Lecture Notes in Mathematics

Edited by A. Dold and B. Eckmann

819

Global Theory of Dynamical Systems

Proceedings of an International Conference Held at Northwestern University, Evanston, Illinois, June 18–22, 1979

Edited by Z. Nitecki and C. Robinson

Springer-Verlag
Berlin Heidelberg New York 1980

Editors

Zbigniew Nitecki
Department of Mathematics, Tufts University
Medford, MA 02155
USA

Clark Robinson
Department of Mathematics, Northwestern University
Evanston, IL 60201
USA

AMS Subject Classifications (1980): 28Dxx, 34Cxx, 34Dxx, 54H20, 58Fxx, 90Dxx

ISBN 3-540-10236-1 Springer-Verlag Berlin Heidelberg New York
ISBN 0-387-10236-1 Springer-Verlag New York Heidelberg Berlin

Printing and binding: Beltz Offsetdruck, Hemsbach/Bergstr.
2141/3140-543210

In Memoriam
Rufus Bowen
(February 23, 1947 - July 30, 1978)

From his first paper on dynamical systems, written in 1968 with O. E. Lanford, until his sudden, tragic death in 1978, Rufus Bowen was a leading figure in the development of dynamical systems theory. His earliest papers on topological entropy and subshifts set the underlying theme for much of his work, relating geometric characteristics of dynamical systems to various apsects of their stochastic behavior. Together with Sinai and Ruelle, he developed the beautiful and comprehensive theory of metric properties of basic sets for axiom A systems, and contributed to the development of the "thermodynamic formalism" for the study of such systems. Bowen gave an integrated exposition of this theory in his first book, Equilibrium States and the Ergodic Theory of Anosov Diffeomorphisms (Springer Lecture Notes in Math. vol. 470, 1975). His contributions to dynamical systems touched diverse problems: the entropy conjecture, zeta functions, maps of the interval, chain-recurrence, symbolic dynamics and ergodic theory for axiom A systems. Early in 1978, working with C. Series, he began to study the ergodic theory of Fuchsian groups. His second monograph, On Axiom A Diffeomorphisms (CBMS Regional Conference Series in Math., vol. 35, 1978) is a concise but incisive introduction to the state of dynamical systems theory in 1977. A complete bibliography of Bowen's prolific work is to be included in a dedicatory volume of the Publications Mathematiques de l'I.H.E.S. (volume 50, 1980).

Those of us who knew Rufus, as (fellow) graduate student at Berkeley in the late 1960's, and then as one of our most brilliant colleagues, remember a tall figure with a mass of red hair, a soft-spoken, alert presence, a gentle friend whose loss is still shocking after eighteen months.

Had he been alive, Rufus would without doubt have been one of the most important participants at the conference of which this volume is a record. We dedicate this collection of papers to a friend and colleague who is sorely missed.

February, 1980

Preface

This volume constitutes the proceedings of the International Conference on the Global Theory of Dynamical Systems, held June 18 - 22, 1979 at Northwestern Univeristy, Evanston, Illinois. The theme of the conference was the qualitative study of smooth maps and flows and its applications. Participants came from Europe, the Americas, and the Middle East.

Most papers in this volume were presented in talks at the meeting. The rest, while not formally presented, are so closely related to the concerns of the conference that they clearly belonged in this collection. All papers in this volume were refereed.

Of the longer morning addresses at the conference, those of Conley, Franks, Palis, Takens and Zeeman are represented by contributions to this volume. In addition, the joint paper of Ruelle and Shub touches on a part of the address delivered by each author. The following is a partial list of references to work presented at the other morning talks:

M. Ratner: Horocycle flows are loosely Bernoulli, Israel J. Math. 31 (1978) 122-132

The Cartesian square of the horocycle flow is not loosely Bernoulli, Israel J. Math. To appear.

D. Ruelle: Ergodic Theory of differentiable dynamical systems, Publ. Math. I.H.E.S. vol. 50. To appear

M. Shub and D. Fried: Entropy, linearity and chain-recurrence. Publ. Math. I.H.E.S. vol. 50. To appear

M. Shub and J. Franks: The existence of Morse-Smale diffeomorphisms. To appear.

S. Smale: The prisoner's dilemma and dynamical systems associated to non-competitive games. Econometrica, To appear.

J. Yorke and J. Mallet-Paret: Snakes: oriented families of periodic orbits, their sources, sinks, and continuation. To appear.

Two types of Hopf bifurcation points: sources and sinks of families of periodic orbits. Annals N.Y. Acad. Sci. (Conf. on Non-Linear Dynamics, December 1979)

Degenerate orbits as clusters of simple orbits, and a bifurcation invariant. To appear.

The organizing committees for the conference included (besides the undersigned) Alan Dankner, Bob Williams, and John Franks - who bore the major burden of initially setting up the conference. We would like to thank the National Science Foundation for its support of this conference through grant MCS 78 - 18180. We would also like to express our appreciation to the many referees whose help was invaluable in selecting papers for inclusion in this volume. And finally, we would like to thank Jeanette Bowden, Melanie Rubin, and Molly Schwarzman for their help in making the conference a success.

Zbigniew Nitecki
Clark Robinson

February, 1980

CONTENTS

MORSE-SMALE FIELDS OF GEODESICS

Daniel Asimov and Herman Gluck

This paper covers the third and final part of a lecture given at the International Conference on Dynamical Systems at Northwestern University during June '79. The first two parts:

I. Geodesic fields on surfaces
II. An exposition of Sullivan's characterization of geodesic fields

are covered in the preceding paper [G]. The language and results of part II are used here.

INTRODUCTION

Two separate threads of thought point to the question: Given a nonsingular Morse-Smale vector field on a smooth manifold, when does there exist a Riemannian metric for which all the orbits are geodesics, ignoring parametrization?

The first thread is the search for structurally stable fields of geodesics. Here are two examples:

1. The suspension of a structurally stable diffeomorphism of a manifold yields a structurally stable flow-with-section, whose orbits can be made into geodesics [G].
2. The orbits of the geodesic flow on the unit tangent bundle UM of any Riemannian manifold M are themselves geodesics for a natural choice of metric on UM. In the special case that M has strictly negative sectional curvatures, this field of geodesics on UM is structurally stable.

But aside from these two types of examples, structurally stable fields of geodesics seem rare. So it is natural to direct attention to the simplest class of structurally stable fields, the Morse-Smale ones, and ask which of these can be made into geodesics.

The second thread is an outgrowth of Sullivan's characterization of geodesic fields [S_2 and G, part II]. His criterion for "geodesibility" can be difficult to verify in practice, so the following question seems appropriate: If a nonsingular vector field V on M is presented almost algorithmically, can one decide, in similar fashion, whether or not the orbits of V can be made into geodesics for some metric on M? Again Morse-Smale fields come to mind, this time as prototypes for "algorithmically presented" fields.

MAIN THEOREM. A nonsingular Morse-Smale field on a closed manifold M is geodesible if and only if it is a suspension (i.e., admits a cross-section). In particular, M must fibre over a circle.

This points even more strongly to the rarity of structurally stable geodesic fields.

We thank William (Bus) Jaco and Sheldon Newhouse for valuable advice, and the National Science Foundation for support.

CONTENTS

1. Nonsingular Morse-Smale fields

In this and the following section we give a brief description of nonsingular Morse-Smale fields for the nonexpert, with reference to [Sm_1 , Sm_2] for more details. The reader conversant with Dynamical Systems should skip to section 3.

Let V be a nonsingular C^∞ vector field on the closed C^∞ manifold M^n, and let $\{\phi_t: -\infty < t < \infty\}$ be the corresponding flow.

The ω-limit set of a point p of M is defined to be

$$\omega(p) = \{q \in M: \phi_{t_n}(p) \to q \text{ for some sequence } t_n \to \infty\} .$$

Replacing $+\infty$ by $-\infty$, we get the α-limit set $\alpha(p)$ of p. All points on the same orbit have the same α-limit set and the same ω-limit set. One views the orbit as being "born" at $\alpha(p)$ and "dying" at $\omega(p)$. The sets $\alpha(p)$ and $\omega(p)$ are each nonempty, compact, connected and invariant under the flow (i.e., a union of orbits).

A closed orbit of V is said to be hyperbolic if the differential of a corresponding Poincaré first-return-map, defined on a local cross section through the orbit, has all its eigenvalues off the unit circle. Any closed orbit can be made hyperbolic by slight perturbation of V. Any closed orbit which is already hyperbolic persists under slight perturbation of V.

If β is a hyperbolic closed orbit, we define the stable manifold of β to be the set

$$W^s(\beta) = \{q \in M: \omega(q) = \beta\} .$$

Similarly, the unstable manifold of β is

$$W^u(\beta) = \{q \in M: \alpha(q) = \beta\} .$$

The stable and unstable manifolds of a hyperbolic closed orbit are each embedded C^∞ manifolds in M [H-P]. Indeed, $W^s(\beta)$ is a copy of either $S^1 \times R^{s-1}$ or $S^1 \tilde{\times} R^{s-1}$ (the nontrivial R^{s-1} bundle over S^1), and similarly for $W^u(\beta)$. Here $s+u = n+1$, and $W^s(\beta)$ and $W^u(\beta)$ meet transversely along β (and perhaps transversely or non-transversely elsewhere).

A point p in M is a <u>nonwandering point</u> of V if for every neighborhood U of p, the set of times $t \in R$ for which $\phi_t(U)$ meets U is unbounded. The <u>nonwandering set</u> $\Omega = \Omega(V)$ consists of all the nonwandering points of V. It is nonempty, compact, invariant under the flow, and contains all closed orbits and all α- and ω-limit sets.

The vector field V is called a <u>nonsingular Morse-Smale field</u> if:

A) V has only finitely many closed orbits, all hyperbolic.

B) If β_1 and β_2 are closed orbits (not necessarily distinct), then $W^u(\beta_1)$ and $W^s(\beta_2)$ meet transversely.

C) The nonwandering set $\Omega(V)$ equals the union of the closed orbits <u>and nothing else</u>.

Two important consequences of this definition are:

D) Each orbit of V is born at some closed orbit and dies at some closed orbit.

E) No orbit of V can be born and die at the <u>same</u> closed orbit (other than that orbit itself).

Nonsingular Morse-Smale fields are structurally stable [Pe,P,P-Sm]. Indeed, they are the simplest structurally stable fields.

2. The corresponding filtration by stable manifolds

> Let V be a nonsingular Morse-Smale vector field on M. If $\beta_1, \beta_2, \ldots, \beta_k$ are all the closed orbits of V, then the corresponding stable manifolds $W^s(\beta_1), W^s(\beta_2), \ldots, W^s(\beta_k)$ provide a <u>filtration</u> of M, i.e., they are disjoint and their union is all of M. A few details about this filtration will be necessary for the proof of the Main Theorem.

For brevity, let W_i denote the stable manifold of β_i, and s_i the dimension of W_i. Let ∂W_i denote the set of limits in M of sequences on W_i whose preimages on $S^1 \times R^{s_i-1}$ or $S^1 \tilde{\times} R^{s_i-1}$ approach infinity. Then, since W_i is <u>embedded</u> in M, its closure is the <u>disjoint</u> union: $\mathrm{Cl}\, W_i = \partial W_i \cup W_i$.

We define an ordering among the closed orbits by writing

$$\beta_i \leq \beta_j$$

if some orbit of V is born at β_i and dies at β_j. We write $\beta_i < \beta_j$ if $\beta_i \leq \beta_j$ but $\beta_i \neq \beta_j$.

Properties of this ordering [Sm1]:

A) It's a partial ordering.

B) $\beta_i \leq \beta_j \implies s_i \leq s_j$.

C) $\beta_i < \beta_j \iff W_i \subset \partial W_j \iff W_i \cap \partial W_j \neq \emptyset$.

D) $\partial W_j = \bigcup_{\beta_i < \beta_j} W_i$.

Thus the stable manifolds $W_1, W_2, \ldots, W_k$ filtrate M in such a way that the closure of W_j consists of all W_i for which $\beta_i \leq \beta_j$.

It will be convenient for the proof to come if we extend our partial ordering of the closed orbits to a total ordering. Begin by listing first all source orbits, i.e., those with stable manifolds of dimension one (themselves). The order among these can be random. Then list all closed orbits with stable manifolds of dimension two. Here the order is chosen to extend the above partial ordering. Then do the same for all closed orbits with stable manifolds of dimension three, and so on, listing the sink orbits (those with stable manifolds of dimension n) last. Suppose that the list $\beta_1, \beta_2, \ldots, \beta_k$ refers to this total ordering. Then if some orbit goes from β_i to β_j, we can be sure that $i \leq j$ (but not conversely). And if $i \leq j$, then $s_i \leq s_j$. In particular, from D) we obtain

E) $\partial W_j \subset \bigcup_{i<j} W_i$.

3. Idea of the proof of the Main Theorem

We want to show that the orbits of the nonsingular Morse-Smale vector field V can be made into geodesics if and only if V admits a cross-section. In doing this, we rely on the theorems of Schwartzman [Sch] and Sullivan [S_2], both discussed in part II of [G].

SCHWARTZMAN'S THEOREM. The nonsingular vector field V on the closed manifold M admits a cross-section if and only if no nontrivial foliation cycle bounds.

SULLIVAN'S THEOREM. Let V be a smooth nonsingular vector field on the smooth manifold M. Then there is a Riemannian metric making the orbits of V geodesics if and only if no nonzero foliation cycle for V can be arbitrarily well approximated by the boundary of a 2-chain tangent to V.

We already know [G, section 2] that the orbits of any nonsingular vector field admitting a cross-section can be made into geodesics. So we assume that V is a nonsingular Morse-Smale vector field which does not admit a cross-section, and plan to show that its orbits can not be made into geodesics.

By Schwartzman's Theorem, some nontrivial foliation cycle c bounds a 2-current σ. The idea of the proof is simply to modify σ through a sequence of operations: "perturb", "puncture", "fracture" and "flow", until it becomes a 2-chain τ, tangent to V, with boundary approximating c. By Sullivan's Theorem, it will be impossible to choose a metric making the orbits of V into geodesics.

We quickly check the truth of the Main Theorem in dimension 2. Indeed, on the torus every field of geodesics admits a cross-section [G, section 3]. So there is nothing to prove. On the Klein bottle, the only fields of geodesics which do not admit cross-sections are those shown in Figure 4 of [G]. These curve fields do not admit a coherent choice of "arrow", hence cannot be the orbits of a vector field. So on the Klein bottle also the Main Theorem is true. Henceforth we work in manifolds of dimension at least 3.

3. Foliation cycles for nonsingular Morse-Smale fields

The proof of the Main Theorem will be phrased in the de Rham-Sullivan language of currents and foliation cycles; we refer the reader to part II of [G] for a brief exposition of these ideas. The first step will be to identify all possible foliation cycles for a nonsingular Morse-Smale flow.

LEMMA. If V is a nonsingular Morse-Smale field on M^n whose closed orbits are $\beta_1, \beta_2, \ldots, \beta_k$, then all foliation cycles for V have the form:

$$a_1\beta_1 + a_2\beta_2 + \ldots + a_k\beta_k \qquad \text{each } a_i \geq 0 .$$

Let c be a foliation cycle for V and μ the corresponding invariant transversal measure [S_1 or G, section 8]. Consider a sink orbit β. Pick a small transverse (n-1)-disk D which contains its image P(D) under the Poincare first-return-map P. By flow-invariance of the transverse measure, $\mu(D) = \mu(P(D))$. Since $P(D) \subset D$, we get $\mu(D - P(D)) = 0$. Iterating forwards, $\mu(D - D \cap \beta) = 0$. Iterating backwards, $\mu = 0$ on $W^s(\beta) - \beta$. Thus on $W^s(\beta)$, μ is an atomic transversal measure, supported on β with some value $a \geq 0$ there.

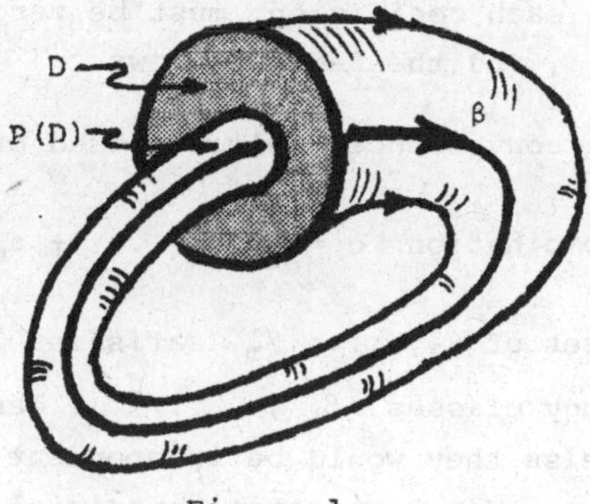

Figure 1

The same argument applies to all sinks of V, then to all closed orbits with stable manifolds of dimension n-1 (this time choosing transverse (n-2)-disks within the stable manifold), and so on, proving the Lemma.

5. We arrange that the coefficients $a_1, \ldots, a_k$ are all integers

Let V be a nonsingular Morse-Smale vector field which does not admit a cross-section. Let c be a nontrivial foliation cycle which bounds, as promised by Schwartzman's Theorem. If $\beta_1, \ldots, \beta_k$ are the closed orbits of V, then $c = a_1\beta_1 + \ldots + a_k\beta_k$, with each $a_i \geq 0$, according to the Lemma of section 4. We will show here that we can select c so that, in addition, the coefficients $a_1, \ldots, a_k$ are all integers.

LEMMA. Let $v_1, \ldots, v_m$ be vectors in a real vector space such that

1) $a_1v_1 + \ldots + a_mv_m = 0$ with each $a_i > 0$,
2) no proper subset of $v_1, \ldots, v_m$ satisfies 1).

Then any dependence relation among $v_1, \ldots, v_m$ is a multiple of 1).

Let $b_1v_1 + \ldots + b_mv_m$ be another dependence relation, and pick the largest ratio, say b_j/a_j, of corresponding coefficients. Then

$$\Sigma_i \left(\frac{b_j}{a_j} a_i - b_i \right) = 0$$

has all coefficients ≥ 0, but the jth coefficient $= 0$. To avoid contradicting condition 2), each coefficient must be zero. Hence $b_j/a_j = b_i/a_i$ for each i, and the Lemma follows.

Now, renumbering for convenience, select closed orbits $\beta_1, \ldots, \beta_m$ such that

1) some linear combination $c = a_1\beta_1 + \ldots + a_m\beta_m$ bounds, with each $a_i > 0$,
2) no proper subset of $\beta_1, \ldots, \beta_m$ satisfies 1).

The rational homology classes $\beta_1, \ldots, \beta_m$ cannot be independent over the rationals, or else they would be independent over the reals, contrary to 1) just above. Hence we get a nontrivial dependence relation

$$b_1\beta_1 + \ldots + b_m\beta_m = 0$$

with each b_i rational. By the above Lemma, the coefficients b_i are proportional to the a_i, hence (changing all signs if necessary) are positive. Multiplying by the least common multiple of the denominators then gives a positive integral dependence.

6. We arrange that c bounds a smoothly embedded surface σ

At this point we have a foliation cycle with positive integral coefficients, which we revert to calling $c = a_1\beta_1 + \dots + a_m\beta_m$, which bounds some 2-current. By the de Rham isomorphism [de R], we know that c also bounds some geometric 2-chain. We want to show in this section that c can be approximated by the boundary of a smoothly embedded surface.

We begin by finding a map $f\colon N^2 \to M$, where N^2 is a compact oriented surface, such that $f(\partial N) = c$. To do this, triangulate M so that the closed orbits $\beta_1, \dots, \beta_m$ appear as subcomplexes. Then choose a polyhedral 2-chain σ in this triangulation satisfying $\partial\sigma = c$. For each 2-simplex of σ, take as many copies as its coefficient in σ. Partially assemble (in abstracto) all these 2-simplexes, identifying edges at most in pairs, to get the required surface N.

Next perturb the map $f\colon N \to M$ so as to make it smooth and in general position. Note that now $f(\partial N)$ approximates, rather than equals, c.

If $\dim M \geq 5$, f will be an embedding, as desired.

If $\dim M = 4$, $f(N)$ will have finitely many points of transversal self-intersection lying in $f(\text{int } N)$. "Pipe" these over the boundary $f(\partial N)$ to get an embedding [Z, Chapter 7].

If $\dim M = 3$, the construction takes a little longer to describe, but is nevertheless routine. We begin with a triangulation of M in which the closed orbits $\beta_1, \dots, \beta_m$ as well as the 2-chain σ appear as subcomplexes. The process described above of "cloning" the 2-simplexes of σ and partially assembling them in abstracto will now be carried out in situ to produce an embedded surface $N \subset M$, with boundary approximating c.

Start by cloning each 2-simplex of σ according to its multiplicity in the chain σ . The clones may each consist of several simplexes, are close to one another, share a common boundary, but are otherwise disjoint, as shown in Figure 2 below. Suppose each 2-simplex of σ oriented so as to appear with positive coefficient in σ . Orient the clones accordingly. The sum of the clones of all the 2-simplexes of σ will then be an algebraic chain whose boundary is c.

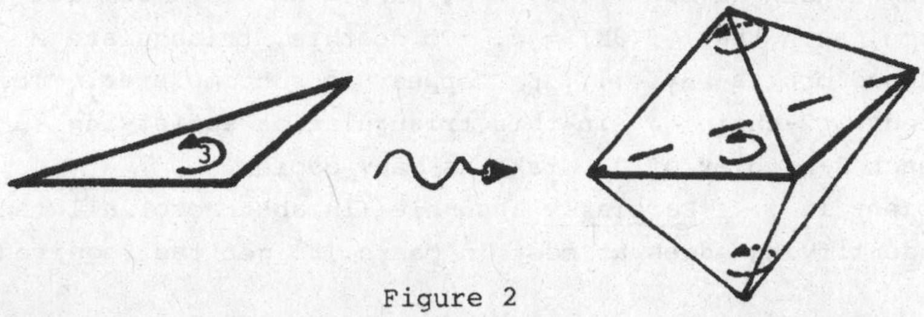

Figure 2

Each edge of the original σ may now appear on the boundary of several cloned 2-simplexes, arranged around it like the pages of an open book, Figure 3 (left). Clone this edge according to the number of adjacent pages, putting each clone in its corresponding page. The cloned edges share a common boundary (2 vertices) but are otherwise disjoint. Then retract each page slightly from the original edge so as to again have the cloned edge on its boundary, Figure 3 (center).

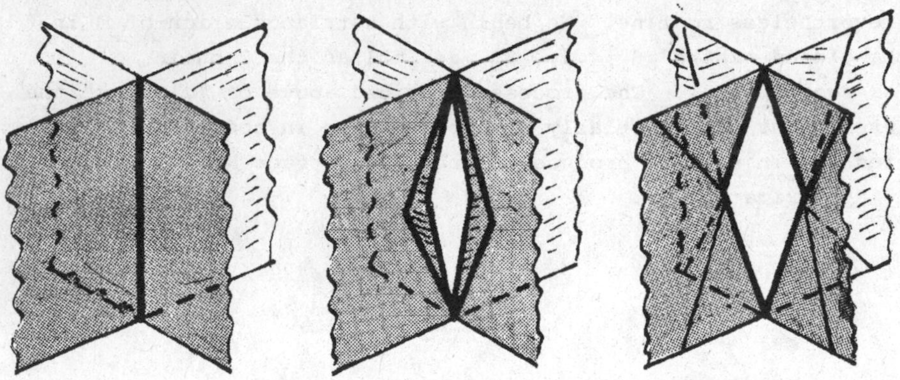

Figure 3

If any pair of pages around a given edge induced opposite orien-
ations on that edge, then the same must be true for some pair of djacent edges. Having retracted slightly from the edge, sew this air of pages together, as in Figure 3 (right), and continue induc-ively. Doing this about each edge of σ then gives a 2-chain whose oundary approximates c.

Each vertex of σ now appears on several "surface fragments" hich are otherwise disjoint. They are arranged around the vertex ike "cones" and "fans". Pull them slightly apart, beginning with n innermost cone or fan, as shown in Figure 4, and continue induc-ively.

Figure 4

What results is a compact oriented surface, piecewise linearly mbedded in M and having boundary approximating the foliation cycle c. Simply smooth this surface along its edges and vertices, call the esult σ again, and we are done.

Many thanks to Bus Jaco for supplying the above argument.

7. <u>We perturb, puncture and fracture σ until each fragment lies in the stable manifold of some sink</u>

At this stage we have in M^n a smoothly embedded surface σ whose boundary approximates the nonzero foliation cycle c. We now modify this surface through a sequence of operations to get a new surface, still called σ, satisfying:

A) The boundary of the new σ approximates the boundary of the old σ, as currents (not as sets), and hence approximates c.

B) Each component of the new σ lies in the stable manifold of some sink orbit.

Let $\beta_1, \ldots, \beta_k$ be a list of all the closed orbits of V, in the order agreed to at the close of section 2. The gradual modification of σ has one step for each closed orbit which is not a sink, and is carried out in this order.

Begin with β_1, a source orbit. Its stable manifold W_1 coincides with β_1 itself. Perturb σ so that it and its boundary become transverse to β_1. Since dim $M \geq 3$, the curves making up $\partial\sigma$ become disjoint from β_1. The perturbed σ meets β_1 transversely in isolated points if $n = 3$, and not at all if $n > 3$.

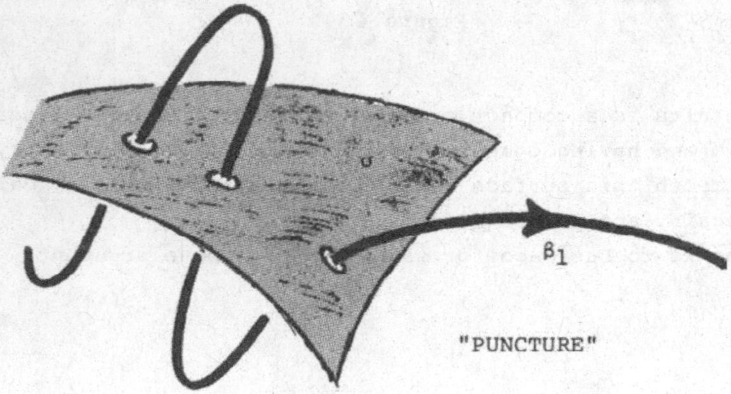

"PUNCTURE"

Figure 5

At each transverse intersection of σ and β_1, remove from σ a small open 2-cell neighborhood of the intersection point. The new punctured σ is now disjoint from β_1, and its boundary can be kept arbitrarily close to the boundary of the old σ, as currents, simply by decreasing the size of the punctures. See Figure 5.

Now let β_j be an orbit which is not a sink, and assume inductively that the surface σ has already been modified so as to be disjoint from all stable manifolds W_i with $i < j$, yet to have boundary arbitrarily close to that of the old σ. Recall E) of Section 2:

$$\partial W_j \subset \bigcup_{i<j} W_i .$$

The surface σ is therefore already disjoint from ∂W_j, and hence meets only a compact portion of W_j. Thus a small perturbation of σ can render it transverse to W_j, yet keep it disjoint from W_i for all $i < j$.

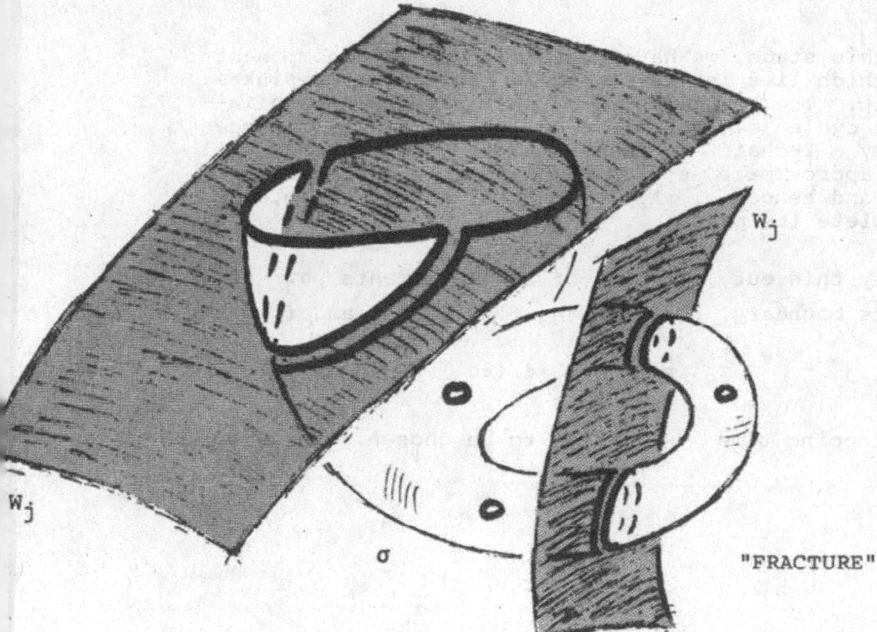

Figure 6

The surface σ now meets W_j transversely along finitely many curves if $s_j = n-1$, at finitely many points if $s_j = n-2$, and not at all if $s_j \leq n-3$.

If $s_j \leq n-3$, we do nothing further to σ at this stage. If $s_j = n-2$, we puncture σ as before to make it disjoint from W_j. And if $s_j = n-1$, we fracture σ along its curves of intersection with W_j, as in Figure 6. That is, we remove from σ a small neighborhood of these curves. The boundary of this new fractured version of σ can be kept arbitrarily close to the boundary of the old σ, as currents, simply by decreasing the width of the fractures.

We iterate this procedure for each closed orbit β_j which is not a sink. At the end we obtain a surface σ, presumably in many fragments, satisfying conditions A) and B).

8. <u>We replace σ by a 2-chain τ tangent to the field, finishing the proof</u>

> At this stage, we have a surface σ, each component of which lies in the stable manifold of some sink orbit. The boundary of σ approximates the foliation cycle c. In this final step, we will replace σ by a 2-chain τ tangent to V. Although τ will not approximate σ, its boundary will approximate $\partial\sigma$, and hence c. Applying Sullivan's Theorem, we complete the proof.

To carry this out, take one of the components σ_j of σ, and keep only its boundary. Turn on the flow $\{\phi_t\}$ and form the tangent 2-chain

$$\tau_j = \bigcup_{0 \leq t \leq t_j} \phi_t(\partial\sigma_j) \quad ,$$

where the stopping time t_j is yet to be chosen. See Figure 7.

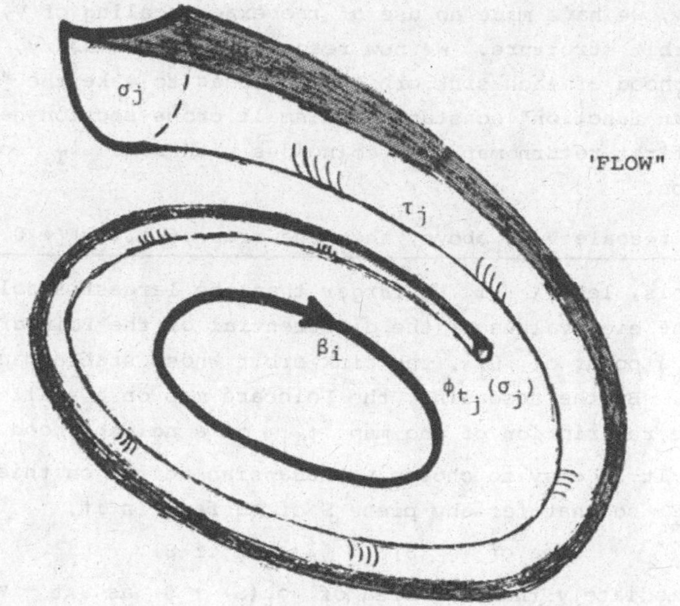

Figure 7

Note that if τ_j is suitably oriented, then

$$\partial\tau_j = \partial\sigma_j - \phi_{t_j}(\partial\sigma_j) .$$

Thus we want to choose the stopping time t_j so large that the 1-cycle $\phi_{t_j}(\partial\sigma_j)$ is approximately zero as a current. That is, we want to show that $\phi_t(\partial\sigma_j) \to 0$ as $t \to \infty$. By the continuity of the boundary map ∂ from 2-currents to 1-currents, this will follow if we show that the 2-currents $\phi_t(\sigma_j) \to 0$ as $t \to \infty$. This in turn would mean that for every bounded set B of 1-forms ω,

$$\phi_t\sigma_j(\omega) = \int_{\phi_t\sigma_j} \omega \to 0 \qquad \text{as } t \to \infty ,$$

uniformly for ω in B. This will certainly follow if we show that the area of $\phi_t(\sigma_j) \to 0$ as $t \to \infty$.

Up to now, we have made no use of the exact scaling of V, but only of its orbit structure. We now rescale V, if necessary, in a small neighborhood of each sink orbit β_i so as to make the "time-of-first-return function" constant on a small cross-section near β_i. The Poincaré first-return map then coincides with some ϕ_{T_i} on such a cross-section.

LEMMA. <u>If we rescale V as above, then the area of $\phi_t(\sigma_j) \to 0$ as $t \to \infty$.</u>

To see this, let $\lambda < 1$ be larger than the largest absolute value among the eigenvalues of the differential of the Poincaré first-return map at a point of β_i, the sink orbit whose stable manifold contains σ_j. By the rescaling, the Poincaré map on a small cross-section is the restriction of the map ϕ_{T_i} of a neighborhood of β_i into itself. It is easy to choose a Riemannian metric on this neighborhood of β_i so that for any piece S of surface in it,

$$\text{area of } \phi_{T_i}(S) \leq \lambda(\text{area of } S).$$

It follows immediately that the area of $\phi_t(S) \to 0$ as $t \to \infty$, and the Lemma is proved.

Thus σ_j has been replaced by a tangent 2-chain τ_j whose boundary approximates $\partial\sigma_j$ as currents. The union τ of the pieces τ_j is then a tangent 2-chain whose boundary approximates the boundary of σ, and hence approximates c. In other words, the nonzero foliation cycle c has been approximated by the boundary of a 2-chain τ tangent to V. By Sullivan's Theorem, no metric on M can make the orbits of V into geodesics, and the proof of the Main Theorem is complete.

REFERENCES

[de R] G. de Rham, VARIÉTÉS DIFFÉRENTIABLES, Hermann, Paris (1960).

[G] H. Gluck, Dynamical behavior of geodesic fields, this volume.

[H-P] M. W. Hirsch and C. C. Pugh, Stable manifolds and hyperbolic sets, GLOBAL ANALYSIS, Proc. of Symposia in Pure Math. XIV Amer. Math. Soc. (1970), 133-163.

[P] J. Palis, On Morse-Smale dynamical systems, Topology 8(1969), 385-404.

[P-Sm] J. Palis and S. Smale, Structural stability theorems, GLOBAL ANALYSIS, Proc. of Symposia in Pure Math. XIV Amer. Math. Soc. (1970), 223-231.

[Pe] M. M. Peixoto, Structural stability on two-dimensional manifolds, Topology 1(1962), 101-120.

[Sch] S. Schwartzman, Asymptotic cycles, Annals of Math. 66(1957), 270-284.

[Sm_1] S. Smale, Morse inequalities for a dynamical system, Bull. Amer. Math. Soc. 66(1960), 43-49.

[Sm_2] ________, Differentiable dynamical systems, Bull. Amer. Math. Soc. 73(1967), 747-817.

[S_1] D. Sullivan, Cycles for the dynamical study of foliated manifolds and complex manifolds, Invent. Math. 36(1976), 225-255.

[S_2] ____________, A foliation of geodesics is characterized by having no tangent homologies, J. Pure and Appl. Algebra 13 (1978), 101-104.

[Z] E. C. Zeeman, Seminar on combinatorial topology, IHES (1963).

SMITH COLLEGE, NORTHAMPTON, MASS.
UNIVERSITY OF PENNSYLVANIA, PHILADELPHIA, PA.

Periodic Points and Topological Entropy of One Dimensional Maps

Louis Block
John Guckenheimer*
Michal Misiurewicz
Lai Sang Young*

This paper is a descendant of the striking theorem of Šarkovskiĭ [Ša] about continuous maps of the interval, stated here as Theorem 2.4. We both give a simpler proof of this theorem and extend it to maps of the circle. Let $f:M \to M$ be a continuous map of a one dimensional manifold into itself. We shall say $x \in M$ has <u>period</u> <u>n</u> if x is a periodic point of minimal period $n: f^n(x) = x$ but $f^i(x) \neq x$ for 0   i   n. Our first problem is to determine, for which m, the existence of points of periods n_1 and n_2 for f implies the existence of points of period m. We answer this question in almost all cases for which $n_1 = 1$. The second problem is to obtain a minimal estimate for the topological entropy of f using only the information that f has points of periods n_1 and n_2. Again we give definitive answers in the case $n_1 = 1$.

Both of these questions have been answered previously for maps of the interval [Ša, Št, M-S, J-R] but the results for maps of the circle appear to be new. Our technique also appears to approach minimal simplicity for the proofs of these results. It relies upon the concepts of <u>f-covers</u> of subintervals and the <u>A-graph</u> of f associated to a partition of M [B, B-F]. Using the existence of points with selected periods, we find a partition A such that the A-graph of f contains a particular subgraph. The existence of other periodic orbits can be deduced from the subgraph. With a useful lemma for calculating the characteristic polynomial of certain matrices, we also obtain estimates for topological entropy. Minimal models for maps with fixed points show that our estimates are sharp.

This paper owes its existence to the International Conference on Dynamical Systems held at Northwestern University in June 1979. We would like to thank the organizers of the symposium for arranging such a stimulating meeting as well as

---

*Research partially supported by the National Science Foundation.

the National Science Foundation and Northwestern University for their support. Evanston may not be Rome, but it has been a center for the development of this field of mathematics. We would like to dedicate this paper to Rufus Bowen and Peter Stefan who each studied one dimensional maps before their untimely deaths.

1. One-dimensional maps and graphs

Let I be an interval, S^1 - a circle ($\{z \in C: |z| = 1\}$), R - the real line. Let $f:S^1 \to S^1$ or $I \to R$ be a continuous map. Further let $A = \{I_1, \ldots, I_s\}$ be a partition of S^1 or I, respectively, into subintervals (we shall call arcs on S^1 also intervals), i.e., a family of closed intervals such that $I_1 \cup \ldots \cup I_s$ is S^1 or I, respectively, and if $i \neq j$ then $I_i \cap I_j$ consists of at most one point. We allow also the degenerate case of S^1 cut at one point and A consisting of one interval with endpoints glued together.

We say that an interval I f-covers J (cf. [B]) if there exists a subinterval K of I such that $f(K) = J$. We say that I f-covers J n times if there exist n subintervals $K_1, \ldots, K_n$ of I with pairwise disjoint interiors such that $f(K_i) = J$ for $i = 1, \ldots, n$. In the degenerate case we require that $f(\text{Int}K)$ (or $f(\text{Int}K_i)$) = IntJ (i.e., S^1 minus the point of cutting).

Definition 1.1. An A-graph of f is an oriented generalized (i.e., possibly with several arrows joining the same vertices) graph with vertices $I_1, \ldots, I_s$ and such that if I_i f-covers I_j n times, but not n+1 times, then there are n (but not n+1) arrows from I_i to I_j.

(Example:

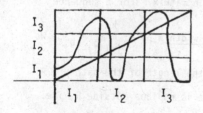

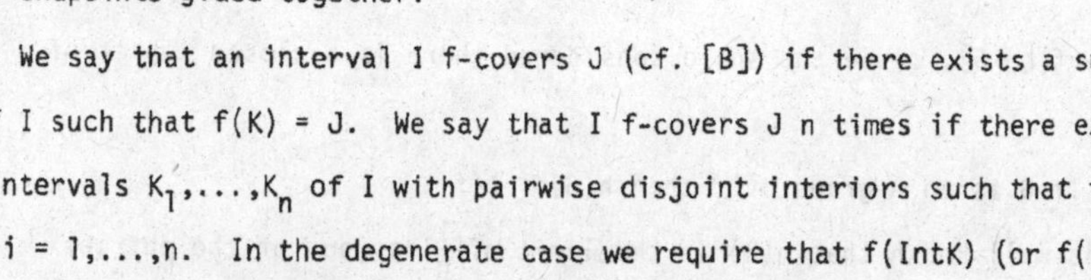

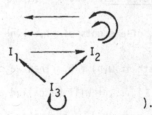

).

We shall use the following simple lemmata ([B]):

Lemma 1.2. If I,J,K are intervals, I f-covers J and $K \subset J$ then I f-covers K. ■

Corollary 1.3. If $f:I \to R$ and $f(K) \supset J$ then I f-covers J.

Proof: $f(I)$ is an interval containing J and I f-covers $f(I)$. ■

Lemma 1.4. If $J_0 \to J_1 \to \dots \to J_{n-1} \to J_0$ is a loop in the A-graph of f then there exists a fixed point x of f^n such that $f^i(x) \in J_i$ for $i = 0,\dots,n-1$. ■

We can associate to our graph G an $s \times s$ matrix $M = (m_{ij})$ such that m_{ij} = (number of arrows from I_i to I_j). We shall call the logarithm of the spectral radius (i.e., of the largest eigenvalue) of M the entropy of G and denote it by $h(G)$.

In the case $f:I \to R$ we define the topological entropy of f as the topological entropy of f restricted to the set $\bigcap_{n=0}^{\infty} f^{-n}(I)$. Notice that in this case we may also use results of [M-S], since the space considered there is a closed subset of an interval.

Lemma 1.5. If G is an A-graph of f then $h(G) \leq h(f)$.

Proof: It is well known that $h(G)$ is the limit of $\frac{1}{n}\log$ (sum of the entries of M^n). But this sum is equal to the number of paths of length n in G. Consequently, it is not larger than $N\left(\bigvee_{i=0}^{n} f^{-1}(A)\right)$. Hence $h(G) \leq h(f,A)$. But by [M-S], $h(f,A) \leq h(f)$. ■

We shall propose a relatively easy way of finding the entropy of G.

Definition 1.6. Let $M = (m_{ij})_{i,j=1}^{n}$ be an $n \times n$ matrix. For a sequence $p = (p_j)_{j=0}^{k}$ of elements of $\{1,\dots,n\}$ we define its width $w(p) = \prod_{j=1}^{k} m_{p_{j-1}p_j}$. We call p a path if $w(p) \neq 0$. Then we call $k = \ell(p)$ the length of the path p. A subset $R \subset \{1,\dots,n\}$ will be called a rome if there is no loop outside R, i.e., there is no path $(p_j)_{j=0}^{k}$ such that $p_k = p_0$ and $\{p_j\}_{j=0}^{k}$ is disjoint from R.

For a rome R we call a path $(p_j)_{j=0}^{k}$ simple if $\{p_0,p_k\} \subset R$ and $\{p_1,\dots,p_{k-1}\}$ is disjoint from R.

For a rome $R = \{r_1,\ldots,r_k\}$ (where $r_i \neq r_j$ for $i \neq j$) we define a matrix function A_R by $A_R = (a_{ij})_{i,j=1}^{k}$ where $a_{ij}(x) = \sum_p w(p)x^{-\ell(p)}$ where the summation is over all simple paths originating at r_i and terminating at r_j. By E we denote the unit matrix (of an appropriate size).

Theorem 1.7. Let $R = \{r_1,\ldots,r_k\}$ (where $r_i \neq r_j$ for $i \neq j$) be a rome. Then the characteristic polynomial of M is equal to $(-1)^{n-k}x^n \det(A_R(x) - E)$.

Proof: We check first the case $R = 1,\ldots,n$. Then $a_{ij} = m_{ij}x^{-1}$. Hence $x^n \det(A_R(x) - E)$ is equal to $\det(M - xE)$, i.e., to the characteristic polynomial of M.

Notice now that any subset of $\{1,\ldots,n\}$ containing a rome is also a rome. Therefore we can use induction, reducing the cardinality of a rome by one at each step.

Hence, it remains to show that if $S = \{s_1,\ldots,s_q\}$ (where $s_i \neq s_j$ for $i \neq j$) is a rome, $s_o \notin S$ and $T = \{s_o\} \cup S$ then $\det(A_S - E) = -\det(A_T - E)$. If $A_S - E = (b_{ij})_{i,j=1}^{q}$, $A_T - E = (c_{ij})_{i,j=0}^{q}$ then we have, by definition, $b_{ij} = c_{ij}+c_{io}c_{oj}$. Since S is a rome, we have $c_{oo} = -1$. Hence, if we multiply the 0-th column of $A_T - E$ by c_{oj} and add it to the j-th column, for $j = 1,\ldots,q$, then we obtain a matrix $(d_{ij})_{i,j=0}^{q}$ such that $d_{ij} = b_{ij}$ for $i,j > 0$, $d_{oo} = -1$ and $d_{oj} = 0$ for $j > 0$. Thus, $\det(A_S - E) = -\det(d_{ij})_{i,j=0}^{q} = -\det(A_T - E)$.■

Notice that if we can find a rome consisting of a single vertex then the characteristic polynomial of M is $(-1)^{n-1}\cdot x^n\cdot\Phi(x)$ where $\Phi(x) = \sum_n$ (number of simple loops of length $n)\cdot x^{-n} - 1$.

We shall have to compare entropies of various graphs. For this we need the following lemma (saying intuitively that shorter loops make larger entropy):

Lemma 1.8. Let $\Phi(x) = \sum_{n=1}^{s} a_n x^{-n} - 1$, $\Psi(x) = \sum_{n=1}^{s} b_n x^{-n} - 1$, a_n and b_n be non-negative integers, $\sum_{n=1}^{s} a_n = \sum_{n=1}^{s} b_n = c > 1$, $\sum_{n=1}^{k} b_n \geq \sum_{n=1}^{k} a_n$ for all k with strict inequality for some k. Then the largest root of $\Phi(x) = 0$ is strictly smaller than the largest root of $\Phi(x) = 0$.

Proof: We have $\Phi(x) = \int_0^c \phi_x(t)dt - 1$, $\Psi(x) = \int_0^c \psi_x(t)dt - 1$, where $\phi_x(t) = x^{-k}$ if $\sum_{n=1}^{k-1} a_n \leq t < \sum_{n=1}^{k} a_n$, $\psi_x(t) = x^{-k}$ if $\sum_{n=1}^{k-1} b_n \leq t < \sum_{n=1}^{k} b_n$. By hypothesis, if $x > 1$ then $\phi_x \leq \psi_x$, and on some subinterval of $[0,c]$ the inequality is strict. Hence, if $x > 1$ then $\Phi(x) < \Psi(x)$. But since $\Phi(1) = \Psi(1) = c-1 > 0$ and $\lim_{x\to\infty}\Phi(x) = \lim_{x\to\infty}\Psi(x) = -1$, the largest roots of $\Phi(x) = 0$ and $\Psi(x) = 0$ are larger than 1.■

Romes as defined here are analogous to cross-sections for flows. For readers familiar with the notion of pressure the above discussion can be rephrased as follows: Recall that for a special flow built under the function ϕ with base transformation $T:X\circlearrowleft$ the entropy of the time-one map is the unique number α satisfying $P(T,-\alpha\phi) = 0$. Solving this equation amounts to finding the largest root of $\det(A_R(x) - E)$, the equation that appears in 1.7. Lemma 1.8 now becomes immediate. For a full shift T and positive functions $\phi_1 \leq \phi_2$, $P(T,-\phi_1) > P(T,-\phi_2)$ if $\phi_1 < \phi_2$ over some cylinder set.

2. The interval

We shall give new, simple proofs of Šarkovskiĭ's theorem ([Ša]) and the estimate of topological entropy of a continuous map of an interval into itself with a periodic point of a given period. This estimate was given in special cases in [J-R] and [M-T]; in the general case it can be easily deduced from the results of Štefan [Št].

Let $f:I \to R$ be a continuous map. We denote an orbit of a point x by Orb x (or $\text{Orb}_f x$).

Lemma 2.1. Suppose that f has a point x of period n, n is odd and $n > 1$, and f has no periodic points of odd periods smaller than n and larger than 1. Let A be the partition of the interval J = [min Orb x, max Orb x] by the elements of Orb x. Then the A-graph of f contains a subgraph of the following form:

(2.1)

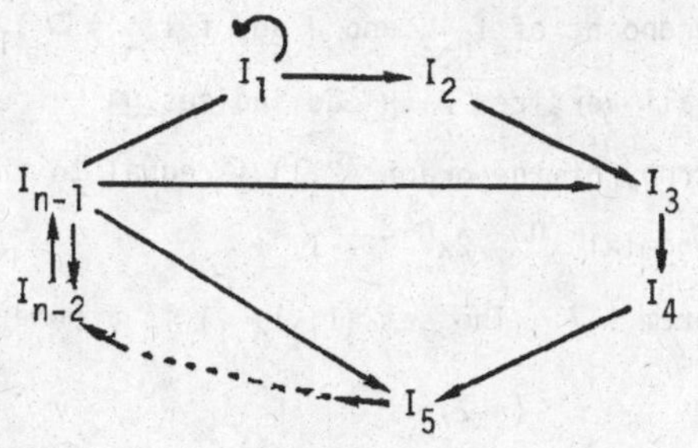

(from I_{n-1} there are arrows to all odd vertices).

Proof. Clearly $f(\max \text{Orb } x) < \max \text{Orb } x$. Hence there exists an element of A whose left endpoint is the largest y Orb x such that $f(y) > y$. Denote this element by $I_1 = [a,b]$. We have $f(a) \geq b$, $f(b) \leq a$ and hence $f(I_1) \supset I_1$. Therefore the sequence $(f^k(I_1))_{k=0}^{\infty}$ is ascending and consequently $f^n(I_1) \supset J$. Since n is odd, there are more elements of Orb x on one side of I_1 than on the other. Therefore some of them must remain on the same side (under the action of f). Clearly, some must change sides. Consequently, $I_1 \subset f(K)$ for some other element K of A. Hence, by Corollary 1.3, we obtain a subgraph of the form

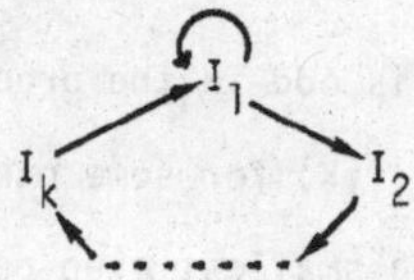

of the A-graph of F. We may assume that the loop $I_1 \to I_2 \to \ldots \to I_k \to I_1$ is the shortest one (except $I_1 \circlearrowright$) from I_1 to itself.

If $k < n-1$ then either the loop $I_1 \to I_2 \to \ldots \to I_k \to I_1$ or the loop $I_1 \to I_2 \to \ldots \to I_k \to I_1 \to I_1$ gives us, by Lemma 1.4, a fixed point z of some f^m, $1 < m < n$, m odd. Since $I_1 \cap I_2$ consists of at most one point and this point is not a fixed point, z is not a fixed point. This contradicts our assumptions. Thus, $k = n-1$.

Since we took the shortest loop from I_1 to itself, there are no arrows from I_i to I_j for $j > i+1$ in the A-graph of f. From this it follows (use induction) that the ordering of the elements of A on the real line must be $I_{n-1}, I_{n-3}, \ldots, I_2, I_1, I_3, \ldots, I_{n-4}, I_{n-2}$ up to orientation. The common endpoint of I_{n-3} and I_{n-1} is mapped onto the common endpoint of I_{n-2} and J and $f(I_{n-1}) \supset I_1$. Therefore there are arrows from I_{n-1} to all vertices with odd indices. ■

Lemma 2.2. The entropy of the graph (2.1) is equal to the logarithm of the largest root of the polynomial $x^n - 2x^{n-2} - 1$.

Proof: We use Theorem 1.7. The set $\{I_1, I_{n-1}\}$ is a rome.

$$\Phi(x) = \begin{vmatrix} x^{-1} - 1 & x^{-(n-2)} \\ x^{-1} & x^{-(n-3)} + x^{-(n-5)} + \ldots + x^{-4} + x^{-2} - 1 \end{vmatrix} =$$

$$= \begin{vmatrix} x^{-1}(1 - x) & x^{-(n-2)} \\ x^{-1} & \dfrac{x^{-(n-3)} - 1}{1 - x^2} - 1 \end{vmatrix} =$$

$$= x^{-1} \frac{x^{-(n-3)} - 1}{1 + x} - x^{-1} + 1 - x^{-(n-1)}, \text{ and hence}$$

$$x^{n-1} \cdot \Phi(x) = \frac{x - x^{n-2}}{1 + x} - x^{n-2} + x^{n-1} - 1 = \frac{x^2 - 2x^{n-2} - 1}{1 + x}. \ ■$$

Lemma 2.3. If $f: I \to R$ has a periodic point of even period then it has a periodic point of period 2.

Proof: We use the fact that n is odd in the proof of Lemma 2.1 only in one place. Namely, we deduced that $I_1 \subset f(K)$ for some other element K of A.

Therefore, if we denote by n the smallest even period of periodic points of f and if $n > 2$ then either the A-graph of f contains a subgraph

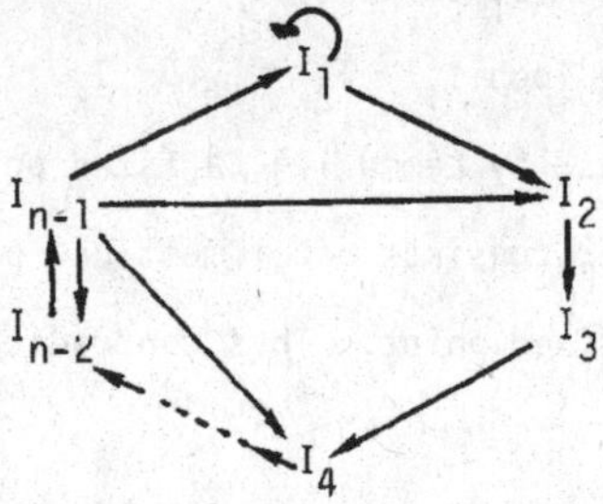

(here n-1 is odd and therefore there are arrows from I_{n-1} to even vertices), or there is no element K of A such that $f(K) \supset I_1$.

In the first case the loop $I_{n-1} \to I_{n-2} \to I_{n-1}$ gives us (by Lemma 1.4) a periodic point of period 2 (the only possible common point of I_{n-1} and I_{n-2} has period n).

In the second case, [min Orb x, min I_1] f-covers [max I_1, max Orb x] and vice versa. Thus, again by Lemma 1.4, there exists a periodic point of period 2.■

Theorem 2.4 (Šarkovskiĭ). Let be the ordering of positive integers:
$3 \lhd 5 \lhd 7 \lhd \ldots \lhd 2\cdot 3 \lhd 2\cdot 5 \lhd 2\cdot 7 \lhd \ldots \lhd 2^2\cdot 3 \lhd 2^2\cdot 5 \lhd \ldots \lhd 2^3 \lhd 2^2 \lhd 2 \lhd 1$.
Let $f: I \to R$ be a continuous map of an interval into the real line. If $n \lhd k$ and f has a periodic point of period n then f has also a periodic point of period k.

Proof: Suppose that $n = 2^m$. Then $k = 2^\ell$, $\ell < m$. The case $\ell = 0$ is trivial. If $\ell > 0$ then consider $g = f^{k/2}$. The map g has a periodic point of period $2^{m-\ell+1}$ and by lemma 2.3 it has a periodic point of period 2. This point has period k for f.

Suppose now that $n = p\cdot 2^m$, p odd, $p > 1$. Then either
(1) $k = q\cdot 2^m$ with q odd, $q > p$, or
(2) $k = q\cdot 2^m$ with q even, or
(3) $k = 2^\ell$, $\ell \leq m$.
In the cases (1) and (2) we use Lemma 2.1 for the map f^{2^m}, and then Lemma 1.4 for one of the loops $I_1 \to I_2 \to \ldots \to I_{n-1} \to I_1 \to I_1 \to \ldots \to I_1 \to I_1$ if $q > p$ or $I_{n-1} \to I_{n-q} \to \ldots \to I_{n-2} \to I_{n-1}$ if $q < p$ and q even. Since the endpoints of I_i have period p for f^{2^m}, we obtain a periodic point of period q for f^{2^m}. If q is even, this point has f-period $q\cdot 2^m$. If q is odd, then either this point has f-period $q\cdot 2^m$ or $q\cdot 2^t$ for some $t < m$. But then we replace n by $q\cdot 2^{m-t}$. For $k = (q\cdot 2^{m-t})\cdot 2^t$, we have case (2) and there is a point of period k.

In case (3), we first use the second case to establish the existence of a periodic point of period 2^{m+1} and then use the results of the first paragraph of the proof.■

Theorem 2.5. Let $f: I \to R$ be a continuous map of an interval into the real line. If f has a periodic point of period $p\cdot 2^m$, where p is odd and $p > 1$ then $h(f) \geq \frac{1}{2^m}\log\lambda_p$, where λ_p is the largest root of the polynomial $x^p - 2x^{p-2} - 1$.

Proof: Use Lemmas 1.5, 2.1 and 2.2 for f^{2^m}. ■

Remark 2.6. The standard examples ([G], [Ša], [Št], [J-R]) show that Theorems 2.4 and 2.5 are in some sense the strongest possible ones. (Use Lemma 1.8.)

3. The Circle

Let $f:S^1 \to S^1$ be a continuous map of a circle into itself. We shall use the standard universal covering $P:R \to S^1$ given by the formula $P(x) = e^{2\pi i x}$.

By F we shall denote the lifting of f to the covering space, $F:R \to R$. It is not defined uniquely, but if F and F' are two liftings of the same map f, then $F = F' + k$ for some integer k. There exists an integer N such that $F(x+1) = F(x) + N$ for all x. We call this N the degree of f and denote it by deg f. Clearly, $\deg (f^n) = (\deg f)^n$.

We shall consider various cases, according to various values of deg f.

(I) $|\deg f| \geq 1$.

In this case f has a fixed point x and the arc J between x and itself (going around the circle) f-covers itself at least $|\deg f|$ times. Therefore we obtain immediately.

Theorem 3.1. If $f:S^1 \to S^1$ is a continuous map of a circle into itself and $|\deg f| > 1$ then f has periodic points of all periods with one exception and $h(f) \geq \log|\deg f|$. The exception occurs when $\deg f = -2$ and there is no point of period 2. ■

(II) $|\deg f| \leq 1$.

We start with the following.

Theorem 3.2. If $f:S^1 \to S^1$ is a continuous map of a circle into itself, $F:R \to R$ is its lifting and F has a periodic point of period n then

(a) if $n \triangleleft k$ then f has a periodic point of period k;

(b) $h(f) \geq \frac{1}{2^m}\log\lambda_p$ if $n = p\cdot 2^m$, p is odd and $p > 1$, where λ_p is the largest root of the polynomial $x^p - 2x^{p-2} - 1$.

Proof: (a) By Theorem 2.4, F has a periodic point x of period k. Then P(x) is f-periodic of period m and $k = m \cdot p$ for some positive integers m,p. We have $F^m(x) = x + q$ for some integer q and $x = F^k(x) = (F^m)^p(x) = x + q\cdot(1 + N + N^2 + \ldots + N^{p-1})$. Hence either $q = 0$ or $N = -1$ and p is even.

If $q = 0$ then $F^m(x) = x$ and therefore $m = k$.

Consider the second possibility: $q \neq 0$ and $N = \deg f = -1$. We have $F^{2m}(x) = x + q\cdot(1 - 1) = x$, and thus $p = 2$. Consider f^2 and its lifting F^2. We have $\deg (f^2) = 1$. F^2 has a periodic point of period n if n is odd or $\frac{n}{2}$ if n is even. If m is even then $n \triangleleft m$ or $\frac{n}{2} \triangleleft m$, respectively. Therefore f^2 has a periodic point of period m. Since m is even, this point has f-period $2m = k$.

We are left with the case of m odd. Suppose first that $m > 1$. We assume that f has a point of odd period $m > 1$ and prove that f has a point of period 2m. Let A be the partition of S^1 by the orbit of period m. Since $\deg f = -1$, f has a fixed point and there is an interval $I \in A$ such that $f(I) \supset I$. Consider the A-graph of f. If there is a loop $I_{i_0} \to I_{i_1} \to \ldots \to I_{i_\ell} = I_{i_0}$ of length $\ell \leq m$ beginning at $I = I_{i_0}$ with $I_{i_j} \neq I_{i_0}$ for $0 < j < \ell$, then the loop $I_{i_0} \to I_{i_1} \to \ldots \to I_{i_0} \to I_{i_0} \to I_{i_0}$ yields a point of period 2m. If there is a loop beginning at I, then there is a loop beginning at I with distinct intermediate vertices (inductively eliminate subloops in a walk) and hence of length at most m. If there is no loop beginning at I, the only arrow ending at I is I↻. To see this note that no pair of points in the orbit of period m are interchanged and $f(I) \supset I$. Hence $f^n(I) = S^1$ for some $n > 0$ and there are paths from I to all other vertices. Eliminating I and the arrows with tail at I leaves us with the A-graph of a map of the interval.

This leaves finally the case of $m = 1$. Here we need to prove that if $\deg f = -1$ and there is a point of period 4, then there is a point of period 2. Let A be the partition of S^1 by an orbit of period 4. We examine the possibilties for the A-graph of f. In all cases with $\deg f = -1$, there will be a loop of length 2 with two different vertices. Let us describe the various cases.

First, f either cyclically permutes points in the orbit of period 4 or it does not. Labelling the points of the orbit cyclically (but not necessarily preserving orientation) we may assume (i) $x_1 \to x_2 \to x_3 \to x_4 \to x_1$ or (ii) $x_1 \to x_3 \to x_2 \to x_4 \to x_1$. Set $I_i = (x_i, x_{i+1})$ where we now take indices mod(4). In case (i), each I_i either f-covers I_{i+1} or f-covers the complement of I_{i+1}. In case (ii), I_1 f-covers I_3 or its complement, I_2 f-covers $I_2 \cup I_3$ or $I_1 \cup I_4$, I_3 f-covers I_1 or its complement, and I_4 f-covers $I_3 \cup I_4$ or $I_1 \cup I_2$.

We now look for the A-graphs which are compatible with deg f = -1. For case (i), there is an i (say i = 1) for which $F(x_{i+1}) < F(x_i)$, F a lift of f. Then I_1 f-covers $I_1 \cup I_3 \cup I_4$. Either I_4 f-covers I_1 and there is a loop of length 2 or we have the subgraph

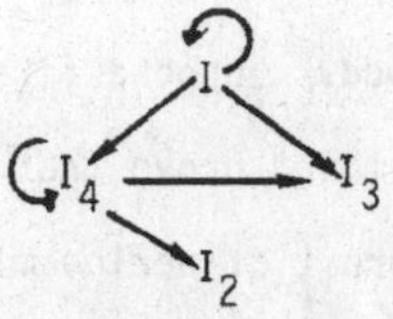

Now I_3 f-covers either I_4 or I_1, so all subcases have a loop of length 2.

For case (ii), there are two subcases. If I_4 f-covers $I_1 \cup I_2$, then both I_1 and I_2 f-cover I_3 or there is a loop of length 2. Since I_3 f-covers I_1 or I_2, there is a loop of length 2 whenever I_4 f-covers $I_1 \cup I_2$. If I_4 f-covers $I_3 \cup I_4$ and there is no loop of length 2, then I_3 f-covers I_1, I_1 f-covers $I_1 \cup I_2 \cup I_4$, and I_2 f-covers $I_2 \cup I_3$. The map with these properties has degree -2 and represents the exception of Theorem 3.1. For all f of degree -1, there is a loop of length 2. This completes the proof of Theorem 3.2(a).

(b) By Theorem 2.5, the entropy of F (restricted to some interval) is at least $\frac{1}{2^m}\log\lambda_p$. By definition, this entropy is equal to $h(F|_X)$ where X is some invariant subset of this interval. But then P(X) is f-invariant and $P|_X$ is bounded one-to-one. Therefore $h(f) \geq h(f|_{P(X)}) = h(F|_X)$.■

In order to apply Theorem 3.2 we need some knowledge about when a lifting of a periodic point is periodic.

<u>Proposition 3.3.</u> Let deg f = 0. Then

(a) f has a fixed point

(b) If x S^1 is a periodic point of f-period n then there exists a periodic point $y \in R$ of F-period n such that $P(y) = x$.

Proof: (a) F is bounded and hence has a fixed point. Its projection to S^1 is a fixed point of f.

(b) Let $z \in R$ be a point such that $P(z) = x$. Take $y = F^n(z)$. Then $F^n(z) = z + k$ for some integer k. Hence $F^n(y) = F^n(z + k) = F^n(z) = y$. The point y cannot have period smaller than n because then $x = P(y)$ would have also smaller period. ■

Proposition 3.4. Let deg f = -1 and let $x \in R$ be a point such that P(x) is a periodic point of f-period n, n odd. Then:

(a) x is a periodic point of F-period n or 2n.

(b) There exists a lifting F' of f such that x is a periodic point of F'-period n.

Proof: We have $F^n(x) = x + k$ for some integer k. Then, since n is odd, $F^{2n}(x) = F^n(x+k) = F^n(x) - k = x$. If x has F-period other than n or 2n then P(x) has period smaller than n. This proves (a).

If we take $F' = F - k$ then we have $F'^n(x) = F^n(x) - k = x$. ■

Let us consider now the case deg f = 1. Fix a lifting F of f. If x is a periodic point of f-period n and $P(y) = x$, then $F^n(y) = y + k$ for some integer k. We shall call the number $\frac{k}{n}$ the rotation number of x and denote it by $\rho(x)$ (or $\rho_F(x)$). It is easy to see that

(a) $\rho(x)$ does not depend on the choice of y

(b) If $F' = F + m$ then $\rho_{F'}(x) = \rho_F(x) + m$

(c) $\rho_{F^m}(x) = m \cdot \rho_{F'}(x)$.

(Notice that we have already used the fact that deg f = 1.)

Denote the set of all rotation numbers of periodic points of f by L (or L_F).

Lemma 3.5. If $a < b < c$ are rational numbers and $a, c \in L$ then also $b \in L$.

Proof: Let $b = \frac{k}{n}$. Consider f^n and its lifting $F' = F^n - k$. If x,y are points with $\rho_F(x) = a$, $\rho_F(y) = c$, respectively, then $\rho_{F'}(x) < 0 < \rho_{F'}(y)$. Hence there exist points $z, t \in R$ such that $P(z) \in \mathrm{Orb}\ x$, $P(t) \in \mathrm{Orb}\ y$ and $F'(z) < z$, $F'(t) > t$. Thus F' has a fixed point. The F-rotation number of its projection to S^1 is equal to b. ■

Lemma 3.6. (i) If $0 \in L$ then f has a fixed point.

(ii) If k and n are relatively prime and $\frac{k}{n} \in L$ then f has a periodic point of period n.

Proof: (i) Let $0 \in L$. There exists a point $x \in R$ such that $F^n(x) = x$ for some n. Then the set $\mathrm{Orb}_F\, x$ is finite and hence F maps the interval [min Orb x, max Orb x] at least onto itself. Thus F has a fixed point in this interval. Its projection to S^1 is a fixed point of f.

(ii) The set L_{F^n} contains an integer k and hence, as above, $F' = F^n - k$ has a fixed point x. We have $f^n(P(x)) = P(x)$ and $\rho_F(P(x)) = \frac{k}{n}$. Since k and n are relatively prime, the period of P(x) cannot be smaller than n. ∎

From Lemmata 3.5 and 3.6 it follows immediately

Theorem 3.7. Let $f:S^1 \to S^1$ be a continuous map of degree 1. If $a,b \in L$, $a < b$, then f has periodic points of all periods n such that for some k, relatively prime with n, $a \leq \frac{k}{n} \leq b$. ∎

Theorem 3.8. (cf. [B]) Let $f:S^1 \to S^1$ be a continuous map of degree 1. If f has a fixed point x and a periodic point y of period $n > 1$ such that $\rho(x) \neq \rho(y)$ then f has periodic points of all periods larger than n.

Now we shall estimate the entropy of f under the assumptions of Theorem 3.8.

Theorem 3.9. Let $f:S^1 \to S^1$ be a continuous map of degree 1. Let f have a fixed point x and a periodic point y of period $n > 1$ such that $\rho(x) \neq \rho(y)$. Then $h(f) \geq \log\mu_n$ where μ_n is the largest zero of the polynomial $x^{n+1} - x^n - x - 1$.

Proof: By Lemma 1.8 we have $\mu_k > \mu_m$ if $k < m$. Therefore we may assume that there is no periodic point z of period smaller than n and such that $\rho(x) \neq \rho(z)$. We may also assume that $\rho(x) = 0$ and $\rho(y) = \pm\frac{1}{n}$ (say $\rho(y) = \frac{1}{n}$).

Since $0 \in L$ and $\rho(y) = \frac{1}{n}$, there exists a fixed point z of F and a point $t < z$ such that $P(t) \in \mathrm{Orb}\, y$, $P((t,z))$ is disjoint from Orb y and $F(t) > t$. Either all points of Orb y are above the diagonal (on the graph) and then $z = x$; otherwise we take two adjacent points of Orb y on opposite sides of the diagonal and then z is between them. Consider two partitions of S^1:A by Orb y and B by Orb y and P(z). Denote the element of A containing P(z) by I_0, and the elements of B having P(z) as an endpoint by I_1 (left) and I_2 (right). Since $F(t) > z$, I_1 f-covers I_2.

There are two possibilities:

(1) I_2 f-covers I_1. Then I_o f-covers itself and we work with the partition A. Since z is a fixed point of F and $\rho(P(t)) = \frac{1}{n}$, after lifting to the covering space, the images of I_o will cover more and more of the half-line $[z,+\infty)$. In particular, after at most n steps they will cover that lifting of I_o which contains z+1. Therefore the A-graph of f contains a subgraph of the form

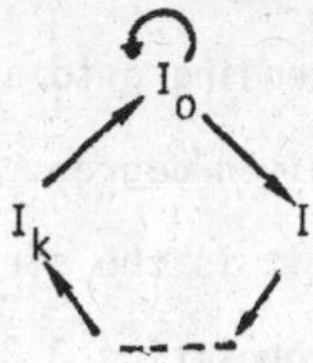

for some $k \leq n + 1$ such that the loop $I_o \to I_3 \to \ldots \to I_k \to I_o$ is not homotopically trivial (i.e., the walk along the loop in the covering space does not lead to the starting place). It is easy to see that then the rotation number of the periodic point of period k-1 obtained from this loop by applying Lemma 1.4 is non-zero. Hence, by our assumption, k = n+1. Since deg f is odd, I_o must be f-covered an odd number of times (f-covered by elements of A, because no image of an endpoint of an element of A can be contained in the interior of I_o). Since all elements of A occur in the subgraph, we get an additional arrow from some I_j to I_o.

Thus we obtain a subgraph in which $\{I_o\}$ is a rome and there is one simple loop of length n, one of length 1, and one additional of length at most n. Therefore, by Lemma 1.8, the entropy of this subgraph is larger than $\log\mu_n$, and consequently, $h(f) > \log\mu_n$.

(2) I_2 does not f-cover I_1. Then I_2 f-covers itself. We work with the partition B. The same arguments as above (we want to cover $I_1 \cup I_2$ in a homotopically non-trivial way) show that the B-graph of f has a subgraph of the form

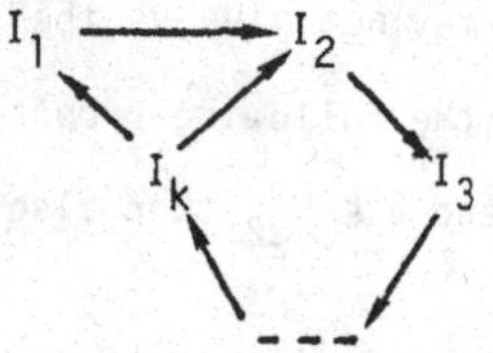

for some $k \leq n + 1$. But the entropy of this subgraph is $\log\mu_{k-1}$ and hence $h(f) \geq \log\mu_n$.■

A question arises, whether our theorems give the strongest possible results. In some cases the answer is affirmative.

If $|\deg f| > 1$ then the example of a map $z \mapsto z^N$ shows that the entropy may be equal to $\log |\deg f|$.

Now let $|\deg f| \leq 1$. Examples corresponding to Theorem 3.2 can be produced by using standard examples $g:I \to I$ and extending g to a map of the circle into itself with the same non-wandering set and a given degree: -1, 0 or 1. Thus, whenever Theorem 3.2 can be applied to a map f, it is the strongest possible result.

Note that if $\deg(f) = 0$ then by Proposition 3.3 we can always apply Theorem 3.2.

Now, suppose that $\deg(f) = 1$, and that f has a fixed point x and a point y of period $n > 1$. There is a lift F of f such that $\rho(x) = 0$. If $\rho(y) = 0$ then Theorem 3.2 applies. If $\rho(y) \neq 0$ then Theorems 3.8 and 3.9 apply. In this case these theorems also give the sharpest possible results. This can be seen by taking a map f of the circle (cf. [B]) such that a lift F of f satisfies the following: $F(0) = 0$, $F(\frac{1}{2n}) = \frac{1}{2n} + \frac{1}{n}$, $F(1 - \frac{1}{2n}) = 1 - \frac{1}{2n} + \frac{1}{n}$, $F(1) = 1$, and F is linear on each interval $[0,\frac{1}{2n}]$, $[\frac{1}{2n},1 - \frac{1}{2n}]$, $[1 - \frac{1}{2n},1]$. Then $\deg(f) = 1$ and $P(\frac{1}{2n})$ is a periodic point with rotation number $\frac{1}{n}$. Thus, the hypothesis of Theorem 3.8 and 3.9 is satisfied, but it is easy to see that f has no periodic points of period k for $1 < k < n$, and $h(f) = \log\mu_n$.

For maps $f:S^1 \to S^1$ of degree 1 without fixed points, our results are not definitive. For example, they do not determine whether periods 5 and 7 imply period 2. There are also problems with the case $\deg f = -1$. If we have a periodic point of odd period then we can apply Proposition 3.4 and then Theorem 3.2. If the period of our points is even then we can take f^2 and apply theorems for degree 1 maps with fixed points. However, this may not give us the strongest possible results for entropy, as illustrated by the following proposition:

<u>Proposition 3.10</u>. If $\deg f = -1$ and $a \in L_{F^2}$ then also $-a \in L_{F^2}$.

Proof: Let $x \in R$ be a point such that $\rho_{F^2}(P(x)) = a$. Denote the f-period of $P(x)$ by n. If n is odd, then, by Proposition 2.4, x is F-periodic of period n or 2n and hence F^2-periodic of period n. Consequently $a = 0$ and $-a = a \in L_{F^2}$. If n is even then $a = \rho_{F^2}(P(x)) = \frac{2}{n}(F^n(x)-x)$ and $\rho_{F^2}(P(F(x))) = \frac{2}{n}(F^{n+1}(x) - F(x))$. But $F^n(x) = x + k$ for some integer k and $F^{n+1}(x) = F(x+k) = F(x) - k$. Hence, $(F^n(x) - x) + (F^{n+1}(x) - F(x)) = k - k = 0$ and therefore $\rho_{F^2}(P(F(x))) = -a$. ■

Some problems remain open, for example:

1) If $\deg f = 1$, $a,b \in L$, then how big is the smallest possible entropy of f and which other periods are present?

2) If $\deg f = -1$ and there exists a periodic point of period n with F^2-rotation number non-integer, then how big is the smallest possible entropy and which periods have to occur for f?

The following table summarizes our results.

deg f	periodic points	lower bound on entropy	best possible estimate on k and periods
$\lvert . \rvert > 1$	all periods except def f = -2	$\log\lvert\deg f\rvert$	yes
0	(same as an interval) Šarkovskiĭ	period $n = 2^m p$, p odd $\Rightarrow h(f) \geq \frac{1}{2^m}\log\lambda_p$ where λ_p = largest root of $x^p - 2x^{p-2} - 1$	yes
-1	period n, n odd $\Rightarrow$ all $n \lhd k$	same as above	yes
	period n, n even (Consider f^2 & use facts about deg 1)		probably no
1	i) $a,b \in L \Rightarrow$ period n if $\exists k$ s.t. $(n,k) = 1$, $a \leq \frac{k}{n} \leq b$		no
	ii) fixed pt x & y of period n and $\rho(x) \neq \rho(y) \Rightarrow$ all periods > n	hypothesis in ii) $\Rightarrow$ $h(f) \geq \log\mu_n$ where μ_n = largest root of $x^{n+1} - x^n - x - 1$	yes

References

[B] L. Block - Periodic orbits of continuous mappings of the circle.

[B-F] R. Bowen and J. Franks - The periodic points of maps of the disk and the interval, Topology 15, 337-342, 1976.

[G] J. Guckenheimer - Bifurcations of maps of the interval, Inventiones Math. 39, 165-178, 1977.

[J-R] L. Jonker, D. A. Rand - A lower bound for the entropy of certain maps of the unit interval, Preprint, University of Warwick.

[M-T] J. Milnor, P. Thurston - Kneading Theory, Preprint, Princeton.

[M-S] M. Misiurewicz, W. Szlenk - Entropy of piecewise monotone mappings. Studia Math. 67, to appear.

[Ša] A. N. Šarkovskiĭ - Coexistence of cycles of a continuous map of a line into itself, Ukr. Mat. Z. 16, 61-71, 1964.

[Št] P. Štefan - A theorem of Šarkovskiĭ on the existence of periodic orbits of continuous endomorphisms of the real line, Comm. Math. Phys. 54, 237-248, 1977.

University of Florida, Gainesville, FL 32611
University of California, Santa Cruz, CA 95064
Institute of Mathematics, University of Warsaw, Warsaw, Poland
Northwestern University, Evanston, IL 60201

Ergodicity of Linked Twist Maps

Robert Burton
Department of Mathematics
Oregon State University

Robert W. Easton*
Department of Mathematics
University of Colorado

§0. Introduction.

The ergodic problem of statistical mechanics is the problem of justifying the replacement of time averages of observables by their phase space averages. This amounts to the problem of showing that a given measure preserving dynamical system is ergodic. Relatively few techniques exist for this purpose. The one we use is discussed in section two below.

In the topological study of dynamical systems the object is to study the phase portrait or orbit structure of the problem. Features of a dynamical system such as the existence of periodic orbits which are of topological interest may not be of statistical interest because they may occur with probability zero. However, we feel that it is important, when possible, to relate topological and statistical properties.

Linked twist mappings provide a class of dynamical systems with interesting topological and statistical properties. Devany [1] has shown that such mappings contain shift automorphisms on finite alphabets as subsystems and hence have positive topological entropy. These subsystems occur however on subsets of measure zero in the domains of these mappings. Nevertheless it is possible that the statistical properties of such a subsystem might be shared by the set of orbits whose alpha and omega limit sets intersect the domain of the subsystem and that this set might have positive measure. Some justification for this speculation is given in [3]. Also in [3] it is conjectured that linked twist mappings are ergodic. It is the purpose of this paper to prove this conjecture.

In section one linked twist mappings are defined and briefly described. Section two recalls relevant results from ergodic theory and outlines the general approach that we will use to establish ergodicity of linked twist mappings. Section three recalls the results of Pesin [5]

* Partially supported by NSF Grant No. MCS 76-84420

which we subsequently use to establish that linked twist mappings are ergodic. In sections four and five we verify the hypotheses necessary to apply Pesin's results. Linked twist mappings in our opinion form an interesting class of examples for which the hypotheses of [5] may be verified. It also follows from Pesin's work that linked twist mappings are Bernoulli.

The study of linked twist maps incorporates some of the difficulties associated with studying so called "ergodic zones" in celestial mechanics, [4]. For a symplectic diffeomorphism of an annulus in the plane which is a sufficiently small (say in the C^3 topology) perturbation of a twist mapping the K.A.M. theorem assures the existence of a set of invariant curves whose union has positive measure. A description of the orbit structure and the measure theoretic properties of perturbations of twist mappings on components of the complement of the set of invariant curves is an important and open problem.

We want to thank Professors Ruelle and Newhouse for pointing out to us the possibility of using Pesin's results in this paper and Professor Robinson for comments which improved the manuscript.

§1. Linked twist mappings

1.1 Definition: Let $A = \{z \in C : 1 \leq |z| \leq 2\}$. A <u>twist mapping</u> of the annulus A is a mapping $\tau : A \to A$ given by the equation

$$\tau(z) = e^{i\theta(|z|)}z$$

where $\theta : [1,2] \to R^1$ is a continuous, increasing function. Let M be a smooth 2-manifold and let $\tilde{A} \subset M$ be an annulus. A map $\tilde{\tau} : \tilde{A} \to \tilde{A}$ is a <u>twist mapping</u> of $\tilde{A}$ provided there exists a homeomorphism $h : \tilde{A} \to A$ such that $h \tilde{\tau} h^{-1}$ is a twist mapping of A.

1.2 Definition: Let A_1 and A_2 be annuli contained in the smooth 2-

manifold M . Suppose that $\tau_1 : M \to M$ and $\tau_2 : M \to M$ are continuous mappings such that $\tau_k(p) = p$ whenever $p \in M - A_k$ for $k = 1,2$. Define $\tau : A_1 \cup A_2 \to A_1 \cup A_2$ by $\tau(p) = \tau_2 \circ \tau_1(p)$. τ is called a <u>linked twist mapping</u> of $A_1 \cup A_2$ provided the τ_1 and τ_2 are twist mappings on A_1 and A_2.

Our goal is to study iterates of linked twist mappings. We will require that the annuli overlap nicely and that the mappings preserve a Borel measure μ . For simplicity we will consider linked twist mappings of annuli which are contained in a torus and which overlap in a square. Under suitable conditions we show that such mappings are ergodic and thus have random orbit structures.

1.3 Definition: $T^2 = R \times R/Z \times Z$ where R and Z respectively denote the groups of real numbers and integers. $R \times R/Z \times Z$ denotes the quotient group formed from the direct product $R \times R$ of R and the subgroup $Z \times Z$. Let $q : R \times R \to T^2$ be the quotient projection. Define

$$A_1 = q([0,1/2] \times R^1)$$

$$A_2 = q(R^1 \times [0,1/2]) .$$

Then A_1 and A_2 are annuli in the torus T^2 . Suppose that for $k = 1,2$ $\alpha_k : R^1 \to R^1$ is a C^2 function with the following properties

(a) $\alpha_k(0) = 0$ and $\alpha_k(t) = n$ for $t \in [1/2,1]$ where $n > 1$ and n is an integer

(b) $\alpha_k(t+1) = \alpha_k(t) + n$ for all t

(c) $\alpha_k'(t) > 0$ whenever $0 < t < 1/2$ and $\alpha'(0) = \alpha'(1/2) = 0$.

Define $\tau_1 : R^2 \to R^2$ by $\tau_1(x,y) = (x+\alpha_1(y),y)$,

$\tau_2 : R^2 \to R^2$ by $\tau_2(x,y) = (x,y+\alpha_2(x))$, and define $\tau = \tau_2 \circ \tau_1$. Thus $\tau(x,y) = (x+\alpha_1(y), y+\alpha_2(x+\alpha_1(y)))$. Define $T : L \to L$ by $T = qT_2qT_1$ where $L = A_1 \cup A_2$. Then T is a linked twist map which preserves the measure μ on L defined by $\mu(E) = m(q^{-1}(E) \cap [0,1] \times [0,1])$ where m denotes Lebesgue measure on R^2. We picture L as the shaded region in the unit square as shown in figure 1. $q : [0,1) \times [0,1) \to T^2$ is 1-1, onto and we will use q^{-1} to determine coordinates on T^2.

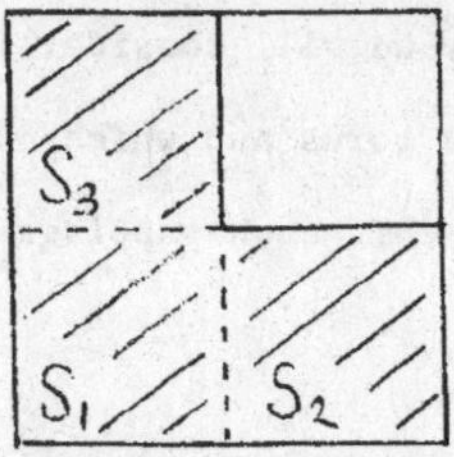

figure 1

Partition L into three sets S_1, S_2, S_3 as follows: Let $S_1 = A_1 \cap A_2$, $S_2 = A_1 - S_1$, $S_3 = A_1 - S_1$.

§2. Results from ergodic theory

This section gives an exposition of results from ergodic theory and stable manifold theory which can be used to prove that a smooth measure preserving diffeomorphism of a smooth manifold is ergodic. Similar ideas are used by Pesin [5] to obtain the results which we quote in section three below. Although we use Pesin's results to conclude that linked twist mappings are ergodic, the approach described here could also be used. For the reader unfamiliar with the ideas developed by Anosov, Sinai, Katok, Pesin and others we hope this section will prove useful.

Let (X,d) be a compact metric space and let μ be a Borel measure on X with $\mu(X) = 1$ and μ positive on open sets. We assume throughout this section that $T : X \to X$ is a homeomorphism of X which preserves μ in the sense that $\mu(T(E))$

$= \mu(T^{-1}(E)) = \mu(E)$ for every Borel set E. For $f \in L^1(X,\mu)$ and $n \geq 0$ define

$$A_n^+(f)(x) = 1/n \sum_{j=0}^{n-1} f(T^j(x))$$

$$A_n^-(f)(x) = 1/n \sum_{j=0}^{-n+1} f(T^j(x)) .$$

Consider A_n^+ and A_n^- as transformations of $L^1(X,\mu)$. We have the following Theorem: (Birkhoff-Kinchin):

a) $P^{\pm}(f) = \lim_{n\to\infty} A_n^{\pm}(f)$ exists in $L^1(X,\mu)$.

b) $P^+(f)(x) = P^-(f)(x)$ for a.e. $x \in X$

c) $\|P^{\pm}(f)\|_1 = \|f\|_1$ where $\| \|_1$ denotes the L^1 norm on $L^1(X,\mu)$.

For a proof of the theorem see [2].

2.1 Definitions: T is <u>ergodic</u> provided $P^+(f)$ is constant almost everywhere for each $f \in L^1(X,\mu)$. The <u>stable</u> and the <u>unstable</u> manifolds of a point $x \in X$ are respectively the sets

$$W^s(x) = \{y \in X : d(T^n(y),T^n(x)) \to 0 \text{ as } n\to\infty\}$$

$$W^u(x) = \{y \in X : d(T^n(y),T^n(x)) \to 0 \text{ as } n\to -\infty\} .$$

The <u>orbit</u> of a point x is the set

$$O(x) = \{T^n(x) : n \text{ is an integer}\} .$$

T is <u>topologically transitive</u> if for each pair of non empty open sets U and V in X there exist an integer n such that $T^n(U) \cap V \neq \phi$.

$f \in L^1(X,R^1)$ is <u>locally constant almost everywhere</u> if there exists a set $Q(f)$ with $\mu(Q(f)) = 0$ such that for each $x \in X - Q(f)$, f is constant almost everywhere on some neighborhood of x.

We show in the following proposition that a transformation T is ergodic provided that it satisfies the following:

Hypothesis A : T is topologically transitive.

Hypothesis B : For each $f \in C^0(X,R^1)$, P^+f is locally constant almost everywhere.

2.2 Proposition: If T satisfies hypotheses A and B then T is ergodic.

Proof: Since $C^0(X,R^1)$ is dense in $L^1(X,R^1)$, and P^+ is continuous and since the constant functions form a closed subspace of $L^1(X,R^1)$ it is sufficient to show that P^+f is constant almost everywhere for $f \in C^0(X,R^1)$. By hypothesis B there exists a set Q of measure zero such that for $x,y \in X - Q$ there exist neighborhoods $V(x)$ and $V(y)$ of x and y such that P^+f is constant almost everywhere on $V(x)$ and on $V(y)$. By hypothesis A there exists n such that $T^n(V(x)) \cap V(y) \neq \emptyset$. Because P^+f is constant on orbits it follows that P^+f is equal to the same constant on $V(x)$, and on $T^n(V(x)) \cap V(y)$ and hence on $V(y)$. It follows that P^+f is constant almost everywhere on X . This completes the proof.

A way to show that a transformation satisfies hypothesis A is to show that for almost every pair of points x,y one has $W^u(x) \cap W^s(y) \neq \emptyset$. A procedure for showing that a transformation satisfies hypothesis B involves examining the "foliations" or partitions of X by the stable and unstable manifolds of points. This procedure is well known to ergodic theorists and is described briefly in [6]. A key observation is

2.3 Proposition: If f is a continuous (hence uniformly continuous) function on X then $P^+(f)$ is constant on the stable manifold of x provided $P^+(f)(x)$ exists. Similarly $P^-(f)$ is constant on the unstable manifold of x provided $P^-(f)(x)$ exists.

Proof: Let $y \in W^s(x)$. Given $\varepsilon > 0$ choose $\delta > 0$ such that

$d(x_1,x_2) < \delta$ implies that $|f(x_1) - f(x_2)| < \varepsilon$. Choose $m > 0$ such that $d(T^k(y),T^k(x)) < \delta$ whenever $k \geq m$. For $n > m$,

$$A_n^+(f)(y) - A_n^+(f)(x) = 1/n \sum_{j=1}^{m} f(T^j(y)) - f(T^j(x))$$

$$+ 1/n \sum_{j=m+1}^{n} f(T^j(y)) - f(T^j(x)).$$

Hence $|A_n^+(f)(y) - A_n^+(f)(x)| \leq \frac{1}{n}| \sum_{j=1}^{m} f(T^j(y)) - f(T^j(x))| + \varepsilon/n$. Therefore $|A_n^+(f)(y) - A_n^+(f)(x)| \to 0$ as $n \to \infty$ and it follows that $P_n^+(f)(y)$ is defined and is equal to $P_n^+(f)(x)$.

Define the <u>stable</u> <u>foliation</u> <u>of</u> <u>X</u> to be the set of equivalence classes of the equivalence relation $\sim$ defined by $x \sim y$ if $d(T^n|x|,T^n|y|) \to 0$ as $n \to \infty$. Similarly define the <u>unstable</u> <u>foliation</u> of X to be the partitioning of X by unstable manifolds. The concept of "absolute continuity" of foliations is needed to relate the geometry of these foliations to the measure μ. For this it is necessary at present to assume that X is a Riemannian manifold and that stable and unstable manifolds are smooth immersed submanifolds of X. For each $k \in X$ define $W_\varepsilon^s(x)$ to be the set of points which can be joined to x by an arc in $W^s(x)$ having arc length less than or equal to ε. We need the stable and unstable foliations to have two properties.

<u>Property 1:</u> Given a set $F \subset X$ with $\mu(F) = 0$ there exists a set F_1 with $\mu(F_1) = 0$ such that if $x \notin F_1$ then there exists $\varepsilon > 0$ such that $W_\varepsilon^u(x) \cap F$ has measure zero relative to the Riemannian measure ν on $W^u(x)$. ($W^u(x)$ inherits from X a Riemannian metric and hence a "volume" measure ν.)

<u>Property 2:</u> Given $B \subset X$ with $\mu(B) = 0$, suppose that $W_\varepsilon^u(x) \cap B$ has measure zero relative to the Riemannian measure on $W^u(x)$. Then there

exists $\delta > 0$ and a neighborhood V of x such that $\mu(V - K(x,\delta)) = 0$ where

$$K(x,\delta) \equiv \bigcup \{W^s_\delta(y) : y \in W^u_\delta(x) - B\} .$$

If the stable and unstable foliations are absolutely continuous according to the definition in [5] then the above properties hold. Further we have

2.4 Proposition: Suppose that the stable and unstable foliations of X relative to the transformation T satisfy properties 1 and 2 . Then T satisfies hypothesis B .

Proof: Let $f \in C^0(X,R^1)$ and define $B(f) = \{x \in X : P^+f(x) \neq P^-f(x)\}$. By property 1 choose a set Q such that $\mu(Q) = 0$ and for each $x \in X - Q$ there exists $\varepsilon > 0$ such that $W^s_\varepsilon(x) \cap B(f)$ has Riemannian measure zero in $W^s_\varepsilon(x)$. By property 2 choose $\delta > 0$ and a neighborhood V of x such that $\mu(V - K(x,\delta)) = 0$. If $z \in K(x,\delta)$, then $z \in W^s_\delta(y)$ for $y \in W^u_\delta(x) - B(f)$. Therefore $P^+f(z) = P^+f(y)$ by proposition 2.2. Hence $P^+f(z) = P^+f(y) = P^-f(y) = P^-f(x) = P^+f(x)$ because $x,y \notin B(f)$. This shows that $P^+f(z) = P^+f(x)$ for almost every point $z \in V$. Thus P^+f is locally constant almost everywhere and hypothesis B is satisfied.

§3. Pesin's Results

Let $f : M \to M$ be a C^2 diffeomorphism of a smooth Riemannian manifold M . For $x \in M$ and $v \in T_xM$ define $\chi^+(x,v) = \lim_{n\to\infty} 1/n \ln(|df^n_x(v)|)$. $\chi^+(x,v)$ is called the <u>exact</u> <u>characteristic</u> <u>Lyapunov</u> <u>exponent</u> of v provided that the limit exists. Suppose that f preserves a smooth measure μ on M (i.e. $\mu = \rho d\nu$ where ρ is a smooth positive function on M and $d\nu$ is the measure on M induced by the Riemannian metric).

3.1 Theorem (Pesin) Suppose that $\mu(\Lambda) > 0$ where

$$\Lambda = \{x \in M : \chi^+(x,v) \neq 0 \text{ for each } v \neq 0 \text{ in } T_x M\} .$$

Then (1) Λ is either a finite or countable union of disjoint measurable sets $\Lambda_0, \Lambda_1, \cdots$

(2) $\mu(\Lambda_0) = 0$, $\mu(\Lambda_n) > 0$ for $n > 0$,

(3) $f(\Lambda_n) = \Lambda_n$, and f restricted to Λ_n is ergodic,

(4) For each $n > 0$, there exists an integer i_n and subsets $\Lambda_n^1, \cdots, \Lambda_n^{i_n}$ of Λ such that $f(\Lambda_n^i) = \Lambda_n^{i+1}$ for $i = 1, \cdots, i_n - 1$ and $f^{i_n} : \Lambda_n^i \to \Lambda_n^i$ is Bernoulli.

(5) A property of the sets Λ_n which Pesin does not state formally in this theorem, but which follows from the way they are constructed is the property that for each $x \in \Lambda_n$, $W^s(x) \subset \Lambda_n$ and $W^u(x) \subset \Lambda_n$. We will use this property later.

§4 Computation of characteristic exponents

In this section we verify that linked twist mappings of the torus defined in section 1 satisfy the hypothesis of Pesin's theorem. Recall that $\tau : R^2 \to R^2$ is a map given by $\tau(x,y) = (x+\alpha_1(y), y+\alpha_2(x+\alpha_1(y)))$. From the properties of α_1 and α_2 we have $q \circ \tau = f \circ q$ where $q : R^2 \to T^2$ is the quotient projection and f is a diffeomorphism of T^2. In this section let

$$\Lambda = \{p \in T^2 : \chi^+(p,v) \neq 0 \text{ for each } v \neq 0, v \in T_p T^2\} .$$

Λ is clearly contained in L (the region pictured in figure 1) because f is the identity map on $T^2 - L$. We will show that $L - \Lambda$ has Lebesgue measure zero. We identify T^2 with the unit square $[0,1] \times [0,1]$ in the plane with opposite sides identified. We also consider tangent vectors to T^2 as real column vectors with two components.

4.1 Definitions: for $0 < \delta < 1/4$ let

$$A_1(\delta) = \{(x,y) \in L : \delta \le y \le 1/2 - \delta\}$$
$$A_2(\delta) = \{(x,y) \in L : \delta \le x \le \tfrac{1}{2} - \delta\}$$
$$A(\delta) = \{(x,y) \in A_1(\delta) : f(x,y) \in A_2(\delta)\}$$
$$E(\delta) = \text{the characteristic function of } A(\delta) .$$
$$Q(\delta) = \{(x,y) \in L : P^+E(\delta)(x,y) \ge \delta\} .$$

$Q(\delta)$ is the set of points in L whose orbits with respect to f hit $A(\delta)$ with frequency greater than or equal to δ.

4.2 Proposition: If $p \in Q(\delta)$ and if $v \in T_pT^2$, then $\chi^+(p,v) \neq 0$ provided $v \neq 0$. Consequently $Q(\delta) \subset \Lambda$.

Proof: $df^n_p = df_{p_{n-1}} \circ \cdots \circ df_{p_0}$ where $p_k = f^k(p)$. Hence df^n_p is a matrix having the form

$$df^n_p = \begin{pmatrix} 1 & a_{n-1} \\ b_{n-1} & 1+a_{n-1}b_{n-1} \end{pmatrix} \cdots \begin{pmatrix} 1 & a_0 \\ b_0 & 1+a_0b_0 \end{pmatrix} ,$$

with $a_i, b_i \ge 0$ and with $a_j, b_j \ge m > 0$ whenever $p_j \in A(\delta)$. The constant m is a lower bound on the derivatives of the functions α_1 and α_2 on the interval $[\delta, 1/2 - \delta]$. Suppose that $v_1, v_2 \ge 0$. Then $|df^n_p(v)|$ is a nondecreasing function of n. If $f^k(p) \in A(\delta)$ then since $a_k, b_k \ge m$ and $v_1, v_2 \ge 0$ it follows that $|df^{k+1}_p(v)| \ge (1+m^2)^{1/2}|df^k_p(v)|$. Since $p \in Q(\delta)$ for sufficiently large n we have $f^k(p) \in A(\delta)$ for at least ℓ integers k between 0 and $n-1$, where ℓ is such that $\ell/n > \delta/2$. Consequently $|df^n_p(v)| \ge (1+m^2)^{\ell/2}|v|$ and therefore

$$1/n \,\ln(|df^n_p(v)|) \ge \delta/4 \,\ln(1+m^2) + \frac{1}{n}\ln(|v|) .$$

It follows that $\chi^+(p,v) > 0$.

Characteristic exponents have the property that $\chi^+(p,sv) = \chi^+(p,v)$ for any real $s \neq 0$. Thus to show $\chi^+(p,v) \neq 0$ for each non-zero $v \in T_pT^2$, it is sufficient to consider in addition to the case where $v_1, v_2 \geq 0$, the case where $v_1 < 0$, $v_2 > 0$. If $df_p^n(v)$ has both components non negative for some $n > 0$, then by the previous argument $\chi^+(p,v) > 0$. So the remaining possibility is where $df_p^n(v)$ has first component negative and second component positive for all $n > 0$. If $w \in T_pT^2$ and $w_1 < 0, w_2 > 0$ then $df_p^{-1}(w)$ also has first component negative and second component positive. Further $|df_p^{-1}(w)| \geq |w|$. Mimicking the earlier argument we get an estimate for n sufficiently large that

$$|v| \geq (1+m^2)^{(\frac{1}{2})\ell} |df_p^n(v)| \quad \text{with } \ell \geq \delta/2n .$$

Hence $1/n \, \ell n(|df_p^n(v)|) \leq -\delta/4 \, \ell n(1+m^2) + 1/n \, \ell n(|v|)$. Therefore $\chi^+(p,v) < 0$ in this case. This completes the proof.

Next we want to verify that Λ has full measure in L . Since $Q(\delta) \subset \Lambda$ it is sufficient to prove the following:

4.3 Proposition: $\bigcup \{Q(\delta) : 0 < \delta < 1/4\}$ has full measure in L .

We prove this proposition with the help of a lemma which is general and has independent interest.

4.4 Lemma: Suppose that X is a compact metric space, μ is a Borel measure on X and T is a homeomorphism of X which preserves μ . Suppose that $Y \subset X$ is a measurable subset of X and define

$$J(Y) = \{x \in X : T^n(x) \in Y \text{ for some } n \geq 0\}.$$

Define $Z(Y) = \{x \in X : P^+E(x) = 0\}$ where E is the characteristic function of Y. Then $\mu(Z \cap J(Y)) = 0$. Hence almost all orbits that intersect Y in forward time intersect Y with positive frequency in n.

Proof: $\mu(Z \cap Y) = \int_Z E d\mu = \int_Z P^+E d\mu = 0$. The second equality follows from the Birkhoff-Kinchin ergodic theorem applied to the dynamical system $(Z, \mu/Z, T/Z)$. The first and third equalities follow from the definitions. Define $Z_k = \{x \in Z : T^k(x) \in Y$ for some $k \geq 0$ and $T^i(x) \notin Y$ for $0 \leq i < k\}$. Since $T^k(Z_k) \subset Z \cap Y$ we have $\mu(Z_k) = 0$ and since $Z \cap Y = \bigcup_{k \geq 0} Z_k$ we have $\mu(Z \cap J(Y)) = 0$. This completes the proof.

Proof of proposition 4.3:

Choose a monotone decreasing sequence $\{\delta_k\}$ of deltas converging to zero. Define

$$U_k = J(A(\delta_k)) - Z(\ A(\delta_k))$$

where the notation of the previous proof is used. (The set $A(\delta_k)$ plays the role of Y in this case.) $U_k \subset \bigcup \{Q(\delta) : 0 < \delta < 1/4\}$ because the orbit of a point of U_k hits $A(\delta_k)$ with some positive frequency $\beta > 0$. Hence it hits $A(\delta_1)$ with frequency at least δ_1 whenever $\delta_1 < \beta$. Therefore the point belongs to $Q(\delta_1)$.

By lemma 4.4 $U = \bigcup \{U_k : k \geq 0\}$. Since $U \subset \bigcup \{Q(\delta) : 0 < \delta < 1/4\}$ by lemma 4. it is sufficient to show that $\bigcup \{J(A(\delta)) : \delta > 0\}$ has full measure in L in order to prove that $\bigcup \{Q(\delta) : 0 < \delta < 1/4\}$ has full measure in L.

Suppose $p \notin \bigcup \{J(A(\delta) : \delta > 0\}$. We consider cases. Case 1: $p \in S_2$. (Recall the partition of L into three squares S_1, S_2, S_3). We must have $f^k(p) \in S_2$ for all $k \geq 0$ because otherwise p would belong to $J(A(\delta))$ for some $\delta > 0$. Therefore p belongs to a set of measure zero, namely the set $W_2 = \{q : f^k(q) \in S_2$ for all $k \geq 0\}$. $\mu(W_2) = 0$

because for $T^k(q)$ always to remain in S_2 implies that $\alpha_1(q)$ is rational and α_1^{-1} of the set of rational numbers has measure zero. Case 2: $p \in S_1$. Then since $p \notin \bigcup\{J(A(\delta)) : \delta > 0\}$ we must have $f^k(p) \in S_2$ for all $k \geq 1$. Hence $p \in f^{-1}(W_2)$. The last case, Case 3, is where $p \in S_3$. Either $p \in W_3 = \{q \in S_3 : f^k(q) \in S_3 \text{ for all } k \geq 0\}$ or $f^{k_1}(p) \in S_1$ for some positive k_1 and $f^k(p) \in S_2$ for all $k > k_1$. Putting these cases together we see that

$$L - \bigcup\{J(A(\delta)) : \delta > 0\} \subset \bigcup\{f^k(W_2) : k \leq 0\} \cup W_3 .$$

Therefore $\mu(\bigcup\{J(A(\delta)) : \delta > 0\}) = \mu(L)$, and the proof is complete.

4.5 Corollary: Λ has full measure in L and Pesin's results apply to show that L is the union of possibly countably many ergodic components.

We show in the next section that L has only one ergodic component with respect to f and hence f is ergodic.

§5 Stable manifold structure and proof of ergodicity

5.1 Theorem: There exists a set $B \subset L$ with $\mu(B) = 0$ such that if $p,q \in L - B$ then $W^u(p) \cap W^s(q) \neq \emptyset$.

Before discussing this theorem we will use it to finish the proof of ergodicity of linked twist maps.

5.2 Theorem: If f is a linked twist map of the torus, then f restricted to L is ergodic. Furthermore f^m restricted to L is Bernouilli for some $m > 0$.

Proof: Combining Pesin's results 3.1 with corollary 4.5 we conclude that L is a finite or countable union of disjoint measurable sets $L = \Lambda_0 \cup \Lambda_1 \cup \cdots$ with $\mu(\Lambda_0) = 0$ such that $f(\Lambda_n) = \Lambda_n$, $\mu(\Lambda_n) > 0$ for $n > 0$, f restricted to Λ_n is ergodic, and some power of f restricted to Λ_n

is Bernoulli. To finish the proof we need to show that $\Lambda_n = \emptyset$ for $n > 1$. By property 5 of theorem 3.1 we have $W^s(p) \cup W^u(p) \subset \Lambda_n$ whenever $p \in \Lambda_n$. If $\Lambda_n \neq \emptyset$ for some $n > 1$, then $\mu(\Lambda_n) > 0$, and by theorem 5.1 we can choose $p \in \Lambda_1 - B$ and $q \in \Lambda_n - B$ such that $W^u(p) \cap W^s(q) \neq \emptyset$. However $W^u(p) \subset \Lambda_1$ and $W^s(q) \subset \Lambda_n$ contradicting the fact that Λ_1 and Λ_n are disjoint. Therefore $\Lambda_n = \emptyset$ and the proof is complete.

A detailed analysis of the stable manifold structure of linked twist maps will be given in a subsequent publication. We will only outline the proof of theorem 5.1 here. By further analysis one can show that the stable and unstable foliations have properties 1 and 2 defined in section two. Hence a proof of the ergodicity of linked twist mappings could be given employing the approach of section two. The smoothness assumption on f necessary to prove that the stable and unstable foliations are absolutely continuous, seems to be a technicality at this point and perhaps linked twist mappings are also ergodic under weaker conditions.

To study the stable manifold structure of f it is convenient to analyse the stable manifold structure of the transformation $\tau : R^2 \to R^2$ which covers f. Let $\overline{G} = \{(u,v) \in R^2 : q(u,v) \in G\}$ where $q : R^2 \to T^2$ is the quotient projection and $G = \bigcup \{J(A(\delta)) : \delta > 0\}$. It can be shown that for $p \in \overline{G}$, the unstable manifold $W^u(p)$ of p with respect to τ is a graph of an increasing function $\gamma : R^1 \to R^1$ such that γ is neither bounded above nor bounded below. Similarly $W^s(p)$ is the graph of an unbounded decreasing function. Consequently for any pair of points p_1, $p_2 \in \overline{G}$, $W^u(p_1) \cap W^s(p_2)$ is non empty. It follows since τ covers f that the same is true for $W^u(r_1) \cap W^s(r_2)$ where $r_k = q(p_k)$. Therefore for any pair of points $r_1, r_2 \in G$, $W^u(r_1) \cap W^s(r_2)$ is non empty. We showed in section four that $\mu(G) = \mu(L)$. Hence choosing $B = L - G$

we have established 5.1 except for the proof that stable and unstable manifolds of points in $\overline{G}$ are unbounded falling and rising curves respectively.

Define an ordering on R^2 by $(u,v) < (u',v')$ provided $u' - u > 0$ and $v' - v > 0$. Then τ preserves this ordering and as well $d\tau$ preserves this ordering. These facts form the basis for a geometric proof that points in $\overline{G}$ have stable and unstable manifolds with the properties described above.

References

1. R. Devaney, Subshifts of finite type in linked twist mappings, Proceedings of the A.M.S. Vol. 71, No. 2, (1978), 334-338.

2. P. Billingsley, Ergodic theory and Information, John Wiley & Sons, Inc. (1965).

3. R. Easton, Chain transitivity and the domain of influence of an invariant set, Lecture Notes in Mathematics, Vol. 668, Springer-Verlag, Inc. (1978) 95-102.

4. M. Henon and C. Heiles, The applicability of the third integral of motion; some numerical experiments, The Astronomical Journal, 69 (1964), 73-79.

5. Y. Pesin, Characteristic Lyapunov exponents and smooth ergodic theory, Russian Math Surveys 32: 4(1977), 55-114.

6. B. Weiss, The geodesic flow on surfaces of negative curvature, Dynamical Systems, Theory and Applications, Lecture Notes in Physics 38, Springer-Verlag (1975), 224-236.

Infinitesimal Hyperbolicity Implies Hyperbolicity

Carmen Chicone and R. C. Swanson

University of Missouri, Columbia, MO 65211

1. Introduction and example.

Let (E, M, π) denote a smooth vector bundle with compact Riemannian base manifold M. A smooth vector bundle flow (Φ^t, ϕ^t) defines a strongly continuous group of operators on the complex Banach space $\Gamma(E)$ of continuous sections of E:

$$\Phi_t^{\#}\eta = \Phi^{-t} \circ \eta \circ \phi^t .$$

The infinitesimal generator L of $\Phi_t^{\#}$ is defined in $\Gamma(E)$ by the formula

$$L\eta = \frac{d}{dt} \Phi_t^{\#}\eta \Big|_{t=0}.$$

We shall treat spectral properties of the flow, e.g. (Φ, ϕ) is <u>hyperbolic</u> if the spectrum of $\Phi_1^{\#}$ is disjoint from the unit circle [14]. Since, formally

$$\Phi_t^{\#} = \exp(tL)$$

and, in practice, the spectrum of L is often easier to compute than the spectrum of $\Phi_t^{\#}$ we seek to express the spectrum of $\Phi_t^{\#}$ as the exponential image of the spectrum of L. This relationship is one theme underlying our results (section 2). Also, since spectral theory is most complete in Hilbert space, we study (section 3) the operators $\Phi_t^{\#}$ and L in the space of square integrable sections.

As an illustration of these concepts we offer the following example, suggested by Moe Hirsch, which evolved over coffee during the conference.

Example (1.1). Consider the flow in the bundle $S^1 \times R \to S^1$ given by

$$\Phi^t(\theta,v) = (\theta + t, e^{\int_\theta^{\theta+t} a(s)ds} v)$$

where $a: S^1 \to R$. One computes

$$\Phi^{\#}_t(\theta, v(\theta)) = (\theta, e^{\int_{\theta+t}^{\theta} a(s)ds} v(\theta + t))$$

and

$$L(\theta, v(\theta)) = (\theta, v'(\theta) - a(\theta)\, v(\theta)).$$

To compute the spectrum $\sigma(L)$ of L, observe that $\sigma(L)$ consists entirely of eigenvalues (L is elliptic). If $v: S^1 \to R$ is an eigenfunction, then

$$v' - av = \lambda v$$

and averaging over S^1 yields

$$\lambda = \frac{1}{2\pi} \int_0^{2\pi} \frac{v'}{v}\, d\theta - \frac{1}{2\pi} \int_0^{2\pi} a(\theta)\, d\theta .$$

Therefore, $\lambda \in \sigma(L)$ has the form

$$\lambda = iN - \text{ave}(a)$$

for $N \in Z$.

The operator $\Phi^{\#}_{2\pi}$ is given by

$$\Phi^{\#}_{2\pi}(\theta, v(\theta)) = (\theta, e^{-2\pi \text{ave}(a)} v(\theta)) .$$

Hence,

$$\sigma(\Phi^{\#}_{2\pi}) = \{e^{-2\pi\, \text{ave}(a)}\}$$

which illustrates the formula

$$\exp \sigma(2\pi L) = \sigma(\Phi^{\#}_{2\pi}) .$$

However, for t and 2π rationally independent, one computes

$$\sigma(\Phi_t^{\#}) = \{e^{-t\,ave(a)} e^{i\theta t} | \theta \in R\}$$

showing that in general

$$\exp t\, \sigma(L) \neq \sigma(\Phi_t^{\#}) .$$

In our example

$$\text{closure}(\exp t\, \sigma(L)) = \sigma(\phi_t^{\#});$$

however, for general semi-groups this relationship is false ([6], p. 665).

The flow Φ^t is hyperbolic if and only if $ave(a) \neq 0$ which is true if and only if $\Phi_t^{\#}$ is hyperbolic for $t \neq 0$, i.e., when $\sigma(\Phi_t^{\#})$ is disjoint from the unit circle. We say that an operator is infinitesimally hyperbolic if its spectrum is disjoint from the imaginary axis. In the example, the flow is hyperbolic if and only if L is infinitesimally hyperbolic. Of course, if the flow is hyperbolic then L is always infinitesimally hyperbolic. We will show in the sequel that for "most" vector bundle flows hyperbolicity is equivalent to infinitesimal hyperbolicity.

It has not escaped our attention that the theory we shall develop can be considered a generalization of Poincaré's stability criterion for a periodic orbit of a differential equation on the plane. In particular, if X is a vector field on R^2 and $\mathcal{O}$ is a periodic orbit of the flow of X with period T, then $\mathcal{O}$ is hyperbolic if and only if

$$\frac{1}{T} \int_{\mathcal{O}} \operatorname{div}(X) \neq 0 .$$

Recall that the flow ϕ^t of X defines a "Poincaré" map from an orthogonal trajectory Σ_p at p on $\mathcal{O}$ to any other orthogonal trajectory Σ_q at $q = \phi^t(p)$ on $\mathcal{O}$ which we denote by

$$\rho: \Sigma_p \to \Sigma_q .$$

If x is a local coordinate for Σ_p and y is a local coordinate for Σ_q with $x(p) = 0$ and $y(q) = 0$ one has

$$\rho'(0) = e^{\int_0^t \operatorname{div}(X)dt} .$$

In our language this says that the tangent flow $(T\phi^t, \phi^t)$ in the bundle TR^2 over the orbit $\mathcal{O}$ projected to the quotient $TR^2/[X]$ is representable as the vector bundle flow in $S^1 \times R$ given by

$$F^t(\theta, v(\theta)) = (\theta + t, e^{\int_\theta^{\theta+t} \operatorname{div}(X)ds} v(\theta)).$$

The analysis of the example discloses that the orbit is hyperbolic if and only if the infinitesimal generator

$$L = \frac{d}{d\theta} - \operatorname{div}(X)(\theta)$$

is infinitesimally hyperbolic; which will be true exactly when

$$\frac{1}{T} \int_0^T \operatorname{div}(X)d\theta \neq 0 .$$

Our main results are as follows:

(A) For a smooth flow f^t, the spectrum of the adjoint representation group $f_t^\#$ is the exponential image of the spectrum of the Lie derivative L_X where $X = \frac{d}{dt} f^t$.

(B) If f^t is measure preserving the analogue of (A) is true in the space of square integrable sections.

(C) The spectrum of $f_t^\#$ on square integrable sections is identical to the spectrum of $f_t^\#$ on the space of continuous sections.

In section 4 we use our results to show that the geodesic flow on the unit tangent bundle of a Riemannian manifold of negative curvature is Anosov.

The complete proofs of any theorems which we do not prove here will appear elsewhere.

2. Spectral mapping theorem

If (Φ^t, ϕ^t) is a smooth vector bundle flow on the bundle (E, M, π), define, as before, the infinitesimal generator L of the induced group $\Phi_t^\#$. For every C_0 semigroup, one has the spectral inclusion ([6], p. 467)

$$\exp(t\,\sigma(L)) \subseteq \sigma(\Phi_t^\#):$$

moreover, the point and residual spectra of the infinitesimal generator exponentiate to give the point and residual spectra of the semigroup. Thus, any element of $\sigma(\Phi_t^\#)$ not accounted for by exponentiation must lie in the approximate point spectrum: $\lambda \in \sigma_{ap}(\Phi_t^\#)$ if and only if for any $\epsilon > 0$ there exists η in $\Gamma(E)$ of norm 1 such that $||\Phi_t^\#\eta - \lambda\eta|| < \epsilon$.

In this section we shall prove that the entire spectrum of the group $\Phi_t^\#$ is accounted for by exponentiation. We denote $\Phi_1^\#$ by $\Phi^\#$.

THEOREM (2.1). If the non-periodic points of the flow ϕ^t are dense, then for $\Phi_t^\#: \Gamma(E) \to \Gamma(E)$

$$\exp(t\,\sigma(L)) = \sigma(\Phi_t^\#)\ .$$

PROOF: in view of the known results and the fact that $L - \lambda$ generates the semigroup $e^{-t\lambda}\Phi_t^\#$, it suffices to prove that $1 \in \sigma_{ap}(\Phi^\#)$ implies $0 \in \sigma_{ap}(L)$.

Choose $\epsilon > 0$ and set $c = \sup\{||\Phi_x^t|| \mid 0 < t \le 1 \text{ and } x \in M\}$. Find $\eta \in \Gamma(E)$ such that $||\eta|| = 2$ and $||\Phi^\#\eta - \eta|| < \epsilon/4c$. If $n > (8c + \epsilon)/\epsilon$, for some integer n, there is a non-periodic point x_0 and a cross-section Σ to ϕ_t with $x_0 \in \Sigma$, $||\eta(x_0)|| \ge 1$, and $\Sigma \times [-n,n]$ a flow box for ϕ^t. Define a smooth bump function $\alpha: \Sigma \to [0,1]$ such that $\alpha(x_0) = 1$ and α vanishes outside some open set in Σ containing x_0. We shall need some additional smooth functions. First, let $f: [0,1] \to [0,1]$ be such that $f[0,1/5] = 0$, $f[4/5,1] = 1$, and $f'(t) < 2$ for $0 \le t \le 1$. Next find a map $g: [-n,n] \to [0,1]$ with the properties

$g(-n) = 0 = g(n)$, $g(0) = 1$ and $g'(t) < 2/n$ for $|t| \leq n$. Finally, define $\beta: M \to [0,1]$ by $\beta(\phi^t\sigma) = \alpha(\sigma)g(t)$ for $\sigma \in \Sigma$ and $|t| \leq n$, and $\beta = 0$, otherwise.

With these auxiliary functions we can construct a smooth vector field ξ with $||\xi|| \geq 1$ and $||L\xi|| < \epsilon$. First, define ξ_1 on the set $T_1 = \{\phi^t\sigma | \sigma \in \Sigma, 0 \leq t \leq 1\}$ such that

$$\xi_1(\phi^t\sigma) = \Phi^t(\eta(\sigma) + f(t)(\Phi^{\#}\eta(\sigma) - \eta(\sigma))) .$$

A calculation now shows

$$L\xi_1(\phi^t\sigma) = \Phi^t(f'(t)(\Phi^{\#}\eta(\sigma) - \eta(\sigma)))$$

and, hence, $$||L\xi_1(x)|| \leq \epsilon/2, \quad \text{for } x \in T_1 .$$

Observe that ξ_1 agrees with η on Σ and on $\phi^1(\Sigma)$, and that $L\xi_1(\phi^t\sigma) = 0$ except for $1/5 < t < 4/5$. Thus, it is possible to carry out the preceding construction in each tubular section $\Sigma \times [k, k+1]$, $-n \leq k < n$. This yields a <u>smooth</u> vector field ξ_1 on the flow box $\Sigma \times [-n,n]$, with the property that $||L\xi_1|| \leq \epsilon/2$. To obtain the desired global vector field, define for $\sigma \in \Sigma$ and $|t| \leq n$,

$$\xi(\phi^t\sigma) = \beta(\phi^t\sigma)\xi_1(\phi^t\sigma) ,$$

and put $\xi = 0$, otherwise.

As defined, ξ is smooth, $||\xi(x_0)|| \geq 1$, and

$$L\xi(\phi^t\sigma) = \beta(\phi^t\sigma)\, L\, \xi_1(\phi^t\sigma) + d\beta/dt\, (\phi^t\sigma)\xi_1(\phi^t\sigma) .$$

Therefore, $||L\xi|| \leq \epsilon/2 + 2||\xi_1||/n$

$\leq \epsilon.$

Q.E.D.

We now specialize to the case of a smooth flow f^t whose adjoint group $f_t^{\#}$ given by

$$f_t^{\#}\eta = Tf^{-t} \circ \eta \circ f^t$$

and associated infinitesimal generator L_X (Lie differentiation in the direction $X = \frac{d}{dt}f^t$) act on the continuous sections of $E = TM/[X]$. The fact that, in this case, the base flow determines the form of the vector bundle map, i.e. $\Phi^t = Tf^t$, allows us to prove a stronger result. We will obtain the spectral mapping Theorem (2.3) without assuming the nonperiodic points are dense.

If U is an open set of periodic orbits define the function $P: U \to (0,\infty)$ which assigns to each point its prime period.

<u>LEMMA</u> (2.2). For almost every periodic orbit $\gamma \subseteq U$, there exists an open tubular neighborhood V of γ such that $V \subseteq U$ and $P|V$ is smooth. Moreover, if η is any smooth section of E supported on V, $L_X\eta = 0$.

<u>PROOF:</u> It is known that P is lower semicontinuous on U and the points of continuity of P are open and dense (Epstein [4], p. 70). If P is continuous at x_0, let V be a tubular neighborhood of the orbit γ_{x_0} such that $P|V$ is continuous. Obviously, there is a one-parameter family of cross-sections S_x along the orbit such that $P|S_x$ coincides with the time of first return map $h_x: S_x \to S_x$, which i smooth. But the flow is smooth so P must be smooth on V.

For the second assertion, since P is smooth on V, we can reparameterize X to a vector field gX on V whose orbits have constant prime period, say 1. Evidently, the flow of gX is simply a suspension of the identity and, hence,

$L_{gX}\eta = 0$, modulo X, for sections of E supported on V. But, modulo X,

$L_X\eta = \frac{1}{g} L_{gX}\eta = 0.$

Q.E.D.

THEOREM 2.3: For the tangent map semigroup $f_t^{\#}: \Gamma(E) \to \Gamma(E)$ with generator L_X, the spectrum is given by

$$\sigma(f_t^{\#}) = \exp(t\sigma(L_X)) .$$

PROOF: We assume that e^{λ} lies in $\sigma_{ap}(f^{\#})$. One possibility is that for any $\epsilon > 0$ there is a section η and a nonperiodic point x_1 with $||\eta(x_1)|| \geq 1$ and $||f^{\#}\eta - e^{\lambda}\eta|| < \epsilon$. If this is the case, the proof proceeds precisely as in the construction given in the proof of Theorem (2.1) when the nonperiodic points were assumed dense. Otherwise, for $\epsilon > 0$, there exists an open set U of periodic orbits such that $||\eta|U|| \geq 1$ and $||f^{\#}\eta - e^{\lambda}\eta|| < \epsilon$ for some smooth section η. We may assume $\epsilon \leq \frac{1}{2}|e^{\lambda} - 1|$. Evidently, by the Lemma there are neighborhoods in U on which L_X vanishes identically. Consequently, $f^{\#}\eta(x) - \eta(x) = 0$ for x in such neighborhoods. But then, the estimate

$$\epsilon \geq ||f^{\#}\eta(x) - e^{\lambda}\eta(x)|| \geq |e^{\lambda} - 1| \cdot ||\eta(x)||$$

yields a contradiction unless $e^{\lambda} = 1$. However, $e^{\lambda} = 1$ is already accounted for, since 0 must lie in the point spectrum of L_X whenever the nonperiodic points are not dense (Lemma 2.2).

Q.E.D.

For completeness, we prove the following folk theorem:

PROPOSITION 2.4: The flow f^t is Anosov if and only if $f^{\#}$ is hyperbolic on $\Gamma(E)$.

PROOF: As in Mather [7], the hyperbolicity of $f^{\#}$ on $\Gamma(E)$ implies the existence of a hyperbolic splitting of the quotient bundle TM/[X]. The required Anosov splitting of the tangent bundle TM now follows from the proof of Theorem 3.1

in Churchill, Franke, and Selgrade [3].

Q.E.D.

COROLLARY 2.5: The flow f^t is Anosov if and only if L_X is invertible in $\Gamma(E)$.

PROOF: By Lemma 2.2 the nonperiodic points are dense. Hence, the result follows from the rotational invariance of the spectrum [7].

REMARKS: (1) While our use of the work "smooth" is taken to mean C^∞, it is clear that all results obtain for C^1 vector fields X. Recent results in David Hart's thesis imply that a reasonable theory also exist for C^0 vector fields. (2) Although (2.3) is false for arbitrary vector bundle flows (Φ^t, ϕ^t) with large sets of periodic orbits (see Section 1), we conjecture that infinitesimal hyperbolicity always exponentiates: if L generates the adjoint semigroup $\Phi_t^\#$ of (Φ^t, ϕ^t) then $\sigma(L) \cap iR = \emptyset$ if and only if (Φ^t, ϕ^t) is hyperbolic.

3. The spectrum in Hilbert space.

Let μ denote a smooth measure on M, the Riemannian base manifold of (E, M, π). The Riemannian metric g on M induces an inner product in $\Gamma(E)$ by complexifying

$$\langle\eta, \xi\rangle = \int_M g(\eta, \xi)d\mu .$$

The completion of $\Gamma(E)$ with respect to the associated norm is the complex Lebesgue space $\Gamma^2(E)$ of square integrable sections.

A smooth flow f^t on M which preserves μ induces a strongly continuous group $f_t^\#$ of operators in $\Gamma^2(E)$, $E = TM/[X]$ $(X = \frac{d}{dt} f^t)$, by

$$f_t^\#\eta = Tf^{-t} \circ \eta \circ f^t .$$

We shall establish suitable versions of (2.3) and (2.4) in $\Gamma^2(E)$ and, in particular, we shall show that f^t is Anosov if and only if L_X, the Lie derivative in the direction X, is invertible as an operator in $\Gamma^2(E)$.

Let (Φ, ϕ) denote either (Tf^1, f^1) acting in $E = TM/[X]$ or, in the case of a measure preserving diffeomorphism f, the map (Tf, f) acting in E = TM.

THEOREM (3.1): $\Phi^\#$ is hyperbolic on $\Gamma(E)$ if and only if $\Phi^\#$ is hyperbolic on $\Gamma^2(E)$. Moreover, in both spaces $\Phi^\#$ is hyperbolic if and only if $1 \notin \sigma(\Phi^\#)$.

To show that hyperbolicity on $\Gamma(E)$ implies hyperbolicity on $\Gamma^2(E)$ one uses the Anosov splitting of $E = E^+ \oplus E^-$ to construct an invariant splitting of $\Gamma^2(E) = \Gamma^2(E^+) \oplus \Gamma^2(E^-)$. The hyperbolic estimates for $\Phi^\#$ on $\Gamma^2(E^\pm)$ follow from the integral estimate

$$\int_M ||\eta||^2 d\mu \le \sup||\eta||^2 \mu(M)$$

and the fact that ϕ preserves μ.

The converse of (3.1) is more subtle; we will indicate the main steps in the proof.

PROPOSITION (3.2): If the non-periodic points of ϕ are dense, then

$$\sigma_{ap}(\Phi^\#, \Gamma(E)) \subseteq \sigma_{ap}(\Phi^\#, \Gamma^2(E)) .$$

The proof requires showing that given an almost invariant C^0 vector field one can construct an almost invariant L^2 vector field. In particular, if ξ_n is a sequence of unit fields in $\Gamma(E)$ such that

$$||\Phi^\#\xi_n - \lambda\xi_n|| \to 0$$

it follows immediately that

$$||\Phi^{\#}\xi_n - \lambda\xi_n||_{L^2} \to 0;$$

however, there is no obvious way to insure that $||\xi_n||_{L^2}$ is bounded away from zero. This difficulty is overcome by making a "Mather type" construction similar to the construction used in the proof of Theorem (2.1).

Using (3.2), we can prove (3.1). Assume $\Phi^{\#}: \Gamma^2(E) \to \Gamma^2(E)$ is hyperbolic. If the non-periodic points of ϕ are not dense, an argument of Mather ([7], p. 481) yields a non-zero section $\eta \in \Gamma(E)$ such that $\Phi^{\#}\eta = \eta$. Clearly, $\eta \in \Gamma^2(E)$, contradicting our assumption.

Thus, we may suppose that the non-periodic points are dense and, by rotational invariance, that $1 \notin \sigma(\Phi^{\#}, \Gamma^2(E))$. By (3.2), $1 \notin \sigma_{ap}(\Phi^{\#}, \Gamma(E))$. Therefore, $\Phi^{\#} - I$ is injective with closed range on $\Gamma(E)$. But this implies that Φ has no bounded orbits (Mañe [8], p. 367). Since ϕ is measure preserving, M equals the nonwandering set of ϕ, and by a result due to Selgrade [12], this implies that Φ has a hyperbolic splitting on E, and $\Phi^{\#}$ is hyperbolic on $\Gamma(E)$.

Q.E.D.

With the outline given above one can prove more:

<u>THEOREM</u> (3.3): $\sigma(\Phi^{\#}, \Gamma(E)) = \sigma(\Phi^{\#}, \Gamma^2(E))$.

In particular, we emphasize that (3.3) implies that for a measure-preserving diffeomorphism or flow f:

(3.4) f is Anosov if and only if $I - f^{\#}$ is invertible on $\Gamma^2(E)$.

In the flow case we also have the analogues of (2.1) and (2.2) for the space $\Gamma^2(E)$

<u>THEOREM</u> (3.5): If f^t preserves a smooth measure μ, then for $f_t^{\#}: \Gamma^2(E) \to \Gamma^2(E)$

$$\exp(t\sigma(L)) = \sigma(f_t^{\#}) .$$

The proof reduces to showing that $1 \in \sigma_{ap}(f^{\#}, \Gamma^2(E))$ implies $0 \in \sigma_{ap}(L, \Gamma^2(E))$. From (3.3) we conclude that $1 \in \sigma'_{ap}(f^{\#}, \Gamma(E))$ and hence there is a sequence ξ_n in the domain of L in $\Gamma(E)$ so that

$$||L\xi_n||_{C^0} \to 0 .$$

But, as in (3.2) we do not know that $||\xi_n||_{L^2}$ is bounded away from zero. Nonetheless, one can find a sequence ξ'_n such that $||\xi'_n||_{L^2} = 1$ and

$$||L\xi'_n||_{L^2} \to 0$$

by using the invariant measure, the sequence ξ_n and making a construction similiar to the construction in the proof of (2.1).

Of course, the main result of this section now follows:

<u>THEOREM</u> (3.6): If f^t preserves a smooth measure μ, then f^t is Anosov if and only if $L_X(X = \frac{d}{dt} f^t)$ is invertible as an operator in $\Gamma^2(E)$.

4. Infinitesimal hyperbolicity for the geodesic flow.

We now consider, as a fundamental example and application, the geodesic flow G^t on the unit tangent bundle $M = T_1N$ of a compact Riemannian manifold (N,g). When the sectional curvatures of g are all negative, the geodesic flow is the classic example of an Anosov flow.

From our perspective, since G_t is volume-preserving, a natural setting for questions of hyperbolicity is the space $\Gamma^2(E) = \Gamma^2(TM/X)$, where X denotes the geodesic vector field. Of course, using (3.6) we can show that G^t is Anosov by proving that L_X is infinitesimally hyperbolic; i.e. L_X is invertible on $\Gamma^2(E)$. The appeal of this approach is evident once one recognizes that the Lie derivative L_X must carry the differential geometric structure associated with the metric g.

The Levi-Civita connection induced by g prescribes a decomposition of TM into horizontal and vertical components. Thus, a real vector field A on M splits as $A = (a,b)$ with a and b vector fields on N. Given $A = (a,b)$ and $B = (c,d)$, two natural metric tensors may be defined on M: the Sasaki metric

$$S(A,B) = g(a,c) + g(b,d)$$

and the Vilms metric

$$V(A,B) = g(a,d) + g(b,c) \ .$$

Results in Section 2 imply that we can replace $\Gamma^2(E)$ by the real Lebesgue space $L^2(E)$, which admits the inner product

$$\langle A,B \rangle = \int_M S(A,B)\,d\mu \ ,$$

and the indefinite inner product

$$(A,B) = \int_M V(A,B)\,d\mu \ .$$

Suppose ∇ is the covariant derivative given by the Levi-Civita connection generated by V. Then, if X is the geodesic vector field one can verify [2] that

(1) $XS(A,B) = S(\nabla_X A,B) + S(A, \nabla_X B)$.

(2) At a point (x,u) in M, the operator given by $\Omega = L_X - \nabla_X$ may be represented in horizontal and vertical components by

$$\Omega = \begin{pmatrix} 0 & -I \\ R & 0 \end{pmatrix} ,$$

where I = identity and R is the curvature operator $R_{(x,u)}(a) = R_x(a,u)u$.

(3) $V(\Omega A,B) = V(A,\Omega B)$.

(4) If the sectional curvatures are bounded below zero, there is a constant $c > 0$ such that

$$(\Omega A,A) \leq -c^2||A||^2 .$$

The hyperbolicity of G^t is then a consequence of the following Hilbert space result:

THEOREM (4.1). Suppose (x,y) is a continuous nondegenerate symmetric form on the Hilbert space H, and suppose D is a densely defined operator in H which is antisymmetric with respect to (x,y). If, in addition, the bounded operator B is symmetric and negative definite, i.e. $(Bx,y) = (x,By)$ and $(Bx,x) \leq -c^2||x||^2$ for some constant $c > 0$ and all $x \in H$, then $D + B$ is invertible in H.

PROOF: First, we show that $L = D + B$ is injective with closed range. If there is a sequence x_n in the domain of D with $||x_n|| = 1$ and $\lim_{n\to\infty} L x_n = 0$, then $\lim_{n\to\infty} (Lx_n,x_n) = 0$. But $(Dx_n,x_n) = 0$, for each n, implies that $\lim_{n\to\infty} (Bx_n,x_n) = 0$ - a contradiction.

To complete the proof notice that if an element z is orthogonal to the range of L, then we have for all x in the domain of D

$$(Lx,z) = (x,L'z) = 0.$$

Therefore, $L'z = D'z + B'z = -Dz + Bz = 0$, where prime denotes the adjoint in the (indefinite) inner product $(\, , \,)$. But this implies $(Bz,z) = 0$ and, hence $z = 0$.

Q.E.D.

COROLLARY 3.2: If the geodesic flow G^t has all negative sectional curves, then L_X is infinitesimally hyperbolic and, hence, G^t is Anosov.

PROOF: The result follows from Theorem (3.5) and Theorem (4.1) with $D = \nabla$ and $B = \Omega$.

Q.E.D.

Questions. E. Hopf observed that the geodesic flow would be hyperbolic even in the presence of small patches of small positive curvature (e.g. see [1]). Can one specify precisely how much positive curvature is allowable? Thus, what is the proof that L_X is invertible when some positive curvature occurs? In this regard, we feel that decomposing L_X will continue to be important. Also, if the hypothesis on the amount of positive curvature takes the form of an average, it seems appropriate to use the Hilbert space $L^2(E)$.

References

1. D. V. Anosov, "Geodesic flows on closed Riemannian manifolds with negative curvature," Proc. Stek. Inst. Math. 90(1967).

2. Carmen Chicone, Tangent bundle connections and the Geodesic flow, Preprint 1978.

3. R. C. Churchill, John Franke and James Selgrade, A geometric criterion for hyperbolicity of flows, Proc. Amer. Math. Soc. 62(1977), 137-143.

4. D. B. A. Epstein, Periodic flows on three-manifolds, Annals of Math. 95(1972), 66-82.

5. J. Franks and C. Robinson, A quasi-Anosov diffeomorphism that is not Anosov, Trans. Amer. Math. Soc. 233(1976), 267-278.

6. E. Hille and R. Phillips, "Functional analysis and semi-groups," Amer. Math. Soc. 1957.

7. J. N. Mather, Characterization of Anosov diffeomorphisms, Indag. Math. 30(1968), 479-483.

8. Ricardo Mañé, Quasi-Anosov diffeomorphisms and hyperbolic manifolds, Trans. Amer. Math. Soc. 229(1977), 351-370.

9. Ricardo Mañé, Persistent manifolds are normally hyperbolic, Bull. Amer. Math. Soc. 80(1974), 90-91.

10. N. Ôtsuki, A characterization of Anosov flows for geodesic flows, Hiroshima Math. Jour. 4(1974), 397-412.

11. R. J. Sacker and G. R. Sell, A note on Anosov diffeomorphisms, Bull. Amer. Math. Soc. 80(1974), 278-280.

12. J. F. Selgrade, Isolated invariant sets for flows on vector bundles, Trans. Amer. Math. Soc. 203(1975), 359-390.

13. R. Swanson and C. Chicone, Anosov does not imply infinitesimal ergodicity, to appear in Proc. Amer. Math. Soc..

14. M. Hirsch, C. Pugh and M. Shub, "Invariant Manifolds," Lecture notes in Math., Springer, Berlin 1977.

A Qualitative Singular Perturbation Theorem

C. Conley

§1. Introduction.

The theorem to be proved here takes the form of a criterion that a compact set which is not an isolating neighborhood for the limit of a (directed) family of flows is for flows close to the limit.

An isolating neighborhood for a flow means a compact set such that no boundary point of the set is on an orbit which is contained in the set. Since all boundary points must then leave the set under the flow, the compactness assures that the same will be true for nearby flows. But in the situation treated here the given set is not an isolating neighborhood in the limit; this explains the adjective "singular".

Motivating the theorem is the view that many significant properties of the flow are reflected in the existence of isolating neighborhoods, or perhaps more accurately, the companion isolated invariant sets (an isolated invariant set is one which is maximal in some isolating neighborhood). This is true in some generality of those properties which are "stable" to perturbation. For example the statement that a flow is Anosov is equivalent to the existence of isolating neighborhoods and the stability of this property is a direct consequence of the stability property of such neighborhoods; also, the criterion that a smooth invariant manifold perturbs to one with a given degree of smoothness can be naturally stated in terms of isolating neighborhoods of (iterated) tangent equations. Further examples involve the existence of special types of solutions such as solutions connecting critical points (structure for shock waves and special traveling wave solutions). More generally, properties that can be derived from the existence of Liapunov type functions are expressible in terms of isolating neighborhoods. In fact, the converse is also true, so one can interpret the present theorem as one asserting the existence of a (local) Liapunov function. (The above remarks are intended to justify the theorem proved here and are needed because the examples included do not.)

In fact, the criterion given in the main theorem is essentially that certain functions exist which are Liapunov functions in some average sense. The first point to watch for is the characterization of the domain in which these functions must be operative in order that the given set be an isolating neighborhood for flows close to the limit.

In this characterization the idea of a Morse decomposition plays the central role. These are similar to the Smale decompositions except that there is no requirement of hyperbolicity on the sets. Because a Morse set is an isolated invariant set it has an "index". The indices of the sets of a decomposition satisfy Morse-Smale type inequalities — hence the name Morse.

The original purpose of the perturbation theorem was to provide a lemma for use in determining the Morse index of an isolated invariant set. The idea is like that in degree theory where one uses the homotopy property to exchange an apparently complicated situation for a simpler one. In that process it must be known that no fixed points cross the boundary of some given open set. In the present case, the open set is replaced by a compact one and the boundary condition is that it be an isolating neighborhood throughout the "homotopy".

Some simple (not to say trivial) examples are included to illustrate the hypotheses and use of the theorem and it is also pointed out that the theorem can be used to show that a relative index (defined in [2]) is well-defined. The questions here grew out of a problem suggested by J. Smoller and in a joint work, the theorem will be more extensively illustrated. Also in a later work, a substantial refinement of the result will be given. This refinement requires an improved development of basic notions which will apply to directed families of flows which do not have a limiting flow; there are too many details to include it here.

The background material required here is developed in [1], but the basic definitions are given in section two for reference. Also the main (qualitative) lemma is proved in this section. This lemma allows the restriction of the domain in which the "Liapunov functions" must operate. In effect, it takes the place of estimates and it is the key point of the theorem. In section three the perturbation theorem is proved and section four contains the examples.

The work here is related to recent work of C. Robinson and J. Murdock, [3], and their ideas are represented here (in different form). Also the work owes a lot to conversations with N. Fenichel whose fine appreciation of the problem led to a much refined application of the qualitative lemma formulated by the author. (The further development also grew from Fenichel's remarks.)

§ 2. A Qualitative Lemma.

2.1 Some Definitions.

The basic definitions used here are listed below along with the necessary results. More detail is found in [1]. (The definition of attractor given below is one of several; in particular, this one places no "internal" restrictions on the set, only that it attract nearby solutions.)

A. Definition 1.

A flow on a topological space X means a continuous function $(x,t) \mapsto x \cdot t$ from $X \times R \to X$ such that for all $x \in X$ and $s, t \in R$, $x \cdot 0 = x$ and $x \cdot (s+t) = (x \cdot s) \cdot t$.

For $Y \subset X$, and $J \subset R$, $Y \cdot J \equiv \{(x,t) \mid x \in Y \text{ and } t \in J\}$.

An invariant set is a set I such that $I \cdot R = I$.

Definition 2.

For $Y \subset X$, $\omega(Y)$ means the maximal invariant set in $cl(Y \cdot [0,\infty))$ (where $cl(Z)$ = closure Z). The corresponding set for the backward flow is denoted $\omega^*(Y)$.

B. Attractors.

Definition.

Let I be a compact invariant set. A set A is called an attractor relative to I if $A = \omega(U)$ where U is a relative neighborhood of A in I. Such a U is called an attractor neighborhood of A.

If A is an attractor relative to I and U is a compact attractor neighborhood then $A^* \equiv \omega^*(I \setminus U)$. A^* is called the repeller dual to A in I.

Remark: A^* is independent of U (subject to the stated condition) and is an attractor for the backward flow.

The strong stability property of attractors is stated in the following lemma wherein the topology on the space of flows is the C-O topology.

Lemma. Suppose I is a compact invariant set and (A, A^*) is an attractor-repeller pair in I. Let U and U^* be neighborhoods in X of A and A^* respectively. Then there are neighborhoods $\mathcal{O}$ and $\mathcal{O}^*$ of $I\setminus U^*$ and $I\setminus U$ respectively and a $T > 0$ such that for sufficiently nearby flows — indicated by $(x,t) \mapsto x:t$ — $\mathcal{O}:T \subset U$ and $\mathcal{O}^*:(-T) \subset U^*$.

Remark: The proof comes from the fact that $\omega(I\setminus U^*)$ is in U together with the definition of C-O topology.

C. Morse Decompositions.

Definition.

Let $\emptyset = A_0 \subset A_1 \subset \ldots \subset A_n = I$ be an increasing sequence of attractors in the compact invariant set I. For $i = 1, \ldots, n$ define $M_i = A^*_{i-1} \cap A_i$. Then the sets M_i are called Morse sets and the (ordered) collection $D = \{M_1, \ldots, M_n\}$ a Morse decomposition.

There is also a need later for the definition $M_{ij} \equiv A^*_{i-1} \cap A_j$.

Let $M(D) \equiv M_1 \cup \ldots \cup M_n$. Then $R(I)$ is defined to be the intersection over all decompositions D of the sets $M(D)$; $R(I)$ is called the chain recurrent set of I.

Remarks. The set $R(I)$ is seen in the criterion of section three to be the domain in which the Liapunov functions must be effective. In the more refined criterion this set will be replaced by a smaller, "higher order" chain recurrent set; though smaller, it requires a larger collection of definitions.

The chain recurrent set is otherwise characterized in terms of approximate orbits ([1]).

The needed results about Morse decompositions are listed in the following lemma.

Lemma.

1) The Morse sets of a decomposition are disjoint.

2) The two set Morse decompositions of I are precisely the attractor-repeller pairs.

3) If $\{M_1, M_2, M_3\}$ is a Morse decomposition of I then so are $\{M_{12}, M_3\}$ and $\{M_1, M_{23}\}$. (This is a special case of the obviously generalized statement.)

4) If $\{M_1, \ldots, M_n\}$ is a Morse decomposition of I then $\{M_i \ldots M_j\}$ is a Morse decomposition of M_{ij}.

5) A finite collection of disjoint invariant sets is a Morse decomposition if and only if it can be ordered, say $M_1, \ldots, M_n$, so that for $x \notin M_1 \cup \ldots \cup M_n$, there exists i and j with $i < j$ and such that $\omega(x) \subset M_i$ and $\omega^*(x) \subset M_j$. (This is proved in [1]; it will not be used here directly but clarifies the definition.

2.2 Behavior Near a Morse Decomposition.

The first lemma, in A of this section, is a step towards the main lemma in B. They are both refinements of the lemma in 2.1B about attractors. The purpose of the somewhat complicated statement becomes more clear in section three where it is applied to prove the singular perturbation theorem which is the aim of this paper.

A. Lemma.

Let $\{M_1, M_2, M_3\}$ be a Morse decomposition of I corresponding to the attractor sequence $\phi = A_0 \subset A_1 \subset A_2 \subset A_3 = I$.

Let U_2 be a neighborhood of M_2 and let V_1 be a neighborhood of M_1.

Then there are neighborhoods W_2 of I and V_{12} of M_{12} such that for all flows $(:)$ sufficiently close to the given one, the following condition is satisfied:

If $x \in V_{12}$ then either

a. $x : [0, \infty) \subset U_2$ or

b. There is a $t' > 0$ such that $x : [0, t'] \subset U_2$ and $x : t' \in U_2 \setminus W_2$ or

c. There is a t' with $x : t' \in V_1$.

Furthermore if W_2 and V_{12} satisfy the conditions and V_{12}' is a neighborhood of M_{12} contained in V_{12} then W_2 and V_{12}' satisfy the conditions. (This is obvious.)

Remark: The lemma says that orbits passing near M_{12} (entering V_{12}) either 1) stay near M_2 (i.e. in U_2) or leave the vicinity of I (W_2) before leaving U_2 or 3). go on down to $M_1(V_1)$. In the application, this means Liapunov functions only have to be effective near Morse sets. This lemma is set up for the induction argument of the next one.

Proof.

1. Choose open neighborhoods $\hat{V}_1$ of M_1 and $\hat{U}_2$ of M_2 with

closures in V_1 and U_2 (respectively) and a neighborhood $\hat{U}_3$ of M_3 so that $\hat{V}_1$, $\hat{U}_2$ and $\hat{U}_3$ have disjoint closures and so that $\mathrm{cl}(\hat{V}_1) \cap M_{23} = \mathrm{cl}\,\hat{U}_3 \cap M_{12} = \phi$. This is possible because M_1, M_2 and M_3 are disjoint and $M_1 \cap M_{23} = M_{12} \cap M_3 = \phi$. (Lemma 2.1C. (3).)

2. Let $C_2 \equiv M_{23} \setminus \hat{U}_2$ and observe that C_2 is disjoint from M_{12} since $M_{12} \cap M_{23} = M_2 \subset \hat{U}_2$. Since (M_2, M_3) is an attractor-repeller pair in M_{23} (2.1C 4).) there is a T_2 and a neighborhood $\mathfrak{O}_2$ of C_2 with closure disjoint from M_{12} such that for sufficiently nearby flows, $\mathfrak{O}_2 : (-T_2) \subset \hat{U}_3$ (2.1B).

Note that $\hat{U}_2 \cup \mathfrak{O}_2$ is a neighborhood of M_{23} and $M_{12} \setminus \hat{U}_2 \cup \mathfrak{O}_2 = M_{12} \setminus \hat{U}_2$ (since $\mathfrak{O}_2$ is disjoint from M_{12}).

3. Let $C_1 = I \setminus \hat{U}_2 \cup \mathfrak{O}_2$. Since $\hat{U}_2 \cup \mathfrak{O}_2$ covers M_{23} and has closure disjoint from M_1, there is a neighborhood $\mathfrak{O}_1$ of C_1 and a $T_1 > 0$ such that for sufficiently nearby flows, $\mathfrak{O}_1 : T \subset \hat{U}_1$ (2.1C 4)., 2). and 2.1B). Observe that $\mathfrak{O}_1 \cup \hat{U}_2 \cup \mathfrak{O}_2$ covers I and therefore $\mathfrak{O}_1 \cup \hat{U}_2$ covers M_{12}.

Define $W_2 = \mathfrak{O}_1 \cup \hat{U}_2 \cup \mathfrak{O}_2$; this is the required neighborhood of I.

4. Choose a neighborhood V_{12} of M_{12} such that for sufficiently nearby flows, $V_{12} : [0, T_2] \subset \mathfrak{O}_1 \cup \hat{U}_2$. This is possible since $M_{12} \cdot R = M_{12} \subset \mathfrak{O}_1 \cup \hat{U}_2$ (and the topology on the space of flows is the C-O topology).

Suppose now that $x \in V_{12}$ and $:$ indicates a flow for which the conditions in 2., 3. and 4. are satisfied.

5. If $x \in \Theta_1$ then $x : T_1 \in \hat{V}_1$ (3.) and the condition c. of the lemma is satisfied. Therefore it can be assumed that x $\hat{U}_2$.

6. Now either $x : [0, \infty) \subset \hat{U}_2$ and condition a. is satisfied or there is a $t > 0$ such that $x : [0, t) \subset \hat{U}_2$ and $x : t \in \partial \hat{U}_2$ (= boundary $\hat{U}_2$). Recall (1.) that $\hat{U}_2$ is open so $x : t \notin \hat{U}_2$ in this case — which is the only one that needs more discussion.

7. If $t \leq T_2$ then by 4., $x : t \in \Theta_1 \cup \hat{U}_2$; since it is not in $\hat{U}_2$, it is in Θ_1 and (3.) $x : (t + T_1) \in \hat{V}_1$ so condition c is satisfied.

8. If $t > T_2$ then $x : t$ cannot be in Θ_2; otherwise $(x:t) \cdot (-T_2)$ would be in $\hat{U}_3$ which is not possible because it is in $\hat{U}_2$ (2. and 1.). Therefore, if $x : t \in W_2 = \Theta_1 \cup \hat{U}_2 \cup \Theta_2$, then it is in Θ_1 and again $x : (t+T_1) \in \hat{V}_1$.

9. If $x : t \notin W_2$, then since $x : t \in \partial \hat{U}_2 \subset U_2$, condition b. is satisfied.

In view of 5. - 9., the lemma is proved.

B. <u>Main Lemma.</u>

Let $\{M_1, M_2, \ldots, M_n\}$ be a Morse decomposition of I corresponding to the attractor sequence $\phi = A_0 \subset A_1 \subset \ldots \subset A_n = I$.

Let U be a neighborhood of $M(D) = M_1 \cup \ldots \cup M_n$.

Then for $i = 1, \ldots, n$ there are neighborhoods V_i of $M_{1i} (= A_i)$ and W_i of I such that for sufficiently nearby flows, the following condition is satisfied.

If $x \in V_i$ then either

a). $x : [0, \infty) \subset U$ <u>or</u>

b). there is a t' such that $x : [0, t'] \subset U$ and $x : t' \in U \setminus W_i$ <u>or</u>

c). there is a t' such that $x : t' \in V_{i-1}$.

Furthermore the sets can be constructed in the order $W_1, V_1, W_2, V_2, \ldots W_n, V_n$ and V_i can be chosen as small as desired without changing the previously constructed sets; in particular, W_i.

Note that if the condition is satisfied, then given $x \in V_n$ and a sufficiently nearby flow, either there is an orbit tail, $x : [t', \infty)$ in U or there is an orbit segment $x : [t', t'']$ and an i with $x : t' \in V_i$, $x : [t', t''] \subset U$ and $x : t'' \in U \setminus W_i$.

<u>Proof.</u> Let $\overline{M}_1 = \emptyset$, $\overline{M}_2 = M_1$ and $\overline{M}_3 = A_1^* = M_{2n}$. Then $\{\overline{M}_1, \overline{M}_2, \overline{M}_3\}$ is a decomposition of I. Let $\overline{U}_2 = U$ and let $\overline{V}_1 = \emptyset$. Then the previous lemma supplies neighborhoods $\overline{W}_2$ of I and $\overline{V}_{12}$ of $\overline{M}_{12}$ $(= \overline{M}_2 = M_1)$ satisfying the condition of the lemma.

Let $W_1 = \overline{W}_2$ and let $V_1 = \overline{V}_{12}$. Then if $x \in V_1$, either $x : [0, \infty) \subset U$ or there is a t' with $x : [0, t'] \subset U$ and $x : t' \in U \setminus W_1$ (the third situation doesn't come up because $\overline{V}_1 = \emptyset$). Thus the condition of the theorem is satisfied for $i = 1$. Note also that $V_1 = \overline{V}_{12}$ can be chosen as small as desired without changing W_1 (last statement in Lemma 2.2A).

Suppose sets U_i and W_i satisfying the condition have been constructed for $i < k$. Let $\overline{M}_1 = M_{1, k-1}$, $\overline{M}_2 = M_k$ and $\overline{M}_3 = M_{k+1, n}$. Let $\overline{V}_1 = V_{k-1}$ and $\overline{U}_2 = U$. Then again there are neighborhoods $\overline{W}_2$

of I and $\overline{V}_{12}$ of $\overline{M}_{12}$ $(= M_{1k})$ satisfying the condition of the lemma. Let $W_k = \overline{W}_2$ and $V_k = \overline{V}_{12}$. Then if $x \in V_k$, either $x : [0, \infty) \subset U$ or for some t', $x : [0, t'] \subset U$ and $x : t' \notin W_k$ or $x : t' \in V_{k-1}$. Again V_k can be reduced as much as needed without changing W_k or the previously constructed sets.

§3. A Singular Perturbation Theorem.

The theorem concerns a family of differential equations on $\mathbb{R}^n$, namely

(1) $$\dot{x} = f(x,\lambda) = f_0(x) + \lambda f_1(x) + \ldots ,$$

wherein f depends smoothly on λ for $\lambda \geq 0$. For each λ these equations are assumed to define a flow (a local flow would be sufficient) which will be indicated by $(x,t) \to x \cdot \lambda \cdot t$ except when $\lambda = 0$ in which case $x \cdot t$ will do.

Let $N \subset \mathbb{R}^n$ be a compact set and let $\overline{I}$ be the set of points x with $x \cdot R \subset N$. If $\overline{I} \cap \partial N = \phi$ then N is an isolating neighborhood for all small enough λ (i.e. orbits through boundary points leave N). The theorem to be proved gives a criterion that N is an isolating neighborhood for small positive λ even if it is not when $\lambda = 0$.

The criterion is stated in terms of functions ℓ which are Liapunov functions in some average sense. In 3.1 some facts about averages of functions on solutions are recalled, and in 3.2 the theorem is proved.

3.1 Averages of Functions on Solutions.

A. Definition.

Let I be a compact invariant set of the flow $(x,t) \to x \cdot t$ and let $g : I \to R$ be a continuous function. Then $\mathrm{Ave}(g, I)$ is the limit as T goes to infinity of the set of numbers $\{\frac{1}{T}\int_0^T g(x \cdot s)ds \mid x \in I\}$. If $\mathrm{Ave}(g, I)$ contains only positive numbers then "g has positive averages on I".

B. Definition. The minimal center of attraction of I, MCA(I), is the set of points $x \in I$ with the following property: if g is a non-negative (continuous) function on I which is positive at x, then Ave(g, I) contains a positive number.

C. Theorem. MCA(I) is a compact invariant set and Ave(g, I) is contained in any interval containing Ave(g, MCA(I)). Also, $MCA(I) \subset R(I)$ (= chain recurrent set of I defined in 2.1C).

D. Theorem.

Given a neighborhood J of Ave(g, I), there is a neighborhood U of I, and a $T > 0$ such that for flows (:) sufficiently near the given one and $t \geq T$, $x:[0,t] \subset U$ implies $t^{-1}\int_0^t g(x:s)ds \in J$.

If g has positive averages on I then U and the neighborhood of the flow can be chosen so that for some constant k, $x:[0,t] \subset U$ implies $\int_0^t g(x:s)ds \geq -k$ (k would generally be positive).

Remarks: The proof is straightforward and will be omitted. Note that Ave(g, I) is, in the above sense, "stable"; but the set MCA(I) which determines Ave(g, I) is not at all stable. For example the equation $\dot{\theta} = \sin^2 \theta$ determines a flow on the circle whose minimal center of attraction contains only two points, but for any $\lambda > 0$, $\dot{\theta} = \sin^2\theta + \lambda$ determines one for which the minimal center of attraction is the whole circle.

3.2 Slow Exit Points.

Let N be a compact set in $\mathbf{R}^n$ and (with reference to equation (1) at the beginning of section three) let $\bar{I}$ be the maximal invariant set of the 0th equation in N.

A. <u>Definition.</u> A point $x \in \bar{I} \cap \partial N$ is called a slow exit point of N if the following conditions are satisfied:

a. There is a compact invariant set, $I \subset \bar{I}$, of the 0th equation such that $\omega(x) \subset I$. Let R be the chain recurrent set of I.

b. There is a neighborhood U of R and a differentiable function $\ell(x) = \ell_0(x) + \ldots + \lambda^m \ell_m(x)$ defined from $cl(U)$ into R. Let $L_0 = \{x \mid \ell_0(x) = 0\}$.

c. $I \cap cl(U) = \bar{I} \cap L_0 \cap cl(U)$, in particular, $\ell_0 | I \cap cl(U) = 0$, and $\ell_0 | \bar{I} \cap cl U \leq 0$.

d. Let $g_j = \nabla\ell_0 \cdot f_j + \ldots + \nabla\ell_j \cdot f_0$. Then for some m, $g_j \equiv 0$ if $j < m$ and g_m has positive averages on R.

A slow entrance point is one that is a slow exit point for the backwards equation.

<u>Remarks.</u> Slow exit points must leave a neighborhood of $\bar{I}$ under the flows corresponding to small positive λ (Lemma B following). The idea is that the orbit from x must pass close to R where ℓ_0 is increasing (on the average) due to the condition on the g_j's. But once ℓ_0 becomes positive, condition c. implies the solution can't be close to $\bar{I}$. The main lemma is used to make sure ℓ_0 becomes positive in the neighborhood U of I.

B. Lemma.

If x is a slow exit point of N, there is a neighborhood $\overline{W}$ of I and a neighborhood $\mathfrak{G}$ of x such that for $x' \in \mathfrak{G}$ and sufficiently small positive λ, $x' \cdot \lambda \cdot [0, \infty)$ is not contained in $\overline{W}$.

Proof:

1. Since g_m has positive averages on R, U can be decreased so that the conditions of Theorem 3.1 D (on averages) are satisfied where J is a compact set of positive real numbers, g_m is g and R is the I of that theorem. In particular there is a λ_1 such that if $\lambda < \lambda_1$ the conclusion applies.

2. Now from a. and b. in the definition of slow exit point, U is a neighborhood of the chain recurrent set of I so contains the sets $\{M_1, M_2, \ldots, M_n\}$ of a Morse decomposition of I (2.1C). Using the main lemma (2.2B) neighborhoods V_i and W_i of M_{1i} and I (respectively) can be constructed so that the conditions of that lemma are satisfied. Choose $\lambda_2 \leq \lambda_1$ so that the lemma applies.

3. Suppose $W_1, V_1, \ldots, W_k$ have been constructed. Then using the next last paragraph of the lemma, V_k will be constructed as in 5 below.

Since W_k is a neighborhood of I and since $I \cap cl(U) = \overline{I} \cap L_0 \cap cl(U)$ (by c. in the definition of slow exit point) $cl[U \cap (\overline{I} \setminus W_k)]$ contains no points of L_0. Furthermore, since $\ell_0 | \overline{I} \cap cl(U) \leq 0$, $\ell_0 | cl[U \cap (\overline{I} \setminus W_k)]$ must be strictly negative, say less than $-5\delta_k$ where $\delta_k > 0$.

4. To find the desired neighborhood of $\bar{I}$, it is useful at this point to choose a neighborhood W_k' of $\bar{I} \setminus W_k$ such that $\ell_0 | W_k' \cap U < -4\delta_k$. Let $\overline{W}_k = W_k \cup W_k'$. Then $\overline{W}_k$ is a neighborhood of $\bar{I}$.

5. Now using the fact that $\ell_0 | I \cap \mathrm{cl}(U)$ is zero (c. again) choose V_k so that $\ell_0 | V_k \cap U > -\delta_k$.

Having constructed the W's and V's in this way, let δ be the minimum of the δ_k's.

6. Since $\omega(x) \subset I \subset V_n$, there is a neighborhood Θ of x such that if λ small enough, say $\lambda < \lambda_3 \le \lambda_2$, then $x' \in \Theta$ implies $x' \cdot \lambda \cdot t \in V_n$ for some t. By the last statement in the main lemma, this implies either some orbit tail of x' is contained in U or there is an orbit segment $x' \cdot \lambda \cdot [t', t'']$ and an i such that $x' \cdot \lambda \cdot t' \in U$. It is now necessary to examine the behavior of the Liapunov function, ℓ, on these orbit segments.

7. Denote the terms in $\nabla \ell \cdot f$ of order higher than m in λ by $\lambda^{m+1} r$. Let K be a bound (in U) for $|r|$ as well as $|\ell_j|$ for $j = 1, \ldots, m$. Then on U, $|\ell_0 - \ell| \le \lambda K + \ldots + \lambda^{m+1} K$. Choose $\lambda_4 \le \lambda_3$ so that $\lambda < \lambda_4$ implies $|\ell_0 - \ell| \le \delta$ on U. It follows that for $\lambda < \lambda_4$ and $y \cdot \lambda \cdot [0,t] \subset U$, $\ell_0(y \cdot \lambda \cdot t) - \ell_0(y) \ge \ell(y \cdot \lambda \cdot t) - \ell(y) - 2\delta$.

Now $\ell(y \cdot \lambda \cdot t) - \ell(y) = \int_0^t \nabla \ell \cdot f = \sum_0^{m-1} \lambda^j \int_0^t g_j + \lambda^m \int_0^t g_m + \lambda^{m+1} \int_0^t r = \lambda^m \int_0^t g_m + \lambda^{m+1} \int_0^t r$ $(g_j \equiv 0$ for $j < m)$.

8. With reference to 1., let J be the interval of positive reals and $T > 0$ and k be the constants guaranteed to satisfy the conditions of Theorem 3.1 D whenever $\lambda < \lambda_1$. Let $\Delta > 0$ be the left end point of

J. Suppose $\lambda < \lambda_4$ $(\leq \lambda_1)$. Then (by 7.) if $t \leq T$, $\ell(y\cdot\lambda\cdot t) - \ell(y) \geq -\lambda^m k - \lambda^{m+1}KT$; if $t \geq T$, $\ell(y\cdot\lambda\cdot t) - \ell(y) \geq \lambda^m \Delta t - \lambda^{m+1}Kt$ is positive and goes to infinity with t. Since ℓ is bounded on U this means that if $0 < \lambda < \lambda_5 = \min(\Delta/K, \lambda_4)$ then U contains no orbit tails of the λ-equation.

9. Choose $\lambda_6 \leq \lambda_5$ so that for $\lambda < \lambda_6$, $-\lambda^m k - \lambda^{m+1}KT > -\delta$ (δ is defined in 5.)

Suppose $0 < \lambda < \lambda_6$ and $x' \in \Theta$. By 6. (and since there are no orbit tails in U) there exists an orbit segment $x'\cdot\lambda\cdot[t', t''] \subset U$ with $x'\cdot\lambda\cdot t' \in V_i$ and $x'\cdot\lambda\cdot t'' \in U \setminus W_i$. Now $\ell_0(x'\cdot\lambda\cdot t'') - \ell_0(x'\cdot\lambda\cdot t') \geq \ell(x'\cdot\lambda\cdot t'') - \ell(x'\cdot\lambda\cdot t') - 2\delta \geq -\delta - 2\delta = -3\delta \geq -3\delta_i$. By 5., this means $x'\cdot\lambda\cdot t''$ is not in the set W_i' constructed in 4. More specifically $\ell_0 | V_i \cap U > -\delta_i$ and $\ell_0 | W_i' \cap U < -4\delta_i$; thus the difference exceeds $3\delta_i \geq 3\delta$. Since $x'\cdot\lambda\cdot t' \in V_i$, $x'\cdot\lambda\cdot t''$ cannot be in W_i'.

But $x'\cdot\lambda\cdot t'$ is not in W_i either, therefore it is not in the neighborhood $\overline{W}_i$ of $\overline{I}$ constructed in 4.

10. Let $\overline{W}$ be the intersection of the $\overline{W}_i$. Then $\overline{W}$ is a neighborhood of $\overline{I}$ and it has been shown that each point of Θ leaves some $\overline{W}_i$ hence $\overline{W}$. This completes the proof of Lemma 3.2B.

C. Theorem.

Let N be a compact set in $\mathbb{R}^n$ and, with reference to equation (1), let $\overline{I}$ be the maximal invariant set (of the 0th equation) in N.

If every point of $\bar{I} \cap \partial N$ is a slow exit or slow entrance point, N is an isolating neighborhood for sufficiently small positive λ.

<u>Proof:</u> For each slow exit or slow entrance point, x, there is a positive $\lambda(x)$, a neighborhood $\mathfrak{G}(x)$ of x and a neighborhood $W(x)$ of $\bar{I}$ such that points in $\mathfrak{G}(x)$ leave $W(x)$ provided $\lambda < \lambda(x)$ and $\lambda > 0$.

A finite number of the $\mathfrak{G}$'s cover $\bar{I} \cap \partial N$; let W_1 be the intersection of the corresponding W's and λ_1 the minimum of the corresponding λ's. Then $\lambda_2 > 0$ and W_1 is a neighborhood of $\bar{I}$.

Now $cl(N \backslash W_1)$ does not meet $\bar{I}$ so points in this set leave N under the 0th equation. Proceeding as in the first paragraph, there is a neighborhood W_2 $cl(N \backslash W_1)$ and a $\lambda_2 > 0$ such that points in W_2 leave N for $\lambda < \lambda_2$.

Observe that $W_1 \cup W_2$ is a neighborhood of ∂N and so all points of ∂N leave N if $\lambda > 0$ is less than $\min(\lambda_1, \lambda_2)$. This proves the theorem.

<u>Remark:</u> As mentioned in the introduction, the statement that N is an isolating neighborhood implies the existence of Liapunov functions which are, however, defined on a much larger domain than the ℓ for the slow exit (entrance) points. The theorem, from this point of view, obtains big Liapunov functions from little ones.

§4 Some Examples.

4.1 Fast-Slow Systems in the Plane.

A. The equations are assumed to be in the form:

$$(1) \qquad \begin{aligned} \mathring{x} &= \lambda g(x, y) \\ \dot{y} &= h(x,y) (= \frac{\partial}{\partial y} H(x,y)). \end{aligned}$$

Assume the equation $h(x,y) = 0$ determines a curve γ in the plane.

Let N be a rectangle in the plane whose horizontal boundaries do not intersect γ. Then $\overline{\Gamma} \cap \partial N$ consists of the two vertical intervals, I_ℓ and I_r, in the left and right hand boundary segments respectively whose end points are the uppermost and lowermost points of γ in the segment. These are invariant sets of the limit equation $(\lambda = 0)$.

The open intervals in $I_\ell \setminus \gamma$ and $I_r \setminus \gamma$ are "directed" by the limit equation: upward if h is positive, downward if h is negative. The components of $I_\ell \cap \gamma$ and $I_r \cap \gamma$, if finite in number, make up the Morse sets of a decomposition of the respective sets. The correct orderings (several are possible in general: cf. 2.1C 5).) are those such that no arrow points from a set with lower index to one with higher index.

Assume the sign of g is non-zero and constant on components of $I_\ell \cap \gamma$. and $I_r \cap \gamma$. It is then seen that a component of $I_r \cap \gamma$ consists of slow exit points if and only if g is positive on the component and there is a permissable ordering so that g is positive on all components with lower index. Similar formulations hold for the slow entrance points of I_r and for I_ℓ.

B. If the horizontal boundaries of N meet γ, say in an isolated point (x_0, y_0), then the Liapunov function to pick is $(y - y_0)$. When the point is isolated, this is not a degenerate situation. Namely, though $\ell = \ell(x,y)$ is zero at (x_0, y_0), $\dot{\ell} = \lambda \frac{\partial h}{\partial x} g + \frac{\partial h}{\partial y} h$ is equal to $\lambda \frac{\partial h}{\partial x} g$. If this is positive on the upper boundary or negative on the lower boundary, the point (x_0, y_0) is both an exit and an entrance point; otherwise the question is more global. In any case the theorem is of no help. (The situation does suggest an improvement taking account of second derivatives of ℓ.)

C. In general for fast slow systems like (1) but in higher dimensions, the set I (e.g. I_ℓ and I_r) in the definition of slow exit point has the form of the set of y's over some one x and the condition for slow exit has a form like that above where the components are replaced by the Morse set of the y-equation and "+" or "-" refers to pointing into or outside of N. (The Morse sets consist of critical points of the y equation provided it is gradient-like as it always is if $y \in \mathbb{R}^1$; otherwise an appropriate average of f determines the "sign".)

D. The popular case of the cubic curve is illustrated in Figure 1. Here N is an annulus however and, as a result of the indicated directions and signs, is an attractor neighborhood for small values of λ. Observe that the sharp corners are cut off with horizontal lines—any (small) amount would do. (The application of the theorem is not quite direct).

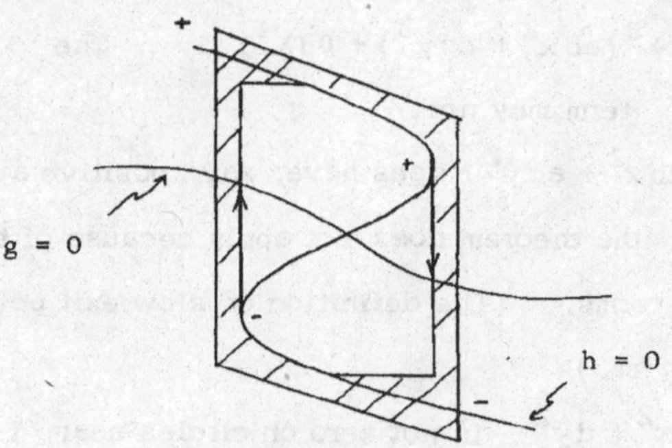

Figure 1

4.2 The Averaging Hypothesis and Use of ℓ_2.

A. The equations are:

$$\dot{x} = y + \lambda a(x,y)y + \lambda^2 b(x,y)x$$
$$\dot{y} = -x - \lambda a(x,y)x + \lambda^2 d(x,y)y.$$

Let N be the unit disk; then $\overline{I} = N$ and $I = \overline{I} \cap N$ is the periodic solution in the boundary.

Let $\ell_0 = \frac{1}{2}(x^2 + y^2)$. Then $\nabla\ell_0 \cdot f = \lambda^2(bx^2 + dy^2)$. The flow in I is (in polar coordinates) $\dot{\theta} = 1$, so if $b(x,y)x^2 + d(x,y)y^2$ has positive average on the unit circle, N is a repeller neighborhood for small positive λ.

B. But the "a" terms indicate that it is really a weighted average that should enter and might swing the result if the average of $bx^2 + dy^2$ is zero. To take advantage of this the time parametrization can be changed by a factor of $(1 + a)^{-1}$ (the property of being an isolating neighborhood has nothing to do with time parametrization).

The new equations are:

$$\dot{x} = y + \lambda^2 bx - \lambda^3 abx + 0(\lambda^4)$$
$$\dot{y} = -x + \lambda^2 dy - \lambda^3 ady + 0(\lambda^4).$$

Now $\ell_0 = \lambda^2(bx^2 + dy^2) - \lambda^3(abx^2 + acy^2) + O(\lambda^4)$. The λ^2 term has zero average but the λ^3 term may not.

However, even if $abx^2 + acy^2$ does have, say, positive average on the circle (i.e. on I), the theorem does not apply because of the presence of the second order terms. (The definition of slow exit points required $g_j \equiv 0$ for $j < m$.)

If the average of $bx^2 + dy^2$ is not zero on circles near I the theorem gives no result - except that there are nearby isolating neighborhoods as in A.

However if it is zero, then there is a function $\ell_2(x, y)$ such that $y\,\partial\ell_2/\partial x - x\,\partial\ell_2/\partial y = -\lambda^2(bx^2 + dy^2)$. Let $\ell = \ell_0 + \lambda^2\ell_2$. Then $\nabla\ell \cdot f = -\lambda^3(abx^2 + ady^2) + O(\lambda^4)$. It follows that if the average of the third order term is not zero then N is an isolating neighborhood for small positive λ.

C. One might hope for a general result in the case where $\mathrm{Ave}(g_j, I) \geq 0$ for $j < m$ and $\mathrm{Ave}(g_m, I) > 0$. However, to apply the theorem, one would have to solve equations like $\nabla\ell \cdot f = g$ where g has zero averages on I. This is not generally possible. One might also expect a counterexample rather than hope for a general result since under perturbation, the set $\mathrm{Ave}(g_j, I)$ may contain negative numbers.

4.3 A Relative Index.

A. Let $\dot{x} = f(x, \mu)$ be a family of equations parametrized by $\mu \in [0,1]$. Suppose $\overline{N} \subset X \times [0,1]$ is compact and let $\overline{N}_\mu = \overline{N} \cap X \times \{\mu\}$. Assume

$\overline{N}$ is an isolating neighborhood for each value of μ. Then all the invariant sets S_μ isolated by $\overline{N}_\mu$ have the same index.

Suppose S_0 is disjoint union of A_0 and A_0^* and S_1 is the disjoint union of A_1 and A_1^*. Then A_0 and A_0^* are both attractors in S_0 as are A_1 and A_1^* in S_1; each set is also the dual repeller of the other in the appropriate S.

The situation of interest is that where A_0 continues to A_1 and A_0^* to A_1^*. That is, there are disjoint compact sets N and N^* contained in $\overline{N}$ which, like $\overline{N}$, provide isolating neighborhoods for each value of μ, but such that $N \cap S_0 = A_0$, $N \cap S_1 = A_1$, $N^* \cap S_0 = A_0^*$ and $N^* \cap S_1 = A_1^*$.

The question is whether $\overline{N}$ contains more solutions than $N \cup N^*$. To get an answer without looking too closely a relative index is defined as follows.

B. Let φ_0 be a function of x defined on $\overline{N}_0$ so that it is positive on A_0 and negative on A_C^*. Likewise let φ_1 be defined on $\overline{N}_1$ so that it is negative on A_1 and positive on A_1^*.

To the x-equation append the equation $\dot{\mu} = \lambda[(1-\mu)\varphi_0+\mu\varphi_1]$ where λ is small and positive. Reasoning as in 4.1, $\overline{N}$, N and N^* will be isolating neighborhoods of the extended equations for all small values of λ.

In this way $\overline{N}$, N and N^* determine isolated invariant sets with indices $\overline{h}$, h and h^* respectively (see [1]). The indices h and h^* are expressible in terms of those of A_0 and A_0^* (or A_1 and A_1^* which have the same indices). Specifically, h is the suspension of $h(A_0)$ (= index of A_0) and h^* equals $h(A_0^*)$.

The statement is that if $\bar{h} \neq h \vee h^*$ then $\bar{N}$ contains more orbits than $N \cup N^*$.

C. In [2] it is seen that such an argument can be used to ascertain the existence of solutions connecting critical points (for a simple example).

The point here is that the relative indices $\bar{h}$, h and h^* are actually well defined; that is the arbitrariness of φ_0 and φ_1, used to define the indices, is of no consequence. Namely if (φ_0, φ_1) and $(\bar{\varphi}_0, \bar{\varphi}_1)$ are different pairs used, then for all values of the parameter $\nu \in [0,1]$, substitution of the pairs $(\nu\varphi_0 + (1-\nu)\bar{\varphi}_0, \nu\varphi_1 + (1-\nu)\bar{\varphi}_1)$ in the appended equation leads to an equation for which $\bar{N}$ is an isolating neighborhood provided λ is small enough and positive. This is a simple application of the main theorem of this paper. The family of equations parametrized by ν then provides the continuation (see [1]) that shows the relative index defined by (φ_0, φ_1) is the same (in fact canonically equivalent to) that defined by $(\bar{\varphi}_0, \bar{\varphi}_1)$.

Such a relative index can be defined for cases where μ ranges over other parameter spaces than the interval of course.

As stated in the introduction, better examples of this perturbation theorem will be presented in a later paper.

1] Conley, C., Isolated invariant sets and the Morse index. N.S.F.C.B.M.S. Lecture Note Series, #38, A.M.S., Providence, 1978.

2] Conley, C. and Smoller, J., Isolated invariant sets of parametrized systems of differential equations, The Structure of Attractors in Dynamical Systems (Eds. N. G. Markley, J. C. Martin and W. Perizzo) Lecture Notes in Mathematics, 668, Springer Verlag, Berlin (1978).

3] Murdock, J. and Robinson, C., Some mathematical aspects of spin orbit resonance II (Preprint).

University of Wisconsin
Madison, Wisconsin 53706

On a Theorem of Conley and Smoller*

Joseph G. Conlon

1. Introduction

We consider the following theorem in [1]:

Theorem 1.1: Let V be a vector field in the plane given by $(\phi(u,v), \psi(u,v))$, where $\phi_v \psi_u > 0$. Assume further that V admits exactly two critical points, both of which are nondegenerate. Then one is a saddle point, the other is a node and there is a unique orbit of V connecting these critical points.

The proof of this result in [1] makes use of an assumption that the curves $\{\phi = 0\}$ and $\{\psi = 0\}$ in the plane are connected. One can easily see that such an assumption is in general unjustified. Here we wish to outline the proof of an n dimensional version of theorem 1.1. For complete details see [2].

First of all let us assume theorem 1.1 is true and that the two critical points are at 0 and $a = (a_1, a_2)$ where a lies in the positive quadrant. Suppose the connecting orbit goes through a point (u_o, v_o) where $u_o < 0$. We consider the change of variables

* Research supported by a University of Missouri summer research grant.

$$(1.1)\qquad e^{w} = u - u_o/2,\quad z = v,$$

which maps the half plane $u > u_o/2$ onto the whole of the (w,z) plane. The transformed vector field $(\Phi(w,z), \bar{\Psi}(w,z))$ has exactly two critical points but no connecting orbit. Evidently the transformed vector field is given by

$$(1.2)\qquad \Phi(w,z) = e^{-w}\phi(e^{w} + u_o/2,z),\quad \bar{\Psi}(w,z) = \psi(e^{w} + u_o/2,z),$$

which satisfies the condition $\Phi_z \bar{\Psi}_w > 0$. Thus $(\Phi(w,z), \bar{\Psi}(w,z))$ satisfies the conditions of theorem 1.1 but not the conclusion. We infer from this that the connecting orbit of our original field $(\phi(u,v), \psi(u,v))$ must lie in the half plane $u \geq 0$. By making similar changes of variable we further conclude that if theorem 1.1 is true then the connecting orbit must lie in the rectangle $B = \{(u,v): 0 \leq u \leq a_1, 0 \leq v \leq a_2\}$. Now we note that the condition $\phi_v \psi_u > 0$ separates into two possibilities. If $\phi_v > 0$, $\psi_u > 0$ then the vector field points inwards on the boundary ∂B of B. On the other hand if $\phi_v < 0$, $\psi_u < 0$ then the vector field points outwards on ∂B.

2. A generalization of the Conley-Smoller theorem

It is clear from the preceding remarks how we might state a generalization of theorem 1.1:

Theorem 2.1: Let $X(x)$ be an n dimensional C^2 vector field defined in the box $B = \{x: 0 \le x_i \le a_i, i = 1, \cdots, n\}$ with critical points at 0 and $a = (a_1, a_2, \cdots, a_n)$ but nowhere else. Suppose the off diagonal terms in the Jacobian matrix $X'(x)$ are all positive for every $x \in B$. Then if the principle eigenvalues of $X'(0)$ and $X'(a)$ are nonzero there is a unique trajectory of the vector field contained in B which joins 0 to a.

The crucial assumption in the statement of Theorem 2.1 is that the off diagonal terms in the Jacobian matrix $X'(x)$ are all positive for $x \in B$. From it we conclude that for all $x \in \partial B$, $X(x)$ points into B. Thus if a trajectory of the vector field starts off inside B from one of the critical points it must remain inside B and consequently go to the other critical point.

The proof separates into two parts; a local part and a global part. First we consider the global part.

Lemma 2.2: Let $X(x)$ satisfy the conditions of theorem 2.1 and $X'(0)$ and $X'(a)$ have principle eigenvalues λ_0 and λ_a respectively. Then either $\lambda_0 < 0 < \lambda_a$ or $\lambda_a < 0 < \lambda_0$.

Proof: Suppose both λ_0 and λ_a are positive. We deform the box B slightly to $\tilde{B}$ as in figure 1 by cutting off the corners of B at 0 and a.

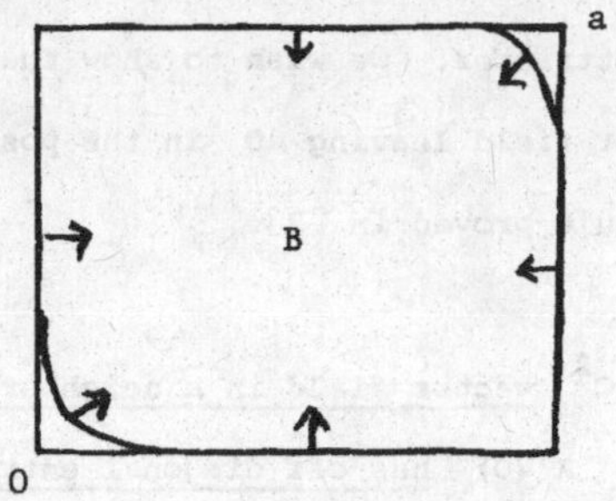

Figure 1

Thus $X(x)$ is nonzero on the boundary $\partial\tilde{B}$ of $\tilde{B}$ and so we can compute the index. In view of the positivity of λ_0 and λ_a the vector field points inwards on $\partial\tilde{B}$ and hence the index is $(-1)^n$. It follows that $X(x)$ has a critical point in the interior of B contradicting an assumption of the theorem. Consequently λ_0 and λ_a cannot both be positive.

Now assume $\lambda_0 < 0$ and $\lambda_a < 0$. Since B is an attractor in this case we can extend the vector field $X(x)$ to the n sphere S^n by adding a repeller at ∞. The Poincaré-Hopf index theorem gives

$$\chi(S^n) = 1 + (-1)^n = \text{index }(0) + \text{index }(a) + \text{index (repeller)},$$

$$\Rightarrow \quad (-1)^n = \text{index }(0) + \text{index }(a),$$

$$\Rightarrow \quad (-1)^n = (-1)^n + (-1)^n,$$

which is a contradiction. Hence at least one principle eigenvalue must be positive. This completes the proof.

Lemma 2.2 can also be proved by using the Rabinowitz global bifurcation theorem [2]. From the lemma we may assume $\lambda_a < 0 < \lambda_0$. Since $\lambda_a < 0$ all eigenvalues of $X'(a)$ are negative and so by the stable manifold theorem a is an attractor. We wish to show that there is a unique trajectory of the vector field leaving 0 in the positive direction. For this we state a result proved in [2].

Proposition 2.3: Let $X(x)$ be a C^2 vector field in a neighborhood of $x = 0$ such that $X(0) = 0$ and $X'(0)$ has off diagonal entries positive. Suppose the principle eigenvalue λ of $X'(0)$ is positive. Then there is a unique trajectory $x(t)$ of the vector field, $-\infty < t \leq 0$, which satisfies $x(t) > 0$, $x(-\infty) = 0$.

Results similar to proposition 2.3 have been proved by Hartman and Wintner [3]. These results require weaker assumptions but do not give uniqueness. It seems likely that Proposition 2.3 is true with just the assumption that $X(x)$ is C'. For the proof in [2] the C^2 assumption is essential.

Proposition 2.3 implies that there is a unique trajectory leaving the critical point 0 which remains in the box B. To complete the proof of theorem 2.1 we must show that the trajectory goes to a. When $n = 2$ this follows from the fact that B cannot contain periodic orbits since this would imply the existence of a critical point in the interior of B. For $n > 2$ one needs the following monotonicity result [4]:

Lemma 2.4: Let $x(t)$ and $y(t)$ be trajectories of $X(x)$ in B such that $x(0) \geq y(0)$. Then $x(t) \geq y(t)$ all $t \geq 0$.

Since the trajectory leaving 0 is initially monotonic, lemma 2.4 implies that it remains monotonic for $t \nearrow -\infty$ and thus must go to a. Also one can easily see from the lemma that every trajectory in B tends to a. This completes the proof of theorem 2.1.

Acknowledgement

I should like to thank Morris Hirsch for pointing out an error in a previous version of theorem 2.1 and James Selgrade for bringing my attention to the importance of monotonicity of solutions. I wish also to thank the referee for supplying the second half of the proof of lemma 2.2.

References

[1] Conley, C. and Smoller, J., Viscosity Matrices for Two-Dimensional Nonlinear Hyperbolic Systems, Comm. Pure Appl. Math., Vol. 23, 1970, 867-884.

[2] Conlon, J., A theorem in Ordinary Differential Equations with an Application to Hyperbolic Conservation Laws, Advances in Math., to appear.

[3] Hartman, P. and Wintner, A., On Monotone Solutions of Systems of Nonlinear Differential Equations, Amer. J. Math., Vol. 76, 1954, 860-866.

[4] Selgrade, J. Mathematical Analysis of a Cellular Control Process with Positive Feedback, Siam J. Appl. Math., Vol. 36, 1979, 219-229.

University of Missouri
Columbia, Missouri 65211

POSITIVELY EXPANSIVE MAPS OF COMPACT MANIFOLDS

Ethan M. Coven and William L. Reddy

It has long been a common practice in topological dynamics to examine the results of differentiable dynamics with an eye towards establishing them in a not-necessarily-differentiable setting. Indeed, one might well argue that this is how topological dynamics arose in the first place. For example, in 1936 H. Bohr and W. Fenchel [1] showed that the classical Poincaré-Bendixson Theorem (in a differentiable flow in the plane, every recurrent orbit is periodic) is true for continuous flows in the plane as well.

In this paper, we consider the dynamics, i.e., those properties invariant under topological conjugacy, of expanding endomorphisms. These maps, the simplest examples of non-invertible hyperbolic maps, were introduced in 1969 by M. Shub [14]. We will show that the dynamics of expanding endomorphisms are shared by a larger class of maps, namely positively expanding maps of compact manifolds.

In his beautiful book [13], D. Ruelle identified those properties of expanding endomorphisms which are sufficient to recover the ergodic theory (invariant measures, equilibrium states, etc.) of these maps. We will show that positively expansive maps of compact manifolds have the properties identified by Ruelle. We will identify the properties of expanding endomorphisms which are sufficient to recover "the rest" of their dynamics and show that positively expansive maps of compact manifolds have them as well.

Finally, we will construct an example of a positively expansive map of the circle which is not an expanding endomorphism. A word of caution: topological dynamics regards two maps as the same if they are topologically conjugate, and our example (as well as every other known example of a positively expansive map of a compact manifold) is topologically conjugate to an expanding endomorphism.

1. Preliminaries and Statement of Results

Let $f : X \to X$ and $g : Y \to Y$ be continuous maps. We say that f and g are <u>topologically conjugate</u> if there is an onto homeomorphism $h : X \to Y$ such that $h \circ f = g \circ h$. If h is not required to be one-to-one, then we say that g is a <u>factor</u> of f.

A <u>dynamical property</u> of a map is one which is invariant under topological conjugacy. The set of dynamical properties of a map includes the topological, but not the differential, invariants of the underlying space. Since invariant measures are preserved by topological conjugacy, the ergodic theory of a map in included in its dynamics.

For (X,d) a metric space, $x \in X$ and $\varepsilon > 0$, we let $S(x,\varepsilon) = \{y \in X | d(x,y) < \varepsilon\}$.

Throughout this paper, the term <u>manifold</u> will mean a connected topological manifold.

A C^1 map $f : X \to X$ of a compact, finite-dimensional, differentiable manifold is called an <u>expanding endomorphism</u> if there is a compatible Riemannian structure, with induced norm $\| \ \|$, and a constant $\lambda > 1$ such that $\|Df^n(v)\| \geq \lambda^n \|v\|$ for all $n \geq 1$ and all tangent vectors v. The prototypical examples are the maps $z \to z^n$ $(n \geq 2)$ of the unit circle.

In [13], D. Ruelle identified the properties of expanding endomorphisms which are sufficient to recover the ergodic theory of these maps. He called maps with these properties "expanding". A continuous map $f : X \to X$ of a compact metrizable space is called <u>expanding</u> if there is a compatible metric d and constants $\delta_0 > 0$ and $\lambda > 1$ such that (i) $d[f(x),f(y)] \geq \lambda d(x,y)$ whenever $d(x,y) < \delta_0$ and (ii) $S(x,\delta_0) \cap f^{-1}(z)$ is a singleton whenever $d(f(x),z) < \lambda\delta_0$. The class of expanding maps includes the expanding endomorphisms as well as the one-sided subshifts of finite type. Among the dynamical properties of expanding maps are the shadowing of pseudo-orbits (which Bowen [2, p. vii] called "the most important dynamical property of Axiom A diffeomorphisms") and the existence of Markov partitions, the key to the above-mentioned ergodic theory. For a fuller discussion of the properties of expanding maps, see [13] or [15].

A continuous map $f: X \to X$ of a metric space (X,d) is called positively expansive [4] if there is a constant $c > 0$ such that if $x \neq y$, then $d[f^n(x), f^n(y)] \geq c$ for some $n \geq 0$. For compact spaces, this concept (although not the expansive constant) is independent of the compatible metric used. The class of positively expansive maps includes the expanding maps as well as all one-sided subshifts. Among the properties of positively expansive maps of compact manifolds is that any such map is a covering map with finite (geometric) degree at least two [8].

Theorem A. Every positively expansive map of a compact manifold onto itself is expanding.

Thus positively expansive maps of compact manifolds have the same ergodic theory as expanding endomorphisms.

The theory of expanding maps is "more general (and thus less rich) than the theory of expanding endomorphisms" [13, p. 149]. Part of the richness which is lost through generalization consists of those properties of expanding endomorphisms whose statements or proofs involve the universal covering space. We list three below. The first two are due to M. Shub [14], the third to M. Hirsch [9].

(1) The universal covering space of X is diffeomorphic to $\mathbb{R}^n$, where n is the dimension of X.

(2) Homotopic expanding endomorphisms are topologically conjugate.

(3) An expanding endomorphism is a factor of the full one-sided k-shift, where $2 \leq k < \infty$ is the degree of the map.

Note that (1) is not a dynamical property and that (3) is. Consider the following not-necessarily-differentiable analogues of (1) - (3).

(1') The universal covering space of X is homeomorphic to $\mathbb{R}^n$, where n is the dimension of X.

(2') Homotopic positively expansive maps of compact manifolds are topologically conjugate.

(3') A positively expansive map of a compact manifold is a factor of the full one-sided k-shift, where $2 \leq k < \infty$ is the degree of the map.

P. Duvall and L. Husch [3] proved (1') for expanding covering maps of compact, finite-dimensional manifolds. We shall show in §2 that no infinite-dimensional compact manifold can support a positively expansive map. Thus (1') holds for positively expansive maps of compact manifolds.

Theorem B. Let $f: X \to X$ be a positively expansive map of a compact manifold onto itself, let $(\tilde{X},p)$ be the universal covering space of X and let $\tilde{f}: \tilde{X} \to \tilde{X}$ be a lift of f. Then $\tilde{f}$ is a homeomorphism and there is a complete metric $\tilde{d}$, compatible with the topology of $\tilde{X}$, and a constant $\lambda > 1$ such that $\tilde{d}[\tilde{f}(\tilde{x}),\tilde{f}(\tilde{y})] \geq \lambda\tilde{d}(\tilde{x},\tilde{y})$ for all $\tilde{x},\tilde{y} \in \tilde{X}$. Furthermore, the deck transformations of $(\tilde{X},p)$ are $\tilde{d}$-isometries.

An examination of the proofs of (2) and (3) reveals that the only properties of expanding endomorphisms used are the conclusions of Theorem B (including obvious corollaries, such as $\tilde{f}$ has a fixed point) and the fact that an expanding endomorphism is a covering map. Thus (2') and (3') hold for positively expansive maps of compact manifolds.

It follows from Theorem A and an examination of the proof of

Theorem 1 of [6] that

(4) The topological entropy of a positively expansive map of a compact manifold is $\log k$, where k is as in (3').

Since the full one-sided k-shift is <u>intrinsically ergodic</u>, i.e., it has a unique (necessarily ergodic) entropy-maximizing invariant measure, and intrinsic ergodicity is inherited by factors with the same topological entropy, it follows from (3') and (4) that

(5) A positively expansive map of a compact manifold is intrinsically ergodic. Furthermore, the unique entropy-maximizing measure is positive on non-empty open sets, vanishes on points, is mixing of all orders and has a Bernoulli natural extension.

The reader is invited to add to this list of dynamical properties of expanding endomorphisms which are shared by positively expansive maps of compact manifolds.

§2. Proof of Theorem A

Lemma 1. *Let f be a continuous map of a compact metrizable space onto itself. Then f is expanding if and only if it is an open map and there is a compatible metric d and constants $\delta_0 > 0$ and $\lambda > 1$ such that $d[f(x),f(y)] \geqq \lambda d(x,y)$ whenever $d(x,y) < \delta_0$.*

Proof. If f is expanding, then it is open [13, p. 144].

Conversely, suppose that the conditions of the lemma are satisfied. By replacing δ_0 with $\frac{1}{2}\delta_0$ if necessary, we may assume that $d[f(x),f(y)] \geqq \lambda d(x,y)$ whenever $d(x,y) < 2\delta_0$. Then $d(x,y) \geqq 2\delta_0$ whenever $x \neq y$ and $f(x) = f(y)$. Thus it suffices to show that there exists $\delta_1 > 0$ such that $S(x,\delta_1) \cap f^{-1}(z) \neq \phi$ whenever $d(f(x),z) < \lambda\delta_1$.

Suppose not. Then for each $n \geqq 1$, there exist x_n and z_n

ith $d(f(x_n),z_n) < \lambda/n$ and $S(x_n,1/n) \cap f^{-1}(z_n) = \phi$. We may assume hat $x_n \to x$ and hence that $f(x_n) \to f(x)$ and $z_n \to f(x)$. Since f s open, $f[S(x,\delta_0)]$ is a neighborhood of $f(x)$. For n large nough, $d(x_n,x) < \delta_0$ and $z_n \in f[S(x, {}_0)]$. Then $z_n = f(y_n)$ for ome $y_n \in S(x,\delta_0)$. If $y_n \in S(x_n,1/n)$, then we have a contradiction. f $y_n \notin S(x_n,1/n)$, then $1/n \leq d(x_n,y_n) < 2\delta_0$ and thus $(f(x_n),z_n) \geq \lambda/n$. This too is a contradiction and so the lemma is roved. □

The proof of the following lemma is a topological version of the roof of J. Mather's result on "adapted metrics" [12].

emma 2. Let f be a continuous map of a compact metrizable space nto itself and let $n \geq 2$. Then f is expanding if and only if f^n is expanding.

Proof. It is clear from Lemma 1 that if f is expanding, then so is f^n.

Suppose that f^n is expanding. Then f^n is open and onto and ence f is open. Let d be a compatible metric on X and let $\delta_0 > 0$ and $\lambda > 1$ be constants such that $d[f^n(x),f^n(y)] \geq \lambda d(x,y)$ henever $d(x,y) < \delta_0$. Define

$$d'(x,y) = d(x,y) + \frac{1}{\mu}d[f(x),f(y)] + \ldots + \frac{1}{\mu^{n-1}}d[f^{n-1}(x),f^{n-1}(y)]$$

here $\mu = \lambda^{1/n}$. Then d' is a metric on X, compatible with the topology of X, and $d'[f(x),f(y)] \geq \mu d'(x,y)$ whenever $d'(x,y) < \delta_0$. □

For the remainder of this section, let $f : X \to X$ be a positively expansive map of a compact manifold with metric d onto itself.

Lemma 3. The manifold X is finite-dimensional.

Proof. Let $X^* = \text{inv lim}(X,f)$ and let $f^* : X^* \to X^*$ be the induced map. Then [4] f^* is an expansive homeomorphism of X^*. If X were infinite-dimensional, then X^* would be too, for there is a neighborhood base for X^* each of whose members is the product of a basic neighborhood in X with a Cantor set. However, R. Mañe [11] has shown that no infinite-dimensional compact space can support an expansive homeomorphism. □

Let $(\tilde{X},p)$ be the universal covering space of X and let Γ be the group of deck transformations of $(\tilde{X},p)$. By [3, Theorem 1], there is a complete metric $\tilde{d}$, compatible with the topology of $\tilde{X}$, and a constant $\varepsilon_0 > 0$ such that

(6) $d[p(\tilde{x}),p(\tilde{y})] = \tilde{d}(\tilde{x},\tilde{y})$ whenever $\tilde{d}(\tilde{x},\tilde{y}) < 2\varepsilon_0$.

(7) Each member of Γ is a $\tilde{d}$-isometry.

Let $\tilde{f}:\tilde{X} \to \tilde{X}$ be a fixed lift of f. Since $p:\tilde{X} \to X$ is uniformly continuous (with respect to the metrics $\tilde{d}$ and d), an argument similar to that in [7] shows that $\tilde{f}$ is positively expansive (with respect to $\tilde{d}$) and is a covering map. Since $\tilde{X}$ is simply connected, $\tilde{f}$ must be a homeomorphism. For $\gamma \in \Gamma$, let $\tilde{f}_{\#}(\gamma) = \tilde{f}\circ\gamma\circ\tilde{f}^{-1}$. Then $\tilde{f}_{\#}$ is a one-to-one homomorphism of Γ into itself.

Lemma 4. <u>Both</u> $\tilde{f}$ <u>and</u> $\tilde{f}^{-1}$ <u>are uniformly continuous</u>.

<u>Proof</u>. Let $\tilde{R}$ be a compact subset of $\tilde{X}$ such that $\Gamma(\tilde{R}) = \tilde{X}$. Choose $\beta > 0$ so that $\tilde{S} = \{\tilde{x} \mid \tilde{d}(\tilde{x},\tilde{R}) \leq \beta\}$ is compact.

Let $\varepsilon > 0$. There exists δ, $0 < \delta < \beta$, such that $\tilde{d}[\tilde{f}(\tilde{x}),\tilde{f}(\tilde{y})] < \varepsilon$ whenever $\tilde{x},\tilde{y} \in \tilde{S}$ and $\tilde{d}(\tilde{x},\tilde{y}) < \delta$. Now let $\tilde{x},\tilde{y} \in \tilde{X}$ and $\tilde{d}(\tilde{x},\tilde{y}) < \delta$. There exists $\gamma \in \Gamma$ such that $\gamma(\tilde{x}) \in \tilde{R}$. Since $\tilde{d}[\gamma(\tilde{x}),\gamma(\tilde{y})] = \tilde{d}(\tilde{x},\tilde{y}) < \delta$, $\gamma(\tilde{y}) \in \tilde{S}$. But $\tilde{f}\circ\gamma = \gamma'\circ\tilde{f}$ for some $\gamma' \in \tilde{f}_{\#}(\Gamma) \subset \Gamma$, so

$$\tilde{d}[\tilde{f}(\tilde{x}),\tilde{f}(\tilde{y})] = \tilde{d}[\gamma'\tilde{f}(\tilde{x}),\gamma'\tilde{f}(\tilde{y})] = \tilde{d}[\tilde{f}\gamma(\tilde{x}),\tilde{f}\gamma(\tilde{y})] < \varepsilon.$$

herefore $\tilde{f}$ is uniformly continuous.

To show that $\tilde{f}^{-1}$ is uniformly continuous, let $\varepsilon > 0$ and let and $\tilde{S}$ be as above. By the standard identification of Γ with (X), the index of $\tilde{f}_{\#}(\Gamma)$ in Γ is the degree of f, which in this ase is finite. Let $\{\gamma_1,\ldots,\gamma_k\}$ be a complete set of coset repre-entatives for $\Gamma/\tilde{f}_{\#}(\Gamma)$ and let $\tilde{T} = \cup_{i=1}^{k}\gamma_i(\tilde{S})$. There exists η, $< \eta < \beta$, such that $\tilde{d}[\tilde{f}^{-1}(\tilde{x}),\tilde{f}^{-1}(\tilde{y})] < \varepsilon$ whenever $\tilde{x},\tilde{y} \in \tilde{T}$ and $(\tilde{x},\tilde{y}) < \eta$. Now let $\tilde{x},\tilde{y} \in \tilde{X}$ and $\tilde{d}(\tilde{x},\tilde{y}) < \eta$. There exists $\gamma \in \Gamma$ uch that $\gamma(\tilde{x}) \in \tilde{R}$. Since $\{\gamma_1\gamma,\ldots,\gamma_k\gamma\}$ is also a complete set of oset representatives for $\Gamma/\tilde{f}_{\#}(\Gamma)$, there exist i, $1 \leq i \leq k$, and $' \in \Gamma$ such that $\gamma_i\gamma = \tilde{f}_{\#}(\gamma')$. Since $\tilde{d}[\gamma_i\gamma(\tilde{x}),\gamma_i\gamma(\tilde{y})] < \eta$, $_i\gamma(\tilde{x}),\gamma_i\gamma(\tilde{y}) \in \tilde{T}$. But $\tilde{f}^{-1}\circ\gamma_i\circ\gamma = \gamma'\circ\tilde{f}^{-1}$, so as in the preceding aragraph, $\tilde{d}[\tilde{f}^{-1}(\tilde{x}),\tilde{f}^{-1}(\tilde{y})] < \varepsilon$. □

The main tool in remetrizing X will be the Metrization Lemma of . H. Frink [5] (also see [10, p. 185]): if there is a sequence $\{U_n\}$ f symmetric subsets of $X \times X$ such that $U_0 = X \times X$, $\cap U_n$ is the iagonal and for all $n \geq 0$, $U_{n+1}\circ U_{n+1}\circ U_{n+1} \subset U_n$, then there is a etric ρ, compatible with the topology of X, such that for all ≥ 1, $U_n \subset \{(x,y)\,|\,\rho(x,y) < 1/2^n\} \subset U_{n-1}$.

We now construct such a sequence. Let $c > 0$ be a common xpansive constant for f and $\tilde{f}$. We may assume that $c < \varepsilon_0$. efine a sequence $\{\tilde{V}_n\}$ of subsets of $\tilde{X} \times \tilde{X}$ as follows.

$$\tilde{V}_0 = \tilde{X} \times \tilde{X}$$

$$\tilde{V}_n = \{(\tilde{x},\tilde{y})\,|\,\tilde{d}[\tilde{f}^i(\tilde{x}),\tilde{f}^i(\tilde{y})] < c \text{ for } 0 \leq i \leq n-1\} \quad (n \geq 1).$$

The following facts are easily established.

(8) $\{\tilde{V}_n\}$ is a nested sequence of open, symmetric neighborhoods of the diagonal of $\tilde{X}$.

(9) $\cap\tilde{V}_n$ is the diagonal of $\tilde{X}$.

(10) $\Gamma(\tilde{V}_n) = \tilde{V}_n$ for all $n \geq 0$.

Lemma 5. (i) <u>For every</u> $m \geq 0$, <u>there exists</u> $\delta > 0$ <u>such that</u> $\{(\tilde{x},\tilde{y}) | \tilde{d}(\tilde{x},\tilde{y}) < \delta\} \subset \tilde{V}_m$. (ii) <u>For every</u> $\varepsilon > 0$, <u>there exists</u> $n \geq 0$ <u>such that</u> $\tilde{V}_n \subset \{(\tilde{x},\tilde{y}) | \tilde{d}(\tilde{x},\tilde{y}) < \varepsilon\}$.

<u>Proof</u>. (i) follows directly from the uniform continuity of $\tilde{f}$.

To prove (ii), let $\tilde{R}$ be a compact subset of $\tilde{X}$ such that $\Gamma(\tilde{R}) = \tilde{X}$. Choose β, $0 < \beta \leq c$, such that $\tilde{S} = \{\tilde{x} | \tilde{d}(\tilde{x},\tilde{R}) \leq \beta\}$ is compact. For each $n \geq 1$, let $\tilde{W}_n = \tilde{V}_n \cap (\tilde{R} \times \tilde{S})$. Since $\cap \tilde{V}_n$ is the diagonal of $\tilde{X}$, $\cap \tilde{W}_n = (\cap \tilde{V}_n) \cap (\tilde{R} \times \tilde{S})$ is the diagonal of $\tilde{R}$. Since $\tilde{R} \times \tilde{S}$ is compact and each $\tilde{W}_n$ is open in $\tilde{R} \times \tilde{S}$, given $\varepsilon > 0$, there exists $n \geq 1$ such that $\tilde{W}_n \subset \{(\tilde{x},\tilde{y}) \in \tilde{R} \times \tilde{S} | \tilde{d}(\tilde{x},\tilde{y}) < \varepsilon\}$. Then $\tilde{V}_n \subset \{(\tilde{x},\tilde{y}) | \tilde{d}(\tilde{x},\tilde{y}) < \varepsilon\}$. □

It follows from (9) and Lemma 5 that

(11) $\cap p(\tilde{V}_n)$ is the diagonal of X.

Lemma 6. <u>For</u> n <u>large enough</u>, $\tilde{f}(\tilde{V}_n) = \tilde{V}_{n-1}$.

<u>Proof</u>. For every $n \geq 1$, $\tilde{f}(\tilde{V}_n) = \tilde{V}_{n-1} \cap \tilde{f}(\tilde{V}_1)$. It follows from the uniform continuity of $\tilde{f}^{-1}$ that there exists $\varepsilon > 0$ such that $\{(\tilde{x},\tilde{y}) | \tilde{d}(\tilde{x},\tilde{y}) < \varepsilon\} \subset \tilde{f}(\tilde{V}_1)$. By Lemma 5, $\tilde{V}_n \subset \tilde{f}(\tilde{V}_1)$ if n is large enough. The result then follows from the fact that the $\tilde{V}_n$'s are nested. □

Lemma 7. <u>If</u> $\tilde{d}(\tilde{x},\tilde{y}) < \varepsilon_0$ <u>and</u> $(p(\tilde{x}),p(\tilde{y})) \in p(\tilde{V}_n)$, <u>then</u> $(\tilde{x},\tilde{y}) \in \tilde{V}_n$.

<u>Proof</u>. There exists $(\tilde{z},\tilde{w}) \in \tilde{V}_n$ such that $p(\tilde{z}) = p(\tilde{x})$ and $p(\tilde{w}) = p(\tilde{y})$. There exist $\gamma,\gamma' \in \Gamma$ such that $\tilde{z} = \gamma(\tilde{x})$ and $\tilde{w} = \gamma'(\tilde{y})$. Then $\tilde{d}[\gamma(\tilde{x}),\gamma'(\tilde{x})] < 2\varepsilon_0$. Since $p\gamma(\tilde{x}) = p\gamma'(\tilde{x})$ and p is one-to-one on $S(\gamma(\tilde{x}),2\varepsilon_0)$, it follows that $\gamma(\tilde{x}) = \gamma'(\tilde{x})$ and hence that $\gamma = \gamma'$. Then $(\tilde{x},\tilde{y}) = (\gamma^{-1}(\tilde{x}),\gamma^{-1}(\tilde{y})) \in \Gamma(\tilde{V}_n) = \tilde{V}_n$. □

Let $N \geqq 1$ be chosen so that

12) $\tilde{f}(\tilde{V}_n) = \tilde{V}_{n-1}$ for all $n \geqq N$.

13) $\tilde{V}_n \quad \{(\tilde{x},\tilde{y}) | \tilde{d}(\tilde{x},\tilde{y}) < \varepsilon_0\}$ for all $n \geqq N$.

By Lemma 5, $\{(\tilde{x},\tilde{y}) | \tilde{d}(\tilde{x},\tilde{y}) < \delta\} \subset \tilde{V}_N$ for some $\delta > 0$ and

$_{N+M} \subset \{(\tilde{x},\tilde{y}) | \tilde{d}(\tilde{x},\tilde{y}) < \delta/3\}$ for some $M \geqq 1$. Define a sequence $\{\tilde{U}_n\}$

f subsets of $\tilde{X} \times \tilde{X}$ as follows.

$$\tilde{U}_0 = \tilde{X} \times \tilde{X}$$

$$\tilde{U}_n = \tilde{V}_{N+(n-1)M} \quad (n \geqq 1).$$

Now let $U_n = p(\tilde{U}_n)$.

emma 8. <u>There is a metric</u> ρ, <u>compatible with the topology of</u> X, <u>uch that</u>

14) $U_n \subset \{(x,y) | \rho(x,y) < 1/2^n\} \subset U_{n-1}$ <u>for all</u> $n \geqq 1$.

<u>roof</u>. To apply Frink's Metrization Lemma we need show only that $_{n+1} \circ U_{n+1} \circ U_{n+1} \subset U_n$ for all $n \geqq 0$. It is easy to show that $_{n+1} \circ \tilde{U}_{n+1} \circ \tilde{U}_{n+1} \subset \tilde{U}_n$ for all $n \geqq 0$. The relevant fact for the nductive step is that $\tilde{U}_{n+1} = \tilde{f}^{-M}(\tilde{U}_n)$.

Let $n \geqq 0$ be fixed and suppose that $(x,y),(y,z),(z,w) \in U_{n+1}$. here exists $(\tilde{x},\tilde{y}) \in \tilde{U}_{n+1}$ such that $p(\tilde{x}) = x$ and $p(\tilde{y}) = y$. Since $(\tilde{x},\tilde{y}) < \varepsilon_0$ and p maps $S(\tilde{y},\varepsilon_0)$ onto $S(p(\tilde{y}),\varepsilon_0)$, there exists $\in \tilde{X}$ such that $p(\tilde{z}) = z$ and $\tilde{d}(\tilde{y},\tilde{z}) < \varepsilon_0$. Then by Lemma 7, $\tilde{y},\tilde{z}) \in \tilde{U}_{n+1}$. Similarly, there exists $\tilde{w} \in \tilde{X}$ such that $p(\tilde{w}) = w$ nd $(\tilde{z},\tilde{w}) \in \tilde{U}_{n+1}$. Then $(\tilde{x},\tilde{w}) \in \tilde{U}_{n+1} \circ \tilde{U}_{n+1} \circ \tilde{U}_{n+1} \subset \tilde{U}_n$ and hence $x,w) \in U_n$. □

emma 9. <u>If</u> $\rho(x,y) < 1/16$, <u>then</u> $\rho[f^{3M}(x),f^{3M}(y)] \geqq 2\rho(x,y)$.

<u>roof</u>. If $\rho(x,y) = 0$, then there is nothing to prove. Suppose that $) < \rho(x,y) < 1/16$. Since X is compact and $\cap U_n$ is the diagonal,

$(x,y) \in U_{n+1} - U_{n+2}$ for some $n \geq -1$. Then $1/2^{n+3} \leq \rho(x,y) < 1/2^{n+1}$ and hence $n \geq 2$.

There exists $(\tilde{x},\tilde{y}) \in \tilde{U}_{n+1}$ such that $p(\tilde{x}) = x$ and $p(\tilde{y}) = y$. By (13) and Lemma 7, $(\tilde{x},\tilde{y}) \notin \tilde{U}_{n+2}$ and thus $(\tilde{f}^{3M}(\tilde{x}),\tilde{f}^{3M}(\tilde{y})) \notin \tilde{U}_{n-1}$. Therefore, $(p\tilde{f}^{3M}(\tilde{x}),p\tilde{f}^{3M}(\tilde{y})) \notin p(\tilde{U}_{n-1})$, i.e. $(f^{3M}(x),f^{3M}(y)) \notin U_{n-1}$. Hence $\rho[f^{3M}(x),f^{3M}(y)] \geq 1/2^n > 2\rho(x,y)$. □

It follows from Lemmas 1, 2 and 9 that f is expanding. This completes the proof of Theorem A.

§3. Proof of Theorem B

We retain the notation from §2.

Let $\tilde{x},\tilde{y} \in \tilde{X}$. By a <u>chain</u> from $\tilde{x}$ to $\tilde{y}$ we mean a finite sequence $(\tilde{x}_0,\ldots,\tilde{x}_m)$ of points in $\tilde{X}$ such that $\tilde{x}_0 = \tilde{x}$, $\tilde{x}_m = \tilde{y}$ and $(\tilde{x}_i,\tilde{x}_{i+1}) \in \tilde{U}_2$ for $0 \leq i \leq m-1$. Define

$$\tilde{\rho}(\tilde{x},\tilde{y}) = \inf\left\{\sum_{i=0}^{m-1} \rho[p(\tilde{x}_i),p(\tilde{x}_{i+1})]\right\}$$

where the infimum is taken over all chains from $\tilde{x}$ to $\tilde{y}$.

Since the set of $(\tilde{x},\tilde{y})$ for which there is a chain from $\tilde{x}$ to $\tilde{y}$ is open-closed, $\tilde{\rho}(\tilde{x},\tilde{y}) < \infty$. Since $\tilde{\rho}(\tilde{x},\tilde{y}) \leq \rho[p(\tilde{x}),p(\tilde{y})]$,

(15) $\tilde{\rho}(\tilde{x},\tilde{y}) = \rho[p(\tilde{x}),p(\tilde{y})]$ whenever $(\tilde{x},\tilde{y}) \in \tilde{U}_2$.

Lemma 10. <u>If</u> $\tilde{\rho}(\tilde{x},\tilde{y}) < 1/16$, <u>then</u> $(\tilde{x},\tilde{y}) \in \tilde{U}_2$.

<u>Proof</u>. Suppose $(\tilde{x},\tilde{y}) \notin \tilde{U}_2$ and let $(\tilde{x}_0,\ldots,\tilde{x}_m)$ be a chain from $\tilde{x}$ to $\tilde{y}$.

If $(\tilde{x}_j,\tilde{x}_{j+1}) \notin \tilde{U}_3$ for some j, then by Lemma 7, $(p(\tilde{x}_j),p(\tilde{x}_{j+1})) \notin U_3$. By (14), $\rho[p(\tilde{x}_j),p(\tilde{x}_{j+1})] \geq 1/16$ and hence $\tilde{\rho}(\tilde{x},\tilde{y}) \geq 1/16$.

Suppose that $(\tilde{x}_i,\tilde{x}_{i+1}) \in \tilde{U}_3$ for all i and let $n = \max\{i \mid (\tilde{x}_0,\tilde{x}_i) \in \tilde{U}_3\}$. Then $n \leq m-1$, $(\tilde{x}_0,\tilde{x}_{n+1}) \in \tilde{U}_3 \circ \tilde{U}_3 \subset \tilde{U}_2$

nd $(\tilde{x}_0,\tilde{x}_{n+1}) \notin \tilde{U}_3$. By Lemma 7, $(p(\tilde{x}_0),p(\tilde{x}_{n+1})) \notin U_3$ and hence $(\tilde{x},\tilde{y}) \geq \rho[p(\tilde{x}_0),p(\tilde{x}_{n+1})] \geq 1/16$. □

emma 11. $\tilde{\rho}$ _is a metric on_ $\tilde{X}$, _uniformly equivalent to_ $\tilde{d}$.

roof. Standard arguments show that $\tilde{\rho}$ is a pseudo-metric. It ollows from (15) and Lemma 10 that $\tilde{\rho}(\tilde{x},\tilde{y}) > 0$ whenever $\tilde{x} \neq \tilde{y}$.

To show that $\tilde{\rho}$ and $\tilde{d}$ are uniformly equivalent, let $\varepsilon > 0$. y Lemma 5, $\tilde{U}_n \subset \{(\tilde{x},\tilde{y}) | \tilde{d}(\tilde{x},\tilde{y}) < \varepsilon\}$ for some $n \geq 3$. We show that $(\tilde{x},\tilde{y}) | \tilde{\rho}(\tilde{x},\tilde{y}) < 1/2^{n+1}\} \subset \tilde{U}_n$. Let $\tilde{\rho}(\tilde{x},\tilde{y}) < 1/2^{n+1}$. Since $n \geq 3$, $/2^{n+1} < 1/16$ and so by Lemma 10, $(\tilde{x},\tilde{y}) \in \tilde{U}_2$. Therefore, $[p(\tilde{x}),p(\tilde{y})] = \tilde{\rho}(\tilde{x},\tilde{y}) < 1/2^{n+1}$ and hence $(p(\tilde{x}),p(\tilde{y})) \in U_n$. But $(\tilde{x},\tilde{y}) < \varepsilon_0$ since $(\tilde{x},\tilde{y}) \in \tilde{U}_2$. Then by Lemma 7, $(\tilde{x},\tilde{y}) \in \tilde{U}_n$.

It follows from (15) that $\tilde{U}_n \subset \{(\tilde{x},\tilde{y}) | \tilde{\rho}(\tilde{x},\tilde{y}) < \varepsilon\}$ if $/2^n < \varepsilon < 1/16$. □

emma 12. $\tilde{\rho}[\tilde{f}^{3M}(\tilde{x}),\tilde{f}^{3M}(\tilde{y})] \geq 2\tilde{\rho}(\tilde{x},\tilde{y})$ _for all_ $\tilde{x},\tilde{y} \in \tilde{X}$.

roof. Let $\tilde{x},\tilde{y} \in \tilde{X}$ and let $(\tilde{x}_0,\ldots,\tilde{x}_m)$ be a chain from $\tilde{f}^{3M}(\tilde{x})$ o $\tilde{f}^{3M}(\tilde{y})$. Then $(\tilde{f}^{-3M}(\tilde{x}_0),\ldots,\tilde{f}^{-3M}(\tilde{x}_m))$ is a chain from $\tilde{x}$ to $\tilde{y}$. t follows from Lemma 9 that for each i,

$$\rho[p(\tilde{x}_i),p(\tilde{x}_{i+1})] = \rho[f^{3M}p\tilde{f}^{-3M}(\tilde{x}_i),f^{3M}p\tilde{f}^{-3M}(\tilde{x}_{i+1})]$$
$$\geq 2\rho[p\tilde{f}^{-3M}(\tilde{x}_i),p\tilde{f}^{-3M}(\tilde{x}_{i+1})].$$

Therefore $\sum\rho[p(\tilde{x}_i),p(\tilde{x}_{i+1})] \geq 2\tilde{\rho}(\tilde{x},\tilde{y})$ and it follows that $\tilde{\rho}[\tilde{f}^{3M}(\tilde{x}),\tilde{f}^{3M}(\tilde{y})] \geq 2\tilde{\rho}(\tilde{x},\tilde{y})$. □

Let $\gamma \in \Gamma$. By (10), $(\tilde{x},\tilde{y}) \in \tilde{U}_2$ if and only if $(\gamma(\tilde{x}),\gamma(\tilde{y})) \in \tilde{U}_2$. Therefore $(\tilde{x}_0,\ldots,\tilde{x}_m)$ is a chain from $\tilde{x}$ to $\tilde{y}$ if and only if $(\gamma(\tilde{x}_0),\ldots,\gamma(\tilde{x}_m))$ is a chain from $\gamma(\tilde{x})$ to $\gamma(\tilde{y})$. Hence

(16) Each member of Γ is a $\tilde{\rho}$-isometry.

Let $\lambda = 2^{1/3M}$ and define

$$\tilde{\rho}'(\tilde{x},\tilde{y}) = \tilde{\rho}(\tilde{x},\tilde{y}) + \frac{1}{\lambda}\tilde{\rho}[\tilde{f}(\tilde{x}),\tilde{f}(\tilde{y})] + \ldots + \frac{1}{\lambda^{3M-1}}\tilde{\rho}[\tilde{f}^{3M-1}(\tilde{x}),\tilde{f}^{3M-1}(\tilde{y})].$$

Then, as in the proof of Lemma 2, $\tilde{\rho}'$ is a metric on $\tilde{X}$ and $\tilde{\rho}'[\tilde{f}(\tilde{x}),\tilde{f}(\tilde{y})] \geq \lambda\tilde{\rho}'(\tilde{x},\tilde{y})$ for all $\tilde{x},\tilde{y} \in \tilde{X}$. It follows from the uniform continuity of $\tilde{f}$ with respect to $\tilde{d}$ (and hence with respect to $\tilde{\rho}$) that $\tilde{\rho}$ and $\tilde{\rho}'$ are uniformly equivalent.

It follows from the definition of $\tilde{\rho}'$ and (16) that each member of Γ is a $\tilde{\rho}'$-isometry.

This completes the proof of Theorem B.

§4. A Positively Expansive Map of the Circle which is not an Expanding Endomorphism

Let $g : \mathbb{R} \to \mathbb{R}$ be a C^1 map with the following properties.

(17) $g(0) = 0, \quad g(1) = 2.$

(18) $g'(0) = g'(1) = 1, \quad g'(x) > 1$ whenever $0 \leq x \leq 1.$

(19) $g(x) = 2n + g(x-n)$ whenever $n \leq x \leq n+1 \quad (n \in \mathbb{Z}).$

For example, let $g(x) = x + x^2(x-2)^2$ on $[0,1]$ and extend g to all of $\mathbb{R}$ by (19).

Let $f : S^1 \to S^1$ be the C^1 map of the circle induced by g. Since $Df^n(v) = v$ for all $n \geq 1$ and all tangent vectors v at $z = 1$, f is not an expanding endomorphism. Since the projection map $p : \mathbb{R} \to S^1$ is uniformly continuous, to show that f is positively expansive, it suffices by an argument similar to that in [7] to show that g is positively expansive.

Choose δ, $0 < \delta < 1/2$, so that

(20) $|g(x) - g(y)| < 1/2$ whenever $|x - y| < 2\delta.$

Choose $\lambda > 1$ so that

21) $g'(x) \geqq \lambda$ whenever $|x-n| \geqq \delta$ for all $n \in \mathbb{Z}$.

emma 13. If $g^j(x) \notin \mathbb{Z}$ for all $j \geqq 0$, then for infinitely many $\geqq 0$, $|g^i(x)-n| \geqq 2\delta$ for all $n \in \mathbb{Z}$.

roof. Suppose not. Let k be the largest integer such that $g^k(x)-n| \geqq 2\delta$ for all $n \in \mathbb{Z}$. There exists $n \in \mathbb{Z}$ such that ither $n < g^{k+1}(x) < n+2\delta$ or $n-2\delta < g^{k+1}(x) < n$. Without loss of enerality we assume the former.

Let $y = g^{k+1}(x) - n$. It is easy to show by induction that $< g^i(y) < 2\delta$ for all $i \geqq 0$. By the Mean Value Theorem, $^{i+1}(y) = g'(t_i)g^i(y)$ where $0 < t_i < g^i(y)$. Since $g^i(y) < 1$, $'(t_i) > 1$. Therefore, the sequence $\{g^i(y)\}$ is increasing. Since t is bounded, it converges, say $g^i(y) \to z \leqq 1$. Then $g(z) = z$. ut $g(z) = g'(t)z$ where $0 < t < 1$ and so $g(z) > z$. This is a ontradiction. □

We now show that δ is an expansive constant for g. Let $x \neq y$. f $g^i(x), g^j(y) \in \mathbb{Z}$ for some $i,j \geqq 0$, then $g^{i+j}(x), g^{i+j}(y) \in \mathbb{Z}$ nd so $|g^{i+j}(x) - g^{i+j}(y)| \geqq \delta$. We may therefore assume that $^j(x) \notin \mathbb{Z}$ for all $j \geqq 0$.

The sequence $\{|g^i(x) - g^i(y)|\}$ is non-decreasing, for $g^{i+1}(x) - g^{i+1}(y)| = |g^i(x) - g^i(y)| \cdot |g'(t_i)|$ for some t_i between $^i(x)$ and $g^i(y)$. By Lemma 13, there are infinitely many i 0 uch that $|g^i(x)-n| \geqq 2\delta$ for all $n \in \mathbb{Z}$. For any such i and any $\in \mathbb{Z}$, if $|g^i(x) - g^i(y)| < \delta$, then

$$|t_i - n| \geqq |g^i(x) - n| - |g^i(x) - g^i(y)| > \delta.$$

Hence $g'(t_i) \geqq \lambda$. If $|g^i(x) - g^i(y)| < \delta$ for all $i \geqq 0$, then $|g^{i+1}(x) - g^{i+1}(y)| \geqq \lambda|g^i(x) - g^i(y)|$ for infinitely many $i \geqq 0$. Then $\{|g^i(x) - g^i(y)|\}$ must be unbounded. This is a contradiction and thus g (and hence f) is positively expansive.

The map f is homotopic to, and hence by (2') topologically conjugate to, the expanding endomorphism $z \to z^2$.

BIBLIOGRAPHY

1. H. Bohr and W. Fenchel, Ein Satz über stabile Bewegungen in der Ebene, Danske Vid. Selsk. Mat.-Fys. Medd. 14 (1936), 1-15. Reprinted in Collected Mathematical Works of Harald Bohr, vol. II, Danish Mathematical Society, Copenhagen, 1952.

2. R. Bowen, On Axiom A Diffeomorphisms, CBMS Regional Conference Series in Mathematics, no. 35, Amer. Math. Soc., Providence, R.I., 1978.

3. P. F. Duvall and L. S. Husch, Analysis on topological manifolds, Fund. Math. 77 (1972), 75-90.

4. M. Eisenberg, Expansive transformation semigroups of endomorphisms, Fund. Math. 59 (1969), 313-321.

5. A. H. Frink, Distance functions and the metrization problem, Bull. Amer. Math. Soc. 43 (1937), 133-142.

6. J. Guinez, Entropie topologique et rayon de convergence de la fonction zêta des endomorphismes dilantes des variétés compactes, C. R. Acad. Sci. Paris Sér. A 270 (1970), 1408-1411.

7. E. Hemmingsen and W. L. Reddy, Lifting and projecting expansive homeomorphisms, Math. Systems Theory 2 (1968), 7-15.

8. E. Hemmingsen and W. L. Reddy, Expansive homeomorphisms on homogeneous spaces, Fund. Math. 64 (1969), 203-207.

9. M. Hirsch, Expanding maps and transformation groups, Proc. Sympos. Pure Math., vol. 14, 125-131. Amer. Math. Soc., Providence, R.I., 1970.

10. J. L. Kelley, General Topology, Van Nostrand, Princeton, N.J., 1955.

11. R. Mañe, Expansive homeomorphisms and topological dimension, Trans. Amer. Math. Soc. 252 (1979), 313-319.

12. J. Mather, Characterization of Anosov diffeomorphisms, Nederl. Akad. Wetensch. Indag. Math. 30 (1968), 479-483.

13. D. Ruelle, Thermodynamic Formalism, Encyclopedia of Mathematics and its Applications, vol. 5, Addison-Wesley, Reading, Mass., 1978.

14. M. Shub, Endomorphisms of compact differentiable manifolds, Amer. J. Math. 91 (1969), 175-199.

15. P. Walters, Invariant measures and equilibrium states for some mappings which expand distances, Trans. Amer. Math. Soc. 236 (1978) 121-153.

WESLEYAN UNIVERSITY

An Algorithm for Finding Closed Orbits

J. H. Curry[1]

National Center for Atmospheric Research[2]

and

Department of Meteorology

Massachusetts Institute of Technology

1. Introduction

In this article we describe a numerical method for finding a point on a periodic solution to an autonomous system of ordinary differential equations. Since, in general it is not difficult to find stable critical elements using the computer, the interest in this procedure lies in the fact that it works equally well for stable and unstable periodic orbits. The idea behind the method was first implemented by Lanford [1] in connection with the Lorenz equations. However, since there is no published record of this method, we are compelled to expose it to a wider audience with the hope that it provides another useful tool to be incorporated into the study of numerical dynamical systems.

We remark that recently Hirsch, Smale, Kellogg, Li and Yorke have studied similar problems to those considered here. For details we refer the reader to [2], [3]. We also note that Robbins [4] has done a detailed analysis of the bifurcations of the closed orbits at high values of r for the Lorenz system of equations.

In Section 2 we outline the mathematics behind the method. Then in Section 3 we use a few of the closed orbits of the Lorenz system of equations as examples of implementations of the method.

---

[1]On leave from the Department of Mathematics, University of Colorado, Boulder, Colorado 80307.

[2]The National Center for Atmospheric Research is sponsored by the National Science Foundation.

2. Mathematical foundation

Denote by

$$\frac{dx}{dt} = X(x) \tag{1}$$

a differential equation in R^n, we shall assume that $X(x)$ is at least C^1 and therefore has a solution $\Phi(x,t)$ such that $\Phi(x,0) = x$. Suppose, in addition, that our evolution equation has a periodic solution γ, and $x_0 \in \gamma$, choose a codimension one section, Σ, which is transverse to γ at x_0. We define a mapping P: $\Sigma \to \Sigma$ which sends a point $y \in \Sigma$ to the first point where the solution curve through y intersects the section Σ again—the mapping P is known as the "first return" map or the Poincaré map. Clearly, P need not be defined for all $y \in \Sigma$, but it is certainly defined for all y in a neighborhood of x_0, and $P(x_0) = x_0$.

There are several standard techniques for numerically finding the zeros of functions, such methods can be found in any book on numerical methods [5]. One such method is due to Newton.

Newton's method can best be illustrated graphically (Fig. 1). Choose initial conditions close to a zero of f, α, compute $f(x_0)$ and $f'(x_0)$, then construct the tangent line to the curve f(x) which passes through $f(x_0)$. This line intersects the x-axis at some value x_1, now repeat the process with $f(x_1)$, $f'(x_1)$. If the initial guess is "reasonable" we obtain a sequence of numbers $\{x_n\}$ such that $|x_n - \alpha| <$ tolerance for all $n > N_0$. The iteration formula for determining the new x_{n+1} is just

$$x_{n+1} = x_n - \frac{f(x_n)}{f'(x_n)} \quad . \tag{2}$$

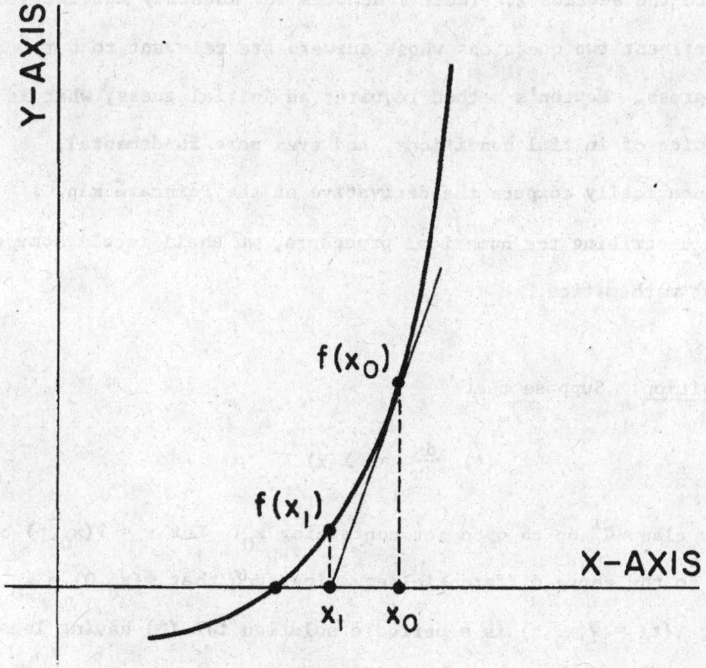

Figure 1 Newton's method for a mapping.

We remark that there are obvious generalizations of Newton's method to higher dimensions.

We are interested in applying Newton's method to the map P-I when restricted to the section Σ. (Here I denotes the identity matrix.) There are at least two questions whose answers are relevant to our further progress. Newton's method requires an initial guess, what is a "good" choice of initial conditions, and even more fundamental, how do we numerically compute the derivative of the Poincaré map.

Before describing the numerical procedure, we shall recall some of the relevant mathematics.

Proposition: Suppose that

$$(*) \quad \frac{dx}{dt} = X(x)$$

and $X(x)$ is class C^1 on an open set containing x_0. Let $x = \Phi(x_0,t)$ be a solution to the above differential equation such that $\Phi(x_0,0) = x_0$. Assume that $\gamma(t) = \Phi(x_0,t)$ is a periodic solution for (*) having least period $p > 0$. Let Σ be a section transverse to γ at x_0. Then there exists a unique real valued function $t = \tau(y)$ which is C^1 in a small neighborhood of x_0 such that $\tau(x_0) = p$ and $\Phi(y,t) \in \Sigma$ when $t = \tau(y)$ and y is in a neighborhood of x_0 contained in Σ.

Proof. The existence of such a τ is an immediate consequence of the Implicit Function Theorem [6].

It is apparent from the above theorem that $P(x) = \Phi(x,\tau(x))$ and the derivative of Φ is

$$D\,\Phi(x,\tau(x)) = D_x\,\Phi(x,\tau(x)) + \frac{d}{dt}\,\Phi(x,\tau(x))\,D\tau(x) \quad . \qquad (3)$$

A key observation is that $D_x\Phi(x,t)$ is the unique solution to the initial value problem

$$\frac{d}{dt}\,H(x,t) = J(\Phi(x,t))\,H(x,t)$$

and (4)

$$H(x,0) = I \quad ,$$

where $J(\Phi(x,t))$ denotes the Jacobian matrix evaluated at $\Phi(x,t)$ and I is the identity matrix.

This initial value problem is called the variational equation. Let us recall a fact from calculus, if h is an implicitly defined function such that $h(x,g(x)) = c$, then

$$\frac{\partial h}{\partial x}\,(x,g(x)) + \frac{\partial h}{\partial y}\,(x,g(x))\,D\,g(x) = 0 \qquad (5)$$

or

$$Dg(x) = -\left[\frac{\partial h}{\partial y}\,(x,g(x))\right]^{-1}\frac{\partial h}{\partial x}\,(x,g(x)) \quad . \qquad (6)$$

With the above observations, we are ready to compute the derivative of P-I. Let $\Phi(x,t) = (\phi_1(x,t), \phi_2(x,t), \cdots, \phi_n(x,t))$ and $\Sigma = \{x = (x_1, \cdots, x_n): x_n = c\}$ we are interested in finding roots of the following system of equations:

$$\begin{aligned}
g_1(x,t) &= \phi_1(x,t) - x_1 \\
g_2(x,t) &= \phi_2(x,t) - x_2 \\
&\vdots \\
g_{n-1}(x,t) &= \phi_{n-1}(x,t) - x_{n-1} \\
g_n(x,t) &= \phi_n(x,t) - c = 0 \quad .
\end{aligned}$$

This problem is equivalent to finding the zeros of $g = (g_1, \cdots, g_{n-1}, 0)$. We will apply Newton's method to g and

$$Dg = \left[\frac{\partial g_i}{\partial x_j}\right] \tag{7}$$

where

$$\frac{\partial g_i}{\partial x_j} = \frac{\partial \phi_i}{\partial x_j} + \frac{\partial \phi_i}{\partial t}\frac{\partial t}{\partial x_j} - \delta_{ij} \quad . \tag{8}$$

We note that three of the quantities on the right-hand side of equation (8) are known (i.e., $\partial\phi_i/\partial x_j$, $\partial\phi_i/\partial t$ and δ_{ij}), the former from solving the variational equation and the latter from evaluating the ith component of the vector field at a particular point.

Hence the only quantity which remains unknown is $\partial t/\partial x_j$, but because of our choice of $g_n(x,t)$ we see that $\partial g_n/\partial x_j = 0$, therefore,

$$\frac{\partial t}{\partial x_j} = -\left(\frac{\partial \phi_n}{\partial x_j}\right) \bigg/ \left(\frac{\partial \phi_n}{\partial t}\right) \quad . \tag{9}$$

We are now ready to locate closed orbits.

3. An example

We consider the following simple system of differential equations

$$\begin{aligned} \dot{x} &= -\sigma x + \sigma y \\ \dot{y} &= -xz + rx - y \\ \dot{z} &= xy - bz \ . \end{aligned} \tag{10}$$

The above system was first studied by Lorenz [7] in his study of the predictability of processes in the atmosphere. In his study Lorenz chose $r = 28$, $\sigma = 10$ and $b = 8/3$, and found what may well be the first example of a "strange attractor." More recently this system of equations has been studied by Guckenheimer, Lanford, Kaplan, Pomeau, Robbins, Rössler, Williams, Yorke and many others.

Several researchers have noted that when $r = 100.0$, $\sigma = 10.0$ and $b = 8/3$. Eq. (9) has at least two stable attracting closed orbits. We have built a computer program which makes use of the mathematics of the previous section. The results of our test are shown in Fig. 2. The orbit in Fig. 2 is not stable for the above parameter values. Indeed, the eigenvalues of the Poincaré map associated with this cycle are (0.00, -405.84). The coordinate of a point on this closed orbit is (-20.217644, -32.763798, 99.0) and the period is 3.3205704.

Many people have considered Lorenz's system of equations for the parameter values which he studied. Williams [8] in his analysis used the theory of kneading sequences to characterize the unstable periodic orbits which are present in the Lorenz attractor. The simplest possible orbit in the Lorenz attractor should, according to Williams, be an "x-y" orbit; that is, a periodic orbit which makes one loop around each of the unstable fixed points before it closes [7].

In Fig. 3 we see the "x-y" orbit of Williams projected onto the axis. The period of the orbit is p = 1.5586522, the eigenvalues of the Poincaré map associated with the "x-y" orbit are (0.00, 4.71). The coordinates of a point on the unstable cycle are (-12.786189, -19.364189, 24.000000).

In Fig. 4 we display the "x-yyy" which is contained in the Lorenz attractor. The period of this orbit is p = 3.0235833, its Poincaré map has eigenvalues (0.00, 16.11), the coordinates of a point on the cycle is (-13.917865, -21.919412, 24.000000).

Recently, Kaplan-Yorke [9] have noted that for R $\approx$ 13.9 the unstable manifold of the origin and its codimension one stable manifold intersect transversally; when this happens a fairly complicated set comes into existence. In [9] it is conjectured that this complicated set will become the Lorenz attractor when $r > 24.74$. We have not been able to verify this conjecture, but we have been able to follow the "x-y" closed orbit for r decreasing from 28 to 15.8, which is well above the parameter value where the attractor is "born."

In Fig. 5 we see the "x-y" orbit when r = 15.8. The period of the orbit has increased to p = 2.7964084, the eigenvalues of DP are (0.00, 30.34), and the coordinate of a point on the cycle is (-10.780017, -6.675985, 24.000000).

We remark that in reporting the eigenvalue for the Poincare map of a closed orbit we have shown only two significant digits after the decimal. Hence the first eigenvalue reported is 0.00. In fact, a typical value for such an eigenvalue is, in the case of the "x-y" orbit, 1.9757636×10^{-7}, which is small.

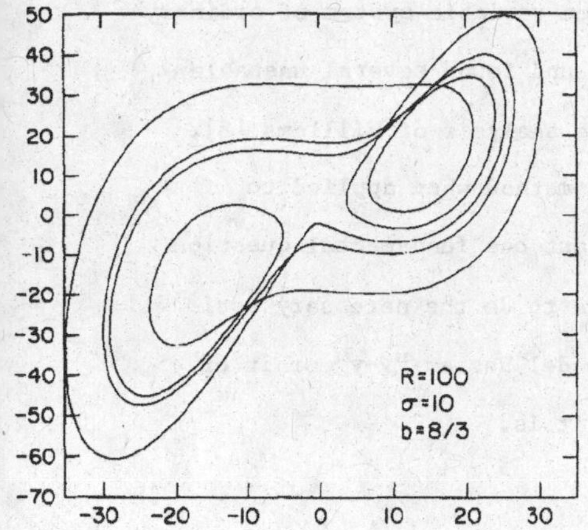

Figure 2 An unstable closed orbit projected onto the (x,y) plane

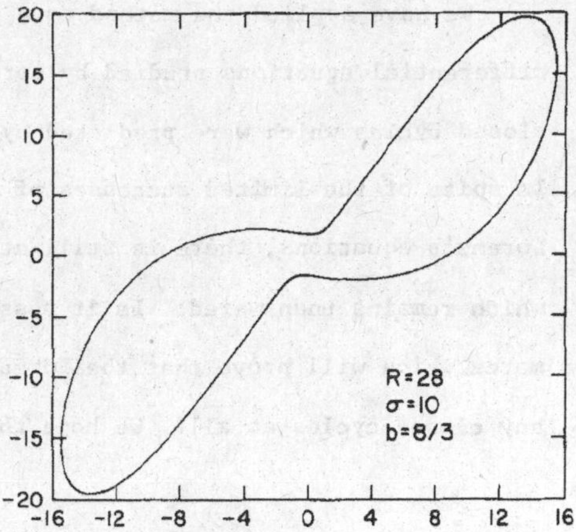

Figure 3 The unstable "x-y" orbit of Williams

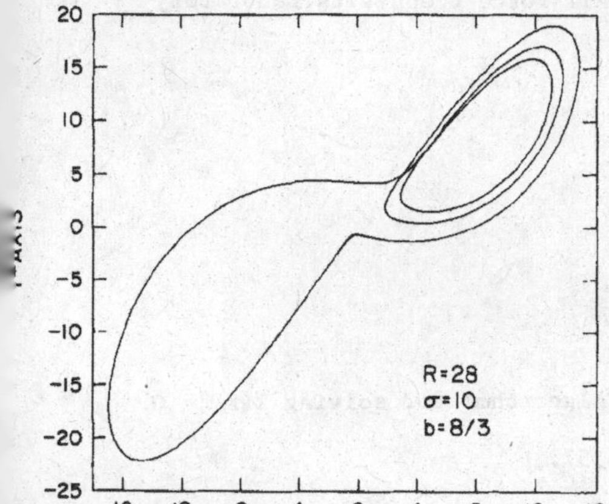

Figure 4 The unstable "x-yyy" orbit

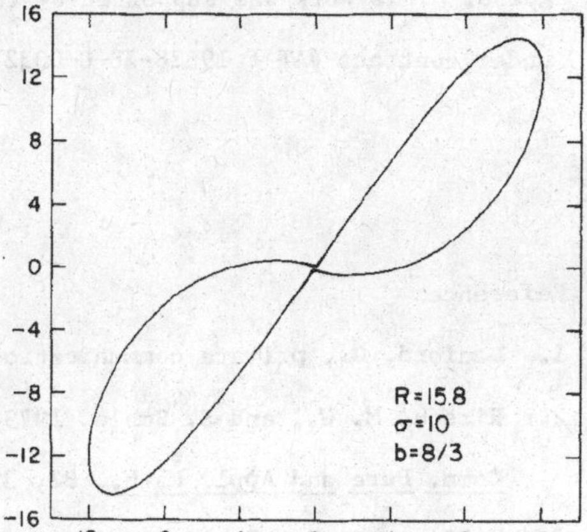

Figure 5 The unstable "x-y" orbit

4. Conclusion

In this article we have described a method which we have found useful for finding a point on a closed orbit. The method works equally well for stable and unstable cycles. Since the algorithm is based in part on Newton's method, it has many of the convergence problems associated with this technique [5].

We have applied the method to a three-variable system of ordinary differential equations studied by Lorenz and found several unstable closed cycles which were predicted by the analysis of Williams [8]. In spite of the limited successes of the method when applied to Lorenz's equations, there is still at least one fundamental question which remains unanswered: Is it possible to do the necessary estimates which will prove that the Lorenz model has an "x-y" orbit or any closed cycles at all? We hope that it is.

Acknowledgments

Thanks to O. E. Lanford for demonstrating that it could be done. Special thanks to B. Williams for providing me with a good initial guess. This work was supported by the Air Force Geophysics Laboratory under contract #AF F 19628-78-C-0032.

References

1. Lanford, O., private communication.
2. Hirsch, M. W., and S. Smale, 1979: Algorithms for solving f(x) = 0. Comm. Pure and Appl. Math., 32, 313-357.
3. Kellogg, R. B., T. Li and J. A. Yorke, 1976: A constructive proof of the Brouwer fixed-point theorem. SIAM J. Numer. Anal., 13, 473-383.
4. Robbins, K. A., 1979: preprint.
5. Dahlquist, G., and A, Bjorck, 1969: Numerical Methods, Prentice-Hall.
6. Hartman, P., 1973: Ordinary differential equations.
7. Lorenz, E. N., 1963: Nonperiodic flow. J. Atmos. Sci., 20, pp. 130-141.
8. Williams, B., 1979: IHES Publication (to appear).
9. Marsden, Chorin and S. Smale, 1977: Berkeley Turbulence Seminar, Springer Lecture Notes in Mathematics.

Linked Twist Mappings are Almost Anosov

Robert L. Devaney*

Introduction. In recent years, a class of homeomorphisms called linked twist mappings have been studied by several different authors for several different reasons. Thurston [T] has encountered such mappings in his study of diffeomorphisms of surfaces. Braun [Br], on the other hand, has shown that linked twist mappings occur as the Poincaré mapping of a surface of section in the classical mechanical system called the Störmer problem. And Bowen [Bo] has used the topological properties of these mappings to show that certain linked twist mappings have positive topological entropy.

Perhaps the most important work on linked twist mappings to date is that of Easton [E]. He has shown that such mappings have non-zero characteristic exponents, and hence admit an ergodic component of positive measure.

Our goal in this paper is more topological in nature. We show that certain linked twist mappings share many of the properties of Anosov diffeomorphisms. More precisely, we prove

Theorem A. Let T be a linked twist mapping with all shears positive or all negative. Then

i. The periodic points of T are dense.

ii. Homoclinic points of T are dense.

iii. T is topologically mixing.

---

* Research partially supported by NSF Grant MCS 79-00430.

It is well known that Anosov diffeomorphisms are structurally stable. Linked twist mappings, however, are not. The reason for this is that linked twist mappings are not quite hyperbolic. Usually, these mappings admit a finite number of periodic intervals which thus destroys any hope of structural stability. This is discussed in more detail in §2.

The most famous examples of Anosov diffeomorphisms — the hyperbolic toral automorphisms — can also be regarded as linked twist mappings in certain cases. This is described in §5. This provides a novel way of looking at such mappings — one which breaks down the complicated hyperbolic toral automorphisms into compositions of simple shear mappings of a torus. Unfortunately, not all of these automorphisms are linked twist mappings the way we have defined them. It is an interesting question as to which such mappings can be written as a linked twist mapping. In §5 we discuss this problem in more detail.

In §1 below, we define a special class of linked twist mappings called toral linked twist mappings. It is these mappings which are related to the hyperbolic toral automorphisms. In §6, however, we define a generalized linked twist mapping. These seem to give a new type of almost-Anosov mapping. Unlike the toral case, we know of no known Anosov diffeomorphism to which these mappings are related.

We would like to acknowledge several interesting and informative discussions with S.E. Batterson, R. Easton, and W. Reynolds while this paper was being written. Reynolds, in particular, supplied me with the examples in §5. Also, several comments by H. Bass and T. Jorgensen were useful in describing the conjugacy classes in $SL_2(\mathbb{Z})$.

§1. Toral Linked Twist Mappings. In this section we will define and give some elementary properties of toral linked twist mappings. Later, in §5, we will give a more general definition.

Let T^2 be the standard torus $\mathbb{R}^2 / \mathbb{Z} \times \mathbb{Z}$, and let A be a closed annulus in T^2 defined by

$$A = \{(x,y) \in T^2 \mid x_0 \leq x \leq x_1, \ y \text{ arbitrary}\}$$

where x_0, x_1 are fixed. For any integer k, a k-twist map F on A is defined by

$$F(x,y) = (x, y + \alpha(x))$$

where $\alpha: [x_0,x_1] \to [0,k]$ is smooth and satisfies

i. $\alpha(x_0) = 0$, $\alpha(x_1) = k$

ii. $\alpha'(x) \neq 0$.

So F fixes both boundaries of A and rotates each circle x = constant by an angle which depends only on x. See Figure 1.

Toral linked twist mappings arise by linking several such annuli and composing their respective twists. More precisely, let $\hat{V}_1,\ldots,\hat{V}_k$ be a collection of disjoint (except possibly at endpoints) closed subintervals of the unit interval, and let $\hat{H}_1,\ldots,\hat{H}_m$ be another such collection. These intervals determine a subset M of T^2 defined by

$$M = \{(x,y) \in T^2 \mid x \in \bigcup_{i=1}^{k} \hat{V}_i \text{ or } y \in \bigcup_{j=1}^{m} \hat{H}_j\}.$$

Clearly, M consists of a union of k "vertical" and m "horizontal" annuli in T^2. See Figure 2.

We denote the individual annuli in M by V_i or H_j, i.e.

$$V_i = \{(x,y) \in T^2 \mid x \in \hat{V}_i,\ y \text{ arbitrary}\}$$
$$H_j = \{(x,y) \in T^2 \mid y \in \hat{H}_j,\ x \text{ arbitrary}\}.$$

For simplicity, we will assume that the $\hat{V}_i$ and $\hat{H}_j$ are arranged in increasing order. Note that V_i and V_{i+1} may intersect only along one or both boundaries.

We now define toral linked twist mappings. They will be piecewise differentiable homeomorphisms of M. For each j, $1 \leq j \leq m$, let $\alpha_j : \hat{H}_j \to [0,h_j]$ be smooth maps satisfying:

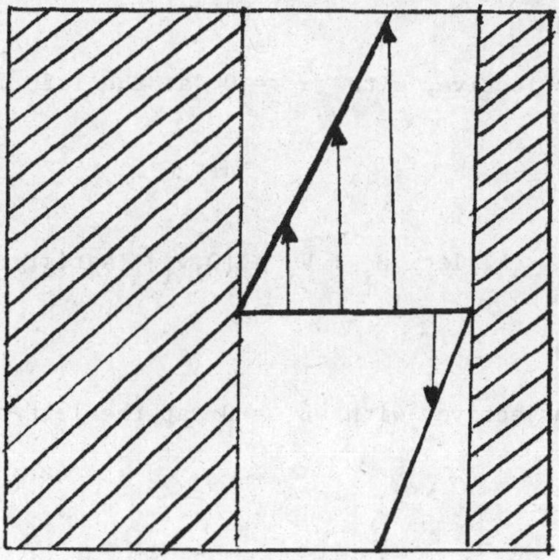

Figure 1. A 1-twist.

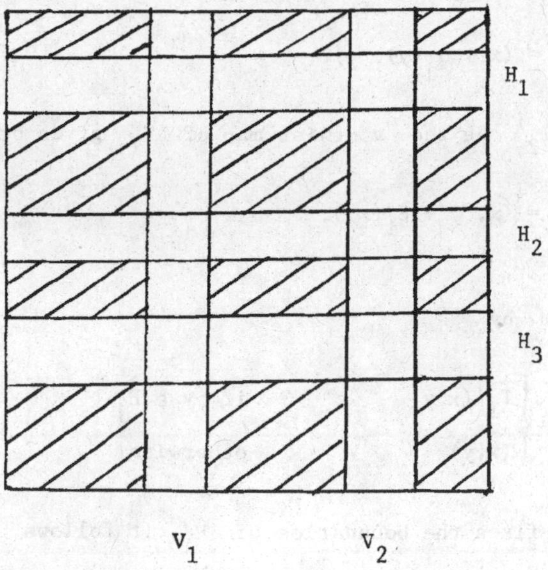

Figure 2. The space M.

i. $\alpha_j'(y) \neq 0$.

ii. α_j is surjective, with $\alpha_j = 0$ at the left hand endpoint of $\hat{H}_j$.

Similarly, for $1 \leq i \leq k$, let $\beta_i : \hat{V}_i \to [0,v_i]$ satisfy:

i. $\beta_i'(x) \neq 0$.

ii. β_i is surjective, with $\beta_i = 0$ at the left hand endpoint of $\hat{V}_i$.

Here, $h_1,\ldots,h_m,v_1,\ldots,v_k$ are integers.

Now let $T_1{}^j : H_j \to H_j$ be the h_j-twist map given by

$$T_1{}^j(x,y) = (x + \alpha_j(y),\ y).$$

Similarly, let $T_2{}^i$ be the v_i-twist map of V_i given by

$$T_2{}^i(x,y) = (x,\ y + \beta_i(x)).$$

Define $T_1 : M \to M$ by

$$T_1(x,y) = \begin{cases} T_1{}^j(x,y) & \text{if } y \in \hat{H}_j \\ (x,y) & \text{otherwise} \end{cases}$$

Since each $T_1{}^j$ fixes the boundaries of H_j, it follows that T_1 is a piecewise smooth homeomorphism of M. We may extend the differential of T_1 to the entire tangent bundle of M as follows. Let (ξ,η) be a tangent vector at (x,y). Let $\gamma(s) = (x,y) + s(\xi,\eta)$ for $s \geq 0$. Define

$$dT_1(\xi,\eta) = \frac{d}{ds}\Big|_{s=0} T_1(\gamma(s))$$

where the derivative with respect to s is the right hand limit. Note that

$$[dT_1(\xi,\eta)]^t = \begin{pmatrix} 1 & \alpha_j'(y) \\ 0 & 1 \end{pmatrix} \begin{pmatrix} \xi \\ \eta \end{pmatrix}$$

if $y \in \operatorname{int} \hat{H}_j$, so dT_1 agrees with the ordinary differential in the interior of H_j. Clearly, dT_1 is the identity mapping in the complement of $\cup H_j$. So it is only on the boundary of H_j where the above definition differs from the ordinary differential.

Clearly, dT_1 is a piecewise linear mapping on each tangent space.

We similarly define $T_2 \colon M \to M$ by

$$T_2(x,y) = \begin{cases} T_2^{\,i}(x,y) & \text{if } x \in \hat{V}_i \\ (x,y) & \text{otherwise} \end{cases}$$

The differential dT_2 is defined exactly as before. So both T_1 and T_2 are piecewise smooth homeomorphisms of M which are twists on the horizontal or vertical annuli and the identity mapping elsewhere. Let T be the composition $T_2 \circ T_1$. T is called a (toral) linked twist mapping.

We close this section with several definitions. Let M_{ij} denote the interior of the rectangle $V_i \cap H_j$. M_{ij} is called the ij^{th} mixing region. Points in M_{ij} move under both T_1 and T_2, whereas other points in M are left fixed by at least one of them. The structure of the set of points whose orbits remain for all time in the mixing regions is the subject of [D]. There it is shown that these points form a hyperbolic set on which T is topologically conjugate to a subshift of finite type.

A point p is said to be mixed by T if $T_1(p)$ lies in a mixing region. So mixed points are acted upon by both T_1 and T_2 in one iteration of T. The point p is said to be eventually mixed if some point on the forward orbit of p is mixed.

Let $\mathcal{M}$ denote the set of eventually mixed points. Observe that if $p \in M - \mathcal{M}$, then the orbit of p lies entirely in one of the horizontal or one of the vertical annuli in M. Therefore such a point lies on one of the T_1- or T_2-invariant circles in this annulus. If such a point never enters a mixing region, it follows that the rotation number of that circle is rational (otherwise the orbit of p would be dense in the circle). Consequently, p is a periodic point, and, in fact, periodic with relatively low period. It follows easily that $M - \mathcal{M}$ consists of a finite number of periodic intervals for T, and thus $M - \mathcal{M}$ has measure zero in M.

The most important type of linked twist mapping for our purposes satisfies the additional hypothesis that all of the shears occur in the same direction. More precisely, a linked twist mapping is said to have <u>all positive shears</u> (resp. all negative shears) if all of the integers $h_1,\ldots,h_m,v_1,\ldots,v_k$ are positive (resp. negative). Equivalently, the derivatives of each α_j and β_i are positive (resp. negative). Our results below apply mainly to these types of linked twist mappings. For simplicity, we shall only consider the positive case; the negative linked twist mappings are handled similarly.

We finally observe that k-twist mappings are area-preserving; hence linked twist mappings also preserve Lebesgue measure in M. Since M is compact, it follows from the Poincaré Recurrence Theorem that all points in M are non-wandering.

§2. Almost Hyperbolicity. In this section we will consider only toral linked twist mappings with all positive shears. The results below clearly extend to the negative case. Later we shall also extend these results to more general types of linked twist mappings.

For each $p \in M$, consider the closed sectors in the tangent space at p, T_pM, defined by

$$S_p^+ = \{(\xi,\eta) \in T_pM \mid \xi\eta \geq 0\}$$
$$S_p^- = \{(\xi,\eta) \in T_pM \mid \xi\eta \leq 0\}.$$

We let S^+ and S^- denote the sector bundles

$$S^+ = \bigcup_{p\in M} S_p^+ \qquad S^- = \bigcup_{p\in M} S_p^-.$$

A mapping is called hyperbolic if, roughly speaking, it preserves a bundle of sectors and expands all vectors in that bundle, while its inverse does the same of the complementary bundle. K-twist mappings almost have this property, as we now show.

Let $p \in \bigcup H_j$. Let $(\xi_0,\eta_0) \in S_p^+$ and suppose $dT_1(\xi_0,\eta_0) = (\xi_1,\eta_1)$. Then

$$\xi_1\eta_1 = \xi_0\eta_0 + \alpha'(p)\,\eta_0^2 \geq 0$$

since $\alpha'(p) > 0$. Hence $(\xi_1,\eta_1) \in S^+_{T_1(p)}$.

Also,

$$|(\xi_1,\eta_1)|^2 = \xi_0^{\ 2} + (1 + \alpha'(p))^2\ \xi_0^{\ 2} + 2\ \alpha'(p)\ \xi_0\ \eta_0$$

$$\geq\ |(\xi_0,\eta_0)|^2$$

with equality only when $\eta_0 = 0$. That is, vectors of the form $(\xi_0,0)$ are mapped to themselves by dT_1, but all other tangent vectors in $S_p^{\ +}$ are expanded. It is only here where hyperbolicity fails.

One has similar estimates for dT_2; dT_2 preserves S^+ over $\cup V_i$ and all vectors are expanded except those of the form $(0,\eta_0)$. In particular, dT_2 expands vectors of the form $(\xi_0,0)$, provided p is also in a vertical annulus. Finally, $dT_1^{\ -1}$ and $dT_2^{\ -1}$ behave similarly on $S_p^{\ -}$.

It follows that the toral linked twist mappings have similar properties. If $p \in M$, then dT maps $S_p^{\ +}$ into $S_{T(p)}^{\ +}$. Moreover, if p is a mixed point, then dT maps $S_p^{\ +}$ properly inside $S_{T(p)}^{\ +}$, and moreover, dT expands every vector in $S_p^{\ +}$. Hence we have genuine hyperbolicity at mixed points. If p is eventually mixed, then there is a positive integer n for which dT^n has this property. Hence it is only over $M - M$ where we do not have eventual hyperbolicity. These non-hyperbolic points are all periodic under T, and the eigenvalues of dT^n at these points are both +1. A mapping with these properties is said to be <u>almost hyperbolic</u>.

§3. <u>Unstable curves</u>. One of the main ingredients in our proof of Theorem A is the notion of an unstable curve in M. Let $\gamma(s)$ be a piecewise smooth curve in M satisfying $\gamma'(s) \neq 0$. $\gamma(s)$ is called an <u>unstable curve</u> if

i. $\gamma'(s) \in \text{int } S^+(\gamma(s))$ for all s for which $\gamma'(s)$ exists.

ii. Both the left and right hand tangents to $\gamma(s)$ lie in the same sector of int $S^+(\gamma(s))$ at all other points.

Stable curves are defined analogously.

We first observe that, if $\gamma(s)$ is an unstable curve, then so are $T_1(\gamma(s))$, $T_2(\gamma(s))$, and $T(\gamma(s))$. Indeed, this follows immediately from the almost hyperbolicity of these mappings. Similarly, the inverses of these mappings preserve stable curves.

We next observe that T stretches an unstable curve in both the horizontal and vertical directions. To see this, let $\gamma(s)$ be an unstable curve and denote by Δx_n (resp. Δy_n) the length of the projection of any lift of $T^n(\gamma(s))$ to the x-axis (resp. y-axis). By the lift of a curve, we mean any piecewise smooth curve in $\mathbb{R}^2$ which projects to the curve under the standard projection $\mathbb{R}^2 \to T^2$. Let $C = \min(\beta_i'(y), \alpha_j'(x))$ for all i,j. Then, by the Mean Value Theorem, we have

$$|\Delta x_n| \geq |\Delta x_0| + nC|\Delta y_0|$$
$$|\Delta y_n| \geq |\Delta y_0| + nC|\Delta x_0|$$

so that, in particular, both $|\Delta x_n|$ and $|\Delta y_n|$ tend to infinity as $n \to \infty$.

We say that a curve $\gamma(s)$ cuts a mixing region vertically if there exists s_0, s_1 such that for $s_0 < s < s_1$, $\gamma(s) \in M_{ij}$ and $\gamma(s_0)$ lies on either the upper or lower boundary of M_{ij}, while $\gamma(s_1)$ lies on the opposite boundary. Similarly, $\gamma(s)$ cuts M_{ij} horizontally if $\gamma(s_0)$ and $\gamma(s_1)$ lie on opposite vertical boundaries of M_{ij}.

The main goal of this section is to prove the following proposition.

Proposition. _Let_ $\gamma(s)$ _be an unstable curve_ (_resp. stable curve_) _in_ M. _Let_ M_{ij} _be any mixing region. Then there exists_ $N > 0$ _such that_ $T^N(\gamma(s))$ _cuts_ M_{ij} _vertically_ (_resp._ $T^{-N}(\gamma(s))$ _cuts_ M_{ij} _horizontally_).

Proof: We prove this for unstable curves; the proof for stable curves is analogous. We need several lemmas.

Lemma 1. Let $\gamma(s)$ be an unstable curve connecting the left and right boundaries of V_j and not lying entirely in one mixing region. Then, for any i, there exists a subcurve of $T(\gamma(s))$ which cuts across M_{ij} vertically.

Proof: Observe first that, since $\gamma(s)$ contains a point fixed by T_1, there is an unstable curve $\gamma_1(s)$ contained in $T_1(\gamma(s))$ which also contains this point and meets the left and right hand boundaries of V_j. This follows immediately from the definition of T_1. We may assume that $\gamma_1(s)$ lies entirely in V_j. Consider $T_2(\gamma_1(s))$. This unstable curve lies entirely in V_j, and one checks easily that $T_2(\gamma_1(s))$ cuts across each mixing region M_{ij} in V_j vertically. QED

Lemma 2. Let $\gamma(s)$ be an unstable curve connecting the upper and lower boundaries of H_i and not entirely contained in a single M_{ij}. Then $T^2(\gamma(s))$ cuts across each mixing region vertically.

Proof: One checks easily that, for each j, $T_1(\gamma(s))$ cuts across each M_{ij} horizontally. This uses the fact that $\gamma(s)$ is not contained entirely in a single mixing region. Let $\gamma_j(s)$ denote a component of $T_1(\gamma(s))$ in M_{ij} which connects the left and right hand boundaries of M_{ij}. Then $T_2(\gamma_j(s))$ is an unstable curve in V_j satisfying the hypotheses of Lemma 1. Therefore, $T(T_2(\gamma_j(s)))$ contains a subcurve which meets each M_{kj} vertically. This completes the proof. QED

We now complete the proof of the proposition. Since the lengths of the projections of $T^n(\gamma(s))$ onto the x- and y-axes grows arbitrarily large with n, it follows that there exists $k > 0$ such that $T^k(\gamma(s))$ meets each of the horizontal and vertical annuli together with all of their boundary circles. Hence there is some subcurve of $T^k(\gamma(s))$, say $\gamma^*(s)$, such that for $s_0 \leq s \leq s_1$, $\gamma^*(s)$ connects the left and right hand boundaries of some V_j, and is not contained entirely in a single mixing region of V_j. By Lemma 1, there is a subcurve of $T(\gamma^*(s))$ which cuts across M_{ij} vertically for any i. Hence $T^{k+1}(\gamma(s))$ cuts M_{ij} vertically for all i.

Now let $\gamma_i(s)$ be any component of $T^{k+1}(\gamma(s))$ which connects the upper and lower boundaries of M_{ij}. As in Lemma 2, one checks easily that $T(\gamma_i(s))$ contains a component which joins the upper and lower boundaries of H_i, but which is not contained in any single mixing region. Lemma 2 applies and shows that T^2 maps this component vertically across any mixing region. QED

§4. Proof of Theorem A. In this section we complete the proof of Theorem A. We first assume that at least one of horizontal twists and one of the vertical twists are k-twists, with $k \geq 2$. This simplifies the proof considerably. In particular, in the mixing region common to these twists, there is a hyperbolic fixed point which we will call p^*. The proof of this is easy and is contained in [D]. Later in the section we will sketch the modifications necessary in case all of the vertical and/or horizontal twists are 1-twists.

Recall that the hyperbolic fixed point p^* admits two invariant curves $W^s(p^*)$ and $W^u(p^*)$ consisting of points which are respectively forward and backward asymptotic to p^* under iteration of T. In our case, the stable "manifold" $W^s(p^*)$ is actually a piecewise smooth curve whose tangent vectors lie in the sector bundle S^- at each point. Similarly, the tangent space to $W^u(p^*)$ is everywhere contained in S^+.

In our case, we may assume that p^* lies in the mixing region M_{11}. Furthermore, it is easy to check that the local stable manifold at p^* is a smooth curve which cuts M_{11} horizontally. Similarly, the local unstable manifold is a smooth curve through p^* which cuts M_{11} vertically. See Figure 3.

A point q in the intersection of $W^s(p^*)$ and $W^u(p^*)$ is called a homoclinic point. Such points play an important role in the

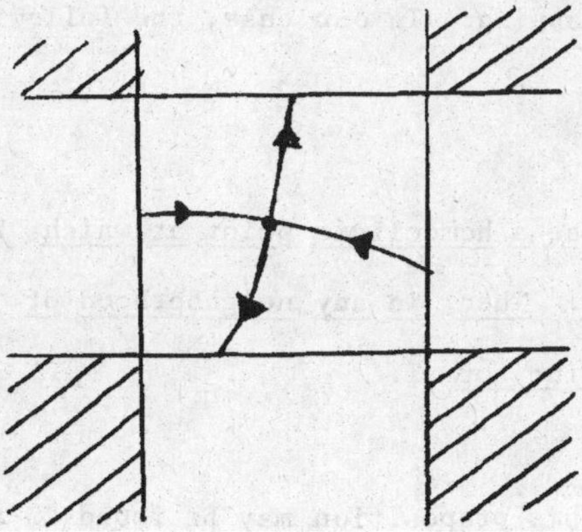

Figure 3. The local stable and unstable manifolds at p*.

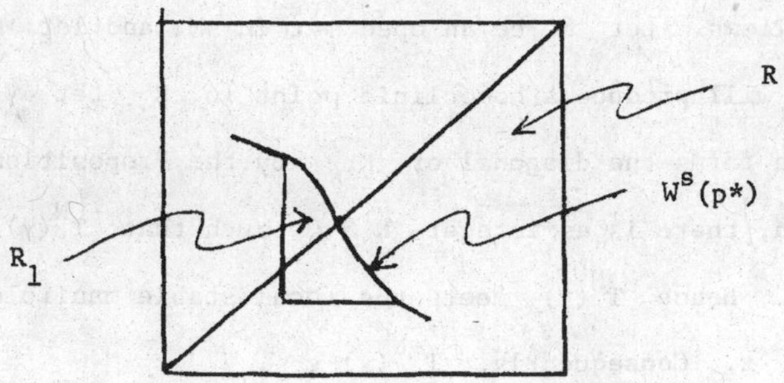

Figure 4. The rectangle R.

orbit structure of a mapping. In our case, the following Theorem of Smale is applicable.

Proposition. Let q be a homoclinic point at which $W^s(p^*)$ and $W^u(p^*)$ intersect transversely. Then, in any neighborhood of q, there is a hyperbolic periodic point for T.

The proof of this proposition may be found in any of [M, N, S2]. In each case, the proposition is proved only for smooth mappings. However it is easy to modify any of the proofs to the piecewise smooth case.

We now proceed to the proof of Theorem A. By the Proposition, it suffices to show that homoclinic points are dense in M. This can be seen as follows. Let N be an open set in M, and let R be a rectangle in N. We will produce a homoclinic point in R. Let γ be an unstable curve which forms the diagonal of R. By the Proposition in the previous section, there is an integer $N > 0$ such that $T^N(\gamma)$ cuts M_{11} vertically. Hence $T^N(\gamma)$ meets the local stable manifold of p^* at some point x. Consequently, $T^{-N}(x) \in \gamma$.

Now consider a small triangular region R_1 in R bounded by a piece of $W^s(p^*)$ containing $T^{-N}(x)$, and by a vertical and horizontal line parallel to the boundaries of the annuli. See Figure 4. Since $W^s(p^*)$ is a stable curve, R_1 is indeed triangular.

Arguments just as above show that any stable curve in R_1 must eventually meet the local unstable manifold of p^*. Hence $W^u(p^*)$ must meet the interior of R_1. Since $W^u(p^*)$ is an unstable curve, it follows that $W^u(p^*)$ must exit R_1 by crossing the hypoteneuse $W^s(p^*)$ at some point q. This point is therefore a homoclinic point. Note that $W^s(p^*)$ and $W^u(p^*)$ meet transversely at q, since the right and left hand tangents to both $W^s(p^*)$ and $W^u(p^*)$ at q lie in distinct sectors. By Smale's result, it follows that there is a periodic point nearby.

Finally, to prove that T is topologically mixing, take two open sets U and V in M. As above, we may assume that $U \cap W^s(p^*) \neq \emptyset$ and $V \cap W^u(p^*) \neq \emptyset$. So there are integers $n > 0$, $k < 0$ such that $T^n(U)$ intersects the local stable manifold in M_{11}, and $T^k(V)$ meets the local unstable manifold. Using the λ-lemma (see [N]), it follows that $T^{n+\alpha}(U) \cap T^{k-\beta}(V) \neq \emptyset$ for any $\alpha, \beta > 0$. This implies that T is topologically mixing.

In the case where one does not have any fixed points in the mixing regions, one can use any of the hyperbolic periodic points guaranteed to exist by [D]. Alternatively, one can use the "corners" of the mixing region. These fixed points are easily seen to admit a one-sided stable or unstable manifolds, and one can manipulate the above proof to find homoclinic points for these curves. One needs a special version of the λ-lemma here to get Smale's result, however.

§5. Hyperbolic Toral Automorphisms. Recall the definition of hyperbolic toral automorphisms. One is given a 2×2 matrix A with integer entries, determinant ± 1, and eigenvalues off the unit circle. Such a matrix induces an automorphism of the two dimensional torus in a natural manner, and this class of mappings is well understood. Our goal here is to detail the relationship between certain of these hyperbolic toral automorphisms and our linked twist mappings.

Suppose first that

$$A = \begin{pmatrix} 1 & n \\ k & nk+1 \end{pmatrix}$$

Then we have

$$A = \begin{pmatrix} 1 & 0 \\ k & 1 \end{pmatrix} \begin{pmatrix} 1 & n \\ 0 & 1 \end{pmatrix} = A_2 A_1 .$$

The matrices A_2 and A_1 also induce mappings on the torus which are respectively k and n-twists. Here the entire torus represents both the vertical and horizontal annulus, so that $A_2 A_1$ is a linked twist mapping.

If the matrix A is linearly conjugate to

$$\begin{pmatrix} 1 & n \\ k & nk+1 \end{pmatrix}$$

via an element of SL(2, $\mathbb{Z}$), then it follows that the induced automorphism is also topologically conjugate to a linked twist mapping. We do not know the conjugacy classes in SL(2, $\mathbb{Z}$). However, the following example due to W. Reynolds shows that not all conjugacy classes include even the more general type of matrix of the form

$$\begin{pmatrix} \pm 1 & n \\ k & \pm(nk+1) \end{pmatrix}$$

Consider the matrix

$$B = \begin{pmatrix} 26 & 45 \\ 15 & 26 \end{pmatrix}.$$

Suppose B is conjugate to the matrix

$$\begin{pmatrix} e & f \\ g & h \end{pmatrix}$$

where $e = \pm 1$. Reducing the conjugacy equation mod 3, one finds that $e \equiv 2 \bmod 3$. However, reducing the equation mod 5, one finds that $e \equiv 1 \bmod 5$. This contradiction shows that the automorphism induced by B is not linearly conjugate to a linked twist mapping.

We conclude this section with one final observation: every hyperbolic toral automorphism is semi-conjugate to a linked twist mapping. Indeed, if

$$A = \begin{pmatrix} a & b \\ c & d \end{pmatrix}$$

and

$$H = \begin{pmatrix} 1 & a-1 \\ 0 & c \end{pmatrix},$$

then

$$H^{-1} A H = \begin{pmatrix} 1 & a+d-2 \\ 1 & a+d-1 \end{pmatrix}$$

which induces a linked twist mapping on the torus. Since $\det H \neq 1$ in general, this only gives a semi-conjugacy.

§6. Generalized Linked Twist Mappings. The toral linked twist mappings defined in §1 share many of the properties of hyperbolic toral automorphisms, and, in fact, certain of these automorphisms are topologically conjugate to linked twist mappings. In a sense, then, these toral linked twist mappings do not represent any substantially new phenomenon. In this section, however, we define a generalized linked twist mapping which does seem to provide a new class of almost-Anosov homeomorphisms.

Consider a collection of horizontal and vertical annuli H_i and V_j for $1 \leq i \leq m$, $1 \leq j \leq n$ defined as in §1. Let $A = [a_{ij}]$ be an $m \times n$ matrix of 0's and 1's. We will modify the manifold on which T is defined via a rule prescribed by A. In the plane, consider the H_i and V_j as subsets of the unit square exactly as in Figure 2. We construct a new collection of linked annuli in $\mathbb{R}^3$ as follows. This collection M_1 will consist of a union $M_H \cup M_V$ where

$$M_H = \{(x,y,0) \in \mathbb{R}^3 \mid (x,y) \in \cup H_i\}$$
$$M_V = \{(x,y,F(x,y)) \in \mathbb{R}^3 \mid (x,y) \in \cup V_j\}.$$

and $F\colon M \to [0,1]$ is a smooth function satisfying $F|M_{ij} = a_{ij}$. So M_H and M_V intersect only in mixing regions where $a_{ij} = 0$. So M_1 has fewer mixing regions than M. Now identify the horizontal and vertical boundaries as in the toral case. The resulting space is no longer a subset of the two-torus; nevertheless, we may define the twist mappings T_1 and T_2 exactly as before, so that $T = T_2 \circ T_1$ is a generalized linked twist mapping.

Theorem A applies to these mappings with minor modifications in the proof.

<u>Note added in proof</u>: Regarding the question of the conjugacy classes in $SL_2(\mathbb{Z})$, Profs. H. Bass and T. Jorgensen have informed me of the following facts. If $A \varepsilon SL_2(\mathbb{Z})$ and $|Tr(A)| \leq 2$, then A has a conjugate of the form

$$\begin{pmatrix} \pm 1 & * \\ * & * \end{pmatrix}.$$

On the other hand, if $|Tr(A)| > 2$, then A^n cannot have such a conjugate unless $n = 1$.

References

[Bo] Bowen, R.: On Axiom A Diffeomorphisms. Proc. CBMS Regional Conf. Math. Ser., No. 35, Amer. Math. Soc., Providence, R.I., 1978.

[Br] Braun, M.: Invariant curves, homoclinic points, and ergodicity in area preserving mappings. To appear.

[D] Devaney, R.: Subshifts of finite type in linked twist mappings. Proc. Amer. Math. Soc. 71 (1978) 334-338.

[E] Easton, R. and R. Burton: Ergodicity of linked twist mappings. This proceedings.

[N] Nitecki, Z.: Differentiable Dynamics. MIT Press, Cambridge, Mass., 1971.

[M] Moser, J.: Stable and Random motions in dynamical systems. Princeton University Press, Princeton, N.J., 1973.

[S1] Smale, S.: Differentiable dynamical systems. Bull. Amer. Math. Soc. 73 (1967) 747-817.

[S2] Smale, S.: Diffeomorphisms with many periodic points. In Differential and combinatorial topology. Princeton University Press, Princeton, N.J., 1965.

[T] Thurston, W.: On the geometry and dynamics of diffeomorphisms of surfaces. To appear.

Tufts University
Medford, MA 02155

SYMBOLIC DYNAMICS, HOMOLOGY, AND KNOTS

by John M. Franks

In this article I want to survey a sequence of results whose roots lie in the classical Euler-Poincaré-Hopf formula. Recall that this formula says that for any flow on a compact manifold with isolated rest points

$$\chi(M) = \Sigma i(p)$$

where $\chi(M)$ is the Euler characteristic, $i(p)$ is the index of the rest point p and the sum is taken over all rest points p of the flow.

In case the flow in question is the gradient of a function with non-degenerate critical points (i.e. a Morse function) then $i(p) = (-1)^{u(p)}$ where $u(p)$ is the dimension of the unstable manifold of p. (The number $u(p)$ is called the Morse index which we hope to avoid confusing with $i(p)$). In the special setting of gradients of Morse functions a much stronger version of the formula was proved by M. Morse (cf. [M]). Morse showed that if $\beta_i = \text{rank } H_i(M)$ and c_i = the number of critical points of Morse index i, then

$$(*) \qquad c_i - c_{i-1} + \dots \pm c_0 \geq \beta_i - \beta_{i-1} + \dots \pm \beta_0$$

for all i and with any choice of coefficients for the homology $H_i(M)$. In the case $i = \dim M$, Morse showed this inequality is an equality and it is in fact just the Euler-Poincaré-Hopf formula mentioned above.

These Morse inequalities form a paradigm for the homological study of dynamical systems. They show that homological invariants of the manifold M (viz. the Betti numbers β_i) restrict the kinds of dynamics (number of critical points of each Morse index) which can occur on M.

This paradigm has another half, however, in the form of a near converse to the result of Morse. The following remarkable result is due to S. Smale [S1].

heorem (Smale). If M is simply connected and has dimension ≥ 6, and $\{c_i\}$ is any set f non-negative integers satisfying (*) for all choices of coefficients for the homlogy of M, then there exists a Morse function on M with exactly c_i critical points ith Morse index i.

The Morse inequalities (*) have been generalized in many ways and found many pplications. (For generalizations applied to dynamical systems see the excellent onograph [C] of C. Conley.) Our approach here will be to pursue those generalizaions related to symbolic dynamics.

In general terms there are two major themes in the theory of smooth dynamical ystems -- statistical or probabilistic methods and topological methods. While often hese two approaches differ greatly in their applications and the kind of information hey give us about a dynamical system, there is a strong unifying element between hem in the form of symbolic dynamics. The ergodic theory of symbol shifts will not e dealt with here. Instead we will consider how symbol shifts arise in smooth dyamical systems and survey some results relating the symbol shift to homological nformation about the system.

We begin with an example of a diffeomorphism of the two sphere S^2 which we hink of as the plane with a point ∞ at infinity added. In the plane we choose a egion X consisting of three disks and two strips and map it as shown in Figure 1.

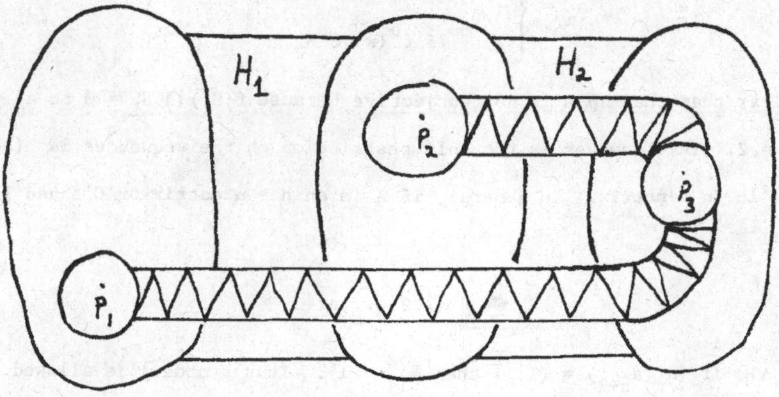

Fig. 1

The map is defined so that the points $\{p_1,p_2,p_3\}$ form a periodic attractor of period 3. We also impose conditions on the behavior of the map on the two strips. Each strip is foliated in two ways -- by horizontal and by vertical line segments. We arrange that f uniformly stretches each horizontal line segment and that f(horizontal line segment) contains any horizontal line segment it intersects. Analogously we arrange that f uniformly contracts vertical line segments and that f(vertical line segment) is contained in any vertical line segment it intersects. Finally we extend the map to all of S^2, making ∞ an expanding fixed point, in such a way that the forward orbit of every point except ∞ enters X.

We now consider the asymptotic behavior of points under the diffeomorphism f. Many points x will satisfy $f^{-n}(x) \to \infty$ as $n \to \infty$ and many will satisfy $f^n(x) \to \{p_1,p_2,p_3\}$ as $n \to \infty$. We consider the closed set Λ of points which do neither. These are the points whose entire orbit lies in the two strips H_1 and H_2. It turns out that each point in Λ is determined by specifying which strip it is in after n iterates of f, $n \in Z$. Thus if we give $\{1,2\}$ the discrete topology and consider the space of bi-infinite sequences $\prod_{-\infty}^{\infty} \{1,2\}$ with the product topology, there is a map

$$h : \Lambda \to \prod_{-\infty}^{\infty} \{1,2\}$$

defined by $h(x) = \underline{a} = (\dots a_{-1},a_0,a_1 \dots)$ where for each $n \in Z$

$$a_n = \begin{cases} 1 & \text{if } f^n(x) \in H_1 \\ 2 & \text{if } f^n(x) \in H_2. \end{cases}$$

It is clear that the map h is not surjective because $f(H_1) \cap H_1 = \emptyset$ so $a_n = 1$ implies $a_{n+1} = 2$. This however is the only restriction on the sequences in $h(\Lambda)$ and we can codify it in a matrix. In general, if A is an $n \times n$ matrix of 0's and 1's we define

$$\Sigma_A \subset \prod_{-\infty}^{\infty} \{1,2,\dots,n\}$$

by $\Sigma_A = \{\underline{a} \mid \forall n,\text{ if } (a_n,a_{n+1}) = (i,j) \text{ then } A_{ij} = 1\}$. Thus symbol j is allowed to follow symbol i if and only if $A_{ij} = 1$. In the example above we have claimed $h(\Lambda) = \Sigma_A$ where $A = \begin{pmatrix} 0 & 1 \\ 1 & 1 \end{pmatrix}$. The proof of this is not difficult (see e.g. (2.4) of [F1]).

There is also a homeomorphism $\sigma : \Sigma_A \to \Sigma_A$ called a subshift of finite type de-ined by $\sigma(\underline{a}) = \underline{b}$ where $b_n = a_{n+1}$. It is clear in our example that the diagram

$$\begin{array}{ccc} \Lambda & \xrightarrow{h} & \Sigma_A \\ \downarrow f & & \downarrow \sigma \\ \Lambda & \xrightarrow{h} & \Sigma_A \end{array}$$

ommutes. We say $f|\Lambda$ is topologically conjugate to σ.

Thus up to homeomorphism the dynamics of orbits in Λ is completely described by he matrix A. In fact one can show that for each $x \in S^2$ there is a y which is either , p_1, p_2, p_3 or is in Λ such that $d(f^n x, f^n y) \to 0$ as $n \to \infty$. Thus we have a good pic-ure of the possible long run behaviors for all points.

The matrix A describing the behavior of $f|\Lambda$ up to homeomorphism is not unique - it depends on the way we have drawn the disks and H_1, H_2, not just the dynamics of . There are many different matrices which correspond to subshifts which are topo-ogically conjugate; i.e. we can find a matrix B and a homeomorphism $\emptyset : \Sigma_A \to \Sigma_B$ uch that

$$\begin{array}{ccc} \Sigma_A & \xrightarrow{\emptyset} & \Sigma_B \\ \downarrow \sigma_A & & \downarrow \sigma_B \\ \Sigma_A & \xrightarrow{\emptyset} & \Sigma_B \end{array}$$

ommutes. The relationship which A and B must have for this to be possible has been uch studied (see [W1]). However, for our purposes we note only that the polynomial et (I-At) is an invariant.

The matrix A is also related to a homological description of the map f in a way hich we now want to describe. A good general reference for this type of analysis is [SS].

We return to Fig. 1 and consider the map f restricted to X. If we denote by Y the union of the three disks in X, and think of the strips H_1 and H_2 as thickened horizontal line segments, then H_1 and H_2, suitably oriented, represent a basis of the homology group $H_1(X,Y)$. The matrix of the map $f_* : H_1(X,Y) \to H_1(X,Y)$ induced by f with respect to this basis is $\tilde{A} = \begin{pmatrix} 0 & -1 \\ 1 & -1 \end{pmatrix}$, which is just A with some minus signs added to reflect the action of f on the orientation of the strips. In this case the $f_* : H_1(X,Y) \to H_1(X,Y)$ is quite different from the maps $f_* : H_*(S^2) \to H_*(S^2)$, though this is not always the case. We can consider, for example, a map of the torus T^2 illustrated in Fig. 2 which is constructed to be similar to f. What is shown in Fig. 2 is a picture of T^2 with a disk D^2 deleted and the image of $T^2 - D^2$ under a diffeomorphism $g : T^2 \to T^2$. The diffeomorphism g will have an attracting fixed point p, a repelling fixed point ∞ in the missing D^2 and a compact invariant set

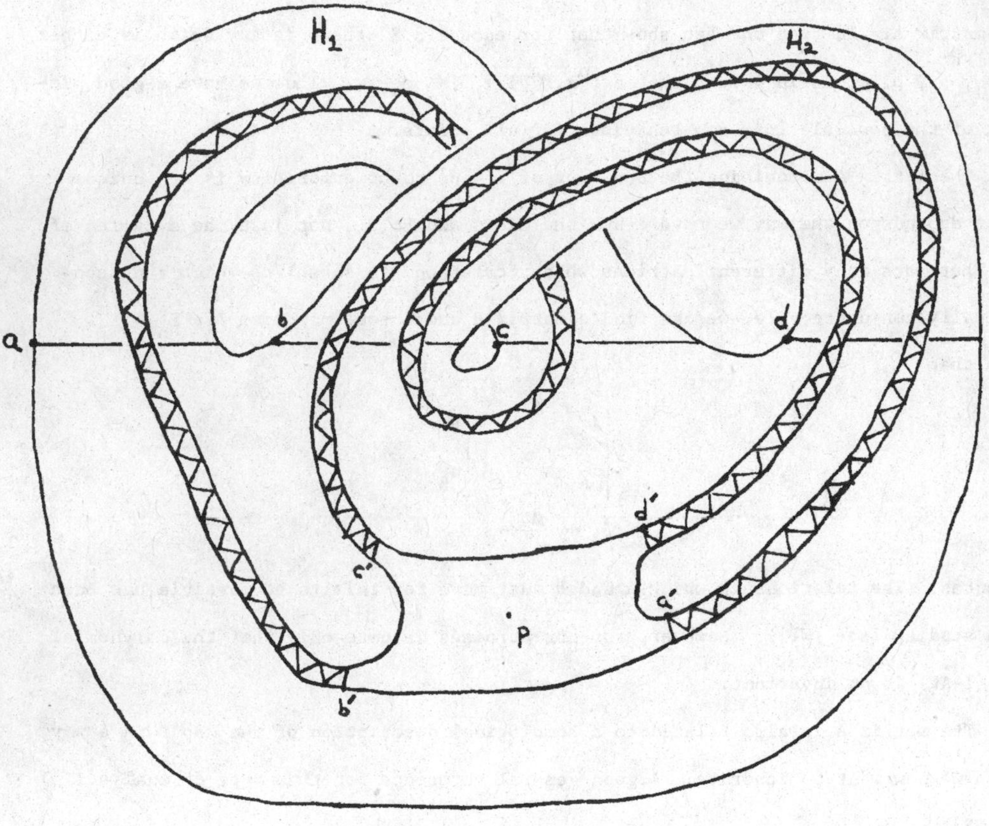

Fig. 2

$\Lambda \subset H_1 \cup H_2$ such that $f|\Lambda$ is topologically conjugate to $\sigma : \Sigma_A \to \Sigma_A$. If we include the signs to reflect how g changes orientations then we have $\tilde{A} = \begin{pmatrix} 0 & 1 \\ -1 & 1 \end{pmatrix}$ and a fairly simple computation shows this is precisely the map $g_* : H_1(T^2) \to H_1(T^2)$ induced by g.

To formally state a theorem relating the signed symbolic matrices $\tilde{A}$ with the aps induced on homology we need some definitions. If $f : M \to M$ is a diffeomorphism nd $x \in M$ then x is said to be chain-recurrent provided that given any $\varepsilon > 0$ there xist points $x = x_1, x_2, x_3, \ldots, x_n = x$ such that $d(f(x_i), x_{i+1}) < \varepsilon$ where d is a fixed etric on M. The set of chain recurrent points $\mathcal{R}$ is a compact invariant set (see or example [C]).

The chain recurrent set $\mathcal{R}$ is said to have a hyperbolic structure or be hyper- olic if the tangent bundle of M restricted to $\mathcal{R}$ is the sum of two Df invariant undles $E^u \oplus E^s$ and if there are constants $C > 0$, $\lambda \in (0,1)$ such that

$$| Df^n(v) | \leq C\lambda^n |v| \qquad \text{for } v \in E^s,\ n > 0,$$

nd

$$| Df^{-n}(v) | \leq C\lambda^n |v| \qquad \text{for } v \in E^u,\ n > 0.$$

If the chain recurrent set $\mathcal{R}$ has a hyperbolic structure then it decomposes nto finitely many disjoint closed invariant pieces $\{\Lambda_i\}$ each with a dense orbit. he Λ_i are called basic sets. Similar definitions can be made for flows.

If Λ_i is zero dimensional then a result of Bowen [Bl] says $f|\Lambda_i$ is topologically onjugate to a subshift of finite type. This is precisely the situation illustrated y the example above.

We wish now to add signs to the (non-unique) matrix A of the subshift of finite ype corresponding to a basic set Λ. Suppose $h : \Sigma_A \to \Lambda$ is the given conjugacy. et $\delta(x) = \pm 1$ according to whether $Df_x : E^u_x \to E^u_{fx}$ preserves or reverses some fixed hosen orientation of E^u (since Λ is totally disconnected $E^u|\Lambda$ is orientable). ince $\delta(x)$ is continuous it is locally constant and we can pick h (and hence A) so hat δ is constant on $C_k = \{x \in \Lambda | h(x)_0 = k\}$ for all k.

efinition. A structure matrix $\tilde{A}$ for Λ is given by $\tilde{A}_{ij} = \delta(C_j)A_{ij}$.

We can now state a generalization of the equality

$$\sum_{i=0}^{h} (-1)^i C_i = \sum_{i=0}^{n} (-1)^i \beta_i$$

rom the Morse inequalities described above.

Theorem [F1]. *Suppose* $f : M \to M$ *is a diffeomorphism whose chain recurrent set* $\mathcal{R}$ *is zero dimensional and has a hyperbolic structure. Then*

$$\prod_{\ell} \det (I-\tilde{A}_{\ell} t)^{(-1)^{u(\ell)}} = \prod_{k} \det (I-f_{*k} t)^{(-1)^k}$$

where $\tilde{A}_{\ell}$ *is the structure matrix of the* ℓ-th *basic set and* $f_{*k} : H_k(M) \to H_k(M)$ *is induced by* f.

If f was the time one map of the gradient flow of a Morse function then both sides of the above equality would consist of powers of (1-t) (the structure matrices would all be the 1×1 matrix (1)). The exponent of the left hand side would be $\Sigma(-1)^i c_i$ and the exponent of the right hand side would be $\Sigma(-1)^i \beta_i$, so this is indeed a generalization of one of the Morse inequalities. In fact, one can formulate a similar generalization of all the inequalities (see [F1]).

We turn now to the investigation of flows and ask if we can find similar results in this setting. We still have a notion of chain recurrence and hyperbolicity whose definitions are quite similar to those given above for diffeomorphisms. Also a result of Bowen [B2] says that if the chain recurrent set $\mathcal{R}$ is hyperbolic and one dimensional then each basic set will admit a cross section such that the first return map is a subshift of finite type $\sigma : \Sigma_A \to \Sigma_A$.

On the other hand there is no interesting analog of f_{*k} since all the maps induced on homology by a flow are the identity map. Also, the matrix A is now even more non-unique since one can choose many different cross sections giving rise to very different subshifts of finite type. In particular, neither det (I-At) nor det $(I-\tilde{A}t)$ will be invariant.

However a result of Parry and Sullivan [PS] shows that the integer det (I-A) is an invariant and by similar methods it is not hard to show det $(I-\tilde{A})$ is also an invariant.

As a beginning on the problem of understanding the flow case we limit our attention to non-singular flows on the three dimensional sphere S^3 which have a hyperbolic chain recurrent set. A great many such flows can be understood by finding partial ross sections which reduce the problem to the study of a map on a surface.

For instance given any map $f : D^2 \to \text{int } D^2$ we can form a flow on $S^1 \times D^2$ called he suspension flow which has $p \times D^2$ as a surface of section with first return map f. e can do this in such a way that the flow is inwardly transverse on the boundary of $^1 \times D^2$ which allows us to patch together with a flow on $D^2 \times S^1$ which has a single epelling closed orbit and is outwardly transverse on the boundary. In this way we an form a flow on $S^3 = (S^1 \times D^2) \cup (D^2 \times S^1)$.

For such flows a description of the dynamics largely reduces to a description f the dynamics of the embedding $f : D^2 \to \text{int } D^2$. In the case that $\mathcal{R}$ is hyperbolic uch maps have been studied considerably. If $\mathcal{R}$ is finite the situation has been early completely analyzed by C. Narasimhan [N]. The more general case with $\mathcal{R}$ zero imensional is not as completely understood but results of Batterson [Ba1] give a airly good picture of the possibilities. Also the case that $\mathcal{R}$ consists of a one imensional attractor has been analyzed in detail by Williams [W2].

In a similar fashion we can describe other examples constructed from maps of urfaces rather than maps of the disk. For example, if we take a map $f : (T^2\text{-int } D^2)$ $\to (T^2\text{-int } D^2)$ homotopic to the one shown in Fig. 2 so the map f_{*1} on homology is iven by the matrix $\begin{pmatrix} 0 & 1 \\ -1 & 1 \end{pmatrix}$, then we can again form the suspension flow which will ave f as the return map on a surface of section. This will be a flow on a three imensional manifold M which is inwardly transverse on the boundary. But M is not simply the product of S^1 and $T^2 - \text{int } D^2$. In fact, it turns out to be diffeomorphic o the complement of a neighborhood of the trefoil knot (Fig. 3). Thus we can again

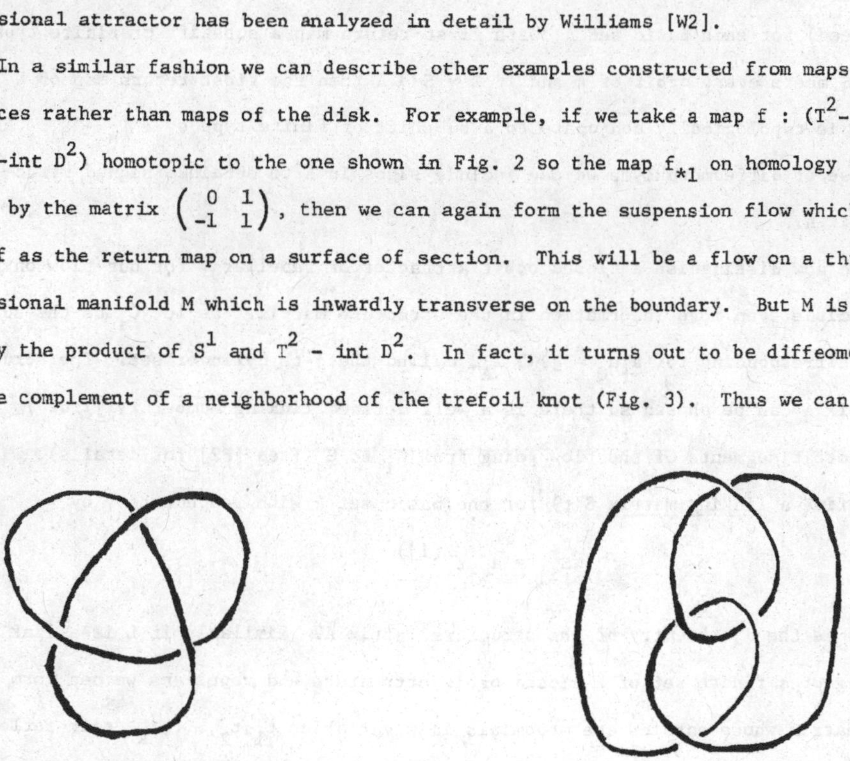

trefoil knot　　　　　figure eight knot

Fig. 3

piece together with a flow on $S^1 \times D^2$ to obtain a flow on S^3 which will have a repelling closed orbit knotted into a trefoil knot.

Diffeomorphisms of the torus in this homotopy class with finite hyperbolic $\mathcal{R}$ have been analyzed by Batterson [Ba2] and are very well understood.

In a similar fashion one can start with a map f which induces a homomorphism on homology given by the matrix $\begin{pmatrix} 2 & 1 \\ 1 & 1 \end{pmatrix}$. Here the situation is more complicated and the knotted closed orbit is in a figure eight knot (see [BW] for an analysis of this and similar cases).

More generally we can ask about non-singular flows the analysis of whose dynamics cannot be reduced to the study of a map of a surface. We will describe the situation when the chain recurrent set $\mathcal{R}$ is one dimensional and has a hyperbolic structure. In this case the flow is called a Smale flow. As mentioned above, a theorem of Bowen [B2] says that there is a surface of section S (with boundary and perhaps disconnected) for each basic set Λ with first return map a subshift of finite type. That is, S meets every orbit of Λ and if $\Sigma = S \cap \Lambda$ then the first return map of Σ to itself is topologically conjugate to a subshift of finite type $\sigma_A : \Sigma_A \to \Sigma_A$. As in the case of diffeomorphisms we can include signs in A to obtain a signed structure matrix $\tilde{A}$.

If we now distinguish a closed orbit attractor or repeller γ for our flow on S^3 we can include even more information in the structure matrix. We let C_i be the subset of S corresponding to $\{\underline{a} | a_0 = j\} \subset \Sigma_A$, called the j-th cylinder set. The structure matrix A can be chosen so there is a well defined linking number $\ell(ij)$ of γ with the orbit segments of the flow going from C_i to C_j (see [F2] for details). We now define a linking matrix S(t) for the basic set Λ with respect to γ by

$$S_{ij} = \tilde{a}_{ij} t^{\ell(ij)}$$

where $\tilde{a}_{ij}$ is the ij-th entry of the structure matrix $\tilde{A}$. Similarly if L is a link consisting of a finite set of k closed orbit attractors and repellers we can form a linking matrix whose entries are monomials in k variables $t_1, t_2, \ldots, t_k$, each reflecting how orbit segments from Λ link the k components of the link L. This construction is very similar to ideas used in different contexts by Williams [W3] and Fried [FR].

The linking matrix depends on the many choices made, but the polynomial det (I-S) n the variables t_i and their inverses is independent of these choices. In fact, . is closely related to a classical invariant of the theory of knots and links alled the Alexander polynomial (see [R] for example).

The following theorem from [F2] expresses this relationship.

heorem. Suppose f_t is a non-singular Smale flow on S^3, L is a link consisting of closed orbits oriented by the flow, each an attractor or a repeller, and $\{S_i\}$ are e linking matrices of the basic sets of index one, with respect to L. Then if > 1,

$$\prod_i \det (I-S_i) = \Delta_L(t_1,\ldots,t_n) \prod_k (1-t_1^{\ell(1,k)} \cdots t_n^{\ell(n,k)})$$

up to multiples of $\pm t_j^{\pm 1}$, $1 \leq j \leq n$, where $\ell(j,k)$ is the linking number of the j-th omponent of L with the k-th component of the set of attractors and repellers not in .

If L is a knot, i.e. n = 1, then

$$(1-t) \prod_i \det (I-S_i) = \Delta_L(t) \prod_k (1-t^{\ell(1,k)})$$

up to multiples of $\pm t^{\pm 1}$.

The equalities in this theorem are valid only modulo multiples of $\pm t_j^{\pm 1}$ since he Alexander polynomial is only defined up to such multiples. We remark also that ny vacuous products in the theorem above are taken to be 1. Thus if L consists of ll attractors and repellers of f_t, we have (modulo $\pm t_j^{\pm 1}$, $1 \leq j \leq n$)

$$\Delta_1(t_1,\ldots,t_n) = \prod_i \det (I-S_i).$$

This theorem is related to the Morse inequality type results cited above. In he trefoil and figure eight knot examples constructed from maps f : (T^2-int D^2) (T^2-int D^2), the Alexander polynomial of the knot is det $(I-f_{*1}t)$. The linking atrices in this case are $\{\tilde{A}_i t\}$, where $\{\tilde{A}_i\}$ are the structure matrices for f and the heorem for Smale flows reduces to the theorem for maps from [F1] which was cited above.

It is also possible to obtain some information without the hyperbolicity assumption, if $\mathcal{R}$ consists of finitely many orbits.

Theorem [F2]. Suppose f_t is a non-singular flow on S^3 and its chain recurrent set consists of finitely many orbits. Then if K is a knotted closed orbit attractor or repeller, its Alexander polynomial $\Delta_K(t)$ has only roots which are roots of unity.

Thus for example the figure eight knot whose Alexander polynomial is $t^2 - 3t + 1$ cannot be an attractor for a non-singular flow unless the chain recurrent set of that flow contains infinitely many orbits.

We close with a very fundamental question for flows on S^3 whose answer is still unknown.

Problem. Given any subshift of finite type $\sigma : \Sigma_A \to \Sigma_A$, is there a non-singular Smale flow on S^3 with the suspension of σ as a basic set?

References

[Ba1] S. Batterson, Constructing Smale diffeomorphisms on Compact Surfaces, to appear in Trans. Amer. Math. Soc.

[Ba2] S. Batterson, The dynamics of Morse-Smale diffeomorphisms on the torus, to appear in Trans. Amer. Math. Soc.

[BW] J. Birman and R. F. Williams, Knotted Periodic Orbits II, in preparation.

[B1] R. Bowen, Topological Entropy and Axiom A, Proc. Sympos. Pure Math. 14 (1970), A.M.S.

[B2] R. Bowen, One dimensional Hyperbolic Sets for flows, Jour. Diff. Eqs. 12 (1972), 173-179.

[C] C. Conley, Isolated Invariant Sets and the Morse Index, CBMS Regional Conference Series 38 (1978).

[F1] J. Franks, A Reduced Zeta function for diffeomorphisms, Amer. Jour. Math. 100 (1978), 217-243.

[F2] J. Franks, Knots, Links and Symbolic Dynamics, to appear.

[Fr] D. Fried, Flow Equivalence, Hyperbolic Systems and a new Zeta function for flows, to appear.

[M] J. Milnor, Morse Theory, Annals of Math. Studies 51, Princeton Univ. Press, Princeton, N.J., 1963.

[N] C. Narasimhan, The Periodic Behavior of Morse-Smale diffeomorphisms on Compact Surfaces, Trans. Amer. Math. Soc. 48 (1979), 145-169.

[PS] W. Parry and D. Sullivan, A topological invariant of flows on one dimensional spaces, Topology 14 (1975), 297-299.

[R] D. Rolfsen, Knots and Links, Publish or Perish Press, Berkeley, 1976.

[SS] M. Shub and D. Sullivan, Homology Theory and Dynamical Systems, Topology 14 (1975), 109-132.

[S1] S. Smale, On the structure of Manifolds, Amer. Jour. of Math. 84 (1962), 387-399.

[W1] R. F. Williams, The classification of subshifts of finite type, Annals of Math 98 (1973), 120-153, and 99 (1974), Errata, 380-381.

[W2] R. Williams, Classification of one dimensional attractors, Proc. Symp. Pure Math. 14 (1970), A.M.S.

[W3] R. Williams, The Structure of Lorenz Attractors, to appear in Publ. I.H.E.S.

Northwestern University
Evanston, IL 60201

Anomalous Anosov Flows

by

John Franks* and Bob Williams*

§1. Introduction

Anosov diffeomorphisms and flows have been much studied since the original paper of Anosov [A]. Their beautiful behavior has led to much work and many conjectures; we answer one of these negatively here:

(1.1) Theorem. There is an Anosov flow ψ_t on M whose chain recurrent set is not all of M. The splitting $E^u \oplus E^s \oplus E^t$ (see definition below) can be

a) $(u,s) = (1,1)$ on a 3-manifold; or

b) any (u,s) with $u \geq 2$, $s \geq 2$.

This settles questions raised in various works on dynamical systems ([S],[F],[N]) and the new Hilbert problems ([H], p. 60, second version of (1)). In addition this question was essentially raised by Anosov [A]. It also contradicts a special case of a theorem of Verjovsky [V], which says that for any codimension one Anosov flow the chain recurrent set is all of M. He uses a special argument when dim $M = 3$ $(u = s = 1)$ which is erroneous, but the higher dimensional cases are presumably correct. Thus his result together with our theorem above cover all possible cases.

We now recall some definitions.

---

*Research supported in part by NSF Grant MCS 79·01080

If f_t is a smooth flow on a compact manifold M, it is said to be structurally stable provided that for any sufficiently close C^1 approximation g_t there is a homeomorphism $h: M \to M$ carrying orbits of f to orbits of g and preserving the sense of orbits. All known examples of structurally stable flows have a hyperbolic chain-recurrent set so we now define these concepts.

A point x of M is called chain-recurrent for f_t provided that corresponding to any ϵ, $T > 0$ there exist points $x = x_0$, $x_1, \ldots, x_n = x$ and real numbers $t_0, t_1, \ldots, t_{n-1}$ all greater than T such that $d(f_{t_i}(x_i), x_{i+1}) < \epsilon$ for all $0 \leq i \leq n - 1$. The set of all such points, called the chain-recurrent set $\mathcal{R}$, is a compact set invariant under the flow.

A compact invariant set K for a flow f_t is said to have a hyperbolic structure provided that the tangent bundle of M restricted to K is the Whitney sum of three bundles $E^s \oplus E^u \oplus E^c$ each invariant under Df_t for all t and that

(a) The vector field tangent to f_t spans E^c.

(b) There are C, $\lambda > 0$, such that

$$\|Df_t(v)\| \leq Ce^{-\lambda t}\|v\| \text{ for } t \geq 0 \text{ and } v \in E^s$$

and

$$\|Df_t(v)\| \geq Ce^{\lambda t}\|v\| \text{ for } t \geq 0 \text{ and } v \in E^u.$$

It is shown in [F-S] that the condition that a flow have hyperbolic chain-recurrent set is equivalent to Axiom A of Smale [S] and the no-cycle property. Results of Smale [S] then show that the chain-recurrent set $\mathcal{R}$ is the union of a finite number of disjoint, compact, invariant pieces called basic sets, each of which contains a dense orbit.

If the entire manifold M possesses a hyperbolic structure then the flow f_t is called an <u>Anosov flow</u>.

One can similarly define hyperbolic set for a diffeomorphism f (there is no E^c and Df must satisfy inequalities obtained from the inequalities above by setting $t = 1$) and Anosov diffeomorphism.

If X is a subset of a hyperbolic set of a flow we define the stable and unstable manifolds $W^s(X)$ and $W^u(X)$ as follows

$$W^s(X) = \{y \mid d(f_t y, f_t(X)) \to 0 \text{ as } t \to \infty\}$$

$$W^u(X) = \{y \mid d(f_t y, f_t(X)) \to 0 \text{ as } t \to -\infty\}.$$

If X is a point x then $W^s(x)$ is called the *strong stable manifold* of x. If X is the orbit containing x then $W^s(X)$ is called the *weak stable manifold* of x. In both these cases $W^s(X)$ is in fact a manifold [HP] and if f_t is Anosov the manifold M is foliated by the strong stable and weak stable manifolds. Stable and unstable manifolds for diffeomorphisms are defined similarly.

In constructing examples, it is useful to have simple criteria to check whether or not a flow is Anosov. The following result essentially due to Mañe [M] is valuable in this regard.

(1.2) *Theorem*. *Necessary and sufficient conditions for a flow ψ_t on M to be Anosov are*:

(1) ψ_t <u>has hyperbolic chain recurrent set</u>

(2) <u>The weak stable and unstable manifolds</u> $W^s(\gamma)$ <u>and</u> $W^u(\gamma)$ <u>intersect transversally for each orbit</u> γ <u>in</u> M

(3) <u>The dimension of</u> $W^s(\gamma)$ <u>is constant, i.e. independent of</u> γ.

Mañe actually proves the analog of this for Anosov diffeomorphisms but the proof for flows is essentially the same (one can also appeal to a combination of results from [HPS]; see for example (2.17), p. 22).

§2. Construction of the main example

Let A be a linear Anosov map on the two-torus T^2 and let f be the DA as introduced by Smale [S]. See also [W]. Briefly, f has the form $q \circ A$, where q is supported in a small neighborhood of a fixed point $\mathcal{O}$ of f, preserves the stable manifold of A, but expands away from $\mathcal{O}$. This is done so that f has $\mathcal{O}$ as a source and has a one dimensional attractor Λ. f in turn has a hyperbolic structure on its chain recurrent set $\Lambda \cup \{\mathcal{O}\}$. The splitting is of type (1,1) on Λ - that is, both the stable and unstable manifolds have dimension one.

Next let M_1, φ_t be the suspension [S] of T^2, f. That is, $M_1 = T^2 \times \mathbb{R}/\sim$ where the equivalence is induced by the map

$$(x,t) \to (fx, t+1).$$

The trivial flow (induced by $\frac{\partial}{\partial t}$) on $T^2 \times \mathbb{R}$ induces in turn the flow φ_t on M_1. Then φ_t has a repelling periodic orbit J corresponding to the source $\mathcal{O}$ in T^2. Now delete a tubular neighborhood of J from M_1. This gives a manifold M_1' with boundary $\partial M_1'$ homeomorphic to T^2 and a flow (also called φ_t) which is inwardly transverse, to the boundary.

The choice of the tubular neighborhood of J is very natural, but critical so we describe it in detail. Note that a neighborhood of J can be considered as $D^2 \times I$ with identifications; we use polar coordinates in D^2, so that the action used to form M_1 as a

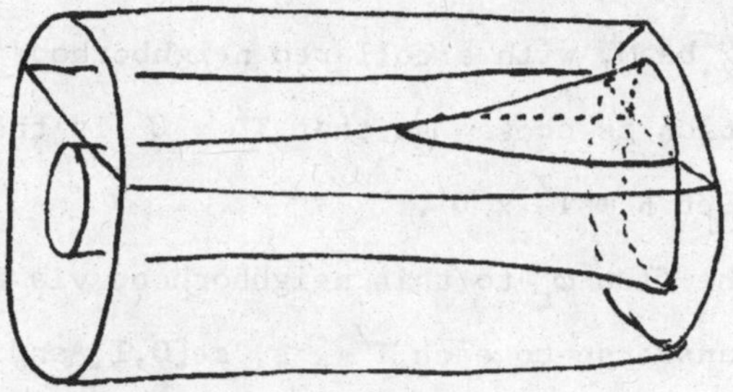

Fig. 1

quotient is generated by

$$(r,\theta,t) \to (Cr,\theta,t+1), \text{ near } J$$

where $C > 1$. Now let $\lambda = \ln C$ and note that points of the form $(\varepsilon e^{\lambda t},\theta,t)$ are invariant under this action; we choose $\varepsilon > 0$ conveniently and this determines the boundary torus B' of M_1'.

Note that the weak stable foliation, W^s, of the attractor Λ_1 of the flow φ_t consists of parallel planes, all parallel to the t-axis; we think of them as horizontal. Thus with one exception, they intersect B' in curves looking roughly like parabolas. The exceptional intersection consists of two horizontal lines. Thus, after the identification $(r,\theta,0) \sim (Cr,\theta,1)$, the foliation is a familiar one with two Reeb components:

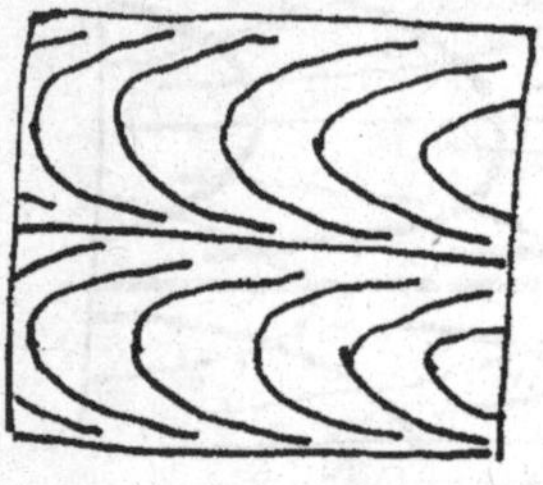

Fig. 2

Next, let M_2 be M_1' with a collared neighborhood $T^2 \times [0,1]$ added. The notation is chosen so that $T^2 \times 0$ is the (new) boundary of M_2; let $B = T^2 \times 0$.

We extend the flow φ_t to this neighborhood via an isotopy so that φ_t is transverse to each $T^2 \times s$, $s \in [0,1]$ and normal, with unit speed for $s \in [0,\frac{1}{2}]$. Then, the stable foliation W^s "propagates" through this neighborhood. It follows that $W^s \cap (T^2 \times 0)$ is smoothly isotopic to $W^s \cap B'$.

Then let $\overline{M}_2, \overline{\varphi}_t$ be a copy of M_2, φ_t with a reversal in sign. That is, $\overline{\varphi}_t$ will flow outwardly normal to $\partial\overline{M}_2 = \overline{B}$.

Then $\overline{\varphi}_t$ has a 2-dimensional repellor with hyperbolic structure, and unstable foliation W^u. Note that the foliation W^u intersects the boundary $\overline{B}$ just as W^s intersects B. Also note that if we now sew M_2 and $\overline{M}_2$ together with essentially any diffeomorphism $\partial M_2 \to \partial\overline{M}_2$ $\varphi_t \cup \overline{\varphi}_t$ yields a smooth flow ψ_t on the resulting manifold M. We do this by the obvious diffeomorphism $\partial M_2 \to \partial\overline{M}_2$ which makes the foliation $W^s \cap B$ and $W^u \cap \overline{B}$ transverse (see Figure 3).

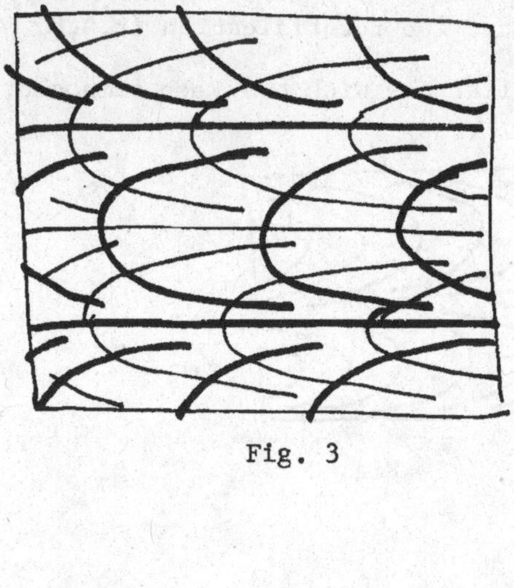

Fig. 3

We now rigorously check transversality by writing down explicit equations for the leaves of the foliation. The torus B' is a quotient of the cylinder $r = \epsilon e^{\lambda t}$ (given in coordinates (r,θ,t) for $D^2 \times I$). A horizontal plane which is distance d from the line $r = 0$ has the equation $r \cos \theta = d$ or $r = d/\cos \theta$. The intersection of the plane and the cylinder has equation $\epsilon e^{\lambda t} = d/\cos \theta$ or taking logarithms $\ln \epsilon + \lambda t - \ln d - \ln(\sec \theta) = 0$. The differential of this gives the one form $\lambda dt + (\tan \theta)d\theta$ defining the foliation (except when $\theta = \pm\pi/2$ when $d\theta$ is the appropriate form). This one form is well defined on the torus B' using the coordinates θ and t mod 1.

The flow defines a diffeomorphism of the torus B' to the torus B which carries $W^s \cap B$. We use this diffeomorphism to define the coordinates θ, t on B, so the foliation is still given by the same one-form $\lambda dt + \tan \theta\, d\theta = 0$. Likewise we have coordinates $\bar{t}$ and $\bar{\theta}$ on $\bar{B}$ and $W^u \cap \bar{B}$ is given by the one form $\lambda d\bar{t} + (\tan \bar{\theta})d\bar{\theta}$. The diffeomorphism gluing B to $\bar{B}$ is given by $t = \bar{t}$, $\theta = \bar{\theta} - \pi/2$ and it is clear that the one forms $\lambda dt + \tan \theta\, d\theta$ and $\lambda d\bar{t} + \tan \bar{\theta}\, d\bar{\theta} = \lambda dt - \cot \theta\, d\theta$ are never parallel (in fact if we rescale t so $\lambda = 1$ they are everywhere perpendicular).

It is now easy to complete part (a) of our theorem using Theorem (1.2). The flow ψ_t satisfies Axiom A (every orbit not in the attractor or repellor passes through B never to return and hence is wandering, not chain-recurrent) and has two basic sets both with two dimensional (weak) stable manifolds. Thus to show ψ_t is Anosov, we need only check transversality of $W^s(x)$ and $W^u(x)$ for each $x \in M$. But it is enough to check for one point on each orbit of ψ_t and each orbit is either contained in a basic set or passes through B. In both cases transversality is clear.

§3. Different Splittings in Higher Dimension

We proceed to prove part (b). To this end, let N,g be an Anosov diffeomorphism with splitting (u,s) = (p,q). We then proceed with our construction, with T^2,f replaced by $T^2 \times N$, $f \times g$. The resulting manifold M_1^* and flow φ_t^* has a repellor J* which is the suspension of N,g.

We now examine a "tubular neighborhood" of J*. A neighborhood U of J* has the form $N \times D^2 \times R/\sim$ where the identification is

$$(x,y,t) \sim (gx,fy,t+1).$$

The flow φ_t^* has an attractor on which the hyperbolic splitting satisfies (u,s) = (p + 1,q + 1) and we must again understand how the weak stable manifold foliation W^{s*} of this attractor intersects the neighborhood U and in particular its boundary. The foliation W^{s*} on U lifts to a foliation on $N \times D^2 \times R$, namely the foliation with leaves of the form $W_g^s(x) \times (W_f^s(y) \cap D^2) \times R$.

The boundary of U,B* is $N \times B$ where B is the toral boundary of the repellor constructed in §2. Thus the foliation W^{s*} intersects B* in a foliation each of whose leaves has the form $W^s_g(x) \times L$ where L is a leaf of the Reeb foliation of B obtained in §2. We now have half of the desired flow, namely a flow on M^*_1 - int U, inwardly transverse to the boundary B* with $W^{s*} \cap B$ as just described and an attractor with splitting $(u,s) = (p+1,q+1)$. The other half on $\overline{M}^*_1$ - int $\overline{U}$ is constructed similarly but using the suspension of $g^{-1} \times f$ instead of $g \times f$. In this way the foliation on the boundary $\overline{B}^* = N \times \overline{B}$ has leaves of the form $W^u_x(g) \times L$. The rest of the construction is done as in §2. The gluing map $B^* = N \times B \to \overline{B}^* = N \times \overline{B}$ is essentially $id \times h$ where $h: B \to \overline{B}$ is the gluing map of §2.

§4. Other Basic Sets

In light of the example of §2 it is natural to ask if it is possible to construct an Anosov flow on a three manifold with more than two basic sets or with basic sets of dimension one. Both of these questions are answered affirmatively by an example which we now construct. In fact, we will produce an example with a closed orbit isolated in the chain-recurrent set.

A similar procedure would lead to Anosov flows with other isolated basic sets, e.g., the suspension of a sub-shift of finite type. Choosing the gluing maps for this latter would be considerably more delicate, as one would have to match two foliations, one of which is not everywhere defined. For this reason we will only treat the simplest case, alluded to above.

(4.1) Proposition. *There is an Anosov flow on a three dimensional manifold with a basic set consisting of a single closed orbit.*

Proof: The idea is to produce a flow on the manifold $S^1 \times$ (disk with 2 holes) which enters on one boundary component and leaves on the other two and with a single closed orbit inside (all other orbits exiting in either forward or backward time). By doing this appropriately we are able to glue together two copies of (M_2, φ) and a copy of $(\overline{M}_2, \overline{\varphi})$ (as in §2) to obtain the desired Anosov flow. We now give the details.

Let D denote the two-dimensional disk with two holes

deleted as shown in Figure 4.

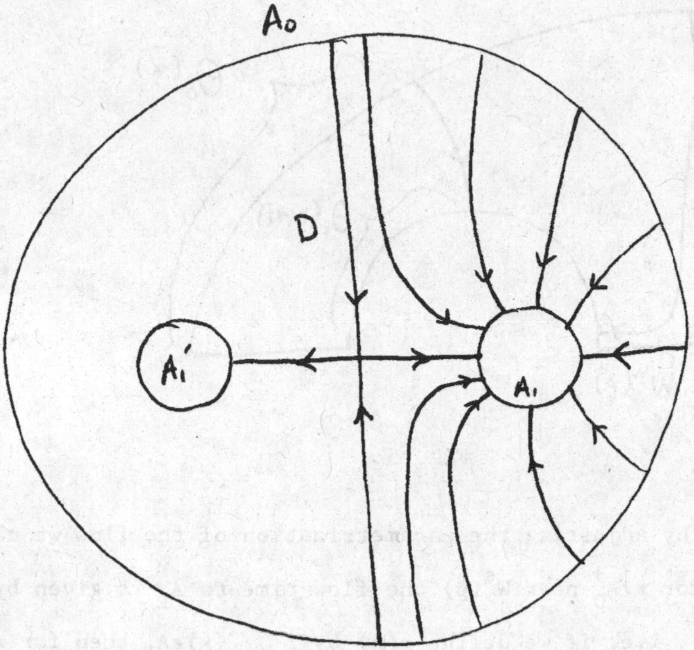

Fig. 4

We endow D with a flow f_t^1 with a single hyperbolic rest point p at its center with a vertical stable manifold and horizontal unstable manifold. The rest of the flow is as sketched and is constructed to be symmetric with respect to reflections through $W^s(p)$ and $W^u(p)$. The boundary components are labelled A_0, A_1, A_1' as shown and the flow is chosen so the exit map r from the right half of A_0 to A_1 is given by $2\theta_0(x) = \theta_1(r(x))$ for all x in the right half of A_0, where $\theta_i(x)$ denotes the angle a ray from the center of A_i to x makes with the horizontal (see Figure 5). Of course $r(x)$ is undefined if $x \in W^s(p)$.

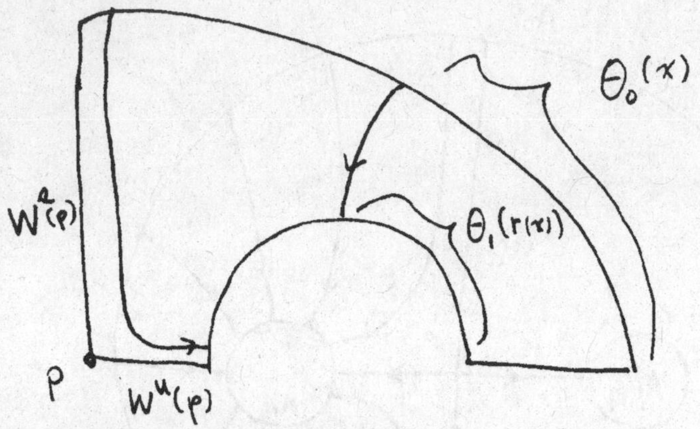

Fig. 5

Finally, by adjusting the parametrization of the flow we can arrange that for $x \in A_0^+$ near $W^s(p)$ the flow time to A_1 is given by $-\ln(\pi/2-\theta_0(x))$, i.e. if we define $\tau(x)$ by $f_{\tau(x)}(x) \in A_1$ then for $x \in A_0^+$ near $W^s(p)$, $\tau(x) = -\ln(\pi/2-\theta_0(x))$. It is easy to see that this is possibl by assuming the flow comes from a linear vector field near $W^s(p) \cup W^u(p)$

Let X denote a vector field on D whose flow we have just described. We want to construct a vector field and flow on $D \times S^1$ which we will do by constructing them on $D \times R$ in such a way that they are periodic of period 1 in $t \in R$. By symmetry it is enough to describe the vector field on $R \times D^+$ where D^+ is the right upper quadrant of D, i.e. those points on orbits of X passing through $\{x \mid x \in A_0 \; 0 \leq \theta_0(x) \leq \pi/2\}$.

Define α on D^+ by $\alpha(x) = \pi/2-\theta_0(y)$ where $y \in A_0$ is the unique point on the same orbit as x ($\alpha(x) = 0$ if $x \in W^u(p)$).

Choose $\alpha_0 > 0$ such that for all $x \in A_0$ with $0 < \alpha(x) \leq \alpha_0$ the exit time $\tau(x) = -\ln(\alpha(x))$. Now choose a bump function $\rho(s) \geq 0$ defined on $[0,\alpha_0]$ such that

1) $\rho \equiv 1$ on a neighborhood of 0 and $\rho \equiv 0$ on a neighborhood of α.

2) The function $-\rho(s)\ln s$ is concave up, i.e. has non-negative second derivative.

This is done by convexly interpolating between $-\ln s$ and 0 (see Figure 6) and dividing the resulting function by $-\ln s$.

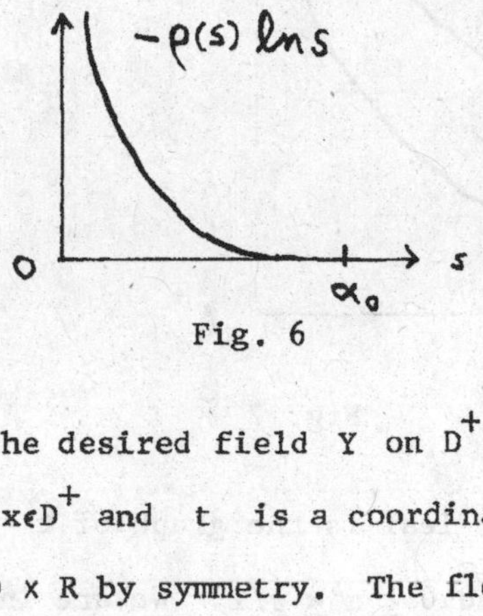

Fig. 6

We now define the desired field Y on $D^+ \times R$ to be $X + \rho(\alpha(x))\frac{\partial}{\partial t}$ where $x \in D^+$ and t is a coordinate on R. We extend Y to all of $D \times R$ by symmetry. The flow generated by Y is given by

$$f_s(x,t) = (f_s^1(x), t + s\rho(\alpha(x))) \text{ if } x \in D^+,$$

provided the right hand side is defined. We now attach $\overline{M}_2$ to

$D \times S^1$ by a diffeomorphism of $\overline{B} = \partial\overline{M}_2$ to $A_0 \times S^1$ which carries the foliation $W^u \cap \overline{B}$ to a foliation of $A_0 \times S^1$ which is as in Figure 7. The figure actually shows the foliation on $A_0^+ \times R$ where $A_0^+ = A_0 \cap D^+$, i.e. the part of the foliation on the first quadrant of $A^0 \times R$. The rest of the foliation is symmetrical.

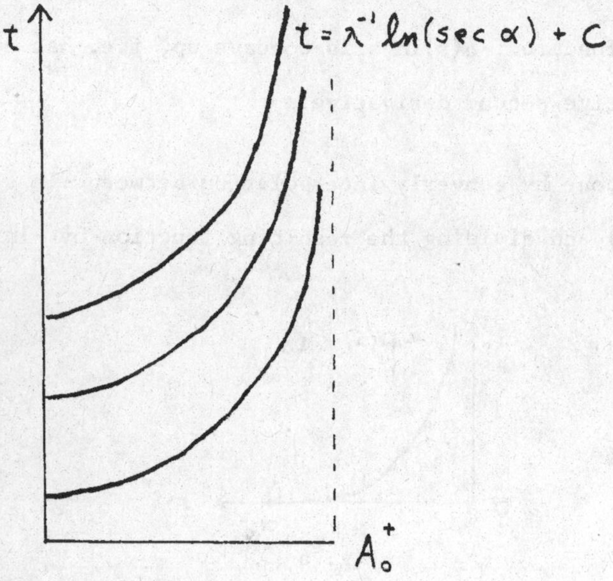

Fig. 7

More precisely each leaf is the graph of $t = \lambda^{-1}\ln(\sec\alpha) + c$ for some constant C and $0 \leq \alpha \leq \pi/2$. We are interested in the foliation of $A_1 \times S^1$ (or $A_1 \times R$) which is the image of this one under the map $A_0^+ \times R \to A_1 \times R$ obtained by flowing along orbits of Y (together with the special leaf $W^u(p \times R)$). Using the formula for $f_s(x,t)$ given above and the fact that the exit time for $x \in A_0$ is

$-\ln(\alpha(x))$ one calculates that the leaves are all graphs of the equations $t = \lambda^{-1}\ln(\sec \alpha_1) - \rho(\alpha_1)\ln \alpha_1 + C$ where $0 \leq \alpha_1 = (\pi-\theta_1)/2 \leq \pi/2$ and C is a constant. By symmetry the foliation on the other half of $A_1 \times R$ is the same. Since by construction the curves $t = \lambda^{-1}\ln(\sec \alpha) - \rho(\alpha)\ln \alpha$ are convex up and have vertical asymptotes at $\alpha = 0, \pi/2$ (or $\theta = 0, \pi$) we have a Reeb foliation of $A_1 \times S^1$ with two Reeb components. The leaves are precisely the intersection of $A_1 \times S^1$ with unstable manifolds. Just as in §2 we can sew a copy of M_2 with the flow φ_t onto $A_1 \times S^1$ in such a way as to preserve transversality of stable and unstable manifolds. The construction and analysis for $A_1' \times S^1$ is precisely the same. The proof that the resulting flow is Anosov is the same as in §2.

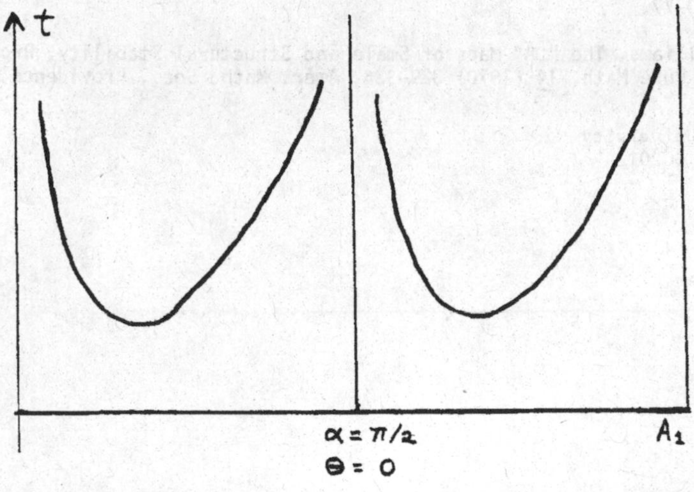

Fig. 8

References

[A] D. V. Anosov, Geodesic Flows on Closed Riemannian Manifolds with Negative Curvature, Proc. of the Steklov Inst. of Math. 90 (1967). English translation AMS, Providence, R.I.

[FS] J. Franke and J. Selgrade, Hyperbolicity and Chain Recurrence, J. Differential Equations 26 (1977), 27-36.

[F] J. Franks, Anosov Diffeomorphisms, Proc. Symposia Pure Math. 14 (1970) 61-93. Amer. Math. Soc., Providence, R.I.

[H] F. Browder (ed.), Proc. Symposia Pure Math. 28 part I Amer. Math. Soc. Providence, R.I.

[HP] M. Hirsch and C. Pugh, Stable Manifolds and Hyperbolic Sets, Proc. Symposia Pure Math 14 (1970) 133-163, Amer. Math. Soc. Providence, R.I.

[HPS] M. Hirsch, C. Pugh and M. Shub, Invariant Manifolds, Springer Lecture Notes in Math 583 (1977).

[M] R. Mañé, Quasi Anosov Diffeomorphisms and Hyperbolic Manifolds. Trans. Amer. Math. Soc. 229 (1977) 351-370.

[N] Z. Nitecki, Differentiable Dynamics, MIT Press, Cambridge Mass. 1971.

[S] S. Smale, Differentiable Dynamical Systems, Bull. Amer. Math. Soc. 73 (1967), 747-817.

[V] A Verjovsky, Codimension One Anosov Flows, Bol. Soc. Mat. Mexicana, 19 (1974) 49-77.

[W] R. F. Williams, The "DA" Maps of Smale and Structural Stability, Proc. Symposia Pure Math. 14 (1970) 329-334, Amer. Math. Soc., Providence, R.I.

Northwestern University
Evanston, IL 60201

Efficiency vs. Hyperbolicity on Tori

by David Fried*

We consider the problem raised by Mike Shub of finding "simplest" iffeomorphisms in an isotopy class [9], that is Axiom A-Strong ransversality diffeomorphisms whose topological entropy is the inimum within the isotopy class. We will show that certain isotopy lasses do not have a "simplest" representative, indicating a con-flict between efficiency (as measured by low topological entropy) nd hyperbolicity.

In fact there are certain homotopy classes on tori T^n for hich no Axiom A representative f has entropy as small as the inear automorphism α. This means that f has exponentially more (isolated) periodic points than α. These classes are those for hich the action $L = \alpha_* = f_*: H_1(T^n, \mathbb{R}) \circlearrowleft$ satisfies

1) 1 is not an eigenvalue of L

2) L has an eigenvalue λ of modulus 1 which isn't a root of unity.

We extend methods Manning used in his study of Anosov diffeomorphisms on tori [7].

We remark that these are the first examples known of isotopy classes without simplest representatives. In [9] Shub constructed isotopy classes in which (as here) the entropy of a fitted diffeo-morphism must exceed the logarithm of the largest homology eigen-value λ, but it isn't known for his examples whether $h(f) \leq \log|\lambda|$ at some f in the isotopy class. In our examples, entropy assumes

Partially supported by the National Science Foundation.

the value $\log|\lambda|$ and this value is minimal.

In Section 1 we will analyze the poles and zeros of the Lefschetz zeta function of an Axiom A diffeomorphism. In Section 2 (which is independent of the other sections) we study cases where simplest representatives do occur on tori and discuss the algebra of toral automorphisms. In Section 3 we prove our main result, the entropy inequality mentioned above.

We thank I.H.E.S. for supporting this research and we're grateful to John Franks for some helpful comments. We especially thank Sheldon Newhouse for encouraging us to study these questions.

Section 1 The Lefschetz Zeta Function for Axiom A Diffeomorphisms

One may count the periodic points of an Axiom A diffeomorphism f algebraically to obtain a sequence of integers $L(f^p)$ determined by the homotopy class of f. Our results are based on measuring the asymptotic behavior of this sequence in 2 ways, using symbolic dynamics on the one hand and homology on the other. This behavior may be described in terms of the innermost poles and zeroes of a rational function $\tilde{\zeta}(f)$ called the Lefschetz zeta function of f [10].

We recall the definition of $\tilde{\zeta}$. The Lefschetz index of a hyperbolic fixed point $gx = x$ was computed by Smale as $\operatorname{ind}(g,x) = (-1)^u \cdot \operatorname{sign} \det(Dg|E^u(x))$, where E^u is the u-dimensional unstable subspace of the tangent space at x. If all the fixed points of g are hyperbolic then $L(g) = \sum_{gx=x} \operatorname{ind}(g,x)$ is the algebraic number f fixed points of g. If $f: M \to M$ is Axiom A then we let

$$(f) = \exp\left(\sum_{n=1}^{\infty} \frac{L(f^p)}{p} z^p \right).$$

The rationality of this formal power eries follows from the Lefschetz Fixed Point Formula [10].

A Lefschetz zeta function $\tilde{\zeta}(f|I)$ for a closed invariant set I ıay be defined by counting over only the periodic points in I. Ve will also use the analogous (Artin-Mazur) zeta functions $\zeta(f)$ ınd $\zeta(f|I)$ obtained by counting all periodic points geometrically, :hat is with weight $+1$ instead of ± 1. If Λ is a basic set for an Axiom A diffeomorphism f, we let $N_p^+(f|\Lambda)$ (resp. $N_p^-(f|\Lambda)$) be the number of points x of period p for $f|\Lambda$ for which $\det(Df^p|E^u(x)) > 0$ (resp. < 0). Then $L(f^p|\Lambda) = (-1^u)(N_p^+(f|\Lambda) - N_p^-(f|\Lambda))$. We note that $\zeta(f|\Lambda)$ is the power series obtained by replacing $L(f^p|\Lambda)$ by $N_p(f|\Lambda) = N_p^+(f|\Lambda) + N_p^-(f|\Lambda)$ in the formula defining $\tilde{\zeta}(f|\Lambda)$. Both $\zeta(f)$ and $\zeta(f|\Lambda)$ are rational functions [6].

Bowen showed that every basic set Λ for an Axiom A diffeomorphism f is the disjoint union of mixing components $M_1, \ldots, M_{n(\Lambda)}$ which are closed sets determined up to a cyclic reordering of the indices. The number $n(\Lambda)$ is the g.c.d. of the periods of the periodic points of $f|\Lambda$. The M_i are cyclically permuted by f and the maps $f^{n(\Lambda)}|M_i$ have periodic points of all sufficiently large periods. The M_i are basic sets for $f^{n(\Lambda)}$.

We will describe three types of mixing basic sets, N, P and R, depending on the behavior of the local unstable orientations of periodic points.

Definition Let Λ be a basic set for the Axiom A diffeomorphism f and let $n = n(\Lambda)$.

1) Λ is type N if $N_p^+(f^n|\Lambda)$ and $N_p^-(f^n|\Lambda)$ are nonzero for sufficiently large p

2) Λ is type P if $N_p^-(f^n|\Lambda) = 0$ for all p

3) Λ is type R if $N_p^+(f^n|\Lambda) = 0$ for p odd.

One may think of N as standing for normal, P for preserving and R for reversing.

Recall the convention that a zero of order k is a pole of order $-k$.

Theorem 1 Every basic set Λ of Axiom A diffeomorphism f belongs to precisely one of the types N,P or R.

The only poles of $\tilde{\zeta}(f|\Lambda)$ on $D_\Lambda = \{|z| \leq e^{-h(f|\Lambda)}\}$ lie on ∂D_Λ and have order $(-1)^{u(\Lambda)}$. If Λ is of type P, these poles occur at $e^{-h(f|\Lambda)}.\xi$ where $\xi^{n(\Lambda)} = +1$. If Λ is of type R, the only poles occur at $e^{-h(f|\Lambda)}.\xi$ for $\xi^{n(\Lambda)} = -1$. If Λ is of type N, no poles occur in D_Λ.

Proof By passing to $f^{n(\Lambda)}$, we reduce to the case $n(\Lambda) = 1$ ($f|\Lambda$ mixing).

Suppose $\mathcal{P}$ is a fine Markov partition for $f|\Lambda$. Then one may orient E^u over each rectangle $R \in \mathcal{P}$ so that when $x \in R_i \cap f^{-1}(\text{int } R_j)$ the value of sign $\det(Df:E^u(x) \to E^u(fx)) = M(R_i,R_j)$ depends only on R_i and R_j. Setting $M(R_i,R_j) = 0$ when $R_i \cap f^{-1}(\text{int } R_j) = \phi$, we obtain a signed transition matrix M.

We associate two subshifts of finite type A and B to M. The transition matrix of A is just $|M|$. The transition matrix of B is given by replacing the entries of M by 2×2 blocks, as follows. 0's are replaced by $\begin{pmatrix} 0 & 0 \\ 0 & 0 \end{pmatrix}$, 1's by $\begin{pmatrix} 1 & 0 \\ 0 & 1 \end{pmatrix}$ and -1's by $\begin{pmatrix} 0 & 1 \\ 1 & 0 \end{pmatrix}$.

<u>Lemma</u> 1) $h(A) = h(B) = h(f|\Lambda)$ (= h, say)

2) $N_p(f|\Lambda) = N_p(A) + o(e^{ph})$

3) $N_p(f|\Lambda) + \tilde{N}_p(f|\Lambda) = N_p(B) + o(e^{ph})$

<u>Proof of Lemma:</u> The natural semiconjugacy π from A to $f|\Lambda$ is bounded-to-one and surjective. This gives $h(A) = h(f|\Lambda)$. But the semiconjugacy from B to A is onto and everywhere two-to-one, which gives 1).

We prove 2) as in [7]. Each $x \in \mathrm{Fix}(f^p|\Lambda)$ contributes +1 to $N_p(A)$ if $x \notin \partial\mathcal{P}$ (since each $y \in \Lambda$ which stays in the interior of the rectangles has only one preimage under π) and a bounded amount in any case. Thus we need only show

(*) $\#(\mathrm{Fix}(f^p|\Lambda) \cap \partial\mathcal{P}) = o(e^{ph})$.

Let μ be the Bowen measure on Λ which is the asymptotic distribution of the periodic points of $f|\Lambda$. It is known that $\mu(\partial\mathcal{P}) = 0$, which shows

$$\frac{\#(\mathrm{Fix}(f^p|\Lambda) \cap \partial\mathcal{P})}{N_p(f|\Lambda)} \to 0. \text{ This implies } \frac{N_p(A)}{N_p(f|\Lambda)} \to 1.$$

Hence A is mixing and Perron-Frobenius theory gives $\frac{N_p(A)}{e^{ph}} \to 1$. So (*) follows.

We show 3) by a similar argument which was suggested by [11]. Observe that $x \in \mathrm{Fix}(f^p)$ contributes $\left\{\begin{smallmatrix}+2\\0\end{smallmatrix}\right.$ to $N_p(f|\Lambda) + \tilde{N}_p(f|\Lambda)$ as $\det(Df^p : E^u(x)) \to E^u(f^p x)) \left\{\begin{smallmatrix}>0\\<0\end{smallmatrix}\right.$. If $x \in \partial\mathcal{O}$ then x contributes this same amount to $N_p(B)$ (by the construction of B) but a bounded amount in any case. Again (*) shows this counting error is $o(e^{ph})$.

Q.E.D. (Lemma)

From the lemma, we see that $\tilde{N}_p(f|\Lambda) = N_p(B) - N_p(A) + o(e^{ph})$. Passing to zeta functions, we find that the rational function $g = \frac{(\tilde{\zeta}(f|\Lambda))^{(-1)^{u(\Lambda)}} \zeta(A)}{\tilde{\zeta}(B)}$ has no poles in int D_Λ. Suppose that g has poles on ∂D_Λ, say at $e^{-h}\xi_i^{-1}$ of order n_i, ξ_i distinct points on $|z| = 1$, $i = 1,\ldots,k$. We would find $o(e^{ph}) = \Sigma n_i(e^h\xi_i)^p + O(e^{p(h-\varepsilon)})$, where the other poles of g have modulus $\geq e^{\varepsilon-h}$. Dividing by e^{ph} gives $\Sigma_i n_i\xi_i^p \to 0$. Since translation by p is recurrent on $G = \overline{\{(\xi_1^p,\ldots,\xi_k^p)\}} \subset T^k$, we have $\Sigma n_i\xi_i^p \equiv 0$ for all p. Thus $\Sigma n_i q(\xi_i) = 0$ for any polynomial q and, since the ξ_i are distinct, we find all n_i are 0. Thus

(**) the poles of $\tilde{\zeta}(f|\Lambda)$ in D_Λ have the same location and order as those of $(\zeta(B)/\zeta(A))^{(-1)^{u(\Lambda)}}$.

We already observed that A is mixing. Perron-Frobenius theory implies that the transition matrix $|M|$ has a unique eigenvalue λ of maximum modulus. This eigenvalue $\lambda = e^h$.

Thus $\zeta(A) = \det^{-1}(I-t|M|)$ has a pole of order 1 at $z = e^{-h}$ and no other pole in D_Λ.

To analyze the poles of $\tilde{\zeta}(f|\Lambda)$ in D_Λ, we must use (**) and examine the mixing properties of B. If B is mixing (this corresponds to Λ of type N) then $\zeta(B)$ has only a pole of order 1 at e^{-h} and $\tilde{\zeta}(f|\Lambda)$ has no poles in D_Λ. If B is nontransitive (Λ of type P) then we see that the transition matrix for B is two copies of $|M|$. Thus $\zeta(B)$ has only a pole of order 2 at e^{-h} and $\tilde{\zeta}(f|\Lambda)$ has only a pole of order $(-1)^u$ at e^{-h}. Finally, if B is transitive and not mixing (Λ of type R) we must have a g.c.d. $d > 1$ to the periods of periodic points in B. Since B has a two-to-one semiconjugacy to the mixing subshift A, we must have $d = 2$. We have $N_{2m}(B) = 2N_{2m}(A)$ and $N_{2m+1}(B) = 0$, so $\zeta(B) = \zeta(A^2) \circ z^2$. Since A^2 is mixing we find that $\zeta(B)$ has only poles of order 1 at $z = \pm e^{-h}$ and that $\tilde{\zeta}(f|\Lambda)$ has only a pole of order $(-1)^u$ at $-e^{-h}$.

It is easy to verify that types N, P and R arise from the 3 cases just described and are exclusive.

Q.E.D.

We observe that if Λ is of a given type for f it is of that type for f^i, $i \neq 0$, with one exception: if Λ is of type R and i is even then Λ is of type P for f^i.

Corollary. The only poles of $\tilde{\zeta}(f)$ on $D = \{|z| \leq e^{-h(f)}\}$ are at points $e^{-h(f)} \cdot \xi \in \partial D$, where ξ is a root of unity. If ξ is a primitive ith root of unity then the order of the pole at ξ is the sum of the coefficients of y^{mi}, $m = 1,2,\ldots$, in the polynomial

$$\sum_{\substack{\Lambda \text{ of type P} \\ h(f|\Lambda)=h(f)}} (-1)^{u(\Lambda)} y^{n(\Lambda)} + \sum_{\substack{\Lambda \text{ of type R} \\ h(f|\Lambda)=h(f)}} (-1)^{u(\Lambda)} (y^{2n(\Lambda)} - y^{n(\Lambda)}) .$$

Proof Since $h(f) = \sup_{\Lambda} h(f|\Lambda)$ and $\tilde{\zeta}(f) = \prod_{\Lambda} \tilde{\zeta}(f|\Lambda)$ one need only check that each $\tilde{\zeta}(f|\Lambda)$ makes the proper contribution to the given polynomial. This follows easily from the preceding theorem. Q.E.D.

Section 2 Ergodic and Hyperbolic Toral Automorphisms

We will give a recipe for constructing matrices $M \in Gl(n,\mathbb{Z})$ that induce ergodic nonhyperbolic toral automorphisms $\bar{M}$. These furnish examples for which the theorem of Section 3 applies. We will also discuss the computational problem of determining the ergodicity and hyperbolicity of a given M. But first we show that simplest representatives do exist in the homology classes on tori mentioned in the introduction.

Given any $M \in Gl(n,\mathbb{Z})$, we may divide the spectrum of M into roots of unity and eigenvalues that are not roots of unity.

Since this is an algebraic condition defined over $\mathbb{Q}$, we find that $\mathbb{Q}^n = Q \oplus E$ where Q and E are M-invariant rational subspaces, the spectrum of $M_e = M|E$ contains no roots of unity (i.e. M_e is ergodic) and the spectrum of $M_q = M|Q$ consists solely of roots of unity (i.e. M_q is quasi-unipotent). We have

Theorem 2. If M_e is hyperbolic (e.g. if $n \leq 3$) then $\overline{M}:T^n \to T^n$ is C^∞ approximable by simplest diffeomorphisms.

Proof Consider the projection $\mathbb{Z}^q$ and $\mathbb{Z}^e$ of $\mathbb{Z}^n$ into Q and E respectively. Note that $\mathbb{Z}^n \subset \mathbb{Z}^q \oplus \mathbb{Z}^e$ is a subgroup of finite index. This gives a finite covering $T^n \to T^q \times T^e$ that is a semiconjugacy from $\overline{M}$ to $\overline{M_e} \times \overline{M_q}$.

By assumption, $\overline{M_e}$ is hyperbolic. By [4, see also 5] the quasi-unipotent toral automorphism $\overline{M_q}$ is C^∞ approximable by a Morse-Smale diffeomorphism f. Lifting $\overline{M_e} \times f$ to $g:T^n \to T^n$ gives a C^∞ approximation g to $\overline{M}$.

We have that g is Axiom A-Strong Transversality since $\overline{M}_e$ and f are. Also $h(g) = h(\overline{M_e} \times f) = h(\overline{M_e})$, since $h(f) = 0$. As $h(\overline{M_e}) = \sum_{\substack{|\lambda|>1 \\ \lambda \in \mathrm{Spec}\overline{M}_e}} \log|\lambda|$ is the logarithm of a homology eigenvalue and the entropy conjecture holds on tori [8], we have that $h(g)$ is minimal in its homology class.

Q.E.D.

We now describe the construction of ergodic nonhyperbolic toral automorphisms $\overline{M}$. Given an irreducible factor p_o of $p(M)$, there is an $\bar{M}$-invariant fibration of T^n over T^j, $j = \deg p_o$, such that the induced map on T^j is a toral automorphism with characteristic polynomial p_o [4]. With this in mind, we ignore the extension problem and assume $p(M)$ is irreducible and ergodic.

We know that two integral matrices M_1 and M_2 with $p(M_1)$ and $p(M_2)$ irreducible are conjugate over $\mathbb{Q}$ whenever $p(M_1) = p(M_2)$. This implies (similar to the argument of the preceding theorem) that there is a finite cover $s: T^n \to T^n$ which is a semiconjugacy from M_1 to M_2. Thinking of $M_1 = M$, $M_2 = C(p(M))$ = the companion matrix of $p(M)$, we will restrict to analyzing only the possibilities for $p(M)$.

So assume p is irreducible, ergodic and nonhyperbolic. Then if λ is a root of p and $|\lambda| = 1$, we have $\lambda^{-1} = \bar{\lambda}$ is also a root of p. It follows that all roots of p occur in pairs ξ, ξ^{-1}, so p is symmetric of degree 2m. We find that $x^{-m}p(x) = q(x + x^{-1})$ where q is a monic integral polynomial of degree m. Moreover q is irreducible and has some (but not all) of its roots in $(-2, 2)$ (ergodicity implies some root of p lies off the unit circle). Conversely, any q with these properties gives an irreducible, ergodic and nonhyperbolic p.

For instance, if $q(x) = x^2 + ax + b$ has noninteger roots $\xi \in (-2,2)$ and η, $|\eta| > 2$, then $p(x) = x^4 + ax^3 + (b+2)x^2 + ax + 1$ is irreducible, ergodic and nonhyperbolic. Taking the companion matrix $C = C(p(M))$ gives examples $\bar{C}: T^4 \to T^4$ to which Theorem 3 below applies, and all such examples on T^4 arise from finite covers of these $\bar{C}$'s.

Finally, we consider the computational problem of deciding if a given integral polynomial p has a root λ on the unit circle. If it does, then λ is a root of $x^{(\deg p)}p(x^{-1}) = p'(x)$ as well. Thus we may pass to the g.c.d. of p and p' and assume that p is symmetric. Dividing out by linear factors $x \neq 1$, we may suppose ± 1 are not roots of p. Then $\deg p = 2m$ is even and $x^{-m}p(x) = q(x + x^{-1})$ where q has a root in $(-2,2)$. Whereas the original problem involved functions of a complex variable, and it can't be established by mere computation that a given complex number lies precisely on the unit circle, we have changed the problem to a computable question about an integral polynomial in a real variable. Thus one may compute whether an element $M \in Gl(n,\mathbb{Z})$ is hyperbolic.

Section 3 The Strict Entropy Inequality on Tori and Related Spaces

In [9] it was shown that the entropy conjecture $(h(f) \overset{?}{\geq} \log|\lambda|)$ holds for Axiom A diffeomorphisms f whenever some homology eigenvalue λ of maximum modulus doesn't cancel out in the Lefschetz formula. For such maps f, the corollary of section 1 imposes a constraint on the possibility that $h(f) = \log|\lambda|$, namely λ must be (real)$\cdot$(root of unity). This will be applied to tori (and spaces which have similar homological properties).

Theorem 3 Suppose $\alpha : T^n \to T^n$ is a toral automorphism with isolated fixed points and some eigenvalue λ, $|\lambda| = 1$, not a root of unity. Then any Axiom A diffeomorphism f homotopic to α satisfies $h(f) > h(\alpha)$.

Proof We use the Lefschetz Fixed Point Formula for f and its iterates. This is concisely expressed in the identity (see [2])

$$\tilde{\zeta}(f) = \frac{\prod_{i \text{ odd}} \det(I - f_{*i} z)}{\prod_{i \text{ even}} \det(I - f_{*i} z)}$$

where $f_{*i} : H_i(T^n, \mathbb{R}) \circlearrowleft$, $i = 0,1,\ldots,n$.

Since the cohomology ring of T^n is an exterior algebra on $H^1(T^n)$, we have $\tilde{\zeta}(f) = \prod_{1 \le j_1 < \ldots < j_i \le n} (1 - \lambda_{j_1} \ldots \lambda_{j_i} z)^{(-1)^{i+1}}$,

where λ_j are the eigenvalues of $f_{*1} : H_1(T^n) \circlearrowleft$. Consider those terms $(1 - \ell z)$ in this product for which ℓ has largest possible modulus. These are necessarily of the form $(1 - P\lambda_{j_1} \ldots \lambda_{j_k} z)$ where $P = \prod_{|\lambda_i| > 1} \lambda_i$ and $1 \le j_1 < \ldots < j_k \le c = c(\alpha)$, where $|\lambda_{j_i}| = 1$.

Letting $Q = \prod_{\substack{1 \le j_1 < \ldots < j_k \le c \\ 0 \le k \le c}} (1 - P\lambda_{j_1} \ldots \lambda_{j_k} z)^{(-1)^k}$ we see

$Q'(0) = -P(1 - \lambda_1) \ldots (1 - \lambda_c) \neq 0$ so $Q \neq 1$. Hence $\tilde{\zeta}(f)$ has radius of convergence $|P|^{-1}$.

From theorem 1 (or [8]) it follows that $h(f) \geq \log|P|$. Since Bowen showed $\log|P| = h(\alpha)$, we need only show $h(f) \neq \log|P|$.

Assume $h(f) = \log|P|$. From Section 1 we see that Q has the same poles as some product $S = \Pi(1\pm(e^{h(f)}z)^n)^{\pm 1}$. Since Q and S agree at $z = 0$, we have $Q = S$. Setting $Q'(0) = S'(0)$ gives $(1-\lambda_1)\dots(1-\lambda_c) \in \mathbb{Z}$. Since $h(f^n) = nh(f) = nh(\alpha) = h(\alpha^n)$, we have

$$(1-\lambda_1{}^n)\dots(1-\lambda_c{}^n) \in \mathbb{Z}, \text{ for all } n.$$

Consider the closure $G = \overline{\{(\lambda_1{}^n,\dots,\lambda_c{}^n)\}}$. G is a compact abelian Lie group. We cannot have G finite since some λ_i is not a root of unity. Thus G contains a nontrivial one-parameter group $\{(e^{ita_1},\dots,e^{ita_c}) | t \in \mathbb{R}\}$, where we may suppose $a_1 \neq 0$. Note that $g = (1-z_1)\dots(1-z_c):G \to \mathbb{Z}$ so $g(\lambda_1 e^{ita_1},\dots,\lambda_c e^{ita_c})$ is a constant k. Since g vanishes when $e^{ita_1} = \lambda_1^{-1}$, we must have $k = 0$. Then some $(1-z_i)$ must vanish on $\{(\lambda_1 e^{ita_1},\dots,\lambda_c e^{ita_c})\}$ which is possible only if $a_i = 0$ and $\lambda_i = 1$. But we assumed α has isolated fixed points so no $\lambda_i = 1$.

Q.E.D.

This theorem holds equally well for nilmanifolds, since the Lefschetz number of a nilmanifold automorphism $\alpha:N\circlearrowleft$ is $L(\alpha) = (1-\lambda_1)\dots(1-\lambda_n)$, where λ_i are the eigenvalues of α_* on the central series [7]. An easy proof of this formula is based on the fibration $T \to N \overset{\pi}{\to} P = N/T$, where T is the torus corresponding to the center of $\tilde{N}$ and P has lower nilpotent degree. It follows that α induces $\alpha_P : P \to P$ and that for each $p \in \mathrm{Fix}(\alpha_P)$ the maps $(\alpha|\pi^{-1}p)$ are conjugate toral automorphisms α_T. One sees $L(\alpha) = L(\alpha_P)L(\alpha_T) = L(\alpha_P \times \alpha_T)$ and the formula follows by induction on the nilpotent degree.

We now recover Manning's result that ergodic nonhyperbolic nilmanifold automorphisms $\alpha : N \circlearrowleft$ aren't homotopic to Anosov diffeomorphisms. If $f \sim \alpha$ <u>were</u> Anosov, then by passing to a finite cover, we may assume that the unstable bundle of f is oriented. Then Shub and Williams showed that $\pm e^{h(f)}$ is a homology eigenvalue of f [12]. But by Theorem 2, $e^{h(f)} > e^{h(\alpha)} \geq$ spectral radius of $(f_* : H_*(N, \mathbb{R}) \circlearrowleft)$, which is a contradiction.

Theorem 2 adapts to oriented manifolds M whose cohomology ring is an exterior algebra on odd-dimensional generators (e.g. M a Lie group) when the action of f_* on the primitive elements is ergodic and nonhyperbolic. For instance, if $M = (S^3)^n$ then any element of $\mathrm{Gl}(n, \mathbb{Z})$ arises as f_{*3} for some diffeomorphism f. If f is Axiom A we find that $h(f) > \log|\lambda|$ for all homology eigenvalues λ. There are some candidates for maps of entropy $\left(\sum_{|\lambda_i|>1} \log|\lambda_i| \right)$, namely maps obtained by quaternion multiplication using automorphisms of the free group on n generators. This suggests that there exist simply connected examples for which efficiency and hyperbolicity clash.

References

1. Bowen, Rufus. Entropy for group endomorphisms and homogeneous spaces, Trans.Amer.Math.Soc.153 (1971), pp. 401-414.

2. Franks, John. Morse inequalities for zeta functions, Ann. of Math. 102 (1975) pp. 55-65.

3. ________. A reduced zeta function for diffeomorphisms, Amer. J. Math. 100 (1978) pp. 217-244.

4. Fried, David and Michael Shub. Entropy, linearity and chain-recurrence. To appear in Publ. I.H.E.S.

5. Halperin, Benjamin. Morse-Smale diffeomorphisms on tori. To appear in Topology.

6. Manning, Anthony. Axiom A diffeomorphisms have rational zeta functions. Bull. London Math Soc. 3 (1971) pp. 215-220.

7. ________. There are no new Anosov diffeomorphisms on tori. Amer. J. Math 96 (1974) pp. 422-429.

8. Misiurewicz, Michael and F. Przytycki, The entropy conjecture on tori, preprint.

9. Shub, Michael. Dynamical systems, filtrations and entropy. Bull. Amer. Math. Soc. 80 (1974) pp. 27-41.

10. Smale, Steve. Differentiable dynamical systems. Bull. Amer. Math Soc. 73 (1967) pp. 747-817.

11. Williams, R. F. The zeta function in global analysis. Proc. Symp. Pure Math XIV pp. 335-340.

12. ________ and Michael Shub. Entropy and stability, Topology 14 (1975) pp. 329-338.

University of California
Santa Cruz, CA 95064

DYNAMICAL BEHAVIOR OF GEODESIC FIELDS

Herman Gluck

This paper covers the first two parts of a lecture given at the International Conference on Dynamical Systems at Northwestern University during June '79:

I. Geodesic fields on surfaces
II. An exposition of Sullivan's characterization of geodesic fields.

The remaining portion of the lecture,

III. Morse-Smale fields of geodesics,

is covered in the following paper [A-G].

INTRODUCTION

For the simplest examples of geodesic fields, consider parallel lines in Euclidean space, parallel circles or winding lines on a flat torus, or a Hopf field of great circles on the 3-sphere. In each case, a certain Riemannian manifold is foliated by geodesics. Many other examples arise naturally in mathematics:

1. Geodesic flows. The orbits of the geodesic flow on the unit tangent bundle UM of any Riemannian manifold M are themselves geodesics for a natural choice of metric on UM. This yields the Hopf fibration of RP^3 when M is a round 2-sphere; structurally stable Anosov flows when M has strictly negative sectional curvatures [A]; flows of positive entropy (hence dynamically complex) when the fundamental group $\pi_1(M)$ has exponential growth [D].

2. Isometric flows. If we start with an isometric flow (i.e., Killing vector field) on a Riemannian manifold M, and then rescale the metric conformally to make the flow of unit speed (Wadsley's trick [Wa]), the orbits become geodesics and the flow remains isometric. For example, rotating the plane about the origin gives a nonsingular isometric flow on the punctured plane, whose orbits are concentric circles.

escaling turns the punctured plane into a cylinder and makes all he orbits geodesics of equal length.

. Contact flows. A contact form on an odd dimensional orientable anifold M^{2n-1} is a one-form ω such that $\omega \wedge (d\omega)^{n-1}$ is ever zero. Example: $\omega = x\,dy + dz$ on R^3. Given such an ω, here is a unique vector field V on M such that $\omega(V) = 1$ and the ie derivative $\mathcal{L}_V(\omega) = 0$. V can be characterized, up to length, s generating the one-dimensional subspace ker $d\omega$ in each tangent pace. V generates the contact flow. It is possible to choose a iemannian metric on M (see section 10) in terms of which the orbits f the contact flow are all geodesics. Indeed, the geodesic flow n the unit tangent bundle of a Riemannian manifold is a contact low, so this example includes example 1 above.

One sharp distinction between contact flows and the more general eodesic fields can be seen by comparing both with flows-with-section: very flow-with-section can be made into a geodesic field; no flow-ith-section is a contact flow.

. Classical Hamiltonian mechanics. Consider a conservative mechan-cal system with configuration space a Riemannian manifold M (whose etric corresponds to kinetic energy), with potential energy : $M \to R$ and with phase space TM, the tangent bundle of M. Roughly peaking, the Principle of Least Action (in the form of Maupertuis-uler-Jacobi) asserts that each orbit $\gamma: R \to M$ of this mechanical ystem is a geodesic in a certain "Jacobi" metric on M, depending nly on the original metric and on the constant total energy E of the rbit γ.

More precisely, first fix a value of $E > \max U$. Then the orbits $\gamma: R \to M$ of energy E lift to orbits $(\gamma,\gamma'): R \to TM$ which fill out a onstant energy hypersurface in phase space. This hypersurface is just a copy of the unit tangent bundle of M in the Jacobi metric orresponding to energy E. The Hamiltonian flow on this hypersurface is, up to a change of time-parameter, just the geodesic flow of example 1. Hence there is a Riemannian metric on the constant energy hypersurface in terms of which the orbits of the Hamiltonian flow are all geodesics.

For regular values of $E < \max U$, the orbits are restricted to a proper subset of M, and the Jacobi metric degenerates on the boundary of this region. Nevertheless, Weinstein [W] has shown that, up to change of time scale, the Hamiltonian flow is a contact flow on the corresponding constant energy hypersurface. By example 3, we get a metric on this hypersurface making all the orbits geodesics.

5. Fields of closed curves. Let M be a closed manifold filled by closed curves. Wadsley [Wa] showed that there is a Riemannian metric on M making these curves geodesics if and only if their lengths are bounded (in some, and hence any, metric). Epstein [E] showed that this bounded length condition is always satisfied in dimension 3. Sullivan [S_1] showed that it can fail in dimensions ≥ 5. Epstein and Vogt [E-V] showed that it can also fail in dimension 4.

The basic questions about geodesic fields seem to be these:

A. Which spaces can be filled by geodesics? That is, given a smooth manifold M, is there a field of curves filling M and a Riemannian metric making them geodesics?

B. Which fields of curves on a given space can be made into geodesics by appropriate choices of metric? What are the dynamical properties of such "geodesible" curve fields?

The first question is treated in [G]. Vanishing of the Euler characteristic is certainly a prerequisite for filling a space by geodesics. In dimension 2, the torus and Klein bottle can clearly be so filled. Using the "open book decomposition" introduced by Winkelnkemper, together with recent results of Quinn and Terry Lawson, it is shown in [G] that every closed odd-dimensional manifold can be filled by geodesics. Furthermore, finitely many disjoint simple closed curves can be specified in advance as "geodesics-to-be". In particular, arbitrary knots and links in 3-manifolds can be included in geodesic fields.

The second question is treated here and in the following aper [A-G] with Dan Asimov. Here we discuss geodesic fields on urfaces, and then give an exposition of Sullivan's theory of oliation cycles and his characterization of geodesic fields, to e used in the following paper. There we look at Morse-Smale ields (dynamically simplest and structurally stable) and show hat only those which are suspensions (i.e., flows-with-section) an be made into geodesics. In particular, there is no Morse-male field of geodesics on the 3-sphere.

I am grateful to Professors Dan Asimov, Eugenio Calabi, 'ozef Dodziuk, David Fried, Alan Kafker and Alan Weinstein for any insights gained in conversation with them. I also thank he National Science Foundation for support.

CONTENTS

I. GEODESIC FIELDS ON SURFACES

1. The Reeb component

The obvious first question is: What could prevent a field G of curves on M from being geodesible? Example: the presence in G of a Reeb component, as shown in Figure 1.

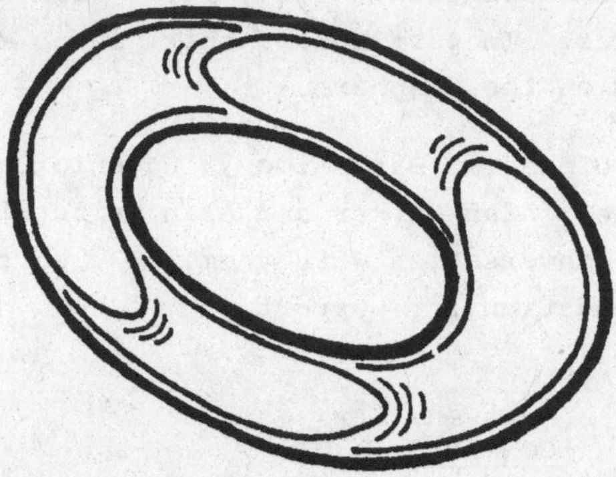

Figure 1

Any field of curves transverse to G must have a limit cycle, by Poincaré-Bendixson. But if G were geodesible, then its orthogonal trajectories would be a constant distance apart, measured along the curves of G, and hence could contain no limit cycle. Conclusion: G is not geodesible. Sullivan's theorem (section 11) displays a sense in which the Reeb component is the prototypical obstruction to geodesibility.

The Reeb component can also appear in several minor variations:

a) The above Reeb annulus, with boundary identified to form a torus;

b) The Reeb annulus, with boundary identified to form a Klein bottle;

c) A Reeb Möbius band (which is double covered by a Reeb annulus);

d) A Reeb Möbius band, with boundary identified to form a Klein bottle.

Each of these will be called a "Reeb component"; none can be made from geodesics for the same reasons as above.

The main result about geodesic fields on surfaces is given by

THEOREM. On a closed surface, any smooth curve field without Reeb components is geodesible.

REMARK. It is natural to interpret "smooth" as meaning of class C^∞. The theorem also works for curve fields of class C^2, if we settle for a Riemannian metric of class C^2. I don't know if one can get a C^∞ metric in this case. If the curve field is only of class C^1, then the theorem is in serious doubt, since we now must deal with the Denjoy flow.

The theorem above will be proved in the course of the next two sections.

2. Manifolds which fibre over the circle; flows-with-section

LEMMA. Flows-with-section are geodesible. Hence any manifold which fibres over the circle can be filled by geodesics in a suitable metric.

A section to a flow on a closed manifold M^n is a closed hypersurface F^{n-1} transverse to the flow and meeting every orbit at least once. Cutting M along F yields $[0,1] \times F$, with the flow lines falling apart into the segments $[0,1] \times$ point. M can then be reassembled by identifying $(1,x)$ with $(0,h(x))$ for some diffeomorphism h of F.

Let g_t be a metric on F which interpolates between a random metric $g = g_0$ and its pullback $h^*g = g_1$, with $g_t = g_0$ for t near 0 and $g_t = g_1$ for t near 1. Then the metric $dt^2 + g_t$ on $[0,1] \times F$ induces a metric on M. Since the segments $[0,1] \times$ point are geodesics on $[0,1] \times F$, the flow lines are geodesics on M.

Now any manifold which fibres over the circle has a flow transverse to the fibres, so can be filled by geodesics in this way.

NOTE that this lemma is not restricted to dimension 2.

3. Proof of Theorem

Let C be a smooth (C^∞) curve field without Reeb components on the closed surface M^2.

A. Suppose M^2 is a torus. It follows from Kneser [K] that the absence of Reeb components guarantees the presence of a smooth closed transversal, meeting every curve of C at least once. Hence C consists of the orbits of a flow-with-section, so is geodesible by the Lemma of section 2.

Indeed, concrete examples of such geodesic fields are easy to visualize. If C has no closed orbits, then it is differentiably equivalent to an irrational flow [D_1-D_4]. Hence its curves appear as geodesics in some flat metric on the torus.

If C has some closed orbits, then it follows again from [K] that finitely or countably many spiral annuli can be removed from

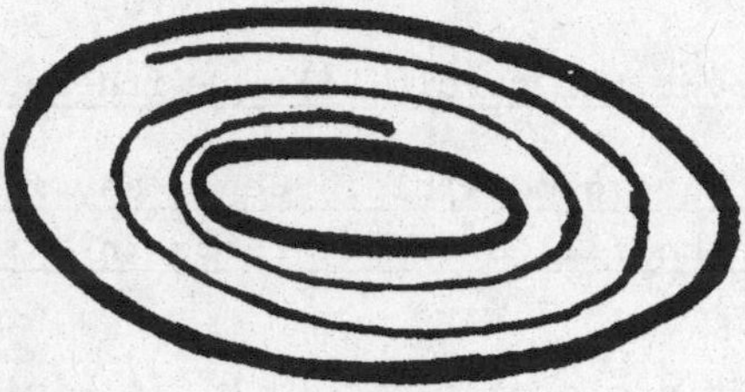

Figure 2

the torus so that only "parallel" closed curves of C remain. In that case we can visualize the surface as a multiply swollen torus of revolution, on which the curves of C are all geodesics.

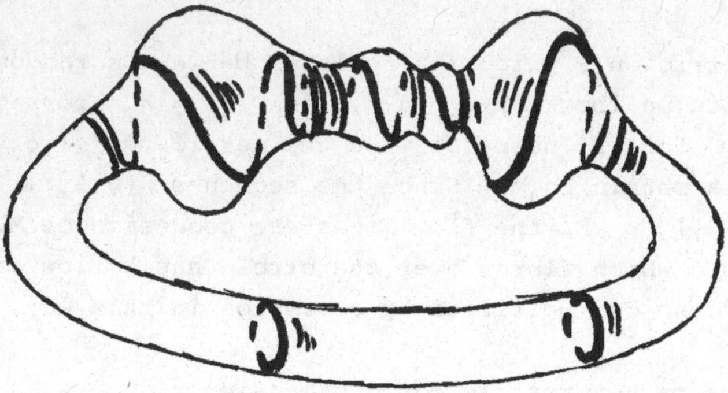

Figure 3

B. Suppose M^2 is a Klein bottle. Any continuous curve field on a Klein bottle must contain a closed curve [K]. So pick one, call it γ, and suppose first that it is one-sided. Cutting the Klein bottle along γ yields a Möbius band, and on it a curve field without Reeb components. Applying [K] again, we easily get a two-sided smooth transversal, meeting every curve of C at least once. Then as before, C consists of the orbits of a flow-with-section, so is geodesible.

If γ is two-sided, then cutting the Klein bottle along it yields an annulus, on which we see a curve field without Reeb components. Just as for the torus, this annulus can be visualized as a multiply swollen surface of revolution, with the curves as geodesics.

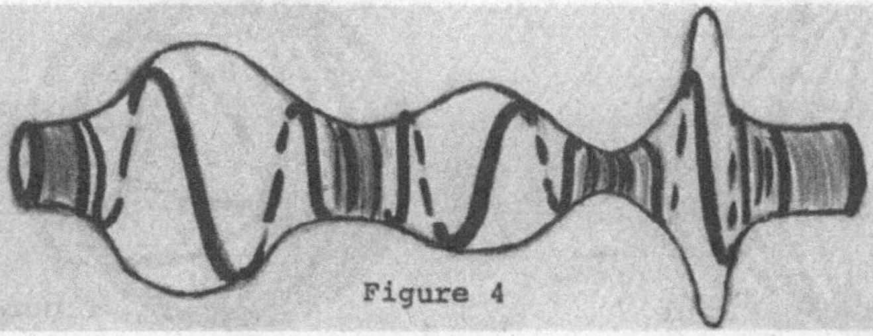

Figure 4

The metric on this annulus can easily be taken so that an orientation reversing isometry between its two boundary curves can be used to recreate the Klein bottle, and induce on it a metric making the curves of C geodesics.

This completes the proof.

4. Geodesic fields with singularities

Still restricting our attention to surfaces, we now look for geodesic fields which may have isolated singularities, thus freeing ourselves from the restriction to Euler characteristic zero.

Two obvious examples in the plane are:

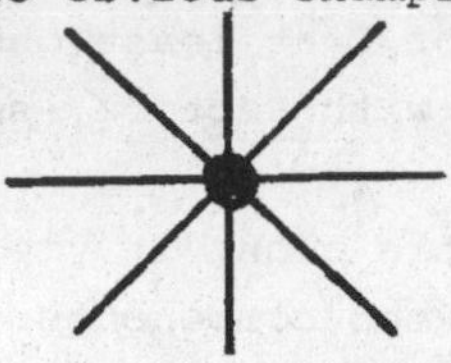

Figure 5: DAY

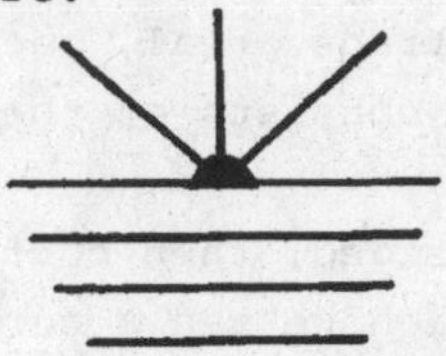

Figure 6: SUNRISE

On a round 2-sphere, we can visualize the fields of geodesics:

Figure 7

Figure 8

On a round projective plane, we can consider the field of great circles through a common point. Diagrammatically:

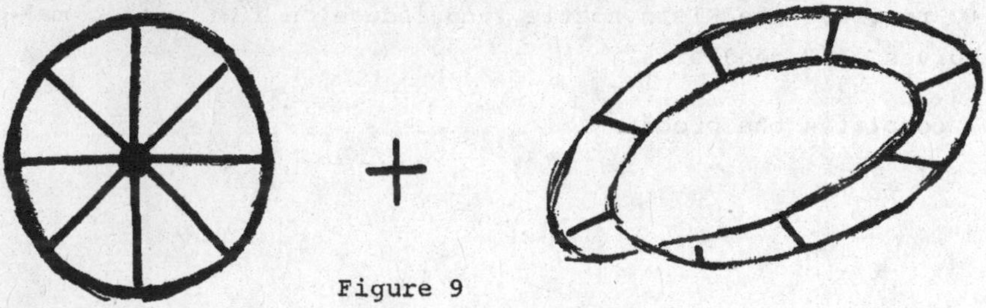

Figure 9

Alan Kafker has shown [Ka] that there is a sense in which Figures 7, 8 and 9 display the <u>only</u> examples of geodesible fields with singularities on closed surfaces. We briefly describe his result.

A word about differentiability: since we are looking for C^∞ etrics which make a given curve field into geodesics, each individual curve in the field must be of class C^∞. But we ask only that he tangent lines to the curve field vary continuously, otherwise he Sunrise singularity of Figure 6 and the beautiful curve field f Figure 8 would be excluded.

If $\{p_1, p_2, \ldots, p_n\}$ is a finite set of points of M and the urve field is only given on their complement, we call the points p_i *singularities* of the field. A singularity p_i is *removable* if he curve field results by deleting p_i from the range of some ther curve field.

Kafker first proves:

A. *Let p be an isolated nonremovable singularity of a eodesic field on a 2-dimensional Riemannian manifold M. Then there s a neighborhood U of p on M and a homeomorphism of U to a eighborhood of the origin in the plane, taking p to the origin nd the curves of the field to those of Figures 5 or 6.*

Note that the Day singularity of Figure 5 has index 1, and he Sunrise singularity of Figure 6 index 1/2. Since the sum of he indices of the singularities equals the Euler characteristic, e get:

B. *A closed surface of Euler characteristic less than zero does ot admit a geodesic field with isolated singularities. If the Euler haracteristic is zero, such a field can have only removable singularities.*

Therefore we can restrict attention to the two-sphere and the rojective plane.

To circumvent certain difficulties, Kafker calls a curve field with isolated singularities *dynamically simple* if all curves "begin" and "end" at singularities. This simplifying hypothesis is involved only in parts 3 and 5 of the following theorem.

C. 1) On a 2-sphere, a geodesic field with two singularities (both Day) is topologically equivalent to the field of great semicircles shown in Figure 7.

2) There does not exist a geodesic field on a 2-sphere with three singularities (one Day and two Sunrises).

3) A dynamically simple geodesic field on a 2-sphere having four singularities (all Sunrises) is topologically equivalent to the field shown in Figure 8.

4) A geodesic field on a projective plane having one singularity (Day) is topologically equivalent to a field of great circles through a single point, diagrammed in Figure 9 .

5) There does not exist a dynamically simple geodesic field on a projective plane having two singularities (both Day).

Two interesting fields of curves on a 2-sphere, each having four Sunrise singularities, and each violating the conditions for dynamic simplicity, are presented in [Ka]. Whether or not they can be made into geodesics is unknown.

D. Let G be a geodesic field with isolated singularities on a 2-sphere or projective plane. If on the 2-sphere, assume no nontrivial recurrence. Then a CLOSED GEODESIC (perhaps pieced together from several curves and singularities) always appears in G.

II. SULLIVAN'S CHARACTERIZATION OF GEODESIC FIELDS

In [S_2], Sullivan gives a geometric characterization of geodesible curve fields, phrased in the DeRham-Sullivan language of currents and foliation cycles [DeR, S_3]. We give a brief description of this language here, and then the proof of Sullivan's theorem. In the following paper [A-G], we use this to understand Morse-Smale fields of geodesics.

. DeRham's theory of currents

Let M^n be a closed C^∞ manifold and let $\mathcal{D}_p$ denote the real vector space of C^∞ p-forms on M. There is a natural C^∞ topology on $\mathcal{D}_p$ which makes it into a locally convex linear space.

A continuous linear functional $c: \mathcal{D}_p \to R$ is called a p-current, or current for short. Examples:

A. $c(\omega) = \int_c \omega$, where c is a p-chain and ω a p-form.

B. a mass distribution of p-chains.

C. $c(\omega) = \int_M \omega \wedge \eta$, where ω is a p-form, η an (n-p)-form and M is oriented. Denote this p-current by c_η .

D. Dirac currents: pick a point x in M and a p-vector V_x at x. Define $c(\omega) = \omega_x(V_x)$.

NOTE that a 0-current is a distribution in the sense of L. Schwartz.

Let $\mathcal{D}_p'$ denote the real vector space of all p-currents on M, that is, the dual space to $\mathcal{D}_p$. Under a natural topology, $\mathcal{D}_p'$ also becomes a locally convex linear space.

The boundary of a p-current c is the (p-1)-current ∂c defined by $\partial c(\omega) = c(d\omega)$ for all (p-1)-forms ω. Thus "boundary" for currents is adjoint to "exterior differentiation" for forms. Currents with zero boundary are called cycles. Examples:

E. integration along a geometric cycle.

F. if η is a closed (n-p)-form, then

$\partial c_\eta(\omega) = c_\eta(d\omega) = \int_M d\omega \wedge \eta = \int_M d(\omega \wedge \eta) = 0$ by Stokes' theorem, since $\partial M = \emptyset$. So c_η is a cycle.

The boundary map $\partial: \mathcal{D}_p' \to \mathcal{D}_{p-1}'$ is continuous.

The strength of these ideas lies in the beautiful theorem of Schwartz [Sc]:

$\mathcal{D}_p$ is also the dual space of $\mathcal{D}_p'$. That is, each continuous linear functional on $\mathcal{D}_p'$ comes from evaluating the currents in $\mathcal{D}_p'$ on some fixed p-form ω.

6. Sullivan's definition of foliation cycles

Let M^n still denote a closed C^∞ manifold, and now suppose that V is a nonsingular C^∞ vector field on M. For each point x in M, let $\delta_x: \mathcal{D}_1 \to R$ denote the Dirac 1-current defined by $\delta_x(\omega) = \omega_x(V_x)$. Then let $\mathcal{C}$ denote the closed convex cone in $\mathcal{D}_1'$ generated by the set of all these Dirac currents. The elements of $\mathcal{C}$ are called foliation currents (referring to the foliation of M by orbits of V). Examples:

A. integration along an arc of an orbit of V.

B. a mass distribution of these.

C. let η be an (n-1)-form such that

1) $\mathcal{L}_V(\eta) = 0$, i.e., η is invariant under the flow of V, and

2) $V \lrcorner\, \eta = 0$, i.e., η kills (n-1)-vectors tangent to V.

Then the 1-current $c_\eta(\omega) = \int_M \omega\wedge\eta$ is a foliation current if M is oriented.

REMARK. Conditions 1 and 2 just above imply $d\eta = 0$. For by the general formula

$$\mathcal{L}_V\eta = d(V \lrcorner\, \eta) + V \lrcorner\, d\eta \quad ,$$

they at least imply $V \lrcorner\, d\eta = 0$. But $d\eta$ is a form of top dimension and V is never zero. It follows that $d\eta = 0$.

A 1-form ω on M is transversal to V if $\omega_x(V_x) > 0$ for all x in M. These are plentiful and can be assembled by partitions of unity. If c is a foliation current and ω a transversal 1-form, then $c(\omega) > 0$.

The cone $\mathbf{C}$ of foliation currents is a compact convex cone, neaning that there exists a continuous linear functional $L: \mathcal{D}_1' \to R$ such that $L(c) > 0$ for all nonzero c in $\mathbf{C}$, and $L^{-1}(1) \cap \mathbf{C}$ is compact. Thus if $\mathcal{X}$ is a closed subspace of $\mathcal{D}_1'$ meeting $\mathbf{C}$ only at 0, then the Hahn-Banach Theorem promises a continuous linear functional $L': \mathcal{D}_1' \to R$ such that $L'(c) > 0$ for all nonzero c in $\mathbf{C}$, and $L'(\mathcal{X}) = 0$.

A foliation cycle is simply a foliation current whose boundary is zero. Examples:

D. integration along a closed orbit of V.

E. a mass distribution of these.

F. the foliation current c_η of example C above is automatically a foliation cycle (by example F of section 5 and the fact that $d\eta = 0$).

Foliation cycles may be thought of as "generalized closed orbits". hey exist, even when (traditional) closed orbits do not. Consider he irrational flow of the torus $S^1 \times S^1$ defined by the vector ield

$$V = \frac{\partial}{\partial x} + \alpha \frac{\partial}{\partial y} .$$

et $\eta = -\alpha\, dx + dy$. Note that $V \lrcorner\, \eta = 0$ and that $d\eta = 0$. o by the formula on the preceding page, $\mathcal{L}_V \eta = 0$ also. Hence by xample F above, we get a nonzero foliation cycle c_η.

7. Existence of nonzero foliation cycles

As before, M^n denotes a closed C^∞ manifold and V a nonsingular C^∞ vector field on M.

SULLIVAN'S THEOREM [S_3]. V has nonzero foliation cycles.

As before, let $\mathcal{C}$ denote the compact, convex cone of foliation currents. Let $Z = \ker \partial : \mathcal{D}_1' \to \mathcal{D}_0'$ denote the cycles. Note that Z is a closed subspace of $\mathcal{D}_1'$ since it is the kernel of a continuous map.

Suppose now that there are no nonzero foliation cycles. That is, suppose $\mathcal{C} \cap Z = \{0\}$. Sullivan records this schematically by the following diagram.

Figure 10

Now by the Hahn-Banach Theorem, there is a continuous linear functional $L: \mathcal{D}_1' \to R$ such that $L(Z) = 0$ and $L(c) > 0$ for all nonzero c in $\mathcal{C}$.

By Schwartz's Theorem (see section 5), L corresponds to a smooth 1-form ω. That is, $L(c) = c(\omega)$ for all 1-currents c. Since $c(\omega) = 0$ whenever c is a cycle, ω must be exact.

Since $c(\omega) > 0$ for all nonzero c in $\mathcal{C}$, if we let c be the Dirac current δ_x, we get $\omega_x(V_x) > 0$ for all x in M. Thus ω is transversal to V.

But these two properties of ω: "exactness" and "transversality to V" are incompatible. For if $\omega = df$, then at a maximum point x of f, $\omega_x = 0$, violating transversality to V at x.

Hence V must have nonzero foliation cycles.

REMARK. Therefore the relative position of Z and $\mathcal{C}$ inside $\mathcal{D}_1'$ may be indicated as follows.

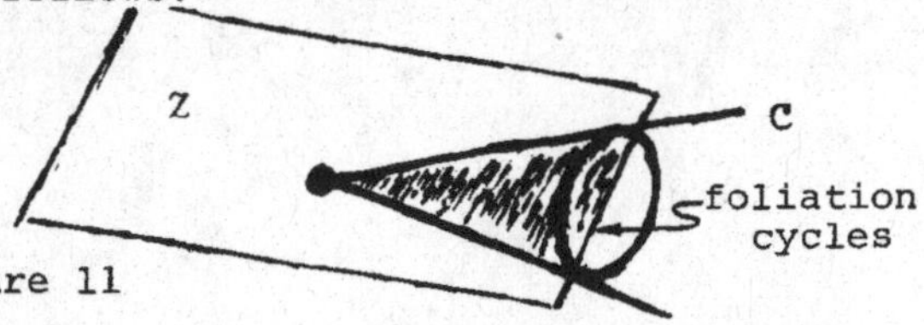

Figure 11

. Foliation cycles and invariant transversal measures

The one-to-one correspondence between foliation cycles and nvariant transversal measures provides an aid to the visualization f foliation cycles in specific examples.

Again let V be a nonsingular vector field on the closed maniold M^n. Let T be a finite union of closed (n-1)-disks in M^n, ransverse to V, such that every orbit of V meets the interior of t least one of these disks.

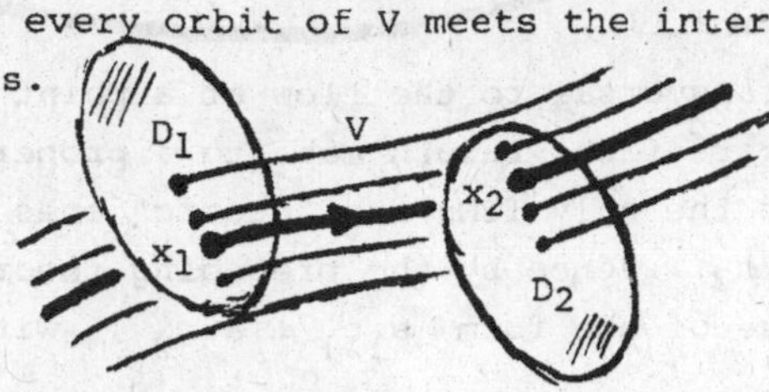

Figure 12

Suppose D_1 and D_2 are two such disks in T, whose interiors ontain points x_1 and x_2 of M which lie on the same orbit of V. he flow along V then determines a homeomorphism germ from a neighorhood of x_1 on D_1 to a neighborhood of x_2 on D_2.

An <u>invariant</u> <u>transversal</u> <u>measure</u> (ITM) for V is a non-negative easure of finite mass on T, which is invariant under all the germs of homeomorphisms described above.

IOTE that only the direction of V, not its speed, figures in this efinition.

ULLIVAN'S THEOREM [S_3]. <u>Given M and V as above, there is a canonical ne-to-one correspondence between foliation cycles and invariant ransversal measures.</u>

The correspondence may be described as follows. Given an ITM, e must define a foliation cycle c. If ω is a 1-form, first 'localize" by writing $\omega = \Sigma\, \pi_i \omega$, where $\{\pi_i\}$ is a partition of nity subordinate to a covering of M by flow boxes for V. In the low box supporting $\pi_i \omega$, integrate $\pi_i \omega$ along each arc of orbit, hen "add up" the answers via the ITM to get the value of $c(\pi_i \omega)$. inally, $c(\omega) = \Sigma\, c(\pi_i \omega)$. The invariance of the measure makes this rocedure well-defined.

EXAMPLE. In the figure below we show an annulus and on it a nonsingular vector field V with c_1 and c_2 as closed orbits and all other orbits spiraling from c_1 to c_2.

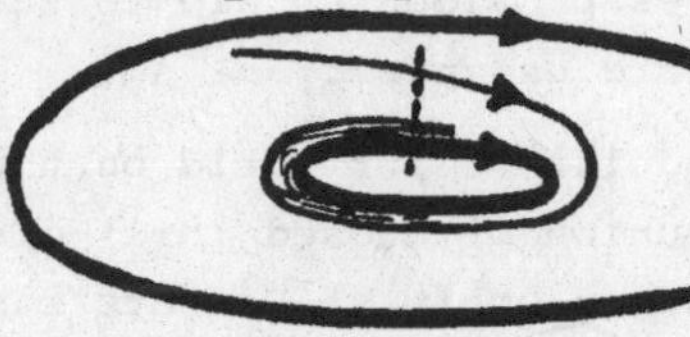

Figure 13

A small transversal to the flow at a point of c_2 is mapped, under the Poincaré first return map, to a proper subset of itself. It follows that the only ITMs are "atomic" ones with nonzero weights just at c_1 and c_2. Hence by the preceding theorem, the only foliation cycles are those of the form $a_1c_1 + a_2c_2$, with a_1 and $a_2 \geq 0$.

9. Schwartzman's Theorem

Sullivan's Theorem in section 7 tells us that the cone of foliation currents has nontrivial intersection with the subspace of cycles. We now ask whether $\mathcal{C}$ meets the subspace $\mathfrak{B}$ of boundaries. Following Sullivan, we indicate the two possibilities schematically by the following diagrams.

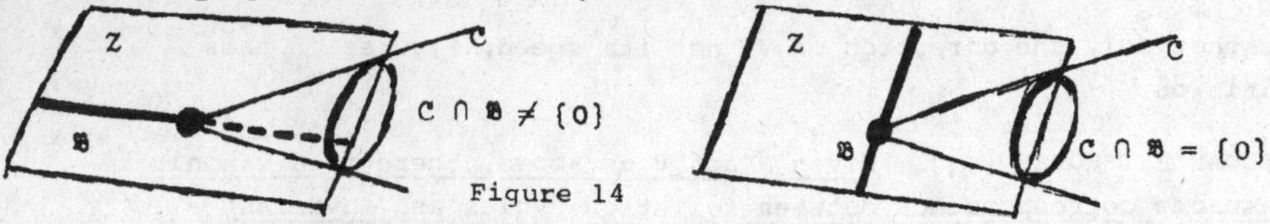

Figure 14

SULLIVAN'S LEMMA. $\mathcal{C} \cap \mathfrak{B} = \{0\}$ if and only if there is a CLOSED 1-form ω transverse to V.

The proof is a blend of the Hahn-Banach and L. Schwartz Theorems, just as in section 7.

CHWARTZMAN'S THEOREM [Sm]. The nonsingular vector field V on the losed manifold M admits a cross-section if and only if no nontrivial oliation cycle bounds.

By Sullivan's Lemma, $\mathcal{C} \cap \mathcal{B} = \{0\}$ if and only if there exists closed 1-form ω transverse to V. By Tischler's theorem [T], his happens if and only if there exists a cross-section to V.

0. Preliminary characterization of geodesic fields

HEOREM. Let M be a smooth manifold and V a smooth nonsingular ector field on M. Then the following conditions are equivalent:

1) There exists a smooth 1-form ω on M such that $\omega(V) = 1$ and $\mathcal{L}_V\omega = 0$ (i.e., ω is invariant under the flow of V).
2) There exists a smooth 1-form ω on M such that $\omega(V) = 1$ and $V \lrcorner d\omega = 0$ (i.e., $d\omega$ vanishes on all 2-planes tangent to V).
3) There exists a Riemannian metric on M making the orbits of V geodesics and V of unit length.

REMARK. The equivalence of 1) and 3) is a standard fact of differential geometry. The equivalence of 1) and 2) was pointed out by Sullivan in [S_2], and is of use in what follows.

Proof. 1) $\Longleftrightarrow$ 2). $\mathcal{L}_V \omega = d(V \lrcorner \omega) + V \lrcorner d\omega = V \lrcorner d\omega$ if $\omega(V) = 1$.

3) $\Longrightarrow$ 1). Suppose M has a Riemannian metric making V of unit length. Define a 1-form ω by $\omega(U) = V \cdot U$, where the dot signifies "inner product". We will see below that in this circumstance

$$\mathcal{L}_V\omega = \nabla_V V \cdot \text{--}$$

as 1-forms, where ∇ denotes covariant derivative. This formula shows that ω is invariant under the flow of V if and only if the orbits of V are geodesics. Hence 3) will imply 1).

To check the formula above, it is sufficient to show that

$$(\mathcal{L}_V \omega)(U) = \nabla_V V \cdot U$$

for all vector fields U invariant under the flow of V, since the value of such a U can still be preassigned at any given point.

We compute:

$$\begin{aligned}(\mathcal{L}_V\omega)(U) &= \mathcal{L}_V(\omega(U)) - \omega(\mathcal{L}_V U) = \mathcal{L}_V(\omega(U)) \\ &= \mathcal{L}_V(V\cdot U) = \nabla_V(V\cdot U) \\ &= (\nabla_V V)\cdot U + V\cdot(\nabla_V U) \\ &= (\nabla_V V)\cdot U + V\cdot(\nabla_U V) \\ &= (\nabla_V V)\cdot U + (1/2)\,\nabla_U(V\cdot V) \\ &= (\nabla_V V)\cdot U \quad .\end{aligned}$$

In this calculation we used the fact that $\mathcal{L}_V = \nabla_V$ on functions, and the fact that $\nabla_V U = \nabla_U V + \mathcal{L}_V U = \nabla_U V$, since U is invariant under the flow of V. This completes the argument that 3) implies 1).

<u>1) $\Longrightarrow$ 3).</u> If 1) holds, define a Riemannian metric on M by requiring that

a) $V\cdot V = 1$

b) V is orthogonal to ker ω

c) arbitrary metric on ker ω.

Notice that for such a metric, $\omega(U) = V\cdot U$. But then by the above computation, $\mathcal{L}_V\omega = 0$ implies $\nabla_V V = 0$, which implies that the orbits of V are geodesics. So 1) implies 3), completing the argument.

COROLLARY [S_2]. <u>Let V be a nonsingular vector field on M. Then there is a Riemannian metric on M making the orbits of V geodesics if and only if there exists a 1-form ω on M satisfying:</u>

$$\omega(V) > 0 \qquad \text{and} \qquad V \lrcorner\, d\omega = 0.$$

Given the metric, let $V' = V/|V|$ and appeal to 3) $\Longrightarrow$ 2) above. Given the 1-form ω, let $V' = V/\omega(V)$ and appeal to 2) $\Longrightarrow$ 3).

REMARK. The convenience of the above Corollary is that it refers only to the direction of V, not to its speed.

. Sullivan's characterization of geodesic fields

IEOREM [S_2]. Let V be a smooth nonsingular vector field on the nooth manifold M. Then there is a Riemannian metric making the :bits of V geodesics if and only if no nonzero foliation cycle or V can be arbitrarily well approximated by the boundary of a -chain tangent to V.

We first give several examples illustrating this theorem, aving the proof until the next section.

KAMPLE 1: The annulus.

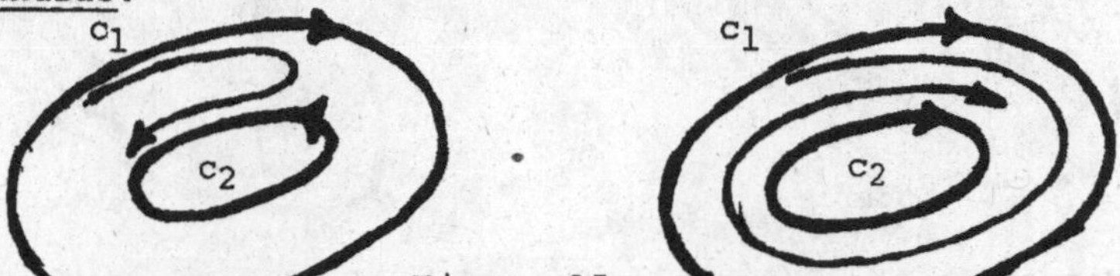

Figure 15

oliation cycles:	$a_1c_1+a_2c_2$ $(a_1,a_2 \geq 0)$	$a_1c_1+a_2c_2$ $(a_1,a_2 \geq 0)$
angent 2-chain:	whole annulus	whole annulus
bserve:	∂(annulus) = c_1+c_2 , a foliation cycle	∂(annulus) = c_1-c_2 , not a foliation cycle
onclude:	not geodesible	
bserve:		no foliation cycle bounds
onclude:		geodesible

EXAMPLE 2: Morse-Smale approximation to the Hopf flow on S^3.

Consider the Hopf flow on S^3, all of whose orbits are great circles and hence geodesics in the standard round metric. Perturb to a flow with just two closed orbits, one repelling and the other attracting. The orbits of this new flow can not be made into geodesics, and we describe how to see this via Sullivan's Theorem.

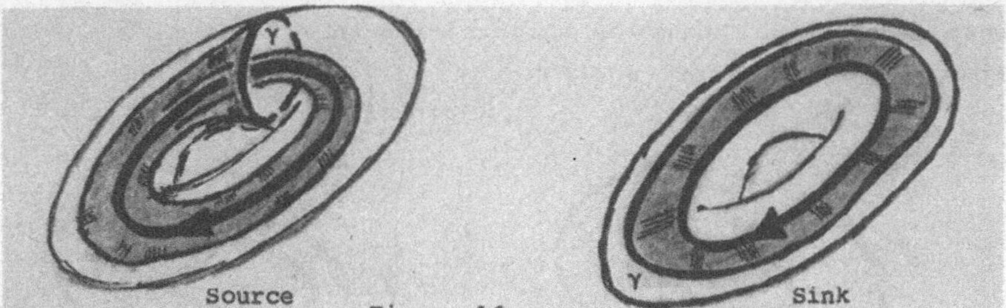

Figure 16

Visualize the 3-sphere split in half by a torus transverse to the perturbed flow: one half is a solid torus neighborhood of the source orbit , the other a solid torus neighborhood of the sink. Pick a simple closed curve γ on this intermediating torus, parallel to the sink as shown. "Turning on the flow" and letting γ move forward in time, a half open annulus tangent to the flow is formed. Its boundary is "sink" - γ . If we view γ on the boundary of the neighborhood of the source orbit, then we can let it flow backwards in time. Doing so, it also traces out an annulus, which appears in the figure as a "French horn" repeatedly swallowing itself. This annulus is also tangent to the flow, and has boundary $\gamma - \gamma_t$, where γ_t denotes the position of γ at time t. As $t \to -\infty$, the curve γ_t is seen encircling the source orbit and sliding around it, becoming ever smaller. Clearly, regarded as a 1-current, $\gamma_t \to 0$ as $t \to -\infty$.

Hence the annulus formed by flowing γ backwards to time t and forward to ∞, and adjoining the sink orbit, is a tangent 2-chain whose boundary is "sink" - γ_t . Thus the foliation cycle "sink" can be arbitrarily well approximated by the boundary of a 2-chain tangent to the flow. By Sullivan's Theorem, it is impossible to select a metric on S^3 making the orbits of this flow into geodesics.

REMARK. Dan Asimov has observed that there is a Reeb component embedded in the perturbed flow, bounded by "source" and "sink". This also shows the impossibility of making the orbits into geodesics.

EXAMPLE 3: Fields of closed curves of unbounded lengths.

Dennis Sullivan [S_1] gave a filling of $S^3 \times S^1 \times S^1$ by closed curves of unbounded lengths. By Wadsley's Theorem [Wa], they can not be made into geodesics. It is therefore instructive to look for the actual tangent 2-chains which, by Sullivan's Theorem, must be present to prevent geodesibility.

Indeed, let $T_1, T_2, T_3, \ldots$ be larger and larger initial segments of the track of a "moving leaf" [E-M-S] whose length becomes unbounded. Then

$$\frac{T_1}{1}, \quad \frac{T_2}{2}, \quad \frac{T_3}{3}, \quad \ldots$$

is a sequence of tangent 2-chains whose initial boundary is (1/n)(initial curve) and whose terminal boundary approaches a fixed leaf (closed curve). Hence this limiting closed curve, regarded as a foliation cycle, can be arbitrarily well approximated by the boundary of a 2-chain tangent to the flow.

12. Proof of Sullivan's characterization

Let V be a nonsingular vector field on the smooth manifold M. Suppose there is a Riemannian metric on M making the orbits of V geodesics. Suppose at the same time that there is a sequence of tangent 2-chains $T_1, T_2, T_3, \ldots$ whose boundaries approach some foliation cycle z.

Define a 1-form ω on M, as usual, by setting $\omega(U) = U \cdot \frac{V}{|V|}$. By the Preliminary Theorem of section 10, $V \lrcorner\, d\omega = 0$. Since the 1-form ω is transversal to V, we have $z(\omega) > 0$, according to section 6. But then we are faced with the contradiction:

$$0 \;=\; T_n(d\omega) \;=\; \partial T_n(\omega) \;\to\; z(\omega) \;>\; 0 \;,$$

since $V \lrcorner\, d\omega = 0$ implies that any tangent 2-chain annihilates $d\omega$. Thus if a Riemannian metric makes the orbits of V geodesics, then <u>no</u> foliation cycle can be arbitrarily well approximated by the boundary of a tangent 2-chain.

Conversely, suppose now that no foliation cycle can be arbitrarily well approximated by the boundary of a 2-chain tangent to the flow. If we let $\overline{\{\partial T\}}$ denote the closed subspace of $\mathcal{D}_1'$ generated by the boundaries of all tangent 2-chains, then the hypothesis reads $\overline{\{\partial T\}} \cap C = \{0\}$, and is illustrated below.

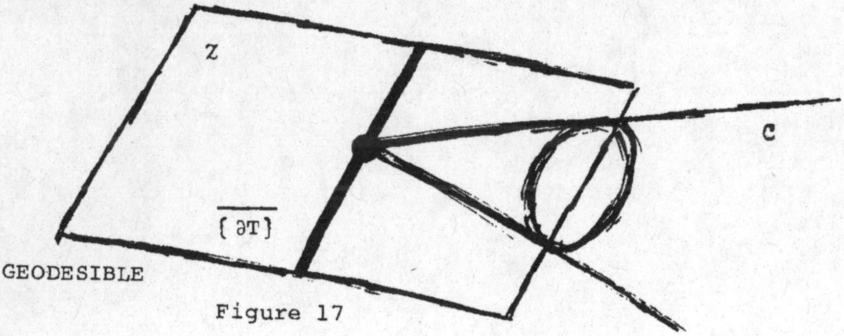

Figure 17

By the Hahn-Banach Theorem, there is a continuous linear unctional $L: \mathcal{D}_1' \to R$ such that

$$L(\overline{\{\partial T\}}) = 0 \qquad \text{and} \qquad L(c) > 0 \quad \text{for all} \quad c \neq 0 \text{ in } \mathbf{C}\,.$$

By Schwartz's Theorem (section 5), this functional corresponds o a smooth 1-form ω such that

$$\partial T(\omega) = 0 \qquad \text{for all tangent 2-chains } T, \text{ and}$$
$$c(\omega) > 0 \qquad \text{for all } c \neq 0 \text{ in } \mathbf{C}\,.$$

The first condition above is equivalent to $T(d\omega) = 0$ for all angent 2-chains T, which is equivalent to $V \lrcorner\, d\omega = 0$. The second ondition is equivalent to $\omega(V) > 0$ at each point of M.

Invoking the Corollary of section 10, we get a Riemannian metric aking the orbits of V geodesics. This completes the argument.

EMARK. By contrast, the non-geodesible case is summarized by ullivan in the diagram:

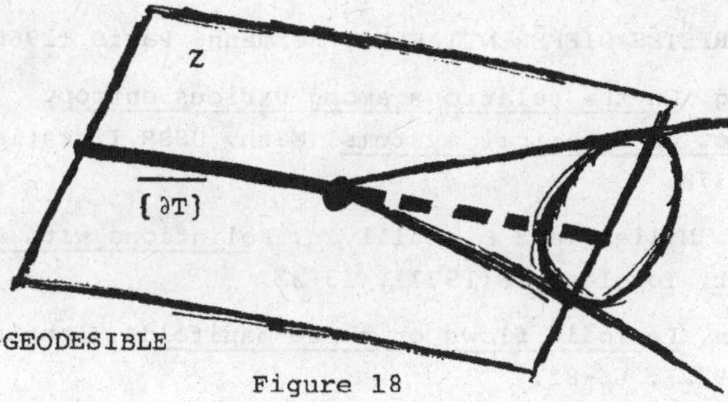

NON-GEODESIBLE

Figure 18

REFERENCES

[A] D. Anosov, GEODESIC FLOWS ON CLOSED RIEMANNIAN MANIFOLDS WITH NEGATIVE CURVATURE, Proc. Steklov Inst. Math. 90(1967).

[A-G] D. Asimov and H. Gluck, Morse-Smale fields of geodesics, this volume.

[D_1] A. Denjoy, Sur les courbes définies par les équations différentielles á la surface du tore, J. Math. Pures Appl. 11(1932), 333-375.

[D_2] ________, Theorie des fonctions sur les characteristiques a la surface du tore, Comptes Rendus Acad. Sci.194(1932), 830-833.

[D_3] ________, Theorie des fonctions sur les characteristiques du tore, Comptes Rendus Acad. Sci. 194(1932), 2014-2016.

[D_4] ________, Les trajectoires á la surface du tore, Comptes Rendus Acad. Sci. 223(1946), 5-8.

[DeR] G. DeRham, VARIÉTÉS DIFFÉRENTIABLES, Hermann, Paris (1960).

[D] E. I. Dinaburg, On the relations among various entropy characteristics of dynamical systems, Math. USSR Izvestia 5(1971), 337-378.

[E-M-S] R. Edwards, K. Millett and D. Sullivan, Foliations with all leaves compact, Topology 16(1977), 13-32.

[E] D.B.A. Epstein, Periodic flows on three-manifolds, Annals of Math. 95(1972), 66-82.

[E-V] D.B.A. Epstein and E. Vogt, A counterexample to the periodic orbit conjecture in codimension 3, Annals of Math. 108(1978), 539-552.

[G] H. Gluck, Can space be filled by geodesics, and if so, how?, to appear.

[Ka] A. Kafker, Geodesic fields with singularities, thesis, U. of Pennsylvania (1979).

[K] H. Kneser, Reguläre Kurvenscharen auf den Ringflächen, Math. Annalen 91(1924), 135-154.

[Sc] L. Schwartz, THÉORIE DES DISTRIBUTIONS, Hermann, Paris (1966).

[Sm] S. Schwartzman, Asymptotic cycles, Annals of Math. 66(1957), 270-284.

[S1] D. Sullivan, A counterexample to the periodic orbit conjecture, Publ. IHES 46(1976), 5-14.

[S2] ________, A foliation of geodesics is characterized by having no tangent homologies, J. Pure and Appl. Algebra 13(1978), 101-104.

[S3] ________, Cycles for the dynamical study of foliated manifolds and complex manifolds, Invent. Math. 36(1976), 225-255.

[T] D. Tischler, On fibering certain manifolds over S^1, Topology 9(1970), 153-154.

[Wa] A. W. Wadsley. Geodesic foliations by circles, J. Diff. Geom. 10(1975), 541-549.

[W] A. Weinstein, On the hypotheses of Rabinowitz' periodic orbit theorems, to appear in J. Diff. Eqs.

UNIVERSITY OF PENNSYLVANIA
PHILADELPHIA, PA.

THE GROWTH OF TOPOLOGICAL ENTROPY FOR ONE DIMENSIONAL MAPS

John Guckenheimer*

This paper can be considered a continuation of [1] and [2] in which dynamical properties of maps of the interval are studied. Here we focus upon the topological entropy or (equivalently) the growth rate as a function on the space of maps. We study the regularity of this function for smooth one parameter families, proving that it is Hölder continuous for some Hölder exponent larger than 0. Examples show that there is no a priori positive lower bound.

We recall from [2] some of the motivation for studying the growth rate function. Let $f_\mu : I \to I$ be a one parameter family of smooth maps with parameter $\mu \in J$. If the growth rate of f_μ changes with μ, then there will be parameter values for which the corresponding map behaves as if it has a "strange attractor." By this we mean that there is a set of positive Lebesque measure which originates trajectories that are not asymptotically periodic. Some of these parameter values can be detected by the fact that the growth rate function is not locally constant at this parameter value. An outstanding question is whether the set of parameter values yielding maps with "strange attractors" has positive measure. For a typical one parameter family, the growth rate defines a Cantor-like function which is constant on many intervals. If this function is sufficiently nice, then the set of parameter values where it is not locally constant will have positive measure. The "strange attractor" behavior corresponding to some parameter values is studied in [4]. These considerations provided an incentive for our interest in the question studied in this paper.

Let us recall the setting in which we work. The class $\mathcal{C} \subset C^3(I,I)$ consists of functions $f : I \to I$, $I = [0,1]$ which satisfy

(1) $f(0) = f(1) = 0$,

(2) f has a single critical point c which is non-degenerate,

(3) the Schwarzian derivative $Sf = \frac{f'''}{f'} - \frac{3}{2}\left(\frac{f''}{f'}\right)^2 < 0$ on $I - \{c\}$.

---

*Research partially supported by the National Science Foundation.

nctions belonging to $\mathcal{C}$ have been classified up to topological equivalence [2]. We e interested particularly in those $f \in \mathcal{C}$ which do not have a neighborhood on which pological entropy is constant. Such functions are characterized by the condition at they are topologically equivalent to one of the piecewise linear functions $(x) = \mu/2 - \mu|x - \frac{1}{2}|$, $\mu \in (1,2]$. There is also a more intrinsic characterization f functions at which topological entropy is locally constant. A <u>central point</u> of is a point $p \in (0,1)$ for which there is an $n > 0$ with $f^n(p) = p$, $Df^n(p) > 0$ and monotone on the interval (p,c). The central point p is <u>restrictive</u> if $f^n(c) \in$ $,p']$ where p' is the point in $I - \{p\}$ with $f(p) = f(p')$.

<u>Proposition</u> [2]: If $f \in \mathcal{C}$ has topological entropy in the interval $\frac{1}{2^{k+1}} \log 2, \frac{1}{2^k} \log 2]$, then f has more than k restrictive central points if and nly if there is a neighborhood of f in $\mathcal{C}$ consisting of functions with the same opological entropy. If $f \in \mathcal{C}$ has zero entropy, then f has a neighborhood of unctions with zero entropy if an only if f has a finite number of periodic orbits.

Note that the presence of central points and restrictive central points is reserved by continuous changes of coordinates. Thus topologically equivalent unctions have the same number of restrictive central points. The piecewise linear aps g_μ introduced above have topological entropy $\log \mu$. If μ is in the interval $\frac{1}{2^{k+1}} \log 2, \frac{1}{2^k} \log 2]$, then g_μ has exactly k restrictive central points. In articular, if $\mu \in (\frac{1}{2} \log 2, \log 2]$ there are no restrictive central points. hese considerations about restrictive central points are the basis for the theorem roved in this paper.

Consider $f \in \mathcal{C}$ whose topological entropy is in the interval $(\frac{1}{2} \log 2, \log 2]$. If such f have restrictive central points, then they have neighborhoods in which all functions have the same topological entropy. In a one parameter family f_μ, we shall study μ intervals throughout which there is a central restrictive point. We shall derive a lower bound for the length of such intervals which depends only on the family f_μ and the period of the central restrictive point. By using this estimate, we obtain an additional estimate for the length of the parameter intervals over which the topological entropy can increase by a given amount. This second estimate yields our theorem. We present these estimates in a pair of lemmas.

<u>Lemma 1</u>: Let $F: I \times J \to I$ be a C^1 map giving a one parameter family of maps $f_\mu(x) = F(x,\mu)$. There is a constant λ such that $|\frac{\partial}{\partial\mu}(f_\mu^n(x))| < \lambda^n$ for all (x,μ,n).

<u>Proof</u>: The chain rule implies $\frac{\partial}{\partial\mu}(f_\mu^n(x)) = \frac{\partial F}{\partial x}(f_\mu^{n-1}(x),\mu)\frac{\partial f^{(n-1)}}{\partial\mu}(x) + \frac{\partial F}{\partial\mu}(f_\mu^{n-1}(x),\mu)$. Let $\lambda = 1 + \sup|\frac{\partial F}{\partial x}| + \sup|\frac{\partial F}{\partial\mu}|$. Then inductively we have $|\frac{\partial}{\partial\mu}(f_\mu^n(x)| < \lambda^n$.

<u>Lemma 2</u>: Let $f \in \mathcal{C}$. There is a constant $\beta > 0$ such that the following holds: if p is a restrictive point with $f^n(p) = p$ and if there is a point $q \in (p,p')$ with $f^n(q) = q$ and $Df^n(q) \leq 0$, then $|p-p'| > \beta^n$.

<u>Proof</u>: From the chain rule, it follows that there is a constant β^{-1} with $|D^2f^n(x)| < \beta^{-n}$ for all $x \in I$ and positive integers n. Consider now an affine change of coordinates τ so that $\tau(p) = 0$ and $\tau(p') = 1$. Then $g = \tau f^n \tau^{-1}$ has a fixed point a 0, $g(1) = 0$, and a fixed point at $\tau(q)$ with $Dg(\tau q) \leq 0$. Now $D^2g(x) = (p'-p)\cdot D^2f^n(\tau^{-1}x)$ because τ is affine with slope $1/(p'-p)$. It follows that $|D^2g(x)| < |p'-p|\beta^{-n}$ for all $x \in [0,1]$. Since p is central, f^n is monotone on (p,c). Since $Df^n(q) \leq 0$, it follows that $|f^n(c)-f^n(p)| \geq |q-p| \geq |c-p|$. For g this means that $|g(\tau(c))| \geq |\tau(c)|$. The mean value theorem implies that there is a $y \in [0,1]$ with $|Dg(y)| \geq 1$. Since $Dg(\tau(c)) = 0$, a second application of the mean value theorem gives an $x \in [0,1]$ with $|D^2g(x)| > 1$. Combined with the inequality above, we find $|p'-p| > \beta^n$ proving the lemma.

In the proof of the theorem, there is a third estimate which is needed to locate values of the topological entropy which correspond to maps with the critical point periodic of period n. Consider an interval $[\rho,\log 2]$, $\rho > 0$. Let $k_{n_1},\ldots,k_{n_2}$ be the values of the topological entropy assumed by maps with its critical point periodic with period n. Denote by M(n) the mesh size of the artition of $[\rho, \log 2]$ by $\{h_{k_i} | k \leq n\}$.

Lemma 3: With the notation of the previous paragraph, there is a constant σ ich that $M(n) < \sigma e^{-n}$.

The proof of this lemma involves the study of the piecewise linear maps g_μ ntroduced earlier and an analysis of the kneading invariants [3] of these maps. e recall the definition and a few basic facts about the kneading invariant. Let be a smooth map of I with the single critical point c. Then the kneading

ıvariant of f is the power series $D(t) = \sum_{i=0}^{\infty} \delta_i t^i$ where $\delta_i = \text{sgn } D(f^i)(f(c))$,

ith the interpretation that sgn0 = 0 if c is periodic. If s is the smallest zero f $D(t)$, then the topological entropy of f is $-\log s$.

We want to examine the kneading invariant of the piecewise linear maps g_μ. or this purpose it is more convenient to work on the interval $[-1,1]$ rather than $0,1]$. With this change of coordinates, $g_\mu(x) = (\mu-1) - \mu|x|$. For $\mu \in (1,2]$, he topological entropy of g_μ is $\log \mu$, so that μ^{-1} is a zero of the kneading

nvariant of g_μ. Denote by $Q_n(\mu)$ the polynomial which is $\frac{g_\mu^n(0)}{(\mu-1)}$ if 0 is not eriodic of period $\leq n$. There are different subintervals of $(1,2]$ with different olynomials Q_n. If $D_n(t)$ denotes the first $(n+1)$ terms of $D(t)$, then we assert hat $D_n(\mu^{-1}) = \delta_n \mu^{-n} Q_n(\mu)$. This can be easily verified inductively. We have $_0 = D_0 = 1$, $D_{n+1}(\mu^{-1}) = D_n + \delta_{n+1}\mu^{-(n+1)}$, and $Q_{n+1}^{(\mu)} = 1 - \mu(\text{sgn}Q_n(\mu))Q_n(\mu)$. Note hat $\delta_{n+1} = -\text{sgn } g_\mu{}^n(0) \cdot \delta_n$ by the chain rule and that $\text{sgn } Q_n(\mu) = \text{sgn } g_\mu^n(0)$. Thus $^{-(n+1)}Q_{n-1}(\mu) = \mu^{-(n+1)} - (\text{sgn } Q_n(\mu))^{-n}(Q_n(\mu))$ and $\delta_{n+1}\mu^{-n+1}Q_{n+1}(\mu) = \delta_{n+1}\mu^{-(n+1)}$ $\delta_n \mu^{-n} Q_n(\mu)$ since $\delta_n = -\text{sgn}Q_n(\mu) \cdot \delta_{n+1}$. Inductively, the right side is $D_{n+1}(\mu^{-1})$.

This formula can be interpreted by saying that $\text{sgn } D_n(\mu^{-1}) = \delta_n \cdot \text{sgn } Q_n(\mu) =$ δ_{n-1}. Thus, inductively again one finds that $\delta_n = -\text{sgn } D_{n-1}(\mu^{-1})$. This makes it lear that $D(\mu^{-1}) = 0$. We find it easier to work with $(1-t)D(t) = L(t)$ than $D(t)$ tself. This can be written $2(1-t) - (1-t^{\ell_1}) + t^{\ell_1}(1-t^{\ell_2}) - g^{\ell_1+\ell_2}(1-t^{\ell_3}) + \ldots$. ere the exponents ℓ_i are characterized by $g^{\ell_1+\ldots+\ell_k}(0) > 0$ and $g_\mu^n(0) < 0$ for $_1+\ldots+\ell_{k-1} < n < \ell_1+\ldots\ell_k$. They mark the number of iterates between successive points in the orbit of 0 landing to the right of 0.

Restrict now attention to $\mu \in (\sqrt{2},2]$. For such μ we assert that $L_\mu(t)$ is strictly decreasing on the interval $[0,\mu^{-1}]$ and that there is an $\alpha < 0$ with $L'_\mu(t) < \alpha$ for all $\mu \in (\sqrt{2},2]$ and $t \in [0,\mu^{-1}]$. Differentiating the formula for $L(t)$ gives $-2+2^{(\ell_1-1)} -2(\ell_1+\ell_2)t^{(\ell_1+\ 2-1)} + 2(\ell_1+\ell_2+\ell_3)t^{(\ell_1+\ell_2+\ell_3-1)} - \ldots$. The function $\ell t^{(\ell-1)}$ is a decreasing function of ℓ for $t \in (0,1/\sqrt{2}]$ provided that $(1-\ell \log\sqrt{2}) < 0$ or $\ell > \frac{2}{\log 2}$. Since $\ell_1 \geq 3 > \frac{2}{\log 2}$, the pair of terms $-2(\ell_1+\ldots+\ell_{2k})t^{\ell_1+\ldots+\ell_{2k}-1} + 2(\ell_1+\ldots+\ell_{2k+1})t^{\ell_1+\ldots+\ell_{2k+1}-1}$ is always negative. Thus, L'_μ truncated after a term with a positive coefficient, is an upper bound for $L'_\mu(t)$.

We now must do a few calculations. If $\ell_1 > 3$, then $\mu > \frac{1+\sqrt{5}}{2}$ and $4\cdot(\frac{\sqrt{5}-1}{2})^3 = 4\cdot(\sqrt{5}-2) < 1$. Therefore $L'_\mu(t) \leq 2(-1+\ell_1 t^{\ell_1-1}) \leq 2(-9+4\sqrt{5}) < 0$. If $\ell_1 = 3$, then there are four additional cases which need to be checked: (a) $\ell_2 = 4$ and $\ell_3 = 6$, (b) $\ell_2 = 4$, $\ell_3 = 5$, $\ell_4 = 6$, and $\ell_5 = 8$, (c) $\ell_2 = 4$, $\ell_3 = 5$, $\ell_4 = 6$, $\ell_5 = 7$, $\ell_6 = 8$ and $\ell_7 = 10$, and (d) $\ell_i = i+2$ for $i = 2,\ldots,9$. We note that the largest block of indices starting with 3 for which $Q_i(\mu) > 0$ must have even length when $\mu > \sqrt{2}$. For $\mu = \sqrt{2}$, the pattern of signs is +-++++... and the monotonicity of the invariant coordinate [3] implies that for $\mu > \sqrt{2}$, the second - sign occurs after an even block of +'s. In each of the cases (a)-(d), we check that the following polynomials are strictly negative on $[0,1/\sqrt{2}]$:

(a) $2(-1 + 3t^2 - 4t^3 + 6t^5 - 7t^6 + kt^{k-1})$, $k = 8$ or 9

(b) $2(-1 + et^2 + 4t^3 + 5t^4 - 6t^5 + 8t^7)$

(c) $2(-t + 3t^2 - 4t^3 - 5t^4 - 6t^5 + 7t^6 - 8t^7 + 10t^9)$

(d) $2(-1 + 3t^2 - 4t^3 + 5t^4 - 6t^5 + 7t^6 - 8t^7 + 9t^8 - 10t^9 + 11t^{10})$

This concludes the proof that $L'_\mu(t) < 0$ for $t \in [0,\mu^{-1}]$, $\mu \in (\sqrt{2},2]$.

The lemma now follows easily for $\mu \in (\sqrt{2},2]$ from the implicit function theorem. If we truncate $D(t)$ with the term of degree$(n-1)$, the remainder is bounded by $t^n/(1-t)$. If we do the truncation with a term whose zero gives a lower estimate ν for μ, then $(\nu-\mu)$ is bounded by the root of $\alpha x - \frac{(\nu-x)^n}{(1-\nu+x)} = 0$ in $[0,\nu]$. This root tends to zero exponentially, so the mesh size of the partition of $[\frac{1}{2}\log 2, \log 2]$ by $\{h_{k_i} | k \leq n\}$ tends to 0 exponentially with n. This is the estimate called for in the lemma.

Almost all of the periodic orbits of maps with topological entropy in the nterval $\left(\frac{1}{2^{k+1}} \log 2, \frac{1}{2^k} \log 2\right]$ have periods which are divisible by 2^k. In partiular, if the critical point is periodic, its period is a multiple of 2^k. The peration of taking the 2^k iterate for the family g_μ has the property that g_μ estricted to a subinterval agrees with $g_{2^k \mu}$ after a linear change of coordinates. herefore, the results we obtained for maps with topological entropy in the nterval $(\frac{1}{2} \log 2, \log 2]$ are also valid for maps in the interval $\left(\frac{1}{2^{k+1}} \log 2,\right.$ $\left.\frac{}{k} \log 2\right]$. Taking k sufficiently large that $\frac{1}{2^{k+1}} \log 2 < \rho$, the lemma is roved. Note that as $k \to \infty$, the mesh sizes of the partitions of $[0, \log 2]$ by $h_{k_i} | k \le n\}$ decrease only linearly with n since the smallest value of k with a $_i \in [0, \frac{1}{2^\ell} \log 2]$ is larger than 2^ℓ.

Theorem. Let $F: I \times J \to I$ be a C^1 map defining a one parameter family of maps $_\mu \in \mathcal{C}$ with positive topological entropy, and let $h(\mu)$ denote the topological ntropy of f_μ. The function h is Hölder continuous (of class C^α for some $\alpha > 0$).

Proof: As a preliminary reduction, we observe that we may assume that the ritical point c_μ of f_μ is a constant c. To make the reduction, we perform a μ-ependent change of coordinates which sends c_μ to c and fixes the boundary of I.

Next we introduce the constant γ obtained from $M(n)$ in Lemma 3 with the roperty that every interval $(\nu, \nu+\varepsilon)$ of values $h(\mu)$ contains a value μ_0 for which he turning point of the function g_{μ_0} is periodic with a period n at most $-\gamma \log \varepsilon$. he proof of the theorem will be completed by estimating the length of the parameter interval in which $h(\mu)$ changes from ν to $\nu+\varepsilon$. This interval is longer than any arameter subinterval containing μ_0 for which $h(\mu)$ remains constant. One such parameter subinterval has endpoints (μ_1, μ_2) at which f satisfies the equations $f^n_{\mu_1}(c) = c$ and $f^{3n}_{\mu_2}(c) = f^{2n}_{\mu_2}(c) = (f^{2n}_{\mu_2}(c))'$. The map $f^n_{\mu_2}$ restricted to the proper domain is topologically equivalent to $g_2 = 1 - 2|x - \frac{1}{2}|$.

We use Lemmas 1 and 2 to estimate $|\mu_2-\mu_1|$. Lemma 2 implies that $|f^{2n}_{\mu_2}(c) - f^n_{\mu_2}(c)| > \beta^n$. Since f has a nondegenerate critical point at c and $f^{2n}_{\mu_2}(c) = (f^n_{\mu_2}(c))'$, there is a constant δ with $|f^n_{\mu_2}(c) - c| > \delta\beta^n$. Now $c = f^n_{\mu_1}(c)$, so Lemma 1 implies that $|\mu_2-\mu_1| \geq |\sup\frac{\partial f}{\partial\mu}|^{-1} |f^n_{\mu_2}(c) - f^n_{\mu_1}(c)| \geq \delta(\beta\lambda^{-1})^n$. Thus we have proved that the length of the parameter interval over which $h(\mu)$ changes values from ν to $\nu+\varepsilon$ is at least $\delta(\beta\lambda^{-1})^{-\gamma\log\varepsilon} = \delta\varepsilon^{-\gamma\log(\beta\lambda^{-1})}$. Therefore h is Hölder continuous with exponent $\alpha = (-\gamma\log(\beta\lambda^{-1}))^{-1}$. This proves the theorem.

We conclude with an example which limits the size of the Hölder exponent α in the Theorem. Let f_μ be a smooth family of maps in $\mathcal{C}$ for which $h(\nu) = \log 2$ and $\frac{\partial}{\partial\mu}(f_\mu(g)) > 0$ where $\mu = \nu$. This implies that ν is the right end point of the parameter interval since when μ is slightly larger than ν, f_μ could not map I into itself. The prototype for this example is $f_\mu(x) = \mu x(1-x)$ with $\nu = 4$. We shall estimate how the topological entropy of f_μ changes as $\mu\to\nu$. For this purpose, there is a sequence of periodic orbits which give easy estimates of both the corresponding topological entropy and the parameter distances from ν.

Let μ_n be a sequence of parameter values for which the map f_{μ_n} has a periodic orbit of period n containing the critical point with $f^i(c) < c$ for $2 \leq i \leq n-1$. The kneading invariant $D_{\mu_n}(t)$ of f_{μ_n} satisfies $(1-t)D_{\mu_n} = 1 - 2x + 2x^n$. The smallest positive root of D_{μ_n} is $\frac{1}{2} + \varepsilon_n$ where $2^n\varepsilon_n \to 1$. This gives $h(\nu) - h(\mu_n) = \log(1 + 2\varepsilon_n) \approx 2^{-(n-1)}$. Next let us estimate the values of $\nu - \mu_n$. Let ϕ_μ be the inverse of the function f_μ restricted to $[0,c_\mu]$. For any $x \in [0,c]$ we have $\phi^n_\mu(x) \to 0$ at the exponential rate $(Df_\mu(o))^{-1}$. The values of the μ_n are given by solving the equations $f^2_\mu(c_\mu) = \phi^{n-2}_\mu(c_\mu)$ for μ. Because ϕ_μ is a contraction near 0, $\frac{\partial}{\partial\mu}\phi^{n-2}_\mu(c_\mu) \to 0$ as $n\to\infty$ uniformly in μ. The ratio $\frac{\phi^{n-2}_\mu(c_\mu)}{\phi^{n-2}_\mu(c_\mu)} \to D\phi_\mu(0)$. Therefore $|\nu - \mu_n|$ will be approximately proportional to $(D\phi_\nu(0))^n$ and the ratio $\frac{|\nu-\mu_{n+1}|}{|\nu-\mu_n|}$ will tend to $D\phi_\nu(0)$. Thus the value of α for which $|h(\nu)-h(\mu_n)|$ is proportional to $|\nu - \mu_n|$ is $\alpha = \frac{\log 2}{\log Df_\nu(0)}$. If $Df_\nu(0)$ is larger than 2 (as in the quadratic example where $Df_\nu(0) = 4$), then $\alpha < 1$. By choosing families f_ν with $Df_\nu(0)$ large, e see that there is no a priori lower bound for α.

EFERENCES

1] J. Guckenheimer, On the Bifurcation of Maps of the Interval, Inventiones Mathematicae 39 (1977), 165-178.

2] J. Guckenheimer, Sensitive Dependence to Initial Conditions for One Dimensional Maps, IHES preprint, May, 1979.

3] J. Milnor and W. Thurston, On iterated maps of the interval I and II, mimeographed, Princeton, 1977.

4] M. Misiurewicz, Absolutely Continuous Measures for Certain Maps of an Interval, IHES preprint, June, 1979.

niversity of California
nta Cruz, CA 95064

Separatrices, Non-isolated Invariant Sets and the Seifert Conjecture

C.S. Hartzman* and D.R. Naugler

1. On the Seifert conjecture.

In 1950, H. Seifert published a paper [24] defining the rotation number of an isotopic deformation of the plane and used it to prove that a continuous vector field with unique integral curves on the 3-sphere S^3, which differs sufficiently little from a vector field tangent to the Hopf fibration, has a least one closed integral curve. In this paper he states, " it is unknown if every continuous vector field of the three-dimensional sphere S^3 contains a closed integral curve." The positive assertion of this statement has subsequently become known as the Seifert conjecture, a term which was popularized by F.W. Wilson Jr. [29], Hsin Chu [3] and C.C. Pugh [16]. The statement of the Seifert conjecture with C^r vector fields replacing continuous vector fields will be called the C^r Seifert conjecture.

In 1974, a counterexample to the C^1 Seifert conjecture was published by P. Schweitzer [20]. Its distinctive feature was the use of the pathogical flow of A. Denjoy [5] on the 2-torus T^2 whose only minimal set is 1-dimensional but not closed. The basic construction consists of erasing the vector field in a solid toroidal annular neighborhood of a point on a closed orbit and replacing it by a vector field containing one of these Denjoy flows in its interior, thus opening closed orbits. Since no C^r vector field on T^2,

*The author thanks the National Research Council of Canada for its support; Grant A8050.

$r \geq 2$, exhibits the pathogies of Denjoy's example, the C^r Seifert conjecture remains open for $r \geq 2$. It should by noted that the techniques used by Schweitzer permit the construction of vector fields with no closed orbits on any 3-manifold M^3. In his paper, Schweitzer has also shown how to construct codimension two foliations without compact leaves on any manifold that admits codimension two foliations.

Schweitzer's counterexample in conjunction with Seifert's original result, lends weight to the question of which flows on 3-manifolds have closed orbits. The investigations into this question follow several lines. Some are extensions and generalizations of Seifert's techniques, some result from unsuccessful attempts to solve the Seifert conjecture and some deal with special flows. There is also a body of research concerned with the existence of closed leaves of foliations whose origin is again the techniques of Seifert. Finally, as Schweitzer's counterexample indicates, careful attention must be paid to the character of minimal sets of flows on 3-manifolds and there is a body of literature concerned with this question.

Among the investigations closely associated with Seifert's techniques are papers by G. Reeb [17] and R. Langevin and H. Rosenberg [13]. Starting with a vector field with all closed orbits on a manifold M^n one obtains on M^n a fibered structure with base V^{n-1}. The crux of Seifert's proof lay in the fact that if his rotation number for a bounded isotopic deformation was zero then the final map in the deformation had fixed points. This was then applied to the base space S^2 of the Hopf fibration of S^3 and closed orbits for small C^0 perturbations of vector fields tangent to the Hopf fibration were deduced. Seifert indicated that if S^3 were replaced by any 3-dimensional fibered manifold as above whose base space is not a torus similar proofs pertain. Reeb reexaming

Seifert's theorem about the rotation number using a C^1 norm indicates a generalization of Seifert's theorem to certain higher dimensional fibered manifolds. Langevin and Rosenberg using Seifert's rotation number theorem are able to prove several results for fibrations $p: E \to B$ where the fibers F are not necessarily circles and E,B,F are all compact manifolds. First, if $\Pi_1(F) = \mathbb{Z}$ and B is a surface with Euler characteristic $\chi(B) \neq 0$ and $\Pi_1(B)$ acts trivially on $\Pi_1(F)$, then any C^0 perturbation of the fibration has a compact leaf. The same result pertains when $B = S^1$ and monodromy is multiplication by -1 in $\Pi_1(F)$. It should be noted that Schweitzer himself has continued the study of foliations (codimension 1) with and without compact leaves [21, 22, 23].

Among the investigations that have led to positive results for special flows are those of the above mentioned Hsin Chu who proves that any almost periodic flow [9] [15] on S^3 has a closed orbit. He also shows that any effective action (i.e., elements acting like the identity are the identity) of the reals on S^3 has exactly two (!) closed orbits. For Hamiltonian systems there are the interesting results of A. Weinstein [27, 28]. Let (P, Ω, H) be a Hamiltonian system (P is a manifold, Ω is a symplectic structure on P and H is a Hamiltonian vector field) and let $p \in P$ be a non-degenerate local minimum of H with $H(p) = 0$. Then for sufficiently small real c, the constant energy surface $H = c$ has at least $\frac{1}{2}\dim P$ geometrically distinct periodic orbits. If $P = \mathbb{R}^{2n}$ he has also shown that if Σ is a compact, convex, regular energy surface for H, then there is a periodic solution on Σ.

We now come to those investigations which, in light of Schweitzer's counterexample, we consider to touch on the heart of the problem of finding somewhat general circumstances under which the Seifert confecture can be positively asserted. This includes work of F.W. Wilson Jr [29], M. Handel [10],

E. Thomas Jr. [26] and A. J. Schwartz [18, 19]. Wilson's work seems to have influenced, to some extent at least, much of the work related to problems raised by the Seifert conjecture. In particular, his considerations led him to ask the following intriguing question; on which manifolds must every non-singular vector field have a minimal set of codimension 2? On 3-manifolds, this reduces to questions about 1-dimensional minimal sets, the subject of study of Handel and Thomas and as we recall, intimately related to the counterexample of Schweitzer. Thomas has shown that if Γ is an isolated (see definition 3.1) minimal set in the interior of M^3 and if some point of Γ is almost periodic, then Γ is periodic. Handel studies one-dimensional minimal sets of M^3 that are not closed orbits, studying behavior which allow for counterexamples to the Seifert conjecture. The result for isolated minimal sets Γ is that Γ must be a surface minimal (i.e., embeddable in a flow on a surface) and that the stable and unstable sets for Γ form the equivalent of a "product structure" for Γ (i.e. Schweitzer's counterexample is typical). If Γ is not isolated, Handel is able to break open periodic orbits with minimal sets which are not surface minimal sets and may not even be 1-dimensional In the course of his work he has answered Wilson's question negatively on S^3.

Lastly we come to the results of Schwartz. Since S^3 can be written as the union of two solid tori $X = D^2 \times S^1$ with their boundaries appropriately identified, he studied flows on the solid torus. Of course, Schweitzer's counterexample, involving only a local construction, pertains in this situation too. The results of Schwartz however, point out a direction in which one might look for tamer behavior.

Theorem 1.1 [18]. Let $\phi(x,t) : X \times \mathbb{R} \to X$ be a flow on X that is asymptotic to $\partial X = T^2$ (i.e., $\phi(x,t) \to T^2$ as $t \to \infty$ for every x). Then T^2 contains a closed orbit.

Theorem 1.2 [19]. Let α be a flow on X such that T^2 is minimal. Then Int X contains a Poisson stable orbit (i.e., $x(t) \subset \omega(x) \cap \alpha(x)$; $\omega(x)$, $\alpha(x)$ are the positive, negative limit sets respectively)

Theorem 1.3 [19]. Let α be a flow on X such that T^2 contains the minimal center of attraction for each $x \in X$. Then T^2 contains a periodic orbit.

Given these last results, we can now say how one might proceed to characterize a relatively broad family of flows on compact M^3 that have periodic orbits. First of all, as Handel's work indicates minimal sets must be tamed. The orbits in minimal sets are among a large class of exceptional orbits that have become known as separatrices [6] [11]. By putting certain fairly natural conditions on the separatrix set one can guarantee that minimal sets are compact manifolds (for flows on M^3, either tori or closed orbits). The tori that may appear can be either isolated or non-isolated. In the separatrix setting, it is the non-isolated torus that causes the difficulty in concluding the existence of a closed orbit on M^3. To this end, it is necessary to study non-isolated invariant sets. Separatrices are examined in the next section and new results are then applied to M^3 to conclude either the existence of a periodic orbit or a flow on a solid torus with the Poisson-stable orbit of Schwartz severely restricted in character. Although we have not been able to conclude the existence of a periodic orbit in all cases, the ones where the Poisson stable orbit is not closed seem so outlandish, we feel it is not likely that they hold.

2. Separatrices

Separatrices are exceptional orbits of a flow which in some sense separate regions of qualitatively different behavior or present obstructions to extending regions of parallelism. They have been recognized since the work of Poincaré but have only recently been defined rigorously. The first rigorous definition was given by L. Markus [14, 14A] in the setting of completely unstable flows. Definitions applicable to more general situations were given by Bhatia and Franklin [1], Elderkin [6] and Hartzman [11].

Two types of separatrices will be defined; the first was studied in [11], the second picks out certain wandering separatrices that the first did not distinguish.

Let M be an n-dimensional smooth manifold and $\phi(t,x)$ a smooth flow on M. Recall that flows ϕ, ψ on manifolds M_1, M_2 are said to be <u>topologically equivalent</u> if there is a homeomorphism $f : M_1 \to M_2$ carring time-sensed orbits of ϕ to orbits of ψ. A topologically embedded submanifold N of M of codimension 1 (i.e. one dimension less than M) is said to be <u>topologically transverse</u> to a flow ϕ on M iff for each $x \in N$ there exists a $\delta > o$ and a neighbourhood U of x in N such that $\phi|(-\delta,\delta) \times U$ is a homeomorphism onto an open set of M. Recall that a topologically transverse submanifold passing through a given non-singular point of a smooth flow can be constructed.

Terminology used is relatively standard (for example dynamical terminology follows [2]). Flows are taken to be smooth, unless otherwise is stated, although C^1 is sufficient for sections 2 and 3.

Definition 2.1 Let ϕ be a flow on M. A subset P of M is said to be admissible iff there is a connected topologically embedded submanifold N of M topologically transverse to ϕ such that P is the smallest invariant subset of M containing N. P is said to generated by N.

Every non-singular point is contained in some admissible region. Admissible regions are invariant open sets

Certain kinds of manifolds admit particularly nice flows. For example, the straight line flow ψ on $\mathbb{R} \times N$ is defined by $\psi(t,(s,x)) = (t+s,x)$. Since distinct decompositions of a space into a product $\mathbb{R} \times N$ are possible [14], a straight line flow may depend on N. The standard periodic flow ψ on $S^1 \times N$ is the flow induced on $S^1 \times N$ induced by identifying S^1 with $\mathbb{R}/\mathbb{Z}$. This depends on the particular decomposition of a space into a product $S^1 \times N$.

Definition 2.2 An admissible region P will be called a region of

(a) type H if ϕ on P is topologically equivalent to the straight line flow ψ on $\mathbb{R} \times N$ for some connected topological manifold N

(b) type Per if ϕ on P is topologically equivalent to the standard periodic flow ψ on $S^1 \times N$ for some connected topological manifold N.

(c) type T' otherwise.

Any admissible region is of one of the types above and these types are mutually exclusive.

Definition 2.3 A region P will be called separatrix admissible of

(a) type H if P is an admissible region of type H and is not contained in a region of type H.

(b) type Per if P is a countable union of an ascending sequence of admissible regions of type Per and is not contained in a region of type Per.

(c) type T' if P is of type T'.

Lemma 2.4 [11]. Every admissible region is contained in a separatrix admissible region.

Definition 2.5 The primary separating points of ϕ consist of all points satisfying :

(a) $x \in \overline{P} - P$ where P is a separatrix admissible region of type H or Per,

(b) $x \in \overline{P} - P$ for some separatrix admissible region P of type T' and x is not an interior point of a region of type H or Per,

or (c) is a singular point.

The closure of the primary separating points is the set of separating points, denoted S (note that S is invariant). An orbit contained in S is called a separatrix.

Definition 2.6. An admissible region P with $P \cap S = \phi$ is said to be of type T if P is of type T' and intersects no regions of type Per or H. A region of type T is also called a transitive region.

The following result shows that non separating behaviour is nice in the sense discussed above.

Theorem 2.7 [11]. The components of the complement of S in M are regions of type H, Per, or T.

The complexity of the above definition arises from the necessity of handling complicated nonwandering sets. Since the boundary of the nonwandering set Ω is a natural dividing line for types of separating behaviour and types of nonseparating behaviour (theorem 2.7) it would seem likely that it consists of separatrices.

Proposition 2.8 The boundary of the nonwandering set is either empty, or consists entirely of separatrices.

Proof: Let $P \in \partial\Omega (\neq \phi)$. If p is a singular point, p is a separatrix. Otherwise, there exists a sequence of wandering points x_i converging to p. Since each x_i is in a region of type H, by lemma 2.4 it is in a separatrix admissible region P_i which must be of type H. If U is any neighbourhood of p, for n large $P_n \cap U \neq \phi$, and hence $(cl(P_n) - P) \cap U \neq \phi$. Thus every neighbourhood of P contains primary separating points so that p is a separating point.

We know that singular points are separatrices as are periodic orbits not contained in open sets of periodic orbits. More generally, minimal sets are separatrices.

Proposition 2.9 A compact minimal set $\Sigma \subseteq M$ is either a periodic orbit, all of M, or consists of separatrices.

Proof: Since $\Sigma \subseteq \Omega \neq \phi$, either $\Sigma \subseteq \partial\Omega$ or $\Sigma \subseteq \text{int}\,\Omega$. If $\Sigma \subseteq \partial\Omega$, by Proposition 2.8 $\Sigma \subseteq S$. If $\Sigma \subseteq \text{int}\,\Omega$, then by Theorem 2.7 Σ either contains a separatrix, hence consists of separatrices, or is contained in a region P of type T containing no separatrices. In the latter case $\Sigma = \overline{P}$, hence has interior. Thus Σ is all of M.

It is conceivable that a component of $M - S$ which is type H may contain an orbit $x(t)$ through a point x whose ω-limit set $\omega(x)$ is contained in a neighborhood which contains no points of $\omega(y)$ for y arbitrarily close to x. For this reason, the definition of separatrix is extended. This extension was originally motivated by definitions of Elderkin [6] and Bhatia and Franklin [1].

<u>Definition 2.10</u> Let I be an invariant set. A nonwandering orbit $\gamma(x)$ not contained in I is said to be a <u>separatrix relative to I</u> if $\omega(x) \subseteq I$, $J^+(x) \subseteq \Omega$ and $J^+(x) \not\subseteq I$ or if $\alpha(x) \subseteq I$, $J^-(x) \subseteq \Omega$ and $J^-(x) \not\subseteq I$. ($J^+(x) = \{y : \{x_n\}, \{t_n\}, t_n \to \infty$ as $n \to \infty$ $|y - x_n| \to 0$ as $n \to \infty$).

We now organize the separatrix set. In [11] it was shown how the separatrix set can be decomposed into a union of disjoint manifolds which can be given structures as differentiable as the flow. In general, the separatrix set can be rather chaotic and the "separatrix manifolds" poorly behaved with respect to one another. We shall make several assumptions which will simplify the structure . The separatrix set is assumed to be decomposed (as in [11]) into disjoint invariant, smooth submanifolds of M, each of which will be called a separatrix manifold of M.

In what follows, the actual decomposition of the separatrix set into manifolds is not important as long as the hypotheses below are satisfied.

If a flow is restricted to a separatrix manifold, it may itself have separatrices, called <u>higher order separatrices</u>.

Example 2.11. Consider $K^2 \times S^1$ represented as $K^2 \times [0,1]$ appropriately identified, where K^2 is the Klein bottle. On $K^2 \times \{0\} \equiv K^2 \times \{1\}$ put a completely peridoic flow parallel to the infinite generator. This flow has two orbits of period 1, γ_1 and γ_2, and all other orbits have period 2. Fill in the remainder of $K^2 \times S^1$ by orbits spiralling from $K^2 \times \{0\}$ to $K^2 \times \{1\}$ along the surfaces $\gamma \times S^1$ for each orbit γ on $K^2 \times \{0\} \equiv K^2 \times$. The surface $K^2 \times \{0\}$ embedded in $K^2 \times S^1$ is a separatrix manifold and the orbits γ_1 and γ_2 are higher order separatrices.

For the rest of this section we will consider, primarily, flows on an n-dimensional manifold M satisfying the follow hypothses;

H1 : Separatrix manifolds are embedded

H2 : There are no higher order separatrices and no singularities.

H3 : There are no n-dimensional regions of separatrices.

Note that the obvious separatrix manifolds of non-singular Morse-Smale flows satisfy these hypotheses.

Flows satisfying these hypotheses have a tamer minimal set structure.

Proposition 2.12. If ϕ is a flow on M^n satisfying hypotheses H1-3 with separatrices and no periodic orbits, then there is a closed separatrix manifold K^ℓ of dimension $\ell \geq 2$ that is a minimal set for the flow. If $n = 3$, $K^\ell = T^2$.

Proof: Let $x(t)$ be a separatrix. Then, $\omega(x)$ is a compact invariant set consisting entirely of separatrices. Let Σ be a compact minimal set contained in $\omega(x)$. Every orbit in Σ is recurrent. Let $y \in \Sigma$ and $P \in y(t)$. No neighborhood of p relative to the induced topology on $y(t)$ is Euclidean so that $y(t)$ is not embedded. Thus $y(t)$ is not on a one-dimensional separatrix manifold, (hypothesis 1) and so must lie on K^ℓ an ℓ-dimensional separatrix manifold, $\ell \geq 2$.

We show that K^ℓ is closed. To do this it will only be necessary to show that $K^\ell \subset \Sigma$. For then, if K^ℓ is not closed, $\mathrm{Cl}K^\ell - K^\ell$ is a non-empty, closed, proper, invariant subset of Σ which is impossible.

Suppose that $K^\ell \subset \Sigma$. Then $K^\ell - \Sigma \neq \phi$ and $K^\ell \cap \Sigma$ consists entirely of recurrent orbits. Let m be any point in $K^\ell - \Sigma$ and $D^{\ell-1}$ be a small transverse disc at m in K^ℓ. The region $P \subset K^\ell$ generated by $D^{\ell-1}$ is an admissible region and must be either type H or type T' since there are no periodic orbits. If P is of type H, then P is contained in a separatrix admissible region P_m of type H. Furthermore, $(\bar{P}_m - P_m) \cap K^\ell \neq \phi$ so that K^ℓ contains separatrices contradicting hypothesis 2. So P must be of type T' and $(\bar{P} - P)^\ell \cap K$ contains no point in the interior of a region of type H. But $(\bar{P} - P) \cap K^\ell \neq \phi$ again leads to separatrices in K^ℓ contradicting hypothesis 2.

The last part of the proposition follows from the Poincaré-Hopf theorem and a result of Kneser [12] showing that every non-singular flow on a Klein bottle has a periodic orbit.

Lemma 2.13. Assume H1-3. If $\Sigma \subset M^n$ is a compact minimal set not a periodic orbit and $\Sigma \neq M^n$, then Σ is a closed separatrix manifold. If $n = 3$, Σ is necessarily a 2-torus.

Proof. This is an immediate consequence of (2.9) and the proof of (2.12).

3. Separatrix structure near a non-isolated invariant set.

Bhatia and Szegö [2; chapter VI] give a remarkable general theorem concerning a flow near a compact invariant set. In particular, they show that a compact invariant set I is either (positively or negatively) asymptotically stable, or there exist points x and y not in the invariant set such that $\omega(x) \cup \alpha(y) \subseteq I$, or every neighbourhood of the invariant set contains an entire orbit not in I. These possibilities are not mutually exclusive.

In this section we classify the behaviour of a flow near a non-isolated invariant set. Both the manifold and invariant sets are assumed to be compact.

Definition: 3.1 An invariant set I is said to be isolated if there exists a neighbourhood of I in which I is the maximal invariant set. An invariant set which is not isolated is said to be non-isolated.

The following useful definition is due to Elderkin and Wilson [7].

Definition 3.2 An invariant set I is said to be solitary if it has a neighbourhood U such that $\omega(x) \subseteq I$ whenever the semi-trajectory $\gamma_+(x) \subseteq U$, and such that $\alpha(x) \subseteq I$ whenever the semi-trajectory $\gamma_-(x) \subseteq U$. U is called a neighbourhood of solitude.

Non-isolated invariant sets have entire orbits contained in any neighbourhood. (which may or may not be asymptotic to the invariant set) Thus non-isolated invariant sets may be solitary or non-solitary - this is a fundamental distinction. Note that a non-solitary invariant set must be non-isolated since each neighbourhood of I contains a limit set not in I.

Proposition 3.3 Let I be a compact non-solitary invariant set. Then one of the following holds:

(a) every neighbourhood of I contains an open set of periodic orbits,

(b) every neighbourhood of I contains a separatrix not contained in I;

(c) for every neighbourhood U of I, U - I has an infinite number of transitive components.

Proof: Since I is non-solitary, every neighbourhood of I contains a limit set which is not in I. If some neighbourhood U of I does not contain any periodic orbits or separatrices, such a limit set contains an orbit which is in a transitive region which is (wholly) contained in U. Choosing a smaller neighbourhood not containing the entire orbit and repeating the argument yields another transitive region which must be disjoint from the first. If there were only a finite number of such regions we could find a neighbourhood of solitude for I, which is a contradition. If these transitive regions were not components of U - I, the closure of one would yield separatrices in $U - I$.

<u>Corollary 3.4</u> Let I be a non-solitary invariant compact submanifold of M. Then either,

(a) every neighbourhood of I contain open sets of periodic orbits,

or

(b) every neighbourhood U of I contains separatrices not in I.

<u>Proposition 3.5</u> Let I be a solitary non-isolated compact invariant set. Then either,

(a) every neighbourhood U of I contains separatrices or separatrices relative to I, not in I, or

(b) every neighbourhood U of I contains open sets or orbits all of whose positive and negative limit sets are contained in I.

Proof Since I is solitary and non-isolated, every neighbourhood U of I contains a point x such that $x \in U - I$, $\gamma(x) \subseteq U$, $\omega(x) \cup \alpha(x) \subseteq I$. If x is a non-wandering point, it is a separating point since it is not contained in any type Per or transitive region (Theorem 2.7)

Thus, if $U - I$ contains no separatrices, then x is wandering. In addition, if $U - I$ contains no separatrices relative to I, then $J^+(x) \cup J^-(x) \subseteq I$. This last condition implies the existence of a neighbourhood V of x such that for every $v \in V$ $\omega(v) \cup \alpha(v) \subseteq I$ [2].

Definition 3.6 An invariant set I is called infinitesimanally elliptic if every neighbourhood of I contains open sets of orbits whose positive and negative limit sets are contained in I.

In Proposition 3.3 we admitted the possibility of invariant sets whose neighbourhoods were highly disconnected. If connected neighbourhoods U of I are not disconnected by I, i.e. if $U - I$ is connected, then I is called one-sided.

Corollary 3.7 Let I be a solitary, non-isolated, one-sided, compact invariant set. Then either,

(a) every neighbourhood U of I contains a separatrix or a separatrix relative to I, not in I, or

(b) I is infinitisimally elliptic and either

(1) every neighbourhood U of I intersects a separatrix or a separatrix relative to I, not in I, or

(2) there is a neighbourhood U of I such that $\alpha(u) \cup \omega(u) \subseteq I$ for all $u \in U$. i.e. I is a positive and negative attractor.

Proof Let U be a neighbourhood of I and suppose that no point in $U - I$ is a separating point, and that there are no separatrices relative to I. By Proposition 3.5, $U - I$ contains open sets of wandering orbits and hence, since I is one-sided, $U - I$ must consist entirely of wandering points.

Let $W \subseteq U$ be a neighbourhood of solitude, and let $A = \{x \in W : \omega(x) \cup \alpha(x) \subseteq I\}$. Since W contains no separating points and there are no separatrices relative to I, $J^+(x) \cup J^-(x) \subseteq I$ for all $x \in A - I$, hence A is open.

If $A \neq W$, there exists $y \in \partial A \cap (U - I)$. For such y we have, without loss of generality, $\omega(y) \not\subset I$, hence $\omega(y) \cap U = \phi$ but $J^+(y) \cap I \neq \phi$. Since $J^+(y)$ is connected, there are separatrices intersecting U.

4. Flow on a solid torus.

We wish to examine the consequences of the previous two sections for flows on a solid torus. For this we will need a theorem of Conley [4] that follows from work of Easton.

Theorem 4.1 Let ϕ be a smooth flow on $\mathbb{R}^3$ which admits an isolated minimal (smooth) torus T^2. Then there is a closed neighbourhood N of T^2 homeomorphic to $T^2 \times [0,1]$ such that each component of ∂N is an embedded torus topologically transverse to ϕ and T^2 is the only invariant set in N.

As a corollary, we have

Corollary 4.2 Let ϕ be a smooth flow on a solid torus X. If ∂X is an isolated, minimal set, then ∂X is asymptotically stable.

Proposition 4.3 If a smooth flow ϕ on the solid torus X has a minimal boundary, then the interior of X contains either a periodic orbit or a separatrix.

Proof: If $X - \partial X$ contains no periodic orbits or separatrices, then by Theorem 2.7 it is a region of type H or T. By Theorem 1.2, $X - \partial X$ contains a Poisson stable orbit since ∂X does not contain a periodic orbit. Thus $X - \partial X$ is not a region of type H, which must consist of wandering points.

If $X - \partial X$ is a region of type T, ∂X is isolated, hence by Corollary 4.2, ∂X is asymptotically stable. However, this implies that no orbit passing near ∂X in $X - \partial X$ is Poisson stable which is a contradiction.

Lemma 4.4 Let ϕ be a flow on X with no closed orbits satisfying hypotheses (1) - (3) with $\partial X = T^2$ minimal. If $C_{T_\alpha} \subset X$ denotes the solid toral region bounded by a minimal torus $T^2_\infty \subset X$, then X contains a minimal torus T^2_∞ such that C_{T_∞} contains no minimal torus in its interior.

Proof Use Zorn's lemma on the set $\{C_T: T^2$ is a minimal torus$\}$ ordered by inclusion and apply (2.18).

Proposition 4.5 Let ϕ be a flow on X satisfying hypotheses (1) - (3) with $\partial X = T^2$ minimal. Then either IntX contains a periodic orbit or every minimal torus in X contains a minimal torus T^2_∞ that is non-isolated relative to C_{T_∞}.

Proof Suppose that the conclusion does not hold. Since there are then no periodic orbits in X, there is a minimal torus $T^2_\infty \subset X$ such that C_{T_∞} contains no minimal torus in its interior. By assumption we may assume that T^2_∞ is isolated relative to C_{T_∞}. Let N be a neighborhood as in (4.1) of T^2_∞ in C_{T_∞}. Since T^2_∞ is either positively or negatively asymptotically stable relative to N and $\partial N \subset C_{T_\infty}$ is transverse to the flow, if $x \in \partial N$

either $\omega(x)$ or $\alpha(x)$ is contained in Int C_{T_∞}. Hence there is a minimal set $\Sigma \subset \text{Int } C_{T_\infty}$. Since Σ is not a periodic orbit, it is a minimal torus (2.13) contradicting the definition of C_{T_∞}.

Corollary 4.6 Let ϕ be a flow on X satisfying hypotheses (1) - (3) with $\partial X = T^2$ minimal. If every minimal torus is isolated, then IntX contains a periodic orbit.

According to the preceding proposition, it only remains to examine flows α on X with $\partial X = T^2$ minimal satisfying hypotheses (1) - (3) and the following conditions:

(a) IntX contains no minimal torus,

(b) $\partial X = T^2$ is non-isolated,

and (c) IntX contains no closed orbit.

By lemma 4.3, we know that IntX contains a separatrix. In fact, an application of (3.7) shows that every neighborhood of T^2 intersects a separatrix or a separatrix relative to T^2. Moreover, $T^2 \subset \omega(x) \cap \alpha(x)$ for all $x \in X$, for otherwise, the usual argument implies that there is a minimal torus contained in IntX. Lastly, one of these orbits is Poisson stable (1.2).

Summarizing, we have the following theorem:

Theorem 4.7 If ϕ is a flow on X with $\partial X = T^2$ minimal satisfying hypotheses (1) - (3), then either

(S_1) there is a periodic orbit in IntX,

or (S_2) every minimal torus in X contains a minimal torus T^2_∞ bounding a toral region C_{T_∞} such that Int C_{T_∞} contains no minimal torus and all of the following hold;

(S_2a) T^2_∞ is non-isolated,

(S_2b) T^2_∞ does not positively and negatively attract every orbit in any neighborhood U, $T^2 \subset U \subset C_{T_\infty}$,

(S_2c) every neighborhood U, $T^2_\infty \subset U \subset C_{T_\infty}$ intersects $S - T^2_\infty$ in a non-empty set; furthermore, if T^2_∞ is non-solitary then U actually contains a separatrix relative to T^2_∞.

(S_2d) $T^2_\infty \subset \omega(x) \cap \alpha(x)$ for every $x \in C_{T_\infty}$.

The last condition severely restricts the behavior of the Poisson stable orbit of Schwartz.

Question: Is it possible for a flow on a solid torus X to have minimal boundary contained in $\omega(x) \cap \alpha(x)$ for all $x \in X$?

<u>Corollary 4.8</u> If ϕ is a flow on a compact manifold M^3 satisfying hypotheses (1) - (3) either

(m_1) there is a periodic orbit,

(m_2) there is a minimal torus T^2 which when bounding a solid torus X satisfies (4.7 (S_2))

or (m_3) there are no separatrices in which case M^3 is the unique minimal set for α.

When $M^3 = S^3$, the last case is the open question - can S^3 be a minimal set for a flow? [25]

Bibliography

1. N. P. Bhatia and L. M. Franklin; Dynamical systems without separatrices, Funcialaj Ekvacioj v.15 (1972) pp. 1-12.

2. N. P. Bhatia and G. Szegö; "Stability theory of dynamical systems," Springer, Berlin, 1970.

3. H. Chu; A remark on the Seifert conjecture, Topology v.9 (1970) pp.275-281.

4. C. C. Conley; Invariant sets which carry a one-form, J. Diff. Eq. v.8 (1970) pp. 587-594.

5. A. Denjoy; Sur les courbes définies par les équations différentielles à la surface du tore, J. Math. Pures et Appl. v.11 (1932) pp.333-375.

6. R. Elderkin; Separatrix structure for elliptic flows, Am. J. Math. v.97#1 (1975) pp.221-247.

7. R. Elderkin and F. W. Wilson Jr.; Solitary invariant sets, in "Dynamical systems, an Int'l. Symp." eds. Cesari et al, v.2, Academic Press 1976.

8. F. B. Fuller; Note on trajectories on a solid torus, Am. J. Math. v.56#2 (1952) pp.438-439.

9. W. A. Gottschalk and G. A. Hedlund; "Topological dynamics," A.M.S. Colloq. Pub. 36 (1965).

10. M. Handel; One dimensional minimal sets and the Seifert conjecture, (to appear).

11. C. S. Hartzman; Separatrices and singular points, Aeq. Math. (to appear).

12. H. Kneser; Reguläre Kurvenscharen auf den Ringflächen, Math. Ann. v.91 (1924) pp.135-154.

13. R. Langevin and H. Rosenberg; Integrable perturbations of fibrations and a theorem of Seifert,"Differential topology, foliations and Gelfand-Fuks cohomology,"ed. P. Schweitzer, Proc. Rio de Janiero 1976 Lecture Notes in Math. #652, Springer.

14. L. Markus; Parallel dynamical systems, Topology v.8 (1969) pp.47-57.

14A. L. Markus; Global structure of differential equations in the plane, TAMS v.26 (1954) pp.127-148.

15. V. V. Nemytskii and V. V. Stepanov; "Qualitative theory of differential equations," Princeton University Press, 1960.

16. C. C. Pugh; the closing lemma, Am. J. Math. v.89 (1967) pp.956-1009.

17. G. Reeb; Sur un théorème de Seifert sur les trajectoires fermées de certains champs de vecteurs, "International Symposium on Nonlinear Differential Equations and Non-Linear Mechanics," eds. J. P. La Salle and S. Lefschetz, Academic Press (1963) pp. 16-21.

18. A. J. Schwartz; Flows on the solid torus asymptotic to the boundary, J. Diff. Eq. v.4 (1968) pp.314-326.

19. A. J. Schwartz; Poisson stable orbits in the interior of a solid torus, in "Topological dynamics," eds. Auslander and Gottschalk, Benjamin, New York, 1968.

20. P. Schweitzer; Counterexamples to the Seifert conjecture and opening closed leaves of foliations, Ann. of Math. v.100 (1974) pp.386-400.

21. P. Schweitzer; Compact leaves of foliations, Proc. Intl. Cong. of Math. Vancouver 1974, v.1 pp543-546, Can. Math. Cong., Montreal 1975.

22. P. Schweitzer; Codimension one foliations without closed leaves,

23. P. Schweitzer; Compact leaves of codimension one foliations, "Applications of topology and dynamical systems," Univ. of Warwick 1973/74 Lecture Notes in Mathematics #468 pp.273-276, Springer, 1975.

24. H. Seifert; Closed integral curves in 3-space and isotopic two-dimensional deformations, PAMS v.1 (1950) pp.287-302.

25. S. Smale; Problems of present day mathematics - dynamical systems, ed. F. E. Browder, in "Mathematical developments arising from Hilbert's problems," Proc. of Symposia in Pure Math. v.XXVIII pt. 1 p.61, A. M. S., Providence, 1976.

26. E. S. Thomas, Jr.; One-dimensional minimal sets, Topology v.12(1973) pp. 233-242.

27. A. Weinstein; Normal modes for non-linear Hamiltonian systems, Inv. Math. v.20 (1973) pp.47-57.

28. A. Weinstein; Periodic orbits for convex hamiltonian systems, Ann. of Math. v.108 (1978) pp.507-518.

29. F. W. Wilson, Jr.; On the minimal sets of non-singular vector fields, Ann. of Math. v.84 (1966) pp.529-536.

Dalhousie University, Halifax, N.S. B3H 4H8 Canada
Mt. St. Vincent University, Halifax, N.S. B3M 2J6

CONSTRUCTION OF INVARIANT MEASURES ABSOLUTELY CONTINUOUS WITH RESPECT TO dx FOR SOME MAPS OF THE INTERVAL

M. V. Jakobson

We study the family of maps $f_\lambda : x \to \lambda x(1-x)$ of the interval $I = [0,1]$ into itself. For the maps corresponding to different values of λ we examine the problem of invariant measures. Especially we are interested in f_λ which admit an invariant measure absolutely continuous with respect to dx.

If $0 < \lambda \leq 4$ the maps under consideration are smooth; if $\lambda > 4$ we consider $x \to \lambda x(1-x) \pmod 1$ which is piecewise smooth.

Let $M_1 = \{\lambda : f_\lambda$ has a periodic sink$\}$. If $\lambda \in M_1$, any f_λ - invariant measure μ_λ is singular with respect to dx. M_1 is an open subset of $\mathbb{R}^+$ and one can suggest that it is a dense one.

Let $M_2 = \{\lambda : f_\lambda$ admit an invariant measure $\mu_\lambda < dx\}$. Then $M_2 \subset \mathbb{R}^+ - M_1$. The first example of $\lambda \in M_2$, $\lambda = 4$, has been considered by J. Neumann and S. Ulam in [1]. L. Bunimovitch in [2] has found a countable set of $\lambda \in M_2$ for the family $x \to \lambda \sin \pi x \pmod 1$ analogous to f_λ. D. Ruelle in [3] has constructed an invariant measure $\mu_\lambda < dx$ for such a $\lambda = 3{,}68\ldots$, that the critical point 1/2 is a preimage of the unstable fixed point. R. Bowen in [4] has found a series of $\lambda \in M_2$ such that 1/2 is a preimage of periodic source. In [5] some $\lambda \in M_2$ have been found such that 1/2 is in the preimage of an invariant unstable Cantor set.

We shall use mes to denote Lebesgue measure in parameter space $\mathbb{R}^+$. The following results assert the hypothesis stated by Ja. G. Sinai and D. Ruelle.

Theorem 1. mes $M_2 > 0$.

Remark. The value $\lambda = 4$ is a Lebesgue point of M_2, i.e. $\forall \varepsilon > 0 \quad \exists \delta > 0$ such that $(1/\delta)$ mes $\{\lambda \in M_2 : 0 < 4 - \lambda < \delta\} > 1 - \varepsilon$. The other Lebesgue points λ of M_2 are studied in [3] - [5]. In order to prove theorem 1 we consider for λ sufficiently close to 4 the induced map introduced in [5]. Then the proof is along the lines of the following theorem 2 which includes a big parameter. In rder to distinguish two cases $\lambda \leq 4$ and $\lambda > 4$, we shall use the notation ϕ_λ or the map $x \to \lambda x(1-x) \pmod 1$.

Theorem 2. $\forall \varepsilon > 0 \quad \exists K_o : \forall K > K_o \quad \text{mes} \{\lambda \in [K,K+4]: \phi_\lambda$ admits an nvariant measure μ_λ absolutely continuous with respect to $dx\} > 4 - \varepsilon$.

We state here the main ideas of the construction of μ_λ. The detailed proof ill be published elsewhere.

. Let $M \subset [K,K+4]$ be the set of λ we are seeking. The central part f the proof of theorem 2 is the construction for $\lambda \in M$ of a special partition λ of [0,1], which we shall call the partition of Adler and Walters (A-W parti- ion), see [6], [7].

Let us fix a positive number $s < 1/6$. The elements of X_λ are intervals $_i(\lambda)$, $i \in Z^+$, which satisfy the next conditions:

1) $\text{int}\ \Delta_i(\lambda) \bigcap \text{int}\ \Delta_j(\lambda) = \emptyset$; $\text{mes} \bigcup_i \Delta_i(\lambda) = 1$.

2) $\forall i \ \exists n_i \in Z^+$ such that $\phi_\lambda^{n_i} | \Delta_i(\lambda)$ is a diffeomorphism of $\Delta_i(\lambda)$ on [0,1].

3) $\inf\limits_{\Delta_i \in X_\lambda} \min\limits_{x \in \Delta_i} |D\phi_\lambda^{n_i}(x)| > 2\lambda^{1-s}$

4) $\sup\limits_{\Delta_i \in X_\lambda} \max\limits_{x \in \Delta_i} \left| \dfrac{D^2\phi_\lambda^{n_i}(x)}{D\phi_\lambda^{n_i}(x)} \right| \cdot |\Delta_i(\lambda)| < C$, where $|\Delta| = \text{diam}\ \Delta$.

The set M and the sets X_λ for $\lambda \in M$ are constructed simultaneously with the help of an inductive construction. M is obtained as $M = \bigcap_{n=0}^{\infty} M_n$, where

$M_o = [K,K+4]$, $M_{n+1} \subset M_n$, $\text{mes}\ M_{n+1} > (1-\varepsilon_{n+1})\ \text{mes}\ M_n$, $\sum_{n=1}^{\infty} \varepsilon_n = 0(\frac{1}{\lambda^t})$, $t > 0$.

At the step n we define for any $\lambda \in M_{n-1}$ a set $X_{n\lambda} \subset [0,1]$, which is the union of a countable number of intervals $\Delta_i^{(k)}(\lambda)$, $k = 1,\ldots n$. The intervals $\Delta_i^{(k)}(\lambda)$, constructed at the step k, do not change at the next steps. The sets $X_{n\lambda}$ satisfy the conditions: $X_{n\lambda} \subset X_{n+1\,\lambda}$; mes $X_{n\lambda} > 1 - \delta_n$, where $\delta_n = o(\varepsilon_n)$.

By definition $X_\lambda = \bigsqcup_{n=1}^{\infty} X_{n\lambda}$.

Let us define the map $T: X_\lambda \to [0,1]$ by $T|\Delta_i(\lambda) = \phi_\lambda^{n_i}$.

Theorem 15 of [7] implies the existence of a T-invariant measure $\nu_\lambda < dx$. The measure μ_λ from theorem 2 is constructed via ν_λ.

2. The graph of the map $\phi_\lambda : x \to \lambda\, x(1-x) \pmod 1$ consists of a number of monotone branches which we denote by $f(\lambda, x)$ and the middle parabola denoted by $h(\lambda, x)$. The domains of the mappings $f(\lambda, x)$, $h(\lambda, x)$ depend continuously on λ. When $\lambda = 4k$, $k \in Z^+$, a new middle branch is born, and the branch $h(\lambda,x)$ which existed for $\lambda \leq 4k$ breaks up into two monotone branches. Without loss of generality we can assume that λ varies from $N_o = 4k_o$ to $4(k_o + 1)$.

We shall denote by $\Delta f(\lambda,x)$ the domain of $f(\lambda, x)$, by $x_{min}(\lambda)$ the end point of the interval $f(\lambda,x)$ nearest to 1/2, and by $x_{max}(\lambda)$ the second end point.

Step 1.

For any $\lambda \in M_o = [N_o, N_o + 4]$ we define $X_1(\lambda) = \{\Delta f(\lambda,x) = [x_{min}(\lambda), x_{max}(\lambda)]$ such that $|x_{min}(\lambda) - \frac{1}{2}| \geq \frac{1}{\lambda^s}\}$. One has the decompositon

$$[0,1] = X_1(\lambda) \cup \delta_1(\lambda) \tag{2.1}$$

where $\delta_1(\lambda) = (\frac{1}{2} - z_2(\lambda), \frac{1}{2} + z_1(\lambda))$ is also the union of domains $\Delta f(\lambda,x)$ and $\Delta h(\lambda,x)$. Since $|Df(\lambda,x)| = |2\lambda(x - \frac{1}{2})|$, we see that whenever $\Delta f \subset X_1(\lambda)$, $f(\lambda,x)$ satisfies $|Df| > 2\lambda^{1-s}$, $|\Delta f| < \frac{1}{2\lambda^{1-s}}$.

ence $\frac{1}{\lambda^s} \le z_1(\lambda) < \frac{1}{\lambda^s} + \frac{1}{2\lambda^{1-s}}$.

In order to construct the set M_1 we enlarge δ_1.

Let us fix some positive $\alpha < s$.

Let $\Delta f(\lambda,x) = [x_{min}(\lambda), x_{max}(\lambda)]$ be a domain such that $|x_{min}(\lambda) - \frac{1}{2}| > \frac{1}{\lambda^{s-2}}$.

The end points $x(\lambda)$ satisfy the condition $\left|\frac{dx}{d\lambda}\right| < \frac{1}{8\lambda^{1-(s-\alpha)}}$.

leanwhile the velocity of the top equals $\frac{d}{d\lambda}(h(\lambda,\frac{1}{2})) = \frac{1}{4}$. This implies that for any $\Delta f(\lambda,x)$ satisfying the condition $|x_{min}(\lambda) - \frac{1}{2}| > \frac{1}{\lambda^{s-\alpha}}$,

nes $\{\lambda : h(\lambda,\frac{1}{2}) \in \Delta f(\lambda,x)\} < \frac{2(1+\varepsilon_1)}{N_o^{1-(s-\alpha)}}$, where $\varepsilon_1 = O(\frac{1}{N_o^{1-(s-\alpha)}})$.

Ve define now M_1 as the union of intervals J_i such that:

1) Any $J_i = [a_i, b_i]$ corresponds to some domain $\Delta f_i(\lambda,x) = [x_{i\,min}(\lambda), x_{i\,max}(\lambda)]$ such that when λ varies in $[a_i,b_i]$, $h(\lambda,\frac{1}{2})$ varies in $[x_{i\,min}(\lambda), x_{i\,max}(\lambda)]$

2) For all $\lambda \in J_i$ $|x_{i\,min}(\lambda) - \frac{1}{2}| > \frac{1}{\lambda^{s-\alpha}}$.

It is easy to check that

$$\text{mes } M_1 > 1 - \frac{2(1+\varepsilon'_1)}{N_o(s-\alpha)}$$

3. Step 2.

a) Construction of $X_2(\lambda)$

We shall use f^1 to denote the branches $f(\lambda,x)$ such that $\Delta f \subset [0,1] - \delta_1$, and g to denote the branches $f(\lambda,x)$: $\Delta f \subset \delta_1$ (the central branch is still denoted by $h(\lambda,x)$).

Let us consider the compositions $f^1 \circ g(\lambda,x)$, $f^1 \circ h(\lambda,x)$. Any domain Δg can be represented in the form

$$\Delta g = \left(\bigsqcup \Delta(f^1 \circ g)\right) \bigsqcup g^{-1}\delta_1 \qquad (3.1)$$

One can choose an interval $\delta_2(\lambda) = \left[\frac{1}{2} - \frac{C_{11}}{\lambda^{2s}}, \frac{1}{2} + \frac{C_{12}}{\lambda^{2s}}\right]$, with $1 < C_{11}, C_{12} < 1 + 0\left(\frac{1}{\lambda^{2-5s}}\right)$, which is the union of domains $\Delta(f^1 \circ g)$ and $g^{-1}\delta_1$.

We shall use g_1 to denote $g|\delta_1 - \delta_2$ and f_1^2 to denote $f^1 \circ g_1$. Using (3.1) one can write $\delta_1 = \left(\bigsqcup \Delta f_1^2\right) \bigsqcup \left(\bigsqcup g_1^{-1}\delta_1\right) \bigsqcup \delta_2$ (3.2)

For any branch g_{1i} we have $g_{1i}^{-1}(\delta_1) = \left(\bigsqcup g_{1i}^{-1}(\Delta f_1^2)\right) \bigsqcup \left(\bigsqcup g_{1i}^{-1} \circ g_1^{-1}\delta_1\right) \bigsqcup$ $\bigsqcup g_{1i}^{-1}\delta_2$, where the sum is taken over all f_1^2 and g_1. Let us denote f_2^2 the branches $f_1^2 \circ g_{1i}$ with the domains $\Delta(f_1^2 \circ g_{1i}) = g_{1i}^{-1}(\Delta f_1^2)$. Then we can rewrite (3.1) as $\delta_1 = \left(\bigsqcup \Delta f_1^2\right) \bigsqcup \left(\bigsqcup \Delta f_2^2\right) \bigsqcup \left(\bigsqcup g_1^{-2}\delta_1\right) \bigsqcup \left(\bigcup g_1^{-1}\delta_2\right) \cup \delta_2$ (3.3)

where g_1^{-2} denote any composition of the form $g_{1i}^{-1} \circ g_{1j}^{-1}$. Proceeding in the same way we obtain the representation

$$\delta_1 = \left(\bigsqcup \Delta f_1^2\right) \cup \dots \cup \left(\bigsqcup \Delta f_k^2\right) \cup \left(\bigsqcup g_1^{-k}\delta_1\right) \cup \left(\bigsqcup g_1^{-1}\delta_2\right) \cup \dots$$

$$\dots \bigsqcup \left(\bigcup g_1^{-(k-1)}\delta_2\right) \cup \delta_2 \qquad (3.4)$$

ith

$$\ell^2 = f_1^2 \circ g_{1i_1} \circ g_{1i_{\ell-1}}, \quad g_1^{-r} = g_{1k_1}^{-1} \circ g_{1k_r}^{-1}.$$

he branches g_1 satisfy the conditions

$$|Dg_1| > \text{const}_1 \cdot \lambda^{1-2s}, \quad |D^2 g_1| < \text{const}_2 \tag{3.5}$$

t follows from (3.5) (see for example [8]) that

$\lim_{k\to\infty} \text{mes}\left(\bigcup g_1^{-k} \delta_1\right) = 0.$ Therefore we get

$$\delta_1 = \left(\bigcup_{k=1}^{\infty} \Delta f_k^2\right) \cup \left(\bigcup_{k=1}^{\infty} g_1^{-k} \delta_2\right) \cup \delta_2 \quad (\text{mod } 0) \tag{3.6}$$

here the equality (mod 0) means that we neglect the sets with zero Lebesgue easure. Using the notation f^2 for all f_k^2 we can write

$$[0,1] = \left(\bigcup \Delta f^1\right) \cup \left(\bigcup \Delta f^2\right) \cup \left(\bigcup_{k=1}^{\infty} g_1^{-k} \delta_2\right) \cup \delta_2 \tag{3.7a}$$

or

$$[0,1] = X_2(\lambda) \cup \left(\bigcup_{k=1}^{\infty} g_1^{-k} \delta_2\right) \cup \delta_2 \tag{3.7b}$$

where $X_2(\lambda)$ is the union of elements of A-W partition constructed at the first and at the second steps.

Remark. (3.1) and (3.6) induce an analogous structure inside δ_2:

$$\delta_2 = \left(\bigcup \Delta f^1 \circ g\right) \cup \left(\bigcup \Delta f^2 \circ g\right) \cup \left(\bigcup g_1^{-n} \circ g^{-1} \delta_2\right) (\text{mod } 0) \tag{3.8}$$

Here take one of g in place of h. Let $h(\frac{1}{2}) \in \Delta f_0^1$. Then for any $f^1 \neq f_0^1$ either there are two monotone branches $f^1 \circ h$ or there is none; the same is true for $h^{-1}\delta_2$; $f_o^1 \circ h$ is the single branch of parabolic type.

b) Construction of M_2.

The intervals $\delta_2, g_1^{-k}\delta_2$ will be called "holes of the range 2" (δ_1 is the unique hole of the range 1).

Let $g_1^{-k}\delta_2 = \delta_\lambda = (x_\lambda, y_\lambda)$ be some hole and let $\delta'_\lambda = (x'_\lambda, y'_\lambda)$,

$x'_\lambda - g_1^{-k}(\frac{1}{2}) = (x_\lambda - g_1^{-k}(\frac{1}{2}))\cdot\lambda^{2\alpha}$, $y'_\lambda - g_1^{-k}(\frac{1}{2}) = (y_\lambda - g_1^{-k}(\frac{1}{2}))\cdot\lambda^{2\alpha}$

be the enlarged hole. If the end of δ'_λ is contained in some $\Delta f^i(\lambda)$, $i = 1,2$, we enlarge δ'_λ once more to $\delta''_\lambda = \delta'_\lambda \cup \Delta f^i(\lambda)$. Let $D_2(\lambda) = \bigcup \delta''_\lambda$ be the union of all enlarged holes. The calculation shows that

$$\text{mes}\,(\delta_2 \cup \bigcup_{n=1}^{\infty} g_1^{-n}\delta_2) < \frac{2}{\lambda^{2s}}(1 + \frac{C_3}{\lambda^s}), \quad \text{and mes } D_2(\lambda) <$$

$$< \frac{2}{\lambda^{2(s-\alpha)}}\cdot(1 + \frac{C_4}{\lambda^s}).$$

Let $J_i \subset M_1$ be one of the intervals constructed at the first step. When λ varies in J_i, $h(\lambda,\frac{1}{2})$ varies in $\Delta f_i^1(\lambda)$ and $f_i^1 \circ h(\lambda,\frac{1}{2})$ varies in $[0,1]$.

We define $M_{2i} \subseteq M_2 \cap J_i$ by the condition:

$$f_i^1 \circ h(\lambda,\tfrac{1}{2}) \in [0,1] - D_2(\lambda) \tag{3.9}$$

It follows from (3.9) that M_{2i} is the union of the intervals

$$J_{irm} = \{\lambda:\ f_i^1 \circ h(\lambda,\tfrac{1}{2}) \in \Delta f_m^r(\lambda),\ r = 1,2,\ m \in \mathbb{Z}^+\}.$$

Comparing the velocity of the top, $\left|\frac{d}{d\lambda}(f_i^1 \circ h(\lambda,\frac{1}{2}))\right|$, with the velocities f the end points $x_m^r(\lambda)$ of the domains $\Delta f_m^r(\lambda)$, we get

$$\text{mes}\,[f_i^1 \circ h(M_{2i})] > 1 - \frac{C_5}{N_o^{2(s-\alpha)}} \tag{3.10}$$

t follows from the definition of branches f^1, that

$$\sup_{f^1} \max_{x,y\in\Delta f^1} \left| \frac{Df^1(\lambda,x)}{Df^1(\lambda,x)} \right| \le \exp\left(\sup_{\Delta f^1} \max_{x,y\in\Delta f^1} \left| \frac{D^2 f^1(x)}{Df^1(x)} \right| \, |\Delta f^1| <\right.$$

$$1 + \frac{1}{\lambda^{1-2s}} \tag{3.11}$$

Hence

$$\text{mes}\, M_{2i} > \text{mes}\, J_i\left(1 - \frac{C_5}{N_o^{2(s-\alpha)}}\right) \frac{1}{1 + \left(\frac{1}{N_o^{1-2s}}\right)} \tag{3.12}$$

Since (3.11) is true for all $J_i \subset M_1$ we get

$$\text{mes}\, M_2 > \text{mes}\, M_1 \left(1 - \frac{C_6}{N_o^{2(s-\alpha)}}\right) > \left(1 - \frac{C_1}{N_o^{(s-\alpha)}}\right)\left(1 - \frac{C_6}{N_o^{2(s-\alpha)}}\right) \tag{3.13}$$

4. Step n+1.

a) We assume that for $\lambda \in M_n$ the interval $[0,1]$ can be represented after the step n in the following form:

$$[0,1] = \left[\bigcup_{i=1}^{n} \left(\cup\, \Delta f^i\right)\right] \cup \left[\bigcup_{k=1}^{\infty} \left(\cup\, \delta_n^{-k}\right)\right] \cup \delta_n \pmod 0 \tag{4.1}$$

where $\Delta f^i \subset X_n$ are the elements of A-W partition,

$$\delta_n = \left(\frac{1}{2} - \frac{C_{n1}}{\lambda^{sn}},\ \frac{1}{2} + \frac{C_{n2}}{\lambda^{sn}}\right),\quad 1 \le C_{n1}, C_{n2} \le 1 + O\left(\frac{1}{\lambda^{tn}}\right),\quad t > 0,$$

and δ_n^{-k} are various preimages of δ_n, such that $\phi_\lambda^k : \delta_n^{-k} \to \delta_n$ are diffeomorphisms (δ_n, δ_n^{-k} are the holes of range n). The interval δ_n can be represented in the form

$$\delta_n = (\bigcup \Delta F^{(n-1)} \circ g) \cup \left[\bigcup_{m=m_o}^{\infty} (\bigcup \delta_n^{-m})\right] \quad (\mathrm{mod}\ 0) \tag{4.2}$$

where we use $F^{(n-1)}$ to denote various compositions of the form $F^{(n-1)} = f^{i_1} \circ \ldots \circ f^{i_r}$, $1 \leq r \leq n-1$, $i_1 \in [1,n]$. All $F^{(n-1)} \circ g$: $\Delta(F^{(n-1)} \circ g) \to [0,1]$ are diffeomorphisms, with exception of one parabolic branch

$$F^{(n-1)} \circ h = f^{i_1} \circ f^{i_2} \circ \ldots \circ f^{i_{n-1}} \circ h.$$

With these notations we proceed at the step n+1 in the following way.

Considering the compositions $f^{(i)} \circ F^{(n-1)} \circ g$, $i \in [1,n]$ we get for any branch $F^{(n-1)} \circ g$ the representation

$$\Delta(F^{(n-1)} \circ g) = \left[\bigcup \Delta(f^i \circ F^{(n-1)} \circ g)\right] \cup \left[\bigcup_{k=0}^{\infty} (\bigcup (F^{(n-1)} \circ g)^{-1} \delta_n^{-k})\right] (\mathrm{mod}\ 0) \tag{4.3}$$

One can find an interval $\delta_{n+1} = [\frac{1}{2} - \frac{C_{n+1,1}}{\lambda^{s(n+1)}}, \frac{1}{2} + \frac{C_{n+1,2}}{\lambda^{s(n+1)}}]$,

$1 < C_{n+1,i} < 1 + O\left(\frac{1}{\lambda^{t(n+1)}}\right)$, composed of the elements of partition (4.3).

Let us denote by g_n the restriction of g on $\delta_n - \delta_{n+1}$ and by f_1^{n+1} the compositions $f^i \circ F^{(n-1)} \circ g_n$. The domains Δf_1^{n+1} will be the first elements of A-W partition constructed at the step n+1. One has

$$\delta_n = (\bigcup \Delta f_1^{n+1}) \cup (\bigcup_{m=m_o}^{\infty} \delta_n^{-m}) \cup \delta_{n+1} \tag{4.4}$$

e shall use the notation G_n for the diffeomorphisms

$^n_\lambda$: $\delta_n^{-m} \to \delta_n$, and $\phi_\lambda^\ell = \phi_\lambda^k \circ F^{(n-1)} \circ g$: $\delta_n^{-\ell} = (F^{(n-1)} \circ g)^{-1} \delta_n^{-k} \to \delta_n$.

sing (4.2) - (4.4) and proceeding as at the step 2 we obtain the branches

$^{n+1}_r = f_1^{n+1} \circ G_{ni_1} \circ \ldots \circ G_{ni_{r-1}}$ and the preimages $(G_{ni_1} \circ \ldots G_{ni_{r-1}})^{-1} \delta_{n+1}^{-m} \subset$

$\subset (G_{ni_1} \circ \ldots \circ G_{ni_{r-1}})^{-1} \delta_n^{-m}$ and get the representation:

$$0,1] = \left[\bigcup_{i=1}^{n+1} (\cup \Delta f^i) \right] \cup \left[\bigcup_{m=1}^{\infty} (\cup \delta_{n+1}^{-m}) \right] \cup \delta_{n+1} \pmod 0 \quad (4.5)$$

ere δ_{n+1}^{-m} are the holes of the range n+1, and all f_r^{n+1} ,r = 1,2,... are

enoted by f^{n+1} . (4.5) induces the representation of δ_{n+1} analogous to

4.2).

b) We assume that M_n is the union of a countable number of intervals

$_{nk}$, such that when λ varies in J_{nk}, $F^{(n-1)} \circ h(\lambda,\frac{1}{2})$ varies in the corre-

ponding interval Δf^i_{nk} (λ,x), i = i(n,k) ε [1,n] and $f^i(\lambda,x) \circ F^{(n-1)} \circ h(\lambda,\frac{1}{2})$

varies in [0,1].

In order to construct M_{n+1} we enlarge all the holes $\delta(\lambda) = \delta_{n+1}^{-k}(\lambda)$,

$\lambda \in M_n$ to $\delta'(\lambda)$, such that diam $\delta' = \lambda^{\alpha(n+1)}$ diam $\delta(\lambda)$, and then, if

$\delta'(\lambda) \cap \Delta f^i(\lambda) \neq \emptyset$, we enlarge $\delta'(\lambda)$ once more to $\delta''(\lambda) = \delta'(\lambda) \cup \Delta f^i(\lambda)$.

Let $D_{n+1}(\lambda)$ be the union of all $\delta''(\lambda)$.

For any J_{nk} we define $M_{n+1} \cap J_{nk}$ as the union of the intervals

$$J_{n,k,n_1,k_1} = \{\lambda: f^i_{nk} \circ F^{n-1} \circ h(\lambda,\tfrac{1}{2}) \ \varepsilon\ \Delta f^j_{n_1k_1} (\lambda,x) \subset [0,1]-D_{n+1}(\lambda)\}.$$

5. Let $f: \Delta \to I$ be a C^2 diffeomorphism of some closed interval Δ on its image I. Then

$$\max_{x,y\in\Delta} \left| \frac{Df(x)}{Df(y)} \right| \le \exp\left(\max_{x\in\Delta} \left| \frac{D^2f(x)}{Df(x)} \right| \cdot |\Delta| \right).$$

We shall use the notation

$$\mu(f) = \max_{x\in\Delta} \left| \frac{D^2f(x)}{Df(x)} \right| \cdot |\Delta|.$$

The possiblity of the inductive construction stated above is based on the following estimations.

Proposition 1. Let f^n_λ, $G_{n\lambda}$, be the diffeomorphisms defined in section 4, s, α the constants defined in sections 1, 2, $c_1 = 1-2s$, $c_2 = \frac{1}{2}(1 - s + \alpha)$, $\gamma = 1 - 3s + \alpha$. Then:

a_{1n}) $|Df^n_\lambda| > 2^n \lambda^{c_1 n}$;

a_{2n}) $\mu(f^n_\lambda) < u_n \cdot \sum_{k=1}^{n} \frac{1}{2^k \lambda^{\gamma k}}$, $u_n < 2$

b_{1n}) $|DG_{n\lambda}| > 2\lambda^{c_2}$

b_{2n}) $\mu(G_{n\lambda}) < \frac{1}{\lambda^{\alpha n}}$

Proposition 2. Let $U_n(\lambda)$ be the union of all holes at the step n: $U_n(\lambda) = \{\delta^{-k}_{n\lambda}\}$, $k = 0,1\ldots$. Then mes $U_n(\lambda) < \left(\frac{2}{\lambda^s}\right)^n$.

Proposition 3. mes $M_n > \prod_{k=1}^{n} \left(1 - v_k \left(\frac{2}{N_o^{s-\alpha}}\right)^k\right)$, $v_k < 2$.

Theorem 2 is the consequence of propositions 1-3 and the following

Proposition 4. For $\lambda \in M \subset [N_o, N_o + 4]$, let T_λ be the map defined in section 1: $T_\lambda|\Delta_i(\lambda) = \phi_\lambda^{n_i}$, and ν_λ the T_λ - invariant measure absolutely continuous with respect to dx, constructed in [7]. Then $\sum_{\Delta i} n_i \nu_\lambda(\Delta_i) < \infty$

d the measure μ_λ is defined by

$$\mu_\lambda (A) = \sum_{\Delta i} \sum_{0 \leq j < n_i} \nu_\lambda \left(\phi_\lambda^{-j} (A \cap \phi_\lambda^j (\Delta_i)) \right).$$

References

. S. Ulam and J. Neumann, On combination of stochastic and deterministic processes. Preliminary report. Bull. Amer. Math. Soc. 53 (1947) 1120.

. L. A. Bunimovitch, About one transformation of the circle. (in Russian) Mat. Zametki 8:2(1970) 205-216.

. D. Ruelle, Applications conservant une mesure absolument continue par rapport à dx sur [0,1]. Comm. Math. Physics 55 (1977) 47-51.

. R. Bowen, Invariant measures for Markov maps of the interval, Comm. Math. Physics 69 (1979) 1-17.

. M. Jakobson, Topological and metric properties of one-dimensional endomorphisms, Soviet Math. Doklady 19 (1978) 1452-1456.

. R. L. Adler, F-expansions revisited, Recent Advances in Topological Dynamics, Lecture Notes in Math. 318 (Springer, 1973) 1-5

. P. Walters, Invariant measures and equilibrium states for some mappings which expand distances, Trans. Amer. Math. Soc. 236 (1978) 121-153.

. H. Brolin, Invariant sets under iteration of rational functions, Arkiv für Mat. 6 (1965) 103-144.

THE ESTIMATION FROM ABOVE FOR THE TOPOLOGICAL ENTROPY OF A DIFFEOMORPHISM

Svetlana R. Katok

1. The topological entropy of a diffeomorphism of a compact manifold is always finite. It follows from the rough estimation [1] which is similar to the earlier estimation of the metric entropy of a diffeomorphism with respect to a smooth invariant measure [2]. Later an exact formula was proved in this case ([3]; the estimation from above belongs to G. A. Margulis). We prove a refined estimation from above for the topological entropy which is similar to the estimation of Margulis. Expressing the right-hand part of our estimation in terms of differential forms we get a relation between the topological entropy and spectral properties of the operator induced by a diffeomorphism in a space of differential forms.

2. Let M be a compact metric space, $T : M \to M$ a homeomorphism of M onto itself. Background material about the topological entropy can be found in [4]. We shall indicate several simple facts concerning the notion of conditional topological entropy which is introduced by M. Misiurewicz [5].

<u>Definition</u>. Let $\mathcal{A},\mathcal{B}$ be two open cover of the space M. The conditional topological entropy $h(\mathcal{A}/\mathcal{B})$ of the cover $\mathcal{A}$ relative to the cover $\mathcal{B}$ is defined by the formula

$$h(\mathcal{A}/\mathcal{B}) = \max_{B\in\mathcal{B}} \log(N_{\mathcal{A}}(B))$$

where $N_{\mathcal{A}}(B)$ is the minimal number of elements of $\mathcal{A}$ which cover the element $B \in \mathcal{B}$.

ROPOSITION 1. Suppose that $\mathcal{B} < \mathcal{C}$, i.e., each element of $\mathcal{C}$ is ontained in an element of $\mathcal{B}$. Then $h(A/\mathcal{B}) \geq n(A/\mathcal{C})$.

ROPOSITION 2. $h(A \vee \mathcal{B}) \leq h(\mathcal{B}) + h(A/\mathcal{B})$.

PROOF: Let us denote the minimal number of elements of a sub-over of the cover A by N_A. Further, let $\mathcal{B}'$ be a subcover of $\mathcal{B}$, hich contains exactly $N_{\mathcal{B}}$ elements. Then

$$N_{A\vee\mathcal{B}} \leq \sum_{B\in\mathcal{B}'} N_A(B) \leq N_{\mathcal{B}} \max_{B\in\mathcal{B}} N_A(B) \quad \text{and}$$

$$h(A\vee\mathcal{B}) = \log N_{A\vee\mathcal{B}} \leq \log N_{\mathcal{B}} + \log \max_{B\in\mathcal{B}} N_A(B) = h(\mathcal{B}) + h(A/\mathcal{B}).$$

ROPOSITION 3. $h(T,A) \leq \lim_{n\to\infty} h(T^nA/A\vee\ldots\vee T^{n-1}A)$.

PROOF: Let us apply the previous proposition n times to the over $A\vee\ldots\vee T^{n-1}A$. We have

$$h(A\vee TA\vee\ldots\vee T^nA) \leq h(A) + h(TA/A) + \ldots + h(T^nA/A\vee\ldots\vee T^{n-1}A).$$

y the definition of the topological entropy and proposition 1

$$h(T,A) = \overline{\lim_{n\to\infty}} \frac{h(A\vee\ldots\vee T^nA)}{n} \leq \lim_{n\to\infty} \frac{\sum_{i=1}^{n} h(T^iA/A\vee\ldots\vee T^{i-1}A)}{n} =$$

$$= \lim_{n\to\infty} h(T^nA/A\vee\ldots\vee T^{n-1}A).$$

ROPOSITION 4. $h(T,A) \leq h(TA/A)$.

PROOF: $h(T^iA/A\vee\ldots\vee T^{i-1}A) = h(TA/A\vee T^{-1}A\vee\ldots\vee T^{-(i-1)}A) \leq h(TA/A)$.

The last inequality follows from proposition 1). Combining this nequality with porposition 3 we obtain

$$h(T,A) \leq \lim_{n\to\infty} (T^nA/A\vee\ldots\vee T^{n-1}A) \leq h(TA/A).$$

3. Let M be an m-dimensional smooth compact Riemannian manifold and $T: M \to M$ be a C^1 diffeomorphism.

THEOREM. $h(T) \le \log \max_{x \in M} \max_{L \subset T_x M} |J(DT_x|_L)| = \alpha(T)$,

where J means the Jacobian and the inner maximum is taken over the set of all linear subspaces of the tangent space $T_x M$.

PROOF OF THE THEOREM: It is known [4] that $h(T) = \lim_{k\to\infty} (T, A_k)$ if A_k is an exhaustive sequence of covers i.e., the maximal diameter of elements of the covers A_k tends to zero as $k \to \infty$. We shall consider a special exhaustive sequence of covers having bounded multiplicities.

Let us choose a positive number ε_0 and a finite set of points $x_1, \ldots, x_s$ M such that the mappings $\exp_{x_i}$, $i = 1, \ldots, s$ are injective on ε_0-balls and the images of these balls cover M. Denote the maximal multiplicity of this cover by N. Let $\delta > 0$ be so small that the images of $(\varepsilon_0 - \delta)$-balls still cover M. Let us denote by $D_y(v,r)$ the ball in the tangent space $T_y M$ of radius r about $v \in T_y M$. Let us fix a sequence of positive numbers $\delta_k \to 0$, $\delta_k < \frac{\delta}{8}$ and cover each of the balls $D_{x_i}(0, \varepsilon_0 - \delta)$, $i = 1, \ldots, s$ by a system of δ_k-balls with the maximal multiplicity bounded by a number R which does not depend on k and i. The images of these δ_k-balls under the action of the corresponding mappings $\exp_{x_i}$ form the cover A_k of M with the maximal multiplicity bounded by the number $C = NR$.

For every element $A \in A_k$ there exist $i \in \{1, \ldots, s\}$ and a tangent vector $v \in T_{x_i} M$ such that $(\exp_{x_i})^{-1} A = D_{x_i}(v, \delta_k)$. Let $\exp_{x_i} v = x_A$. Suppose that the number $\varepsilon_0 > 0$ is chosen so small that

$$D_{x_A}(0, \frac{\delta_k}{2}) \subset (\exp_{x_A})^{-1} A \subset D_{x_A}(0, 2\delta_k) \tag{1}$$

EMMA. There exists a constant C' such that for every integer n here exists a positive integer $k(n)$ such that for $k > k(n)$

$$h(A_k/T^n A_k) \le \alpha(T^n) + C'$$

PROOF OF THE LEMMA: Let us fix an integer n and choose k to provide the mapping T^n be sufficiently close to a linear mapping n each element A of the cover A_k. To be more precise, let us onsider a set A and the corresponding point x_A. Condition (1) mplies that:

$$(\exp_{T^n x_A})^{-1} B \subset (\exp_{T^n x_A})^{-1} T^n \exp_{x_A}(D_{x_A}(0,2\delta_k)) \quad \text{where} \quad B = T^n A.$$

e can choose the number k so large (and consequently δ_k so small) hat

$$(\exp_{T^n x_A})^{-1} T^n \exp_{x_A}(D_{x_A}(0,2\delta_k)) \subset DT^n D_{x_A}(0,3\delta_k)$$

hence

$$(\exp_{T^n x_A})^{-1} B \subset DT^n D_{x_A}(0,3\delta_k) \qquad (2)$$

esides that, we shall claim the diameter of the set $DT^n D_{x_A}(0,3\delta_k)$ o be less than $\frac{\delta}{2}$. Condition (2) expresses our demand to the apping DT^n to be close to a linear map. Let us estimate now $(A_k/T^n A_k) = \max\limits_{B \in T^n A_k} \log N_{A_k}(B)$. Denote by B' the set $\exp_{T^n x_A} U_{2\delta_k}((\exp_{T^n x})^{-1} B)$ where $U_\alpha(E)$ is an α-neighborhood of the et E. By (2) we have

$$(\exp_{T^n x_A})^{-1} B' = U_{2\delta_k}((\exp_{T^n x_A})^{-1} B) \subset U_{2\delta_k}(DT^n D_{x_A}(0,3\delta_k)) \qquad (3)$$

nd the diameter of the set at the right hand part of this formula is ess than δ.

Let $A' = \{A' \in A_k, \quad A' \cap B \neq \emptyset\}$. Obviously the inclusion $' \in A'$ implies that $A' \subset B'$. Consequently,

$$\sum_{A' \in \mathcal{A}'} v(A') \leq C\, v(B')$$

where v is a Riemannian volume on M and C is the maximal multiplicity of the cover $\mathcal{A}_k$. Thus

$$N_{\mathcal{A}_k}(B) \leq |\mathcal{A}'| \leq C \frac{v(B')}{\min\limits_{A' \in \mathcal{A}'} v(A')} .$$

Compactness of M and our choice of the number ε_0 guarantee that the ratio $\frac{v(A')}{v(A'')}$ for every two elements $A', A'' \in \mathcal{A}_k$ bounded from positive constant. In particular $v(A') > C_1 v(A)$ so that

$$N_{\mathcal{A}_k}(B) \leq \frac{C}{C_1} \cdot \frac{v(B')}{v(A)} . \tag{4}$$

Now we shall estimate the volume $v(B')$. Let $\sigma : T_{x_A} M \to T_{T^n x_A} M$ be an isometry. Then $v = DT^n \cdot \sigma^{-1} : T_{x_A} M \to T_{x_A} M$ is a linear operator in $T_{x_A} M$. It can be represented in the form $V = U \cdot S$ where U is an isometric operator and S is a positively definite symmetric operator with eigenvalues $\lambda_1, \ldots, \lambda_m$, where $\lambda_1 \geq \lambda_2 \ldots \lambda_l > 1 \geq \geq \lambda_{l+1} \geq \ldots \geq \lambda_m$. Condition (3) shows that there exist constants C_2, C_3 such that

$$v(B') \leq C_2 \bar{v}((\exp_{T^n x_A})^{-1} B') \leq C_3 (3\delta_k)^m \prod_{i=1}^{m} (\tfrac{2}{3} + \lambda_i) \tag{5}$$

where $\bar{v}$ is a volume in a tangent space. On the other hand

$$v(A) \geq C_4 \bar{v}((\exp_{x_A})^{-1} A) \geq C_5 \delta_k^m \tag{6}$$

Combining inequalities (4), (5), (6) we get an estimation for $N_{\mathcal{A}_k}(B)$:

$$N_{\mathcal{A}_k}(B) \leq C_6 \prod_{i=1}^{m} (\tfrac{2}{3} + \lambda_i) \leq C_7 \prod_{i=1}^{m} \lambda_i .$$

The value $\prod_{i=1}^{l} \lambda_i$ is equal to $|J(DT^n x|_L)|$ where L is a

ıbspace of $T_n x$ generated by the eigenvectors of S with eigen-
ılues $\lambda_1,\ldots,\lambda_l$. Thus we obtain the estimation

$$h(A_k/T^n A_k) \le \log \max_{x\in M} \max_{L\subset T_x M} |J(DT_x^n|_L)| + \log C_7 .$$

ımma is proved.

Now we can finish the proof of the theorem. Let us fix a posi-
.ve integer n and $\varepsilon > 0$ and choose k according to the lemma.
›reover, k can be chosen so large that

$$nh(T) = h(T^n) < h(T^n, A_k) + \varepsilon .$$

›oposition 4.1 implies that
$T^n, A_k) = h(T^{-n}, A_k) \le h(T^{-n}A_k/A_k) = h(A_k/T^n A_k)$, so that
ı$(T) < h(A_k/T^n A_k) + \varepsilon$. By the lemma we have $h(A_k/T^n A_k) \le \alpha(T^n) + C'$.
ıt $\alpha(T^n) \le n\alpha(T)$, that $h(T) \le \alpha(T) + \frac{C'+\varepsilon}{n}$. Since n can be
ıosen arbitrary large, $h(T) \le \alpha(T)$. The theorem is proved.

4. Let us denote by $\Omega^k(M)$ the space of all continuous real-
.lued differential antisymmetric k-forms on M with the norm

$$\|\omega\| = \max_{x\in M} \max_{\substack{v_1,\ldots,v_k\in T_x M \\ \det(v_1,\ldots,v_k)=1}} |\omega(v_1\ldots v_k)|$$

ıc diffeomorphism T induces a linear operator $T_k^\# : \Omega^k(M) \to \Omega^k(M)$.
ıt us denote the direct sum $\bigoplus_0^m \Omega^k(M)$ by $\Omega(M)$ and $\bigoplus_0^m T_k^\#$ by $T^\#$.
ıe proof of the following proposition is routine.

<u>:OPOSITION 5.</u> $\lim_{n\to\infty} \frac{\alpha(T^n)}{n} = \log s(T^\#)$

ıere $s(T^\#)$ is a spectral radius of the operator $T^\#$. The next fact
ıllows immediately from our Theorem and proposition 5.

<u>ROLLARY.</u> $n(T) \le \log s(T^\#)$.

Remark. K. Krzyzewski has generalized our result from diffeomorphisms to arbitrary C^1 mappings of smooth manifolds. His proof used the definition of the topological entropy through ε-separated sets (see [4]).

REFERENCES

1. S. Ito, An estimate from above for an entropy and the topological entropy of a C^1-diffeomorphism, Proc. Japan Acad. 46:3(1970), 226-230.

2. A. G. Kušnirenko, An estimate from above for the entropy of classical dynamical system, Soviet Mathematics, Doklady, 161, N1, (1965), 360.

3. Ya. B. Pesin, Lyapunov characteristic exponents and smooth ergodic theory, Russian Math. Surveys, 32, N4, 1977.

4. E. I. Dinaburg, On the relations among various entropy characteristics of dynamical systems, Math. USSR-Izvestija, 5, N2, 1971, 337-378.

5. M. Misiurewicz, Topological conditional entropy, Studia Mathematica, LV(1976), 175-200.

University of Maryland
College Park, MD 20742

Ergodicity in (G,σ) - extensions

H.B. Keynes* and D. Newton

§0 Introduction.

In this paper we study a more general situation than group extensions (or skew-extensions), which reduces to affine-type extensions when the integers are the acting group. Namely, we consider an extension $\pi : (X,T) \xrightarrow{(G,\sigma)} (Y,T)$, where σ is an action of T on a compact (not necessarily abelian) group G by automorphisms, satisfying the relationship $(gx)t = \sigma^t(g)(xt)$; such extensions are called (G,σ) - extensions (or, simply, σ - extensions). If m is an invariant measure on Y with some dynamical property, the question arises as to whether the Haar lift $\tilde{m}$ enjoys the same property.

Our approach is to use the dynamics of (G,σ) to simplify the analysis, with no assumptions other than G compact and T abelian. We note a decomposition for $L^2(X,\tilde{m})$ (Theorem 1.4), which gives rise to the notion of a weak - $\bar{\gamma}$ - function, for γ an irreducible representation of G. The key observation (Theorem 2.1) is that an $L^2(X,\tilde{m})$ - eigenfunction can be decomposed into a sum of weak - $\bar{\gamma}$ - functions, where the γ's involved have a stabilizer subgroup of finite index. This yields a necessary and sufficient condition for ergodicity of $\tilde{m}$ in terms of weak - $\bar{\gamma}$ - functions, generalizing the result in [7], and yielding a result similar to one of W. Parry [10]: ergodicity and weak-mixing of $\tilde{m}$ is reduced to the corresponding question on a naturally defined factor system which is an equicontinuous - σ - extension of (Y,T).

*Research supported by NSF MPS 75-05250.

Turning to equicontinuous - σ - extensions in §3 , we show that there is a compact subgroup K of the automorphism group of G , and a direct product K - extension $(X \times K, T)$ of (X,T) which, at the same time, is a $K \cdot G$ extension of (Y,T) , and gives a commuting triangle

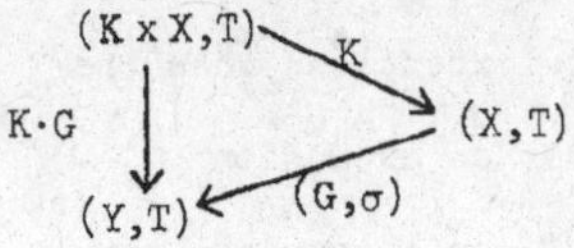

Noting that G is a normal subgroup of $K \cdot G$ we also construct a G - extension $(K \times X, T) \to (K \times Y, T)$, and show that if G is a connected abelian group then the ergodicity of $\tilde{m}$ is equivalent to the ergodicity of certain Haar lifts over this latter extension.

§1 Preliminaries.

In this paper, G will denote a compact Hausdorff group with dual object $\Sigma(G)$ and T an abelian Hausdorff group. An automorphism transformation group (G,T) requires that each transition map $\sigma^t : G \to G$ is an automorphism; we frequently will use the notation (G,σ) . If $\overline{\gamma} \in \Sigma(G)$ and γ a representation in $\overline{\gamma}$, set $\gamma t = \gamma \circ \sigma^t$. Since σ^t is a group automorphism it respects equivalence classes and we may unambiguously define $\overline{\gamma} t = \overline{\gamma t}$. If G is abelian this leads to an induced automorphism group action on the character group $\Gamma(G)$, and a natural subgroup for $T = Z$, by taking those characters with finite orbits under the induced action. To generalize, let $\overline{\gamma} \in \Sigma(G)$ and let $T_{\overline{\gamma}}$ be the closed stabilizer subgroup of $\overline{\gamma}$; $T_{\overline{\gamma}} = \{t \in T : \overline{\gamma} t = \overline{\gamma}\} = \{t \in T : \gamma t \sim \gamma\}$ where '$\sim$' denotes equivalence of representations.

Definition 1.1. We say that $\bar{\gamma}$ (or γ) is T-cofinite, if $T_{\bar{\gamma}}$ has finite index in T. Set $\Sigma_F = \{\bar{\gamma} \in \Sigma(G): \bar{\gamma}$ is T-cofinite$\}$, and put $G_F = \mathrm{Ann}\,\Sigma_F = \bigcap_{\bar{\gamma}\in\Sigma_F} \ker\gamma$. Now Σ_F is T-invariant, and G_F is a T-invariant closed normal subgroup of G. So we consider the induced automorphism group $(G/G_F,T)$.

The proof of the major property will be deferred until later (Corollary 2.5) although we state it now.

Theorem 1.2. $(G/G_F,\sigma)$ is the maximal equicontinuous factor of (G,σ).

Proof. A relatively straightforward argument yields equicontinuity of $(G/G_F,\sigma)$. Maximality will be shown in Corollary 2.5.

To say that $\pi: (X,T) \to (Y,T)$ is a (G,σ)-extension, we mean that X is a compact Hausdorff space with a right action of T, xt, and a free left action of G, gx, related by $(gx)t = \sigma^t(g)(xt)$ $(t \in T, g \in G, x \in X)$. T then induces an action on the orbit space X/G and we assume that (Y,T) is isomorphic to $(X/G,T)$ and $\pi: (X,T) \to (Y,T)$ satisfies the property that $\pi^{-1}(y)$ is a G-orbit for each $y \in Y$.

We denote by $M(X,T)$ the set of regular Borel probability measures on X which are invariant under T and we will suppose $M(X,T)$ is non-empty; then $M(X,T)$ is a compact convex set and its extreme points $E(X,T)$ are the ergodic measures. $\mathcal{B}(X)$ will denote the σ-algebra of Borel subsets of X. If $m \in E(Y,T)$ then we will denote by $P_m(X,T)$ the subset of $M(X,T)$ consisting of those $\mu \in M(X,T)$ for which $\pi^*\mu = m$, where $\pi^*\mu(B) = \mu(\pi^{-1}B)$, $B \in \mathcal{B}(Y)$. It is known ([6], Theorem 2.3.2) that $P_m(X,T)$ is a simplex and its set of extreme points is the set $P_m(X,T) \cap E(X,T)$, which we denote $E_m(X,T)$. If π is a G-extension, that is, $\sigma^t = \mathrm{id}$ for all $t \in T$, then $E_m(X,T)$ is a compact set [6], and so the measures in $P_m(X,T)$ have ergodic decompositions.

We note that there is always at least one measure in $P_m(X,T)$ namely, the Haar lift $\tilde{m}$ (or $\tilde{m}(G)$) of m given by $\tilde{m}(f) = \int_X \int_G f(gx)\, d\lambda\, d\pi^{-1}m \quad (f \in C(X))$. Here λ is normalized Haar measure on G and $\pi^{-1}m$ is the measure, on the σ-algebra $\pi^{-1}B(Y)$, defined by $\pi^{-1}m(\pi^{-1}B) = m(B)$. Since $\tilde{m}$ is G-invariant, in $L^2(X,\tilde{m})$ we have continuous unitary representations U and V of T and G respectively, defined by $U_t f(x) = f(xt)$, $V_g f(x) = f(g^{-1}x)$ $(f \in L^2(X,\tilde{m}))$, and satisfying, $U_t V_g = V_{\sigma^{t^{-1}}(g)} U_t$ $(g \in G,\ t \in T)$. Since G is compact the representation V decomposes $L^2(X,\tilde{m})$ into finite dimensional V-invariant subspaces. This decomposition is not unique but has a degree of uniqueness expressed by Theorem 1.4. First a definition.

Definition 1.3. Let $\Sigma^{n-1} = \{v \in C^n : \|v\|_n = 1\}$ and let $\gamma: G \to U(n)$ be an irreducible representation of G. A Borel function $f: X \to \Sigma^{n-1}$ is called a γ-function if $V_{g^{-1}} f(x) = f(gx) = \gamma(g) f(x)$ $(x \in X,\ g \in G)$.

Theorem 1.4. Let V be the continuous unitary representation of G on $L^2(X,\tilde{m})$ described above. Then for each $\bar{\gamma} \in \Sigma(G)$ there is a V-invariant subspace $M_{\bar{\gamma}}$ such that:

1. V restricted to $M_{\bar{\gamma}}$ decomposes into a direct sum of irreducible representations each of which is in $\bar{\gamma}$.

2. If M is any V-invariant subspace on which V decomposes into irreducible representations each of which is in $\bar{\gamma}$, then $M \subset M_{\bar{\gamma}}$.

3. $L^2(X,\tilde{m}) = \bigoplus_{\bar{\gamma}} M_{\bar{\gamma}}$,

4. $M_{\bar{\gamma}} \neq \{0\}$ <u>for all</u> $\bar{\gamma} \in \Sigma(G)$,

5. $U_t M_{\bar{\gamma}} = M_{\bar{\gamma}t}$,

6. M_1 <u>may be identified with</u> $L^2(Y,m)$.

Proof.

The existence of the $M_{\bar{\gamma}}$'s and 1,2, and 3 follow from a general result on representations ([2], Theorem 27.44). 4 follows from the existence of Borel γ-functions ([7], Lemma 1.3) . 5 follows from the relation $V_g U_t = U_t V_{\sigma^t(g)}$ and the uniqueness in 2 . Finally, 6 follows because M_1 consists of all the V - invariant functions in $L^2(X,\tilde{m})$ and $L^2(Y,m)$ may be thought of as the V - invariant functions in $L^2(X,\tilde{m})$.

<u>Definition 1.5</u>. We say that a non-zero function $f \in L^2(X,\tilde{m})$ is a <u>weak-$\bar{\gamma}$ - function</u> if $f \in M_{\bar{\gamma}}$.

If γ is 1 - dimensional then a γ - function is in $L^2(X,\tilde{m})$ and is a weak - $\bar{\gamma}$ - function. Conversely, in this case, a weak - $\bar{\gamma}$ - function f satisfies $f(gx) = \gamma(g)f(x)$ a.e. $\tilde{m}$. However its modulus need not be non-zero a.e. and so we cannot recover a γ - function from it. For dimensions $n > 1$, we may write a γ - function f in the form $f = \begin{bmatrix} f_1 \\ \vdots \\ f_n \end{bmatrix}$.

Each of the components f_i is then a weak - $\bar{\gamma}$ - function. If f is a weak - $\bar{\gamma}$ - function, then one can choose a basis $f_1,\ldots,f_n$ of an n - dimensional V - invariant subspace generated by f so that $\bar{f} = \begin{bmatrix} f_1 \\ \vdots \\ f_n \end{bmatrix}$ satisfies $\bar{f}(gx) = \gamma(g)\bar{f}(x)$ a.e. $\tilde{m}$. Again $\bar{f}$ need not have non-zero a.e. norm.

To conclude this section we make some remarks about $P_m(X,T)$. Let $M(G,\sigma)$ and $E(G,\sigma)$ denote respectively the invariant and ergodic regular Borel probability measures for (G,σ). Then $\lambda \in M(G,\sigma)$, and if $\nu \in P_m(X,T)$, $\mu \in M(G,\sigma)$, we put $\mu * \nu(f) = \iint_{XG} f(gx)\,d\mu\, d\nu \quad (f \in C(X))$.

It is easy to verify that $\mu * \nu \in P_m(X,T)$ and that for any $\nu \in P_m(X,T)$ we have $\lambda * \nu = \tilde{m}$. If π is a G-extension then for any $\nu \in E_m(X,T)$ we have $P_m(X,T) = \{\mu * \nu\colon \mu \in M(G) \equiv M(G,id)\}$ and $E_m(X,T) = \{\delta_g * \nu\colon g \in G\}$, where δ_g is the atomic measure with mass 1 at g, [6]. Note that $E(G,id) = \{\delta_g\colon g \in G\}$. For non-trivial T-action on G, these characterizations do not hold, as simple examples show. The problem appears to lie in the fact that $\nu \in E_m(X,T)$, $\mu \in E(G,\sigma)$ does not necessarily imply that $\mu * \nu \in E_m(X,T)$. A simple sufficient condition for this is given by

<u>Proposition 1.6</u>. <u>If</u> $\mu \in E(G,\sigma)$, $\nu \in E_m(X,T)$ <u>and if</u> $(G \times X, T, \mu \times \nu)$ <u>is ergodic, then</u> $\mu * \nu \in E_m(X,T)$.

<u>Proof</u>.

Define $p\colon G \times X \to X$ by $p(g,x) = gx$. Then p is a transformation group homomorphism since $p(\sigma^t(g), xt) = \sigma^t(g)(xt) = (gx)t = p(g,x)t$, and $(p^*)(\mu \times \nu) = \mu * \nu$, giving the result.

§2 Ergodicity of (G,σ) - extensions.

Let $\pi:(X,T) \to (Y,T)$ be a (G,σ) - extension and let $m \in E(Y,T)$. Our aim is to discuss the ergodicity of $\tilde{m}$, the Haar lift of m. All our results follow from Theorem 2.1 which states how the 1 - dimensional subrepresentations of U (i.e., eigenfunctions of $(X,T,\tilde{m})$) occur.

Theorem 2.1. Let $\lambda : T \to K$ be a continuous homomorphism, K the circle group. If there is a function $f \in L^2(X,\tilde{m})$ satisfying $U_t f = \lambda(t)f$, $t \in T$, $f \neq 0$, then $f \in \bigoplus_{\bar{\gamma} \in \Sigma_F} M_{\bar{\gamma}}$ and there is a weak - $\bar{\gamma}$ - function h, some $\bar{\gamma} \in \Sigma_F$, satisfying $U_t h = \lambda(t)h$, $t \in T_{\bar{\gamma}}$.

Conversely, if there is a weak - $\bar{\gamma}$ - function h satisfying $U_t h = \lambda(t)h$, $t \in T_{\bar{\gamma}}$, $\bar{\gamma} \in \Sigma_F$, then there is a function $f \in \bigoplus_{\bar{\gamma} \in \Sigma_F} M_{\bar{\gamma}}$ satisfying $U_t f = \lambda(t)f$, $t \in T$, $f \neq 0$.

Proof.

Let $P_{\bar{\gamma}}$ denote the projection of $L^2(X,\tilde{m})$ onto $M_{\bar{\gamma}}$. Let f satisfy $U_t f = \lambda(t)f$. Write

$$f = \sum_{\bar{\gamma} \in \Sigma(G)} P_{\bar{\gamma}} f .$$

Then

$$U_t f = \sum_{\bar{\gamma} \in \Sigma(G)} U_t P_{\bar{\gamma}} f = \sum_{\bar{\gamma} \in \Sigma(G)} \lambda(t) P_{\bar{\gamma}} f .$$

Since $U_t P_{\bar{\gamma}} f \in M_{\bar{\gamma}t}$, it follows that

$$U_t P_{\bar{\gamma}} f = \lambda(t)\, P_{\bar{\gamma}t} f .$$

Hence $\|P_{\bar{\gamma}} f\| = \|P_{\bar{\gamma}t} f\|$. Thus if $\bar{\gamma}$ has an infinite orbit under T then we must have $P_{\bar{\gamma}} f = 0$. Also, if $t \in T_{\bar{\gamma}}$, then $U_t P_{\bar{\gamma}} f = \lambda(t) P_{\bar{\gamma}} f$.

Thus $f = \sum_{\bar{\gamma} \in \Sigma_F} P_{\bar{\gamma}} f \in \bigoplus_{\bar{\gamma} \in \Sigma_F} M_{\bar{\gamma}}$, and, if we choose a $\bar{\gamma} \in \Sigma_F$ with $P_{\bar{\gamma}} f \neq 0$ and put $h = P_{\bar{\gamma}} f$, then h is a weak - $\bar{\gamma}$ - function and $U_t h = \lambda(t)h$ for $t \in T_{\bar{\gamma}}$.

Conversely, suppose h is a weak - $\bar{\gamma}$ - function, $\bar{\gamma} \in \Sigma_F$, such that $U_t h = \lambda(t)h$, $t \in T_{\bar{\gamma}}$. Since $T_{\bar{\gamma}}$ has finite index in T we can choose a set of coset representatives of $T_{\bar{\gamma}}$ in T , $t_1, \ldots, t_n$. Put

$$f = \sum_{i=1}^{n} \lambda(t_i^{-1}) U_{t_i} h \quad .$$

Then $U_t f = \lambda(t) f$, $t \in T$, and $f \neq 0$ since the $U_{t_i} h$ are non-zero elements in orthogonal subspaces. Finally $f \in \bigoplus_{t_i} M_{\bar{\gamma} t_i} \subset \bigoplus_{\bar{\gamma} \in \Sigma_F} M_{\bar{\gamma}}$.

The proof is complete.

As a corollary of this result we get a necessary and sufficient condition for ergodicity of $\tilde{m}$.

<u>Corollary 2.2</u>. <u>Let</u> $\pi : (X,T) \to (Y,T)$ <u>be a</u> (G,σ) - <u>extension,</u> $m \in E(Y,T)$. <u>Then</u> $\tilde{m} \in E(X,T)$ <u>if and only if there are no</u> $T_{\bar{\gamma}}$ - <u>invariant weak</u> - $\bar{\gamma}$ - <u>functions</u>, $\bar{\gamma} \in \Sigma_F$, $\bar{\gamma} \neq 1$.

<u>Proof</u>.

Suppose there are no $T_{\bar{\gamma}}$ - invariant weak - $\bar{\gamma}$ - functions, $\bar{\gamma} \in \Sigma_F$, $\bar{\gamma} \neq 1$. Let $U_t f = f$. Then $f = \sum_{\bar{\gamma} \in \Sigma_F} P_{\bar{\gamma}} f$ and each non-zero $P_{\bar{\gamma}} f$ is a $T_{\bar{\gamma}}$ - invariant weak - $\bar{\gamma}$ - function. Since these can only exist for $\bar{\gamma} = 1$ we get $f = P_1 f$. Thus f is a T - invariant and G - invariant function, hence constant by the ergodicity of m . So $\tilde{m} \in E(X,T)$.

Now suppose $\tilde{m} \in E(X,T)$. Let h be a $T_{\bar{\gamma}}$ - invariant weak - $\bar{\gamma}$ - function, $\bar{\gamma} \in \Sigma_F$. Let $t_1, \ldots, t_n$ be a set of coset representatives for $T_{\bar{\gamma}}$ in T

and put $f = \sum_{i=1}^{n} U_{t_i} h$. Then $U_t f = f$ and hence is constant. Thus h is a constant and $\bar{\gamma} \equiv 1$.

We will now briefly indicate what happens in the case of trivial T-action on G . In this case $\Sigma_F = \Sigma(G)$ and $T_{\bar{\gamma}} = T$ for all $\bar{\gamma}$. A T-invariant weak-$\bar{\gamma}$-function has the property that the V-invariant subspace generated by it consists of T-invariant functions. So we can find a set of n T-invariant weak-$\bar{\gamma}$-functions $f_1, \ldots, f_n$ such that

$$\begin{bmatrix} f_1(gx) \\ \cdot \\ \cdot \\ \cdot \\ f_n(gx) \end{bmatrix} = \gamma(g) \begin{bmatrix} f_1(x) \\ \cdot \\ \cdot \\ \cdot \\ f_n(x) \end{bmatrix} \quad \text{a.e.} \tilde{m} , \ g \in G \quad .$$

Now the norm of $\begin{bmatrix} f_1(x) \\ \cdot \\ \cdot \\ \cdot \\ f_n(x) \end{bmatrix}$, as a vector function, is a T-invariant G-invariant function on X and hence is a non-zero constant, a.e.. Thus dividing by the norm we obtain a T-invariant function which is a γ-function a.e.. We can thus deduce Theorem 2.1 of [7], from the above Corollary 2.2 and Lemma 1.4 of [7].

In general it is difficult to see how to obtain characterizations in terms of γ-functions rather than weak-$\bar{\gamma}$-functions. If we allow some extra hypotheses we can get

Corollary 2.3. Let G be abelian and let $m \in E(Y, T_{\bar{\gamma}})$ for all $\bar{\gamma} \in \Sigma_F$. Then $\tilde{m} \in E(X,T)$ if and only if there are no $T_{\bar{\gamma}}$-invariant γ-functions, $\bar{\gamma} \in \Sigma_F$, $\bar{\gamma} \neq 1$.

Proof.

We need only show that the existence of a $T_{\bar{\gamma}}$ - invariant weak - $\bar{\gamma}$ - function implies the existence of a $T_{\bar{\gamma}}$ - invariant γ - function.

So let f be a $T_{\bar{\gamma}}$ - invariant weak - $\bar{\gamma}$ - function. Then f satisfies $f(gx) = \gamma(g)f(x)$ a.e. $\tilde{m}$. Thus $|f|$ is a G - invariant $T_{\bar{\gamma}}$ - invariant function. Since $m \in E(Y,T_{\bar{\gamma}})$ it follows that $|f|$ is a non-zero constant a.e. Thus $f/|f|$ is an $L^2(X,\tilde{m})$ γ - function which is $T_{\bar{\gamma}}$ - invariant and hence in equal a.e. to a $T_{\bar{\gamma}}$ - invariant (Borel) γ - function [7], Lemma 1.4].

Note that if $T = Z$, (2.3) holds without any additional assumptions on m .

Recalling that $(G/G_F,\sigma)$ is equicontinuous, the extension $\pi : (X,T) \to (Y,T)$ can be decomposed into two extensions

$$\pi_1 : (X,T) \to (X/G_F,T)$$

$$\pi_2 : (X/G_F,T) \to (Y,T)$$

where π_1 is a (G_F,σ) - extension and π_2 is an equicontinuous - $(G/G_F,\sigma)$ - extension. Here the adjective equicontinuous refers to the property of $(G/G_F,\sigma)$ and does not mean that $(X/G_F,T)$ is equicontinuous. Then Theorem 2.1 and Corollary 2.2 give

Corollary 2.4. Let $m \in E(Y,T)$. Then $\tilde{m}(G) \in E(X,T)$ if and only if $\tilde{m}(G/G_F) \in E(X/G_F,T)$.

In addition, if f is an eigenfunction of T on $L^2(X,\tilde{m}(G))$, then f is G_F - invariant and hence is an eigenfunction for T on $L^2(X,\tilde{m}(G/G_F))$.

Proof.

We can identify $L^2(X,\tilde{m}(G/G_F))$ as the set of G_F - invariant functions in $L^2(X,\tilde{m}(G))$. Since this set contains $\bigoplus_{\bar{\gamma}\in\Sigma_F} M_{\bar{\gamma}}$, these results follow directly from 2.1 and 2.2.

Corollary 2.5. $(G/G_F,\sigma)$ is the maximal equicontinuous factor of (G,σ) .

Proof.

Note that (G,σ) is a (G,σ) - extension of the trivial one point flow. Suppose $\pi:(G,\sigma)\to(Y,T)$ is a homomorphism with (Y,T) equicontinuous. Since T is abelian $C(Y)$ is generated by eigenfunctions of T . Let f be such an eigenfunction. Then $\pi^* f = f\circ\pi\in C(G)$ is an eigenfunction for the action of T on G . It follows, from Corollary 2.4, that, as a $L^2(G,\lambda)$ function, $\pi^* f$ is G_F - invariant. But λ is a supported measure so $\pi^* f$ is spatially G_F invariant and hence can be regarded as belonging to $C(G/G_F)$. Since this is true for any eigenfunction it follows that π factors through $(G/G_F,\sigma)$. Since $(G/G_F,\sigma)$ is equicontinuous it is therefore the maximal equicontinuous factor of (G,σ) .

We now consider the applications of these results to a (G,σ) - extension of the one point flow (so $X=G$). If we put $f(t) = et$, where e is the identity of G , then $gt = (ge)t = \sigma_t(g)(et) = \sigma_t(g)f(t)$. Thus each $t\in T$ acts as an affine transformation of G . We refer to (G,T) as an affine action. The group property of the T - action implies that the function $f:T\to G$ satisfies $f(t\cdot t_1)=\sigma_{t_1}(f(t))\cdot f(t_1)$. If $f(t) = e$ for all $t\in T$ then we have $(X,T) = (G,\sigma)$. If H is an σ - invariant closed normal subgroup of G , then the induced action

$(G/H,T)$ is also an affine action.

Many of the results that follow are known in the case $T=Z$.

Corollary 2.6. Let (G,T) be an affine action. Then (G,T) is ergodic relative to Haar measure if and only if $(G/G_F,T)$ is minimal. In particular, if (G,T) is minimal then it is ergodic relative to Haar measure.

Proof.

We know that (G,T) is ergodic relative to Haar measure on G if and only if $(G/G_F,T)$ is ergodic relative to Haar measure on G/G_F . Since $(G/G_F,T)$ is an equicontinuous - $(G/G_F,\sigma)$ - extension of the one point flow it is equicontinuous. Since Haar measure is supported it follows that $(G/G_F,T)$ is ergodic if and only if it is minimal.

In the special case $f(t) = e$, Corollary 2.6 gives

Corollary 2.7. Let (G,σ) be an automorphism group action. Then (G,σ) is ergodic relative to Haar measure on G if and only if $\Sigma_F = \{1\}$, that is, if and only if no nontrivial irreducible unitary representation of G has a finite T - orbit.

If we examine the minimality condition of Corollary 2.6 we obtain

Corollary 2.8. Let (G,T) be an affine action. Then (G,T) is ergodic relative to Haar measure if and only if the smallest closed subgroup generated by G_F and $\{f(t):t\in T\}$ is G .

Proof.

$(G/G_F,T)$ is minimal if and only if the orbit of $e.G_F$ is dense in G/G_F . In other words, if and only if the set of cosets $\{f(t)\cdot G_F:t\in T\}$ is dense in G/G_F . That is, if and only if G is the closed subgroup generated by G_F and $\{f(t):t\in T\}$.

Corollary 2.8 has been shown by H. Hoare and W. Parry [3] for an abelian semigroup action on a connected abelian group.

Applying the ergodicity criteria of Corollary 2.2 we obtain

Corollary 2.9. Let (G,T) be an affine action with G abelian. Then (G,T) is ergodic relative to Haar measure if and only if $\gamma\in \Sigma_F, \gamma\neq 1$, implies $\gamma(f(t))\neq 1$ for some $t\in T_\gamma$.

Proof.

In this situation a weak - $\bar{\gamma}$ - function is simply a non-zero multiple of the character γ . A character γ is T_γ - invariant if and only if $\gamma(f(t)) = 1$ for all $t\in T_\gamma$. The result now follows from Corollary 2.2.

Finally we give one corollary and an example in the special case $T=Z$. We suppose the actions of Z on X,G,Y are given by generators ϕ,τ,ψ respectively and we will denote the factor of ϕ on X/G_F by ϕ_F . R.K. Thomas [12] showed that if ψ is a K - automorphism and ϕ is weakly mixing, then ϕ is a K - automorphism. With our characterization of eigenfunctions and this result we get

Corollary 2.10. Let ψ be a K - automorphism. Then ϕ is a K - automorphism if and only if ϕ_F is a K - automorphism. In particular, if τ is ergodic then ϕ is a K - automorphism.

Proof.

Theorem 2.1 implies that ϕ is weakly mixing if and only if ϕ_F is weakly mixing. The corollary then follows from Thomas' theorem. If τ is ergodic then $G_F = G$ and $\phi_F = \psi$.

Finally we give an example to show that for any compact abelian G and any automorphism τ there is a Bernoulli automorphism (Y,ψ) and a weakly mixing (X,ϕ) which is a (G,τ)-extension of (Y,ψ).

<u>Example 2.11</u>. Let $Y = G^Z = \prod_{-\infty}^{\infty} G_i$, where each G_i is a copy of G. Let ψ denote the shift on Y. Then (Y,ψ) is a Bernoulli group automorphism. Let m be Haar measure on Y. Put $X = Y \times G$ and define $\phi : X \to X$ by

$$\phi(y,g) = (\psi(y), y_0\tau(g)) \quad .$$

Here $y = \{y_i\}_{i=-\infty}^{\infty}$. G acts on X in the obvious way and (X,ϕ) is a (G,τ)-extension of (Y,ψ). If $\alpha \in \Gamma(G)$ then an α-function on X is of the form: $f(y,g) = h(y)\alpha(g), h \in L^2(Y,m)$. We will show that (X,ϕ) is weakly mixing. By Theorem 2.1 and a technique similar to that used in Corollary 2.3 to get α-functions instead of weak-$\bar{\alpha}$-functions, β is an eigenvalue for (X,ϕ) if and only if there is an $\alpha \in \Gamma(G)$, $\alpha \circ \tau^n = \alpha$ and an α-function f such that

$$f(\phi^n(y,g)) = \beta^n f(y,g)$$

or, putting $f(y,g) = h(y)\alpha(g)$,

$$h(\psi^n(y))\alpha(y_{n-1}\tau(y_{n-2})\cdots\tau^{n-1}y_0\cdot\tau^n g) = \beta^n h(y)\alpha(g) \quad .$$

Since $\alpha\cdot\tau^n = \alpha$, this reduces to

$$h(\psi^n y)\alpha(y_{n-1}\tau(y_{n-2})\cdots\tau^{n-1}y_0) = \beta^n h(y) \quad .$$

We denote

$$\gamma_o(y) = \alpha(y_{n-1}\cdot\tau(y_{n-2})\cdots\tau^{n-1}y_o) \in \Gamma(G^Z) \quad .$$

Now $h \in L^2(G^Z)$ and so we may write

$$h = \sum_{\gamma\in\Gamma(G^Z)} k_\gamma \cdot \gamma \,, \quad \Sigma|k_\gamma|^2 < \infty \quad .$$

The equation $h(\psi^n y)\gamma_o(y) = \beta^n h(y)$ becomes

$$\sum_\gamma k_\gamma \cdot (\gamma\circ\psi^n)\cdot\gamma_o = \sum_\gamma \beta^n k_\gamma \cdot \gamma \quad .$$

Since $|\beta| = 1$ we have $|k_\gamma| = |k_{(\gamma\circ\psi^n)\cdot\gamma_o}|$. Thus $k_\gamma \neq 0$ if and only if for some $m,k > 0$, $m \neq k$, we have

$$(\gamma\circ\psi^{mn})\cdot(\gamma_o\circ\psi^{(m-1)n})\cdots(\gamma_o\circ\psi^n)\cdot\gamma_o = (\gamma\circ\psi^{kn})\cdot(\gamma_o\cdot\psi^{(k-1)n})\cdots(\gamma_o\circ\psi^n)\cdot\gamma_o \quad .$$

Assume $m > k$. Since $\gamma \in \Gamma(G^Z)$ then, either $\gamma = 1$, or there is a finite index set I such that $\gamma(y) = \prod_{i\in I} \alpha_i(y_i)$, $\alpha_i \neq 1$, $\alpha_i \in \Gamma(G)$. Suppose $\gamma \neq 1$. Let $i_o = \min I$, $i_1 = \max I$. Looking at independence of coordinates in the above equation we get $i_o + kn = kn$ and $i_1 + mn = i_1 + kn$, which is impossible. Therefore $\gamma = 1$ and so $k_\gamma = 0$ for all $\gamma \neq 1$. This means that h is a constant function and $\gamma_o(y) = \beta^n$. This can only happen if $\beta^n = 1$ and hence $\gamma_o \equiv 1$. Again independence of coordinates implies $\alpha = 1$. Thus $n = 1$, $\beta = 1$ and so (X,ϕ) is weakly mixing and thus a K-automorphism.

It is natural to conjecture that (X,ϕ) is in fact Bernoulli. If (G,τ) were ergodic this does follow from a recent paper of D.A. Lind [9].

§ Equicontinuous - (G,σ) - extensions.

Let (Z,T) denote an equicontinuous topological transformation group. It is well known that if (Z,T) is minimal then it is uniquely ergodic. Our purpose first is to briefly discuss the invariant measures in the non-minimal case. These results may be known but we have been unable to locate a reference.

Let $\mathcal{R}$ denote the orbit closure relation of (Z,T). Then $Z/\mathcal{R}$ is a compact Hausdorff space and the elements of $Z/\mathcal{R}$ are the minimal subsets of (Z,T). Since each of them supports precisely one ergodic invariant Borel probability measure for (Z,T) and all such measures are obtained in this way, we have a bijection $\eta\colon Z/\mathcal{R} \to E(Z,T)$. If we provide $Z/\mathcal{R}$ with the quotient topology and $E(Z,T)$ with the weak topology, then one can show

Theorem 3.1. *η is a homeomorphism. In particular $E(Z,T)$ is a compact subset of $M(Z,T)$ in the weak topology.*

A consequence of this result is that all measures in $M(Z,T)$ have ergodic decomposition.

We now wish to examine the ergodic measures associated with an equicontinuous - (G,σ) - extension $\pi\colon (X,T) \to (Y,T)$. One feature of a group extension is that if $m \in E(Y,T)$ and the Haar lift $\tilde{m}$ is ergodic then it is the only ergodic measure which projects to m. This remains true for an equicontinuous - (G,σ) - extension.

Proposition 3.2. *If $m \in E(Y,T)$ and $\tilde{m} \in E_m(X,T)$, then $P_m(X,T) = \{\tilde{m}\}$.*

Proof.

Let $\nu \in E_m(X,T)$. Then $\tilde{m} = \lambda * \nu$, where λ is Haar measure on G. Now (G,σ) equicontinuous and $\lambda \in M(G,\sigma)$ implies that there is a measure

$\beta \in M(G/\mathcal{R})$, namely $\tau * \lambda$, such that , with obvious notation,

$$\lambda(f) = \int_{G/\mathcal{R}} \mu_F(f)\,d\beta(F) \quad (f \in C(G)) \text{ , and so } \tilde{m} = \Big(\int_{G/\mathcal{R}} \mu_F d\beta\Big) * \nu = \int_{G/\mathcal{R}} \mu_F * \nu\, d\beta \ .$$

Now each $\mu_F * \nu$ belongs to the simplex $P_m(X,T)$ and, by assumption, $\tilde{m}$ is extreme in that simplex. It follows then that $\mu_F * \nu = \tilde{m}$ for β - almost all F . However the mapping $F \to \mu_F * \nu$ is continuous and, since the support of β is $G/\mathcal{R}$, it is constant on a dense set. Therefore $\mu_F * \nu = \tilde{m}$ for all F . Since point mass at the identity is such a μ_F we get $\nu = \tilde{m}$ and hence $P_m(X,T) = \{\tilde{m}\}$.

Corollary 3.3. _If (Y,T) is uniquely ergodic with invariant measure m , then (X,T) is uniquely ergodic with invariant measure $\tilde{m}$ if and only if $\tilde{m}$ is ergodic._

In the case that $T = Z^m$ and G is abelian, the results of 3.2 and 3.3 can be extended to distal - (G,σ) - extensions using the inverse system decomposition of (G,σ) into equicontinuous - extensions, see [8] . In the case $T = Z$, $\sigma^{-1} = \tau$ and τ^P unipotent, this had been noted by P. Walters [14] . Since τ^P unipotent implies τ distal this observation includes Walters' result.

Our main aim in this section is to show the very close relation between equicontinuous - (G,σ) - extensions and group extensions. We will now exhibit a direct product group extension of (X,T) such that the resulting transformation group is a group extension of (Y,T) .

Since (G,σ) is equicontinuous, the pointwise and uniform closures in $\mathrm{Aut}(G)$ of $\{\sigma^t \mid t \in T\}$ coincide and the resulting set K_σ is a compact abelian group on which T acts minimally via the group of translations $\{\sigma^t \mid t \in T\}$ (indeed , (K_σ,T) is just the enveloping group of (G,σ)) . Form the direct product with action $(k,x)t = (k\sigma_t, xt)$.

We now construct a group H such that $K_\sigma \times X$ is a free H-space, the actions of T and H commute and $(K_\sigma \times X/H,T)$ is isomorphic to (Y,T). Let H be the semi-direct product $K_\sigma \cdot G$, with the multiplication given by $(k_1,g_1)(k_2,g_2) = (k_1k_2,k_2(g_1)g_2)$. We can identify G with $\{(1,g): g \in G\}$ and K_σ with $\{(k,e): k \in K_\sigma\}$, using obvious notation; then G is normal in H and $K_\sigma \cong H/G$. We define an action of H on $K_\sigma \times X$ by $(k,g)(k_1,x) = (kk_1,k_1(g)x)$.

It is clear that $K_\sigma \times X$ is a free H-space under this action, and one directly verifies that the actions of T and H commute. We note finally that two elements (k,x) and (k_1,x_1) of $K_\sigma \times X$ belong to the same H-orbit if and only if x and x_1 belong to the same G-orbit of X. Thus we can identify (Y,T) and $(K_\sigma \times X/H,T)$. This gives us the following commuting diagram of extensions

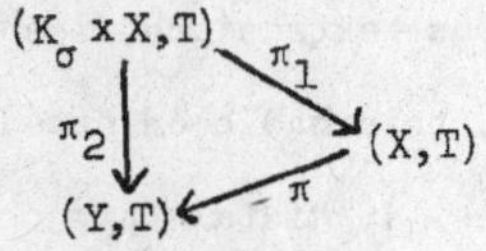

where π_1 is projection onto the second coordinate, and hence a K_σ-extension, π_2 is an H-extension and π is the original (G,σ)-extension.

To complete the picture, since G is a normal subgroup of H we can split the extension π_2 into a G-extension and an $H/G \cong K_\sigma$-extension. Thus

$$(K_\sigma \times X) \xrightarrow{\pi_3} (K_\sigma \times X/G,T) \xrightarrow{\pi_4} (Y,T)$$

We note that the G-action on $K_\sigma \times X$ induced by H is given by $g(k,x) = (k, k(g)x)$. Thus two points are in the same G-orbit if and only if they have the same K_σ coordinate and their X coordinates are in the same G-orbit. So we can identify $K_\sigma \times X/G$ with $K_\sigma \times Y$. Under this identification the T-action on $K_\sigma \times Y$ is the product action and π_4 is projection onto the second coordinate. Thus we obtain the following commutative diagram

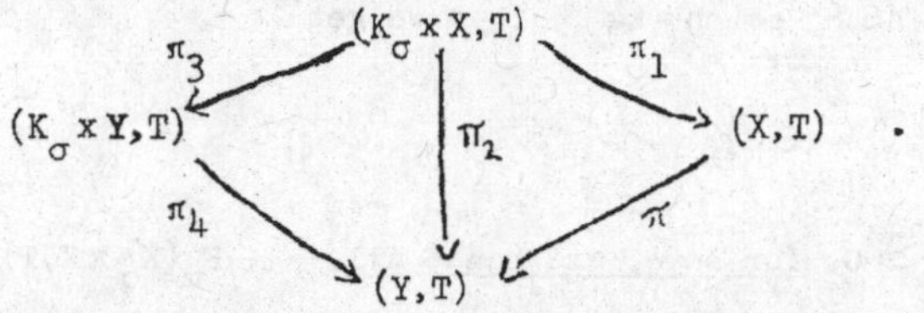

Now let $m \in E(Y,T)$. We will denote the Haar lift of m through the extension π by $\tilde{m}$. If μ denotes Haar measure on K_σ and if $\nu \in E_m(X,T)$, then the Haar lift of ν through π_1 is $\mu \times \nu$. It follows that the Haar lift of m through π_2 is $\mu \times \tilde{m}$. The Haar lift of m through π_4 is $\mu \times m$ and if $\omega \in P_m(K_\sigma \times Y, T)$ we will denote its Haar lift through π_3 by $\tilde{\omega}$. Thus we have $\widetilde{\mu \times m} = \mu \times \tilde{m}$. We recall the following property of a group extension.

<u>Proposition 3.4.</u> [6]. <u>Let</u> $\pi : (X,T) \to (Y,T)$ <u>be a</u> G-<u>extension,</u> $m \in E(Y,T)$ <u>and</u> $\nu \in E_m(X,T)$. <u>Then</u> $E_m(X,T) = \{g\nu : g \in G\}$, <u>where</u> $g\nu(B) = \nu(g^{-1}B)$, $B \in \mathcal{B}(X)$.

Again, letting $\pi : (X,T) \to (Y,T)$ be a G-extension, $m \in E(Y,T)$ and $\nu \in E_m(X,T)$, we recall that the stabilizer of ν is defined by $S_\nu = \{g \in G : g\nu = \nu\}$. We note that S_ν is a closed subgroup of G and that $S_{g\nu} = gS_\nu g^{-1}$.

We are interested in the application of Proposition 3.4 to our situation of a (G,σ) - extension $\pi:(X,T)\to(Y,T)$. Let $m\in E(Y,T)$, $\nu\in E_m(K_\sigma\times X,T)$. Then $E_m(K_\sigma\times X,T)=\{h\nu : h\in H\}$ and so

$$E_m(X,T) = \pi_1^* E_m(K_\sigma \times X,T)$$
$$= \{\pi_1^*(h\nu) : h\in H\} \quad .$$

However π_1^* is a K_σ- extension and so $\pi_1^*(k\nu)=\pi_1^*(\nu)$ for all $k\in K_\sigma$. Since we can write each $h\in H$ as $h=kg$, then we get

$$E_m(X,T) = \{\pi_1^*(g\nu) : g\in G\} \quad .$$

__Proposition 3.5.__ __If__ $S_\nu\supset G$ __for some, and hence all,__ $\nu\in E_m(K_\sigma\times X,T)$, __then__ $\tilde{m}\in E_m(X,T)$.

__Proof.__

First we note that since G is normal in H then $S_\nu\supset G$ implies $hS_\nu h^{-1}=S_{h\nu}\supset G$ for all $h\in H$. Now if $S_\nu\supset G$ then $g\nu=\nu$ for all $g\in G$ and so $\{\pi_1^*(g\nu):g\in G\}=\{\pi_1^*(\nu)\}$. Thus $P_m(X,T)$ has exactly one element which must be $\tilde{m}$. So $\tilde{m}\in E_m(X,T)$.

In §1 we noted that for a general (G,σ) - extension it is not always the case that $E_m(X,T)=\{\mu_o*\nu : \mu_o\in E(G,\sigma)\}$, for some fixed $\nu\in E_m(X,T)$. However in the situation here we can give a sufficient condition for this to hold.

__Proposition 3.6.__ __Let__ $\nu\in E_m(X,T)$ __satisfy__ $\mu\times\nu\in E_m(K_\sigma\times X,T)$. __Then__

$$E_m(X,T) = \{\mu_o * \nu : \mu_o\in E(G,\sigma)\} \quad .$$

Proof.

Put $\nu_o = \mu \times \nu \in E_m(K_\sigma \times X,T)$. Then $E_m(X,T) = \{\pi_1^*(g_o\nu_o) : g_o \in G\}$.
Now for $f \in C(X)$

$$\pi_1^*(g_o\nu_o)(f) = \int_{K_\sigma}\int_X f(\pi_1[(1,g_o)(k,x)])d\nu(x)d\mu(k)$$

$$= \int_X \int_{K_\sigma} f(k(g_o)x)d\mu(k)d\nu(x) \quad .$$

Now recalling that (K_σ,T) is the enveloping group of (G,σ) and using notation suggested by Theorem 3.1 we get

$$\pi_1^*(g_o\nu_o)(f) = \int_X\int_G f(gx)d\mu_{g_o}\, d\nu$$

$$= (\mu_{g_o} * \nu)(f) \quad .$$

Letting g_o run through G , $\pi_1^*(g_o\nu_o)$ runs through $E_m(X,T)$ and μ_{g_o} runs through $E(G,\sigma)$. Thus we get the desired result.

Remark 3.7. We note that the condition $S_\nu \supset G$ for $\nu \in E_m(K_\sigma \times X,T)$ is equivalent to $\widetilde{\pi_3^*\nu} = \nu$. That is, if $S_\nu \supset G$ for some, and hence every, $\nu \in E_m(K_\sigma \times X,T)$, then the Haar lift of each $\omega \in E_m(K_\sigma \times Y,T)$ is ergodic, and conversely. Regarding Proposition 3.6 the condition $\mu \times \nu \in E_m(K_\sigma \times X,T)$ implies, but is not implied by, the condition $\mu \times m \in E_m(K_\sigma \times Y,T)$.

We will now give a necessary and sufficient condition in terms of stabilizers which indicates how close Proposition 3.5 is to being necessary and sufficient.

Proposition 3.8. The following statements are equivalent:

1) $\tilde{m} \in E_m(X,T)$;

2) for all $\nu \in E_m(K_\sigma \times X,T)$ we have $\{h\nu : h \in H\} = \{k\nu : k \in K_\sigma\}$

3) for all $\nu \in E_m(K_\sigma \times X,T)$ we have $H = K_\sigma S_\nu$.

Proof.

1) $\Rightarrow$ 2). If $\tilde{m} \in E_m(X,T)$ than $E_m(X,T) = \{\tilde{m}\}$ and so for any $\nu \in E_m(K_\sigma \times X,T)$ we have $\pi_1^*(\nu) = \pi_1^*(h\nu) = \tilde{m}$ for all $h \in H$. Since π_1 is a K_σ - extension it follows that for each $h \in H$ there is a $h \in K_\sigma$ with $k\nu = h\nu$. Thus we get 2).

2) $\Rightarrow$ 3) Since for each $h \in H$ there is a $k \in K_\sigma$ such that $h\nu = k\nu$ it follows that for each h there is a k such that $h \in kS_\nu$, thus $H = K_\sigma S_\nu$.

3) $\Rightarrow$ 1) Let $\nu \in E_m(K_\sigma \times X,T)$. Then $E_m(K_\sigma \times X,T) = \{h\nu : h \in H\}$. So if $h = ks$, $k \in K_\sigma$, $s \in S_\nu$, then

$$\pi_1^*(h\nu) = \pi_1^*(ks\nu) = \pi_1^*(s\nu) = \pi_1^*(\nu) .$$

Thus $E_m(X,T) = \{\pi_1^*(\nu)\}$. So $\pi_1^*(\nu) = \tilde{m}$ and $\tilde{m} \in E_m(X,T)$.

In a semidirect product $H = K_\sigma \cdot G$ the condition $H = K_\sigma S_\nu$ for some closed subgroup S_ν does not always imply $S_\nu \supset G$. However if we place conditions on K_σ and G then we do get this result.

Theorem 3.9. Let G be connected and K_σ zero-dimensional. Then $\tilde{m} \in E_m(X,T)$ if and only if $S_\nu \supset G$ for all $\nu \in E_m(K_\sigma \times X,T)$ if and only if $\tilde{\omega} \in E_m(K_\sigma \times X,T)$ for all $\omega \in E_m(K_\sigma \times Y,T)$.

Proof.

From Propositions 3.5, 3.8 and Remark 3.7 we have to show that if $H = K_\sigma S_\nu$ then $S_\nu \supset G$.

Let $g \in G$ and put

$$K(g) = \{k \in K_\sigma : g \in kS_\nu\}$$

We will show that $K(g)$ is a coset of $K_\sigma \cap S_\nu$ in K_σ. Let $k_1, k_2 \in K(g)$ Then there are $s_1, s_2 \in S_\nu$ such that $g = k_1 s_1 = k_2 s_2$. Thus $k_2^{-1} k_1 = s_2 s_1^{-1} \in S_\nu \cap K_\sigma$. Hence $K(g)$ is contained in a coset of $K_\sigma \cap S_\nu$. Now let $k \in K(g)$ and let $k_1 \in k \cdot (K_\sigma \cap S_\nu)$, say $k_1 = ks$. Then $g = ks_1 = k_1 s^{-1} s_1 \in k_1 s_\nu$ and so $k_1 \in K(g)$.

Thus we have a mapping $\eta : G \to K_\sigma / K_\sigma \cap S_\nu$, $\eta(g) = K(g)$, and we claim that η is continuous. To see this, if U is open in $K_\sigma / K_\sigma \cap S_\nu$ and $\tau : K_\sigma \to K_\sigma / K_\sigma \cap S_\nu$ is the canonical map, then $\eta^{-1}(U) = \{g \in G : K(g) \subset \tau^{-1}(U)\} = \{g \in G : g \in \tau^{-1}(U) S_\nu\} = G \cap \tau^{-1}(U) S_\nu$. Since $\tau^{-1}(U) S_\nu$ is open in H, $\eta^{-1}(U)$ is open in G. Moreover, since K_σ is zero-dimensional, so is $K_\sigma / K_\sigma \cap S_\nu$. As G is connected, $\eta(G) = \{\eta(e)\} = \{K_\sigma \cap S_\nu\}$, and thus $G \subset (K_\sigma \cap S_\nu) \cdot S_\nu = S_\nu$.

The condition that K_σ be zero-dimensional is not so stringent as it might at first appear. In fact we have

Remark 3.10. If G is abelian, then K_σ is zero-dimensional.

Proof.

By its definition, K_σ is a subgroup of the automorphism group of G. By Iwasawa's theorem [4], that automorphism group is totally disconnected and so K_σ is totally disconnected hence zero-dimensional.

If the acting group T is Z, then K_σ is a compact zero-dimensional monothetic group and these have been classified (see [1], p. 408): if G is abelian, then K_σ is either a finite cyclic group or $K \cong \Delta_{\underline{a}}$ for some sequence $\underline{a}$ (see [1] p. 109 for the definition of $\Delta_{\underline{a}}$).

We now apply Theorem 3.9 to get

<u>Corollary 3.11</u>. <u>If</u> G <u>is a connected abelian group, then</u> $\tilde{m} \in E_m(X,T)$ <u>if and only if</u> $S_\nu \supset G$ <u>for all</u> $\nu \in E_m(K_\sigma \times X,T)$ <u>if and only if</u> $\tilde{\omega} \in E_m(K_\sigma \times X,T)$ <u>for all</u> $\omega \in E_m(K_\sigma \times Y,T)$.

We note in the case $T=Z$, with the actions on Y,X,G generated by ψ,ϕ,τ respectively, that if (Y,ψ,m) is totally ergodic and G is connected abelian then $\tilde{m} \in E_m(X,\phi)$ if and only if $\mu \times \tilde{m} \in E_m(K_\sigma \times X, R_\tau \times \phi)$ (R_τ denotes translation by τ on K_τ). This follows by noting that if $\tilde{m}$ is ergodic, $\mu \times m$ is ergodic iff $\mu \times \tilde{m}$ is ergodic.

We end by briefly considering minimality for equicontinuous-(G,σ)-extensions with abelian G and $T=Z$, and look for similar observations to the results on ergodicity.

Recall that (X,ϕ) is a topologically simple (G,τ)-extension of (Y,ψ) if for each $\gamma \in \Gamma(G)$ there is a continuous $\gamma^{\perp}$function f.

The first result is the topological analogue of Corollary 2.3.

<u>Theorem 3.12</u>. <u>Let</u> (X,ϕ) <u>be a topologically simple equicontinuous</u>-(G,τ)-<u>extension of</u> (Y,ψ). <u>Suppose that</u> X <u>is connected and</u> (Y,ψ) <u>is minimal</u>. <u>Then</u> (X,ϕ) <u>is minimal if and only if given a continuous</u> γ-<u>function</u> f, $f\phi^n = f$ <u>for some</u> $n \geq 1$ <u>implies</u> $\gamma \equiv 1$.

Proof.

Suppose (X,ϕ) is minimal and f is a continuous γ-function with $f\phi^n = f$. Then, since (X,ϕ) is totally minimal, f is a constant, β say, and so $\beta = f(g^{-1}x) = \gamma(g)f(x) = \gamma(g)\beta$. Thus $\gamma(g) = 1$ for all $g \in G$.

Conversely, if (X,ϕ) is not minimal let M be a proper minimal subset. Put $H = \{g : gM = M\}$; then H is a closed proper subgroup of G and so there is a $\gamma \in \Gamma(G)$ with $\gamma|_H = 1$, $\gamma \not\equiv 1$. Let f be a continuous γ-function and define $\bar{f} : Y \to S^1$ by $\bar{f}(\pi x) = f(gx) = \gamma(g)f(x)$, where g is chosen so that $gx \in M$. One directly verifies that $\bar{f}$ is continuous. Since τ is equicontinuous, Theorem 1.4 implies that $\gamma\tau^n = \gamma$ for some $n \geq 1$. Since $gx \in M$ if and only if $\tau(g)\phi(x) \in M$ it follows that

$$\frac{\bar{f}\pi\phi^n(x)}{\bar{f}\pi(x)} = \frac{f(\tau^n(g)\phi^n(x))}{f(gx)} = \frac{\gamma\tau^n(g)}{\gamma(g)}\,\frac{f\phi^n x}{f(x)} = \frac{f\phi^n(x)}{f(x)} .$$

Then $\frac{f}{\bar{f}\pi}$ is a ϕ^n-invariant γ-function, $\gamma \not\equiv 1$, as required

The next result is similar to Corollary 3.11

Theorem 3.13. Let (X,ϕ) be an equicontinuous (G,τ)-extension of (Y,ψ) with X connected. Then (X,ϕ) is minimal if and only if $(K_\tau \times X, R_\tau \times \phi)$ is minimal (R_τ denotes translation by τ in K_τ).

Proof.

Since (K_τ, R_τ) is equicontinuous and K_τ is a group then the characters $\gamma \in \Gamma(K_\tau)$ form a complete set of eigenfunctions for (K_τ, R_τ). Since K_τ is zero-dimensional, Remark 3.10, it follows that each $\gamma(K_\tau)$ is finite, $\gamma \in \Gamma(K_\tau)$, and hence all the eigenvalues of (K_τ, R_τ) have finite order.

Now suppose (X,ϕ) is minimal. Then, since X is connected, (X,ϕ) has no eigenvalues of finite order other than 1. It follows then that (X,ϕ) and (K_τ, R_τ) have no common factors and hence $(K_\tau \times X, R_\tau \times \phi)$ is minimal by [5, Corollary 3.4].

The converse is clear.

References

[1] E. Hewitt, K. Ross: Abstract Harmonic Analysis Vol. 1. Springer-Verlag, 1963.

[2] Ibid, Vol. II, Springer-Verlag, 1970.

[3] H. Hoare, W. Parry: Semi-Groups of Affine Transformations, Quarterly Jour. Math. 17, 106-111 (1966).

[4] K. Iwasawa: On some types of topological groups, Ann. of Math. (2) 50, 507-558 (1949).

[5] H.B. Keynes: Disjointness in transformation groups, Proc. Amer. Math. Soc. 36, 253-259, (1972).

[6] H.B. Keynes, D. Newton: The structure of ergodic measures for compact group extensions, Israel Jour. Math. 18, 363-389, (1974).

[7] H.B. Keynes, D. Newton: Ergodic measures for non-abelian compact group extensions, Compositio Math. 32, 53-70, (1976).

[8] H.B. Keynes, D. Newton: Minimal (G,τ) - extensions, Pacific Journal Math. 77 (1978), 145-163.

[9] D.A. Lind: The Structure of Skew Products with Ergodic Group Automorphisms, to appear: Israel Jour. Math.

[10] W. Parry: Ergodic properties of affine transformations and flows on nilmanifolds, Amer. J. Math. 91, 757-771 (1969).

[11] D. Rudolph: If a finite extension of a Bernoulli shift has no finite rotation factors, it is Bernoulli, Preprint.

[12] R.K. Thomas: Metric properties of transformations of G - spaces, Trans. Amer. Math. Soc. 160, 103-117 (1971).

[13] W. Veech: Finite group extensions of irrational rotations, Israel Jour. Math. 21, 240-25 (1975).

[14] P. Walters: Some transformations having a unique measure with maximal entropy, Jour. Lond. Math. Soc. 43, 500-516 (1974).

University of Minnesota, Minneapolis, Minn. 55455
University of Sussex, Brighton, England

A Probabilistic Version of Bowen - Ruelle's Volume Lemma

by
Yuri Kifer
Institute of Mathematics, Hebrew University of Jerusalem
Jerusalem, Israel

Introduction

Let M be a compact n-dimensional Riemannian manifold and B a C^2-class ctor field on M, generating a dynamical system S^t such that

$$d(S^t x)/dt \Big|_{t=0} = B(x).$$

t us consider the Markov diffusion process x_s^ε with the generator of the rm $L^\varepsilon = \varepsilon^2 L + B$, where L is a nondegenerate elliptic differential operator the second order. The process x_s^ε is called a small random perturbation of e dynamical system S^t.

We shall study in this paper the asymptotic behavior as $t \to \infty$ and $\varepsilon \to 0$ the probability $P_x^\varepsilon \{\text{dist}(x_s^\varepsilon, S^s x) \le \delta, \ 0 \le s \le t\}$, where $P_x^\varepsilon\{\cdot\}$ enotes the probability of the event in brackets for the process x_t^ε starting at . Under some hyperbolicity assumptions on the dynamical system S^t we shall prove, rovided δ is small enough that the asymptotic in question turns out the same as n Volume lemma (Lemma 4.2 in [1]).

The results of the present paper Theorem 2.1 and also Proposition 5.1 are interesting lso in the frame of the study of parameters of dynamical systems which can be approximated y some probabilistic parameters of their small random perturbations and thus can be onsidered as stable with respect to such perturbations (see also [2] and [3]).

. Assumptions and the main theorem.

Let the dynamical system S^t has the compact invariant set $\Lambda \subset M$ which is basic hyperbolic set (see [1]). The tangent bundle restricted to Λ can be ecomposed into the Whitney sum of three DS^t-invariant continuous subbundles

$$T_\Lambda M = B + E^s + E^u,$$

here DS^t is the differential of S^t, and there are constants $c, \lambda > 0$ so that

(i) $\|DS^t v\| \leq ce^{-\lambda t}\|v\|$ for $v \in E^s$, $t \geq 0$ and

(ii) $\|DS^{-t} v\| \leq ce^{-\lambda t}\|v\|$ for $v \in E^u$, $t \geq 0$.

For any $x \in \Lambda$ let $D_t(x)$ be the determinant of the linear map $DS^t : E^u_x \to E^u_{S^t_x}$, provided inner products in tangent spaces are induced by the Riemannian metric. The main result of the present paper is the following theorem.

Theorem 2.1. Let $x \in \Lambda$ and there exists

$$\text{(2.1)} \qquad \lim_{t\to\infty} \frac{1}{t} \ln D_t(x) = \Delta_x,$$

then for some $\delta_0 > 0$ independent of x,

$$\text{(2.2)} \qquad \lim_{\varepsilon\to 0} \lim_{t\to\infty} \frac{1}{t} \ln P^\varepsilon_x \{\text{dist}(x^\varepsilon_s, S^s_x) \leq \delta,\ 0 \leq s \leq t\} = -\Delta_x,$$

provided $\delta \leq \delta_0$.

This theorem can be proved also for a single trajectory of the dynamical system S^t, provided this trajectory satisfies some kind of uniform hyperbolicity conditions. Theorem 2.1 seems to be true also if the hyperbolicity is replaced by some more weak assumption similar to [4]. If there exist the exact Ljapunov exponents (see [4]) for the trajectory $\{S^s_x,\ s_0 \leq s < \infty\}$, then Δ_x is equal to the sum of the positive Ljapunov exponents.

Set

$$\varphi(x) = -\left.\frac{d \ln D_t(x)}{dt}\right|_{t=0}$$

then also

$$\varphi(S^t_x) = -\frac{d \ln D_t(x)}{dt}$$

and therefore

$$\ln D_t(x) = -\int_0^t \varphi(S^s_x) ds .$$

The function $\varphi(x)$ is Holder continuous on Λ (see [1]) and by the ergodic heorem it follows that the limit (2.1) exists almost everywhere with respect o any Borel invariant measure of the dynamical system S^t and Δ_x is equal o a constant almost everywhere with respect to any ergodic measure. If Λ s an attractor then by [1] this limit equals the entropy of the dynamical ystem S^t with respect to some Gibbs measure almost everywhere relative to his measure, which according to [2] is stable to random perturbations..

. Topological lemmas. We shall need the following result which is alled the shadowing property.

Lemma 3.1. There exist $K_0, \rho_0, \delta_0 > 0$ such that for any set of oints $x_1,\ldots,x_m$ with the properties:

i) $\mathrm{dist}(x_i,\Lambda) \leq \rho_0$; $i = 1,\ldots,m$, and

ii) $\mathrm{dist}(S^1x_i,x_{i+1}) \leq \delta$; $\delta \leq \delta_0$; $i = 1,\ldots,m-1$,

here exist some point y such that

$$\mathrm{dist}(y,\Lambda) \leq \rho_0 \text{ and } \mathrm{dist}(x_i,S^iy) \leq K_0\cdot i\cdot\delta \quad (3.1)$$

The proof of this lemma is obtained by the method of [5] (see also Lemma 5.1 f [2], Lemma 4.1 of [3] and the paper [6]).

We shall use also the following lemma.

Lemma 3.2 (a) There exist $\gamma_1,K_1,K_2, \delta_0 > 0$ such that if $x \in \Lambda$, $y \in M$ nd

$$\mathrm{dist}(S^iy,S^ix) \leq \delta \leq \delta_0, \quad i = 1,\ldots,[t], \quad (3.2)$$

here $[t]$ is the integral part of t,

hen

$$\max \{\mathrm{dist}(S^sy,S^{s+\tau}x), \mathrm{dist}(S^{t-s}y, S^{t-s+\tau}x)\} \leq \quad (3.3)$$

$$\leq K_1 e^{-\gamma_1 s} \max \{\mathrm{dist}(y,x), \mathrm{dist}(S^ty,S^tx)\}$$

<u>for some</u> $\tau = \tau(x,y) \leq K_2\delta$ <u>provided</u> $0 \leq s \leq t/2 - \tau$;

(b) <u>There exist</u> γ_2, $K_3 > 0$ <u>such that if</u> $x \in \Lambda$, $y \in M$, $z \in M$ <u>and</u>

$$\text{(3.4)} \quad \text{dist}\,(S^i_y, S^i_x) \leq \delta_0 \ \ \underline{\text{and}}\ \ \text{dist}\,(S^i_z, S^i_x) \leq \delta_0 ,$$

$i = 1,\ldots,[t]$ <u>then</u>

$$\text{(3.5)} \quad \max_{0\leq s\leq t} \{\text{dist}\,(S^s_y, S^s_z)\} \leq$$

$$\leq K_3 \max\{\text{dist}(y,z), \text{dist}(S^t_y, S^t_z)\}^{\gamma_2} .$$

The proof of (a) is contained in the proof of Lemma A.2 of [1]. The statement of the item (b) follows immediately from (a) in the same way as in Lemma 4.3 of [3].

4. <u>Proof of Theorem 2.1.</u>

Let

$$\text{(4.1)} \quad Q^\varepsilon(\delta;t;x,y) = P^\varepsilon_x \{\text{dist}(x^\varepsilon_s, S^s_y) \leq \delta , \ 0 \leq s \leq t\} .$$

Then by the Markov property of the process x^ε_t we obtain

$$\text{(4.2)} \quad Q^\varepsilon(\delta;t,x,x) \leq Q^\varepsilon(\delta;[\tfrac{t}{k}]k;x,x) = E^\varepsilon_x \ \chi_{\text{dist}(x^\varepsilon_s, S^s_x) \leq \delta ,\ 0 \leq s \leq k}$$

$$\times E^\varepsilon_{x^\varepsilon_k} \chi_{\text{dist}(x^\varepsilon_s, S^{s+k}_x) \leq \delta ,\ 0 \leq s \leq k} \cdots$$

$$\cdot \cdots E^\varepsilon_{x^\varepsilon_{([\frac{t}{k}]-1)k}} \chi_{\text{dist}(x^\varepsilon_s, S^{s+([\frac{t}{k}]-1)k}_x) \leq \delta ,\ 0 \leq s \leq k} \leq$$

$$\leq \prod_{\ell=0}^{[\frac{t}{k}]-1} \max_{x_\ell \in U_\delta(S^{\ell k}_x)} Q^\varepsilon(\delta;k,x_\ell,S^{\ell k}_x),$$

here $0 < k \leq t$ is an integer, E_x^ε is the expectation of the process x_s^ε starting at x, χ_A is the indicator of the event A, $\prod$ denotes the product and $U_\delta(z)$ is the ball of radius δ with the centre at z.

at

$$(4.3)\quad J^\varepsilon(\delta;m,v,w) = \int_{U_\delta(S_w^1)} \cdots \int_{U_\delta(S_w^m)} p^\varepsilon(1,v,z_1)p^\varepsilon(1,z_1,z_2)\ \cdots\ p^\varepsilon(1,z_{m-1},z_m)\ dz_1,\ldots,dz_m\ ,$$

here $p^\varepsilon(s,x,y)$ is the transition density of the process x_s^ε. It is clear that the following inequality is true

$$(4.4)\quad Q^\varepsilon(\delta;k,x_\ell,S_x^{\ell k}) \leq J^\varepsilon(\delta;k,x_\ell,S_x^{\ell k}).$$

On the other hand, by the same arguments, we get

$$(4.5)\quad Q^\varepsilon(\delta;t,x,x) \geq \geq \prod_{\ell=0}^{[\frac{t}{k}]} \min_{x_\ell \in U(S_x^{\ell k})} Q^\varepsilon(\delta;k,\ x_\ell,S_x^{\ell k}).$$

e can also write

$$(4.6)\quad Q^\varepsilon(\delta;k,x_\ell,S_x^{\ell k}) = J^\varepsilon(\varepsilon^{\alpha_1};k;x_\ell,S_x^{\ell k}) - R_1(\varepsilon,x_\ell),$$

where $\alpha_1 > 0$ and

$$(4.7)\quad R_1(\varepsilon,x_\ell) \leq k \max_{0\leq s,\ z\in U_{\varepsilon^{\alpha_1}}(S_x^s)} P_x^\varepsilon\{\max_{0\leq u\leq 1} \operatorname{dist}(x_u^\varepsilon,S_x^{s+u}) > \delta\}$$

$$\leq K_4 k \exp(-\gamma_3\delta^2/\varepsilon^{2(1-\alpha_1)}),$$

for some $K_4,\ \gamma_3 > 0$. The second inequality follows easily from [7].

Next

(4.8) $J^\varepsilon(\delta;k,x_\ell,S^{\ell k}_x) =$

$$= \int_{U_\delta(S^{\ell k+1}_x) \,\cap\, U_{\varepsilon^{\alpha_1}}(S^1_{x_\ell})} \quad \int_{U_\delta(S^{\ell k+2}_x) \,\cap\, U_{\varepsilon^{\alpha_1}}(S^1_{z_1})} \cdots$$

$$\cdots \int_{U_\delta(S^{(\ell+1)k}_x) \,\cap\, U_{\varepsilon^{\alpha_1}}(S^1_{z_{k-1}})} \quad p^\varepsilon(1,x_\ell,z_1)\, p^\varepsilon(1,z_1,z_2) \cdots$$

$$\cdots p^\varepsilon(1,z_{k-1},z_k)\, dz_1 \ldots dz_k + R_2(\varepsilon,x_\ell) =$$

$$= J(\varepsilon,x_\ell) + R_2(\varepsilon,x_\ell),$$

where the first and the second equalities in (4.7)
are just the definitions of $R_2(\varepsilon,x_\ell)$ and $J(\varepsilon,x_\ell)$, respectively.

From [7] it follows that

(4.8) $R_2(\varepsilon,x_\ell) \le K_5\, k \exp(-\gamma_4/\,\varepsilon^{2(1-\alpha_1)})$

for some K_5, $\gamma_4 > 0$.

Notice that the integration in $J(\varepsilon,x_\ell)$ is over ε^{α_1}-pseudo orbits i.e.
the sequences of points satisfying (ii) of Lemma 3.1 with $\delta = \varepsilon^{\alpha_1}$ and $m = k$.
Therefore by Lemma 3.1 there exist trajectories of the dynamical system S^t which
are close to corresponding pseudo orbits i.e. (3.1) holds. According to Lemma 3.2
all such trajectories will be ε-close to the trajectory of the point x except
for their ends of the length not more than $\tilde{k}(\varepsilon)$, where

(4.9) $\tilde{k}(\varepsilon) = \gamma_1^{-1} \ln \dfrac{\delta K_1}{\varepsilon}$.

Therefore we have for $k > 2\tilde{k}(\varepsilon) + 3$

(4.9) $J(\varepsilon,x_\ell) \leq$

$$\leq \sup_{v\in U_{\varepsilon^{\alpha_2}}(S_x^{\ell k+\tilde{k}(\varepsilon)})} J^\varepsilon(\varepsilon^{\alpha_2};k-2\tilde{k}(\varepsilon),v,S_x^{\ell k+\tilde{k}(\varepsilon)}),$$

where $\alpha_2 > 0$ such that $\varepsilon^{\alpha_2} > k\,\varepsilon^{\alpha_1} + \varepsilon$.

If ε is small enough we can actually take

(4.10) $\alpha_1 = 9/10,\ \ \alpha_2 = 3/4,\ \ k = [\ \varepsilon^{-1/10}\ ]$.

In order to estimate the right hand side of (4.9) we approximate and replace each density in the integral $J^\varepsilon(\varepsilon^{\alpha_2};k-2\tilde{k}(\varepsilon),v,\ S_x^{\ell k+\tilde{k}(\varepsilon)})$ by some Gaussian density in the same way as in [2] and [3] (see Lemma 4.1 in [2] and Lemma 6.2 in [3]).

After integration this leads to the following inequality (see (5.4) in [2] and §§ 5 and 6 in [3]),

(4.11) $J^\varepsilon(\varepsilon^{\alpha_2};\ k-2\tilde{k}(\varepsilon),v,S_x^{\ell k+\tilde{k}(\varepsilon)}) \leq$

$$\leq (1+K_6\,\varepsilon^{\gamma_3})^{k-2\tilde{k}(\varepsilon)}\ D^{-1}_{k-2\tilde{k}(\varepsilon)}\ (S_x^{\ell k+\tilde{k}(\varepsilon)})$$

for some $K_6, \gamma_3 > 0$ independent of ε. Therefore by (4.1) - (4.4) and (4.8)-(4.11) we obtain, provided ε is small enough and δ satisfies the conditions of Lemma 3.1 and 3.2, that

(4.12) $$\limsup_{t\to\infty} \frac{1}{t} \ln P_x^\varepsilon\{\mathrm{dist}(x_s^\varepsilon, S^s x) \leq \delta\ ,$$

$$0 \leq s \leq t\} \leq K_7\,\varepsilon^{\gamma_4} - \liminf_{t\to\infty} \frac{1}{t}\ln D_t(x),$$

for some $K_7, \gamma_4 > 0$.

On the other hand, in the same way as in [2] and [3] by the substitution of Gaussian densities for the densities in the integral $J^\varepsilon(\varepsilon^{\alpha_1};k,x_\ell,S_x^{\ell k})$ we obtain that

$$(4.13) \qquad J^{\varepsilon}(\varepsilon^{\alpha_1}; k, x_\ell, S\overset{\ell k}{x}) \geq$$

$$\geq (1 - K\varepsilon^{\gamma_5})^k D_k^{-1}(S\overset{\ell k}{x}) , \quad \text{if} \quad x_\ell \in U_\varepsilon(S\overset{\ell k}{x}) ,$$

for some $K_8, \gamma_5 > 0$. Thus by (4.1), (4.5) - (4.7), (4.10) and (4.13) we get

$$(4.14) \quad \liminf_{t\to\infty} \frac{1}{t} \ln P_x^\varepsilon \ \text{dist}\,(x_s, S\overset{s}{x}) \leq \delta , 0 \leq s \leq t\} \geq$$

$$\geq - (K_9\varepsilon^{\gamma_6} + \limsup_{t\to\infty} \frac{1}{t} \ln D_t(x)) ,$$

for some $K_9, \gamma_6 > 0$. The relations (4.12) and (4.14) complete the proof of (2.2) , provided (2.1) holds.

5. Related problem. Consider the probability

$$(5.1) \qquad Q^\varepsilon(\delta; t; x) = P_x^\varepsilon\{\text{dist}(x_j^\varepsilon, \Lambda) \leq \delta , \quad 0 \leq s \leq t\}$$

with dist $(x , \Lambda) < \delta$. It is known (see [3] Lemma 3.1) that for δ small enough there exists the limit

$$(5.2) \quad \lambda^\varepsilon = \lim_{t\to\infty} \frac{1}{t} \ln Q\,(\delta; t; x) ,$$

where λ^ε is the principal eigenvalue of the Dirichlet problem for the operator L^ε in the domain $U_\delta(\Lambda) = \{z : \text{dist}(z, \Lambda) \leq \delta\}$.

On the other hand let us consider the topological pressure $P(\Lambda)$ which according to [1] and some additional considerations has the following representation

$$(5.3) \qquad P(\Lambda) = \limsup_{t\to\infty} \frac{1}{t} \ln \text{volume}\ \{z : \text{dist}(S\overset{s}{z}, \Lambda) \leq \delta, \quad 0 \leq s \leq t\}$$

independently of δ , provided δ is small enough. The methods of [2], [3] and the present paper together with [1] are sufficient to justify the following proposition.

Proposition 5.1. Let in (5.3) be the exact limit then

$$\lambda^{\varepsilon} \underset{\varepsilon\to 0}{\to} P(\Lambda) .$$

If Λ is an attractor then by [1] $P(\Lambda) = 0$. On the other hand in the ame way as in [8] it follows easily that $\lambda^{\varepsilon} \to 0$. Thus in this particular ase Proposition 5.1 follows. When Λ is a point or a circle Proposition 5.1 ollows from [3].

References

1. R. Bowen and D. Ruelle,, The ergodic theory of Axiom A flows, Invent. Math. 29 (1975), 181-202.

2. Yu. I. Kifer, On small random perturbations of some smooth dynamcial systems, Mathe. USSR Izvestija 8(1974), 1083-1107.

3. ___________, On the principal eigenvalue in a singular perturbation problem with hyperbolic limit points and circles, PReprint, 1979.

4. Ja. B. Pesin, Ljapunov characteristic exponents and smooth ergodic theory, Russian, Math. Surveys, 37: 4 (1977).

5. C. Robinson, Stability theorem and hyperbolicity in dynamical systems, The Rocky Mount. J. of Math. 7(1977), 425-437.

6. J. Franke and J. Selgrade, Hyperbolicity and chain recurrence, J. Diff. Equat. 26(1977), 27-36.

7. D. G. Aronson, The fundamental solution of a linear parabolic equation containing a small parameter, Ill. J. Math. 3(1959), 580-619.

8. A. Friedman, The asymptotic behavior fo the first real eigenvalue of a second order elliptic operator with small parameter in the highest derivatives, IndianaUniv. Math. J. 22(1973), 1005-1015.

Periodically forced relaxation oscillations

Mark Levi

§1. Introduction

Our aim is to apply some recent results and methods of the theory of dynamical systems to qualitative analysis of a Van der Pol-type system with forcing

$$\varepsilon\ddot{x} + \varphi(x)\dot{x} + \varepsilon x = bp(t), \tag{1}$$

where ε is a small but fixed parameter, $\varphi(x)$ (the damping) is negative for $|x| < 1$ and positive elsewhere, $p(t)$ is a periodic forcing of period T and b belongs to some finite interval $[b_1, b_2]$ of length of order 1 (independent of ε), to be specified later. One can choose φ,p close (in some sense) to $\varphi_0 = \mathrm{sgn}(x^2-1)$, $p_0(t) = \mathrm{sgn}\sin\frac{2\pi}{T}t$ *), see Fig. 1.

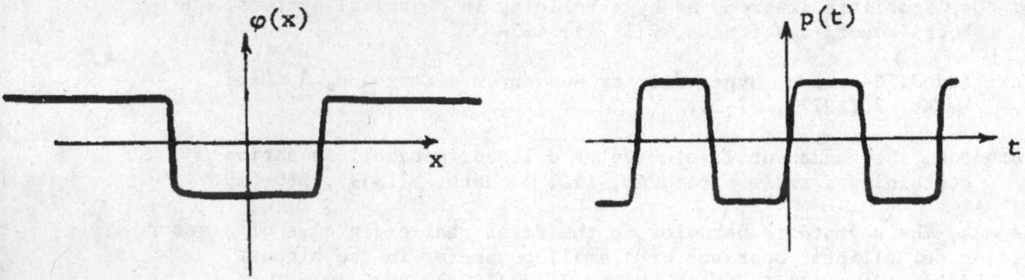

Fig. 1

Such an equation describes oscillations of the current in a triode circuit with a feed-back and with a periodic external

*This specific choice of φ,p is inessential for the qualitative behavior of the system; however, it allows a rigorous and complete analysis for "most" b. The choice $\varphi(x) = x^2-1$ corresponds to the classical Van der Pol left-hand side.

orcing; some biological systems also obey such an equation.

For $b = 0$ we have an autonomous system whose behavior has been ell-known for over half a century.

Making $b \neq 0$, however, complicates the behavior drastically. n the early 1940's it was observed experimentally that the equaion (1) has a periodic solution of a period much larger than that f a forcing term, namely, an integer multiple (around 400 for ertain ε) of T. This effect was used in electronics to obtain ow-frequency oscillations. An interest in this problem was timulated by a puzzling observation: for some values of b the ystem possessed two periodic solutions of <u>different</u> periods; in act, the experiments showed that the intervals of b for which here is one or two observable periodic solutions, alternate, i.e. or b increasing, the system admits alternately only one or only wo stable periodic regimes.

The significance of two periodic solutions with different eriods was noticed by Cartwright and Littlewood [1], who observed hat it implies existence of the so-called strange attractor*) - n attractor which is neither a point nor a curve, previously not nown to arise in differential equations.

Most interestingly, Cartwright and Littlewood had discovered a subfamily of solutions, which exhibits a "random" behavior.**) ([1],[4],[5]).

---

*The term "strange attractor" was not used by Cartwright and Littlewood, but was introduced by Ruelle and Takens in a different context about two decades later. The term attractor is used in the sense of Conley since the whole invariant set is not nonwandering.

**We will see later how this family fits within the attractor.

Their analysis, quite involved, was considerably simplified by Levinson [7], who chose $\varphi(x)$ so as to make (1) piecewise linear, so that the solutions could be analyzed using the explicit formulae on each linearity interval.

These classical results described a certain subfamily of solutions of (1). It remained unclear, however, how do the other solutions behave, and, most importantly, what is the geometrical reason for such a behavior, and how is this geometry deduced from the forms of the equation (1). Also, what kind of bifurcations occur as b changes?

These questions will be answered in the following order. First we state the results (§2); their informal justification is given in §3 and §4. More specifically, in §3 we reduce the study of eq. (1) to that of an annulus map and describe its qualitative behavior, bifurcations, etc., assuming that this map has a certain simple form. This assumption is justified on intuitive level in §4; that is, we describe the behavior of the flow of (1). This description is the basis (and the main difficulty) of our analysis.

Our attack on the problem consists therefore of two main parts: 1) determination of the form of Poincare map associated with eq. (1); 2) Deduction of the properties of the high iterates of this map using the form found in 1).

Much of the analysis in part 2) uses some recent results in the theory of dynamical systems - notably, the concept of horseshoe map of Smale [14],[15], see also Moser [8], bifurcation theory

Newhouse and Palis [9],[10],[11], etc. It should be noted
hat eq. (1), which arose in electronics, was a major incentive
n the development of the theory; we apply this theory back to
he equation.

2. Qualitative properties of the system - the results.

.1. Assumptions

Prior to stating the results, we indicate the assumptions and ntroduce some notations.

We assume that $\varphi(x)$ is even: $\varphi(x) = \varphi(-x)$, and that $p(t)$ atisfies a symmetry property $p(t + \frac{T}{2}) = -p(t)$, T being the eriod of $p(t)$. To be specific, we take for φ, p the functions $gn(x^2-1)$, sgn sin $\frac{2\pi}{T}$ t correspondingly, smoothed near their iscontinuities so as to preserve their symmetry properties (and eriodicity of p). Any functions uniformly close to these will o too.*) Assume also, that $p(t) > 0$ for $0 < t < \frac{T}{2}$. Introduce

$$= \int_0^{T/2} p(t)dt, \quad \Phi(x) = \int_0^x \varphi(\xi)d\xi \text{ (see Fig. 2)}$$

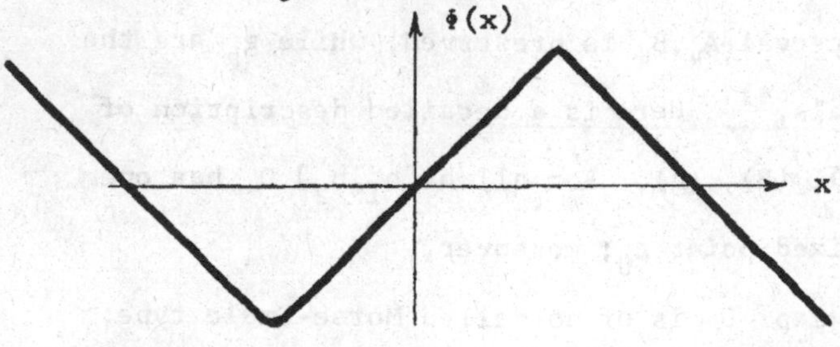

Fig. 2

*This class can be extended (see remark in [3]).

To specify the interval $[b_1,b_2]$, we fix $b_1 > 0$ to be a small constant (say, $b_1 = \frac{1}{100}$), choose $b_2 = \frac{2m-b_1}{\overline{p}}$ and assume that period T is long enough - as it turns out in the proof (given in [3]), it suffices to have $\frac{Tb_1}{2} > \frac{2m-b_1}{\overline{p}} = b_2$. Finally, instead of looking at eq. (1) we consider an equivalent system

$$\begin{aligned} \dot{x} &= \frac{1}{\epsilon}(y - \Phi(x)) \\ \dot{y} &= -\epsilon x + bp(t), \end{aligned} \qquad (2)$$

and describe the nonautonomous flow (2) by sampling the positions of the solutions $(x(t),y(t))$ at discrete times nT - in other words, we look at the Poincare map $D\colon (x,y)_{t=0} \to (x,y)_{t=T}$.

2.2. Qualitative properties of the system

If the above assumptions hold, then for $\epsilon > 0$ small enough, the following (including the classical results) holds.

The range $[b_1,b_2]$ of b-values consists of the alternating subintervals A_k,B_k separated by thin gaps g_k of small (with ϵ) total length, such that the qualitative behavior of the map D throughout each interval A_k,B_k is preserved, while g_k are the bifurcation intervals.*) Here is a detailed description of what happens in (A), (B), (g). For all $b \in [b_1,b_2]$ D has one totally unstable fixed point z_0; moreover,

(A) for $b \in A_k$, the map D is of so-called Morse-Smale type; more specifically, D has exactly one pair of periodic points

*For a simple geometrical explanation of such alternating behavior see §3.2 (Fig. 6), and beginning of §3.3.

of period (2n-1) with an integer $n = n(k) \sim \frac{1}{\varepsilon}$ constant throughout each A_k. One of these points is a sink, another a saddle, see Fig. 3.

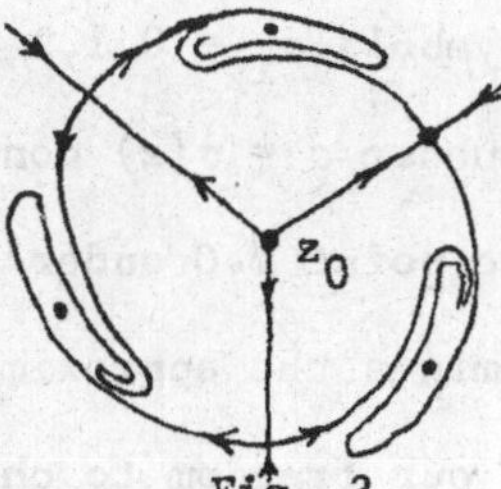

Fig. 3

any point which lies off the stable manifold of the saddle (except for z_0) tends to the sink.

A more interesting case is

(B) for $b \in B_k$, the invariant set of D consists (besides z_0) of two sink-saddle pairs of periods 2n + 1, 2n - 1 correspondingly, and of an invariant hyperbolic Cantor set C, to which the saddles belong; symbolically the situation is depicted on Fig. 4. The set C can be thought of as the set of the points which are undecided to which of the two sinks to tend for future iterates, and which stay away from z_0 and ∞ for all negative iterates by D.

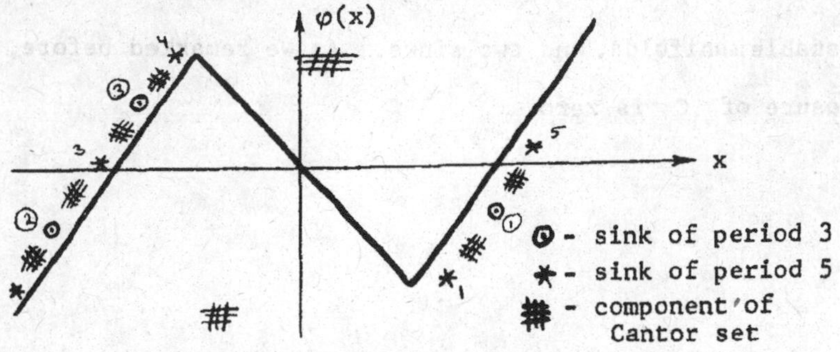

Fig. 4

A more precise description of this Cantor set is given in §3 in terms of a certain return map. Here we indicate only that each point z in C can be represented by a certain sequence $\sigma = \sigma(z) = (\ldots \sigma_{-1}\sigma_0\sigma_1 \ldots)$ of symbols $\sigma_i = 0,1,2$ or 3 with some pairs $\sigma_i\sigma_{i+1}$ forbidden. Each sequence $\sigma = \sigma(z)$ contains the information on the behavior of the point $z \in C$ under the iterations by D: namely, the j-th symbol determines the approximate position of j-th iterate $D^j z$. In particular, our freedom to choose σ_j arbitrarily (within the restriction mentioned above) reflects in the "random" behavior of a sequence of iterates $D^j z$, where $z = z(\sigma)$ corresponds to σ.

As a consequence of this description, we obtain infinitely many periodic points of D, since there are infinitely many periodic sequences.

We remark that the measure of C is zero - this is the answer to a conjecture of Littlewood. In particular, iterates $D^j z$ of almost all points z tend to one of the two sinks, which explains why the Canotr set is not observed experimentally.

Attractor

The attractor consists of the Cantor set C with its unstablemanifolds, and two sinks. As we remarked before, measure of C is zero.

Rotation numbers.*

An interesting phenomenon related to the stochasticity is he existence of the full interval of rotation numbers - namely, he set of rotation numbers is exactly a closed interval $[\frac{2\pi}{2n+1}, \frac{2\pi}{2n-1}]$. In other words, for any number r in this inter- al there is a point $z = z(r) \in$ Cantor set, whose rotation number s r. Here n is the integer in the expression for the periods f the two sinks. (Here b still belongs to B_k).

Structural Stability

Both cases: $b \in A_k$ and $b \in B_k$ correspond to D structurally table - the above described behavior is not pathological in that t cannot be destroyed by small perturbations of the system (1).

This follows by application of the structural stability heorems of Palis [12] for $b \in A_k$ and Robbin [13] for $b \in B_k$.

g) $g \in g_k$: Bifurcations

As b crosses the gap g_k, a complicated sequence of bifur- ations occurs. Perhaps the most interesting feature of these ifurcations is occurence (for some b) of infinitely many table periodic points. Despite their stability, they would be ery hard to detect on computer due to their high period and mall basin of attraction.

---

*Definition. A real number r is called a rotation number of a ap $D: \mathbb{R}^2 \to \mathbb{R}^2$ with respect to a fixed point z_0 of D if for some $\neq z_0$ $r = \lim_{n \to \infty} \frac{\arg(D^n z - z_0)}{n}$. In other words, r is an average angle if exists) by which a point is related by application of D.

Classical results.

We point out, that the existence of alternately one and two sink-saddle pairs was shown by Cartwright, Littlewood and Levinson. The family of the solutions, found by Levinson for $b \in B_k$, corresponds to the sequences containing no 0's and 2's. The above description shows, that in addition to Levinson's periodic solutions, there are infinitely many others.

Remark 2.1. In analyzing the case $b \in B_k$ we use the concept of the horseshoe map (Smale); for its description see [8].

The bifurcations ($b \in g_k$) are analyzed by applying recent results of Newhouse and Palis [9], [10], [11].

§3. Reduction to the annulus map; its analysis

3.1. Reduction to the annulus map.

It is proven in [3] that there is a rectangular region r in the (x,y)-plane (see Fig. 5), such that an iterate of each point $z \neq z_0$

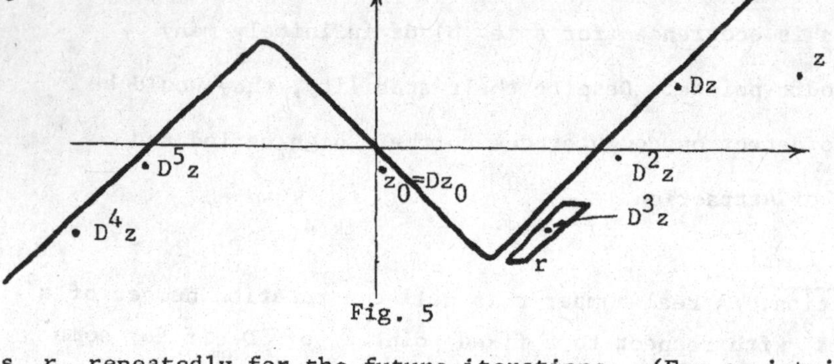

Fig. 5

enters r repeatedly for the future iterations. (For an intuitive explanation see §4). It suffices, therefore, to study map D restricted to r. In fact, we make an additional simplification:

ıstead of studying map D, we analyze the return map M: r → r, efined for each $z \in r$ as $D^j z$, with $j > 0$ being the first integer for ıich $D^j z \in r$ again. Clearly, j depends on z, which makes the eturn map M discontinuous: two nearby points may require a fferent number of iterations to come back to r. This discon- ınuity is removed, however, if within r we identify any pair points z, Dz into one. Now, r is chosen in such a way, that s upper side is mapped onto the lower one (and no two points ıside correspond to each other under D). Identification of the o sides of r makes M continuous, while r becomes an ınulus. Summarizing, we reduced D to an annulus map M; the ıly information lost by this reduction is the number of steps it ıkes to come back to r under iterations by D. This informa- on is easily recoverable from some additional properties of D; e details can be found in [3].

Our aim now is to describe the form of the map M and then e it to analyze the behavior of its high iterates. As it turns ıt, the symmetry properties of the damping $\varphi(x)$ and of the ›rcing p(t) (see Sec. 2.1) reflect in the fact that M: r → r can represented as a second iterate of another map N: r → r of a ımpler form than M: $M = N \cdot N = N^2$; it suffices, therefore, to udy N: r → r.

.2. Properties of the annulus map N.

Analysis shows that r is an extremely thin ($\sim e^{-\frac{1}{\epsilon^2}}$) ınulus, i.e. is nearly a circle. Therefore, a two-dimensional ıpping N: r → r can be represented by a circle map (one-dimensional) to a high degree of accuracy.*)

To describe properties of N, we treat r as a circle and normalize its length to be 1. Map N is such, that there exists a short (for ε small) arc Δ which is stretched by N to the length between 1 and 2; say, it is 1.5, see Fig. 6.

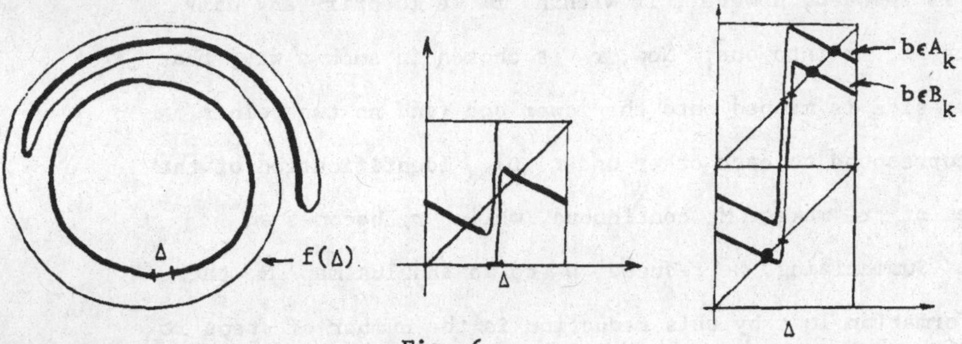

Fig. 6

The rest of the circle is deformed in the simplest possible way: it is reversed in direction and somewhat contracted. In addition to having this simple form, N depends nicely on the amplitude b: in essence, increasing b causes the image N(r) to rotate clockwise.

3.3. Analysis of N.

As an immediate consequence of the above description we recover the classical result on the alternating appearance of one and two sink-saddle pairs. Namely, as b grows, N(r) rotates clockwise, i.e. the graphs on Figs. 6b,c move downwards, which causes alternately one and two pairs of intersections of the graph with the bisectors (which correspond to the fixed points).

---

*In the case of our map N, the properties of N can be recovered completely for most values of b from the 1-dimensional information.

ne intersections where the slope of the graph is >1, correspond
o saddles, while the ones with the |slope| <1 correspond to sinks
f map N. These fixed points of N are the periodic points of
he Poincare map D and can be shown to have periods $2n \pm 1$
$n \sim \frac{1}{\varepsilon}$) correspondingly for each pair.

Below we state without proof the results of the analysis of
. They are easily seen to imply the results of §2.

Range $[b_1, b_2]$ consists of the alternating intervals A_k, B_k
eparated by short gaps g_j, such that the qualitative behavior
f N persists as b ranges in A_k or B_k, while g_j are the bifur-
ation intervals. More precisely,

A) if $b \in A_k$, N is a Morse-Smale type map. More exactly, every
oint not on the stable manifold of the saddle tends to the sinks
Fig. 7,a). This picture translates into Fig. 3 for map D.

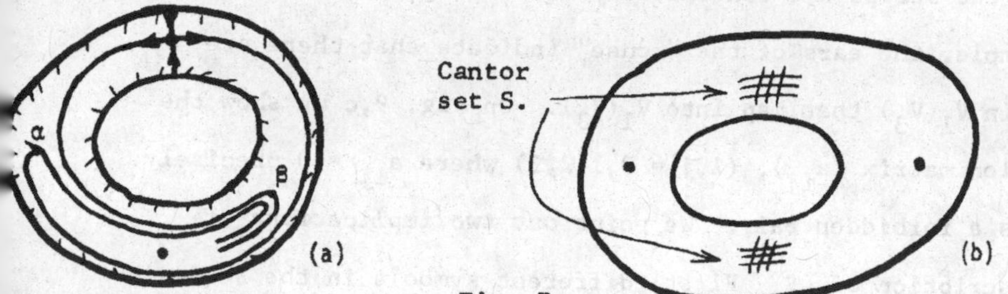

Fig. 7

A more interesting case is

(B) for $b \in B_k$ the map N has an invariant Cantor set S (to
which the two fixed saddle points belong), and two sinks (Fig. 7,b).
S can be thought of as a very complicated watershed - it separates
the basins of attraction of the two sinks.

To completely describe N, it remains to specify its
behavior on the set S. Here is this description.

Each point $z \epsilon S$ can be represented uniquely by a biinfinite sequence $\sigma = (\ldots\sigma_{-1}\sigma_0\sigma_1\ldots)$ of symbols σ_i which can take on one of four values 0,1,2, or 3. Also, any combination of these symbols

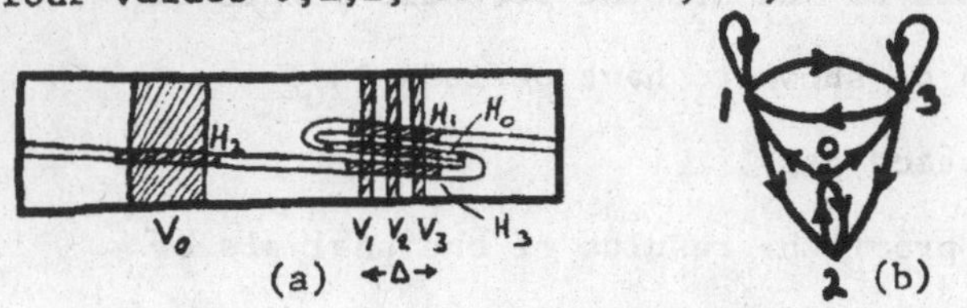

$$\begin{pmatrix} 0 & 1 & 1 & 1 \\ 0 & 1 & 1 & 1 \\ 1 & 0 & 0 & 0 \\ 0 & 1 & 1 & 1 \end{pmatrix}$$

(c)

Fig. 8

except for 00,10,21,22,23,30 can occur. j-th symbol σ_j $(j > 0)$ determines in which of the four vertical strips v_{σ_j} on Fig. 8 $N^j z$ lies. We note that the fact that a pair st is forbidden means that a point in V_s cannot map into V_t - for example, no point in V_0 remains in V_0 (i.e. $V_0 \cap H_0 = V_0 \cap N(V_0) = \emptyset$), and no point in V_2 maps into V_1, V_2 or V_3. The permitted transitions between the strips are conveniently shown on graph in Fig. 8,b. For example, the ears of the "mouse" indicate that there are points in $V_1(V_3)$ that map into $V_1(V_3)$. On Fig. 9,c we show the transition matrix (a_{ij}), $(i,j = 0,1,2,3)$ where $a_{ij} = 0$ precisely if ij is a forbidden pair. We point out two implications of this description of S. First, different symbols in the sequence σ can be prescribed independently of each other (as long as the forbidden pairs are avoided); this is the meaning of randomness in our deterministic system.

Second, to the periodic sequences there correspond periodic points of N; thus we have infinitely many of the latter.

<u>Remark 3.1</u>. The sequences consisting of symbols 1 and 3 only, correspond to the family of solutions of equation (1) described

y Levinson. Note, that 1 and 3 can occur in an arbitrary combi-ation.

emark 3.2. Sequences $\sigma^1 = (\ldots 111 \ldots)$

and $\sigma^3 = (\ldots 333 \ldots)$

orrespond to the two fixed saddle points of N.

emark 3.3. Map N is structurally stable for both $b \in B_k, A_k$.

g) When b passes through a gap g_j from A_k to B_k, the simple ituation of (A) undergoes a complicated sequence of bifurcations. ts onset can be seen from Fig. 7: as b is increasing, the fold of the unstable manifold of the saddle will move clockwise and ill become tangent to the stable manifold of the saddle. This eads to the bifurcations which had been studied by Newhouse and alis. In particular, for some values of $b \in g_j$ there are infinitely any stable periodic points of N. Another implication of their esults is the existence of infinitely many intermediate open sub-ntervals of g_j where N is structurally stable.

4. Analysis of the flow.

Recall, that our system is of the form

2a) $$\dot{x} = \frac{1}{\varepsilon}(y - \Phi(x))$$

2b) $$\dot{y} = -\varepsilon x + bp(t)$$

e describe the flow heuristically, using pictures. Unfortunately, he rigorous description is considerably more complex. It can be ound in [3].

We start ($t = 0$) with a fat annulus on the plane - a large disc with a small disc deleted (Fig. 9,a). If chosen properly, the inner disc expands, whereas the outer boundary contracts as shown in stages on Figs. 1a,b,c,d.

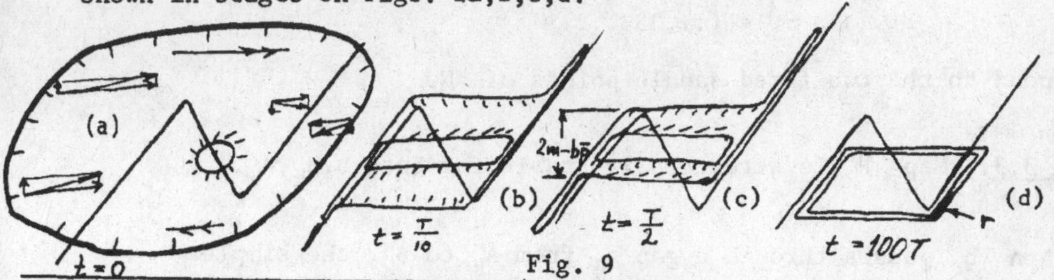

Fig. 9

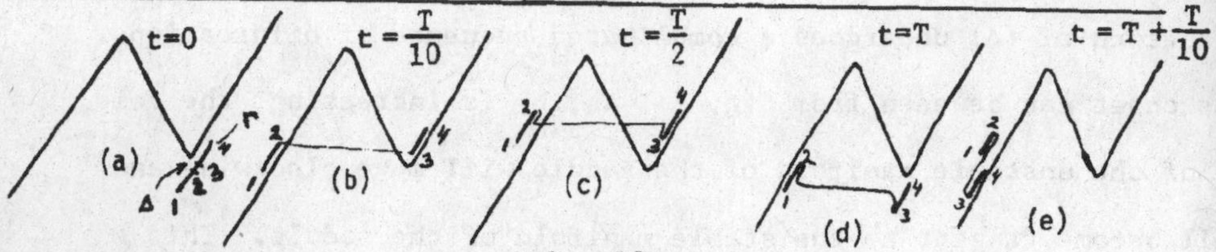

Fig. 10

The evaluation shown in Fig. 10 is determined by the following properties of equation (2).

- by (2a) the flow contracts strongly in the horizontal direction towards the two parts of the curve $y = \Phi(x)$ with the positive slope, and strongly expands near the part of $\Phi(x)$ with negative slope. This explains transition from (a) to (b) on Fig. 9.

- by (2b), the points oscillate in the vertical direction, up to an error $O(\varepsilon)$: integrating, we obtain

$$y(t) = y(0) + b \int_0^t p(\tau)d\tau - \varepsilon \int_0^t x(\tau)d\tau.$$

he amplitude of these oscillations is $b\bar{p}$ ($\bar{p} = \int_0^{T/2} p\, dt$). The erm $-\epsilon x$ in (2b) defines a small vertical shear (see arrows on ig. 9,a) in addition to the vertical oscillations. It is a comination of these three factors (expansion - contraction, shear nd oscillations) that leads to the evolution shown on Fig. 9.

After many periods the fat annulus shrinks into the thin ne (Fig. 9,d), which oscillates up and down between the horizontal ines $y = \pm m$ ($\pm m$ are the extrema of $\Phi(x)$); in addition (due to hear), points in this annulus circulate clockwise and thus enter epeatedly a slice r of the annulus; r is chosen so that its pper side maps onto the lower one by D.

Evolution of r is shown on Fig. 10. The main feature here s the fact that two different pieces of r separated by a short arc" Δ undergo a different evolution during one period; due to hear they have different vertical positions with respect to each ther after they both end up on the same side of $y = \Phi(x)$, ig. 10,e.

Treating r' as a line, if we plot (-vertical coordinate) of point in r at $t = T + \frac{T}{2}$ against its vertical coordinate at $= 0$, we will obtain a graph similar to the one on Fig. 6,c. A ore detailed description of the flow is given in [3].

References

[1] M. L. Cartwright and J. E. Littlewood, On nonlinear differential equations of the second order: I. The equation $\ddot{y} - k(1-y^2)\dot{y} + y = b\lambda k \cos(\lambda t + \alpha)$, k large, J. London Math. Soc., Vol. 20 (1945), pp. 180-189.

[2] J. E. Flaherty and F. C. Hoppensteadt, Frequency entrainment of a forced Van der Pol oscillator, Studies in Applied Mathematics 18 (No. 1) (1978), 5-15.

[3] M. Levi, Qualitative analysis of the periodically forced relaxation oscillations. Ph. D. Thesis, NYU, 1978.

[4] J. E. Littlewood, On non-linear differential equation of second order: III. Acta Math. Vol. 97 (1957), pp. 267-308.

[5] J. E. Littlewood, On non-linear differential equation of second order: IV, Acta Math. Vol. 98 (1957), pp. 1-110.

[6] J. E. Littlewood, Some problems in real and complex analysis, Heath, Lexington, Mass., 1968.

[7] N. Levinson, A second order differential equation with singular solutions, Ann. Math. Vol. 50, No. 1, 1949, pp. 127-153.

[8] J. Moser, Stable and random motions in dynamical systems, Princeton University Press (Study 77), 1973.

[9] S. Newhouse, Diffeomorphisms with infinitely many sinks, Topology 13 (1974), 9-18.

[10] S. Newhouse, The abundance of wild hyperbolic sets and non-smooth stable sets for diffeomorphisms, IHES, January 1977.

.1] S. Newhouse and J. Palis, Cycles and bifurcation theory, Asterisque 31 (1976), 43-141.

.2] J. Palis, On Morse-Smale dynamical systems, Topology, 8, No. 4 (1969), 385-404.

.3] J. Robbin, A structural stability theorem, Ann. Math. 94 (1971), 447-493.

.4] S. Smale, Differentiable dynamical systems, Bull. Amer. Math. Soc. 73 (1967), 747-817.

.5] S. Smale, Diffeomorphisms with many periodic points, Differential and Comb. top. (ed. S. Cairns), Princeton University Press, 1965, 63-80.

rthwestern University
anston, IL 60201

MODULI OF STABILITY FOR DIFFEOMORPHISMS

W. de Melo and J. Palis

In a very natural way, the notion of modulus of stability has been brought up in recent papers in the context of bifurcation theory of dynamical systems [5], [8], [10], [11] and also in the study of holomorphic vector fields near a singularity [1], [2].

In the present paper we introduce a class of diffeomorphisms (of a compact manifold M) whose modulus of stability is one. That is, these diffeomorphisms are not structurally stable, but the different conjugacy classes of the nearby diffeomorphisms can be expressed by a finite number of (real) one-parameter families. Thus, the dynamics (orbit structure) of these diffeomorphisms and the nearby ones can be well described. There is a converse, to appear latter, to the result discussed here: for diffeomorphisms in a generic or residual subset of arcs of diffeomorphisms and whose limit sets are hyperbolic to have modulus of stability one they must belong to the class we present below.

We wish to thank Floris Takens for useful conversations.

Let us recall some basic definitions and give the precise statement of our result. Throughout the paper, M is a C^∞ compact manifold without boundary and $\mathrm{Diff}(M)$ indicates its set of C^∞ diffeomorphisms with the usual C^∞ topology. Given $f,g \in \mathrm{Diff}(M)$, we say that they are conjugate if there is a homeomorphism h of M such that $hf = gh$. An element $f \in \mathrm{Diff}(M)$ has finite modulus of stability if there are an integer $k \geq 0$ and a finite number of

-parameter families $\mathfrak{F}_1,\ldots,\mathfrak{F}_\ell$ such that any diffeomorphism sufficiently near f is conjugate to an element of $\mathfrak{F}_i$, for some $\le i \le \ell$. Notice that each $\mathfrak{F}_i$ consists of only one diffeomorphism f $k = 0$. When k can be taken to be one, but not zero, we say that has modulus of stability one. A point $y \in M$ is called a limit oint of f if for some point $x \in M$ and sequence $n_i \in \mathbb{Z}$, with $n_i| \to \infty$, $\lim f^{n_i}(x) = y$. A periodic point p of period k is yperbolic if $df^k(p)$ has no eigenvalue with norm one. In this case, he set of points whose positive orbits have the orbit of p as their imit set is an injectively immersed submanifold of M, denoted by $^s(p)$. Similarly, if we take negative orbits, we get the unstable anifold $W^u(p)$. The diffeomorphism f is called Morse-Smale if its et of limit points $L(f)$ is finite and hyperbolic and satisfies the ransversality condition; i.e., all pairs of stable and unstable maifolds of the periodic orbits meet transversally. When $L(f)$ is inite and hyperbolic, we say that $L(f)$ has a cycle if there are a equence of periodic points $p_1,p_2,\ldots,p_{k+1}$ and a sequence of noneriodic points $y_1,y_2,\ldots,y_k$ such that p_{k+1} is in the orbit of p_1 nd $y_i \in W^s(p_i) \cap W^u(p_{i+1})$ for $1 \le i \le k$. Notice that if f is Iorse-Smale then $L(f)$ has no cycles.

Let p be a hyperbolic fixed point for f and let α^* be the argest modulus of the eigenvalues of $df(p)$ which are inside the nit circle. We call α the weakest contracting eigenvalue at p f $|\alpha| = \alpha^*$, α has multiplicity one and all other eigenvalues λ of $df(p)$ different from the complex conjugate $\bar{\alpha}$ of α satisfies $\lambda| \ne \alpha^*$. The definition of the weakest expanding eigenvalue is siilar and, if p has period k, we take the eigenvalues of $df^k(p)$. If the weakest eigenvalue $\alpha(\bar{\alpha})$ of f at p is defined, there is an f-invariant submanifold $W^{ss}(p)$, called the strong stable manifold, which is tangent to the subspace of T_pM defined by the other eigenvalues of $df(p)$ with norm less than one. Moreover, there is a uniquely defined foliation $\mathcal{F}^{ss}(p)$ of $W^s(p)$ with smooth leaves such that $W^{ss}(p)$ is a leaf and such that f maps leaves to leaves (see [3]). If the weakest expanding eigenvalue is defined, there are a uniquely defined strong unstable manifold $W^{uu}(p)$ and a uniquely defined strong unstable foliation $\mathcal{F}^{uu}(p)$ in $W^u(p)$. We say that p is s-critical if there is some periodic point q such that $W^u(q)$ intersects some leaf of $\mathcal{F}^{ss}(p)$ non-transversally; u-critical is defined similarly.

Our main purpose here consists in defining a subset $\mathcal{G} \subset \mathrm{Diff}(M)$ and showing that if $f \in \mathcal{G}$ then f has modulus of stability one. As mentioned before, $\mathcal{G}$ should exhaust all diffeomorphisms with modulus of stability one, under the assumptions that they appear in generic arcs of diffeomorphisms and their limit set are finite and hyperbolic. This fact will be published somewhere else.

The set $\mathcal{G}$ is defined as follows. A diffeomorphism $f \in \mathcal{G}$ if

1) $L(f)$ is finite, hyperbolic and has no cycles,
2) there is a pair of periodic points p, q such that $W^s(p)$ and $W^u(q)$ have a (unique) orbit of non-transversal intersection; along all the other orbits, the stable and unstable manifolds of periodic orbits meet transversally,
3) for this pair of periodic points p, q we have that the weakest expanding eigenvalue at q is defined, the weakest contracting eigenvalue at p is defined and they are both real. Moreover, p is not s-critical and q is not u-critical.

We will further restrict the set $\mathcal{G}$ imposing the following generic (open and dense) conditions:

4) along the orbit of tangency, $W^s(p)$ and $W^u(q)$ have a second order contact (see [7]); we say that $W^s(p)$ and $W^u(q)$ meet quasi-transversally along this orbit,

5) there are C^2 coordinate systems near p and q in which f can be expressed as linear isomorphisms (see [15], [16]),

6) let $W^{cu}(p)$ be an invariant manifold tangent at p to the direct sum of the expanding subspace and the weakest contract-ng subspace; similarly for $W^{cs}(q)$. We demand that $W^u(p)$ is trans-ersal to $W^{cs}(q)$ and $W^s(q)$ is transversal to $W^{cu}(p)$.

heorem. If $f \in G$, then f has modulus of stability one.

This result will follow from a sequence of lemmas. We now set ome preliminary concepts.

If $f \in \mathrm{Diff}^\infty(M)$ is such that $\Omega(f) = \mathrm{Per}(f)$ is hyperbolic nd has no cycles we can define a relation of partial order in the set f periodic orbits of f as follows: $\theta(x) < \theta(y)$ iff $W^u(\theta(x)) \cap W^s(\theta(y)) \neq \emptyset$ and $\theta(x) \neq \theta(y)$. Let $x,y \in \mathrm{Per}(f)$. We say that the ehavior of x with respect to y is one if $\theta(x) < \theta(y)$ and there s no $z \in \mathrm{Per}(f)$ with $\theta(x) < \theta(z) < \theta(y)$. We say that the behav-or is k if there is a sequence of periodic orbits $\theta(x) = \theta(x_1) < \ldots < \theta(x_k) = \theta(y)$ such that $\theta(x_i)$ has behavior one with respect o $\theta(x_{i+1})$. Let p be a hyperbolic periodic point of a diffeomor-hism f. A fundamental domain for the stable manifold of the orbit f p is a compact set $D^s(p,f) \subset W^s(\theta(p))$ which is diffeomorphic o a closed annulus and is such that an orbit of each point $\in W^s(\theta(p)) - \theta(p)$ has one and at most two points in $D^s(p,f)$. he set $\partial_{ex} D^s(p,f) = \{y \in D^s(p,f);\ f^k(y) \in D^s(p,f)\}$ is a sphere and he boundary of $D^s(p,f)$ is equal to $\partial_{ex} D^s(p,f) \cup \partial_{in} D^s(p,f)$ here $\partial_{in} D^s(p,f) = f^k(\partial_{ex} D^s(p,f))$, k being the period of p. t is clear that if the behavior of x with respect to y is one hen $W^u(\theta(x)) \cap D^s(f,\theta(y))$ is a compact set.

Now, let $f \in G$. Since $\Omega(f) = \mathrm{Per}(f)$ is hyperbolic and has o cycle we can order the periodic orbits of f as follows:

$$\mathrm{Per}(f) = \theta(p_1) \cup \ldots \cup \theta(p_k) \cup \theta(p) \cup \theta(q) \cup \theta(q_1) \cup \ldots \cup \theta(q_\ell)$$

where $\theta(p)$ is the periodic orbit whose unstable manifold meets the stable manifold of $\theta(q)$ non-transversally; $i < j$ whenever $\theta(p_i) < \theta(p_j)$; there is no $i = 1,\ldots,k$ with $\theta(p) < \theta(p_i)$ or $\theta(q) < \theta(p_i)$; for every $j = 1,\ldots,\ell$ we have either $\theta(p) < \theta(q_j)$ or $\theta(q) < \theta(q_j)$; $i < j$ whenever $\theta(q_i) < \theta(q_j)$; there is no $j=1,\ldots,\ell$ with $\theta(q_j) < \theta(p)$.

Since $\Omega(f)$ is hyperbolic and has no cycles, f is Ω-stable [12], [14]. This implies that, if $\mathfrak{h}$ is a small neighborhood of f, we can write for every $g \in \mathfrak{h}$:

$$\mathrm{Per}(g) = \theta(p_1(g)) \cup \ldots \cup \theta(p_k(g)) \cup \theta(p(g)) \cup \theta(q(g)) \cup \\ \cup\, \theta(q_1(g)) \cup \ldots \cup \theta(q_\ell(g))$$

where $p_i(g)$, $p(g)$, $q(g)$, $q_j(g)$ are the hyperbolic periodic points of g which are near p_i, p, q, q_j respectively.

We may assume that all periodic points of f (and thus of g near f) are fixed; all arguments we present below work as well for periodic orbits with simple and obvious modifications. Also, frequently we indicate $p(g)$, $q(g)$, $p_i(g)$, $q_j(g)$ simply by (p,g), (q,g), (p_i,g) and (q_j,g).

If we take the neighborhood $\mathfrak{h}$ of f small enough then, by [15] or [16], every $g \in \mathfrak{h}$ is C^2 linearizable in neighborhoods of $\theta(p(g))$ and $\theta(q(g))$. Furthermore the weakest contracting (resp. expanding) eigenvalue is defined at p (resp. q) and we can choose C^2 center unstable (resp. stable) manifolds $W^{cu}(\theta(p(g))$ (resp. $W^{cs}(\theta(q(g)))$ depending continuously on g in compact parts.

<u>Lemma 1</u>. We can choose the neighborhood $\mathfrak{h}$ so small that the following holds for every $g \in \mathfrak{h}$:

1) $W^{cu}(p(g))$ is transversal to $W^s(q(g))$;
2) $W^{cs}(q(g))$ is transversal to $W^u(p(g))$;
3) $p(g)$ is not s-critical
   $q(g)$ is not u-critical

4) if $x,y \in Per(g)$, $x \neq p$ or $y \neq q$ then $W^u(x)$ is transversal to $W^s(y)$.

roof. The first two statements can be proved in an entirely similar ay as in the proof of the openess of the transversality condition or Morse-Smale diffeomorphisms [7], [12]. One uses the fact that he invariant manifolds (including the center-unstable and center-table ones) vary continuously with g in the C^1 sense on compact arts. One then proceeds by induction on the periodic orbits. Si-ilarly for the third statement since the strong stable foliation of (g) and the strong unstable foliation of $q(g)$ vary continuously ith g on compact neighborhoods of p and q. To prove (4), we irst observe that off a neighborhood of the orbit of z the argument s the same as for (1) and (2). Let us now suppose by contradiction hat there are sequences $f_n \to f$ and $z_n \to z$ such that $W^u(p_i, f_n)$ s not transversal to $W^s(q_j, f_n)$ in z_n for some $1 \leq i \leq k$ and $\leq j \leq \ell$. By taking a subsequence if necessary, we may assume for he tangent spaces that $T(W^u(p_i,f_n))_{z_n} \to \mathcal{T} \supset T(W^{cu}(p,f))_z$ and $(W^s(q_j,f_n))_{z_n} \to \mathcal{T}' \supset T(W^{cs}(q,f))_z$, the last inclusions being a con-equence of the non-criticality assumptions. Then $\mathcal{T}$ and $\mathcal{T}'$ would ot be transversal, which is contradiction since $W^{cu}(p,f)_z$ and $^{cs}(q,f)_z$ are. This finishes the proof of Lemma 1.

emma 2. Each $f \in \mathcal{G}$ has a neighborhood $\mathcal{N}$ such that $\mathcal{N} - \mathcal{G}$ is ontained in the set of Morse-Smale diffeomorphisms and has two con-ected components.

roof. As we noticed before, since $\Omega(f)$ is hyperbolic and has no ycles, $\Omega(g)$ is also hyperbolic and has no cycles for g near f. rom Lemma 1, it follows that if $g \in \mathcal{N} - \mathcal{G}$ then $W^s(x)$ and $W^u(y)$ re transversal for all $x,y \in Per(f)$. Thus if $g \in \mathcal{N} - \mathcal{G}$ then g s Morse-Smale. Let now $\mathcal{G}^r$ be the set of C^r diffeomorphisms, $2 \leq r < \infty$, in a small C^r neighborhood $\mathcal{N}^r$ of f defined by: $g \in \mathcal{G}^r$ iff $W^s(q,g)$ and $W^u(p,g)$ have a non-transversal orbit of intersection. Due to a result of Sotomayor, as explained in [6], $\mathcal{G}^r$ is a codimension one imbedded submanifold of $\mathrm{Diff}^r(M)$. We have that $\mathcal{H}^r - \mathcal{G}^r$ has two components and, as above, if $g \in \mathcal{H}^r - \mathcal{G}^r$ then g is Morse-Smale. This implies that $\mathcal{H} - \mathcal{G}$ has two components each one made of Morse-Smale diffeomorphisms.

It follows from the above lemma and from the stability of Morse-Smale diffoomorphisms that thore arc at most two topological equivalence classes of Morse-Smale diffeomorphisms in a small neighborhood of f.

Let $\lambda(g)$ (resp. $\mu(g)$) be the weakest contracting (resp. expanding) eigenvalue at $p(g)$ (resp. $q(g)$). From [8] it follows that if $g,\bar{g} \in \mathcal{H} \cap \mathcal{G}$ are conjugate then $\frac{\log|\mu(g)|}{\log|\lambda(g)|} = \frac{\log|\mu(\bar{g})|}{\log|\lambda(\bar{g})|}$.

Hence our result follows from the following:

<u>Main Lemma</u>. If $f \in \mathcal{G}$ there is a neighborhood $\mathcal{H}$ of f such that $g,\bar{g} \in \mathcal{H} \cap \mathcal{G}$ are conjugate whenever $\frac{\log|\mu(g)|}{\log|\lambda(g)|} = \frac{\log|\mu(\bar{g})|}{\log|\lambda(\bar{g})|}$.

Before proving the Main Lemma, we need some more definitions and results.

<u>Definition</u>. Let z be a hyperbolic periodic point of a C^∞ diffeomorphism f. An <u>unstable foliation</u> (or unstable tubular family) for f at z is a C^0 foliation $\mathcal{F}^u(z,f)$ in a neighborhood V^u of $W^u(\theta(z))$ satisfying the following properties:

1) the leaves are C^r, $r \geq 2$, discs varying continuously in the C^r topology;
2) the foliation is f invariant; namely, $f(F^u(x)) \subset F^u(f(x))$ where $F^u(x)$ is the leaf through the point x;
3) each leaf intersects $W^s(\theta(z))$ transversally at a unique point;
4) the leaf through z is $W^u(z)$.

Let $z,w \in \mathrm{Per}(f)$ be such that $z < w$. We say that the unstable foliation $\mathcal{F}^u(f,z)$ is compatible with the unstable foliation $\mathcal{F}^u(f,w)$

f each leaf of $\mathcal{F}^u(f,w)$ that intersects some leaf of $\mathcal{F}^u(f,z)$ is ontained in this leaf and the foliation $\mathcal{F}^u(f,w)$ restricted to a eaf of $\mathcal{F}^u(f,z)$ is a C^r foliation. We write $\mathcal{F}^u(f,z) < \mathcal{F}^u(f,w)$. imilarly we define stable foliations and compatibility of stable oliations.

n [13] it was proved the existence of a compatible system of unstable oliations for all periodic points of a Morse-Smale diffeomorphism. his family of foliations was used to construct a conjugacy between a orse-Smale diffeomorphism and a small perturbation of it. Here we ill need a similar construction which was introduced in [8].

efinition. Let p be a hyperbolic periodic orbit of a C^∞ diffeo-orphism f such that the weakest contracting eigenvalue is defined t p. Let $W^{cu}(\theta(p))$ be a center-unstable manifold. A center-nstable foliation $\mathcal{F}^{cu}(p,f)$ is a decomposition of a neighborhood V^u of $W^u(p)$ in a union of C^1 discs called the leaves of $\mathcal{F}^{cu}(p,f)$, atisfying the following properties:

1) if $F^{cu}(x) \cap F^{cu}(y) \neq \emptyset$ then either $F^{cu}(x) = F^{cu}(y)$ or $F^{cu}(x) \cap F^{cu}(y)$ is one of the connected components of $W^u(p)$;
2) $\mathcal{F}^{cu}(p,f)$ is f-invariant;
3) for every $x \in V^u$, $F^{cu}(x)$ intersects transversally each leaf of the strong-stable foliation at a unique point;
4) the mapping $x \mapsto T_x F^{cu}(x)$ is continuous.

We observe that a center-unstable foliation of p is not ecessarily a foliation in V^u because two different leaves may ntersect along $W^u(p)$. But from the invariance of the foliation and he continuity of the field of tangent planes to the leaves it fol-ows that the tangent planes of two leaves at a point of intersection ust coincide. On the other hand it is clear that a center-unstable oliation $\mathcal{F}^{cu}(p,f)$ defines a C^0 foliation in $V^u - W^u(p)$. Sim-ilarly we define a center stable foliation.

Lemma 3. Let $f \in \mathcal{G}$ as before. Then for each g in a small neighborhood $\mathfrak{N}$ of f we can construct a family of unstable foliations $\mathcal{F}^u(p_1(g)),\ldots,\mathcal{F}^u(p_k(g))$ and a center-unstable foliation $\mathcal{F}^{cu}(p(g))$ such that $\mathcal{F}^u(p_i(g))$ is compatible with $\mathcal{F}^u(p_j(g))$ for every $j<i$ and each leaf of $\mathcal{F}^u(p_i(g))$ which intersects a leaf of $\mathcal{F}^{cu}(p(g))$ contains this leaf.

Proof. The construction of a compatible system of foliations $\mathcal{F}^u(p_1(g)),\ldots,\mathcal{F}^u(p_k(g))$ is in [13] and more recently in [8]. To construct a center-unstable foliation, we need first to construct a center foliation which will be the restriction of $\mathcal{F}^{cu}(p(g))$ to $W^s(p(g))$. Following [8], we indicate how this can be done. Consider a fundamental domain A in $W^s(p(g))$ whose exterior boundary $\partial_{ex}A$ is transversal to $W^u(p_i(g))$ for all $1\le i\le k$. In a neighborhood U of $W^u(p_k(g))\cap\partial_{ex}A$ we construct a vector field X along the leaves of $\mathcal{F}^u(p_k(g))$ which is of class C^1 restricted to each leaf. Near $W^u(p_k(g))\cap\partial_{in}A$, X is simply defined as $Dg_x(X(x))$ for $x\in U$. We can now extend X along the leaves of $\mathcal{F}^u(p_k(g))$ near $W^u(p_k(g))\cap A$ so that X is C^1 along each leaf and varies continuously in the C^1 sense with the leaves. We proceed by considering $p_{k-1}(g)$. Either $W^u(p_{k-1}(g))\cap A$ is compact and we reason as before or else $W^u(p_{k-1}(g))\cap A$ accumulates on $W^u(p_k(g))\cap A$. In the last case, X is already defined along the leaves of $\mathcal{F}^u(p_{k-1}(g))$ in a neighborhood V of $W^u(p_k(g))\cap A$. This follows from the fact that $\mathcal{F}^u(p_k(g))$ is compatible with $\mathcal{F}^u(p_{k-1}(g))$. Then $W^u(p_{k-1}(g))\cap(A-V)$ is compact and thus we can apply the previous argument to have X defined along the leaves of $\mathcal{F}^u(p_{k-1}(g))$. By induction on the $p_i(g)$, we have X defined in a neighborhood of A. We can define X on all of $W^s(p(g))$ by simply taking $X(g^n(x)) = Dg_x^n(X(x))$, $x\in A$, and $X(p(g))=0$. The integral curves of X give us the center foliation $\mathcal{F}^c(p(g))$. We now construct an unstable foliation $\mathcal{F}^u(p(g))$ compatible with $\mathcal{F}^u(p_i(g))$, $1\le i\le k$. A leaf of $\mathcal{F}^{cu}(p(g))$ is obtained as the union of the leaves of $\mathcal{F}^u(p(g))$

ırough the points of a given leaf of $\mathcal{F}^c(p(g))$. In this way we otain the center-unstable foliation as desired.

For $g \in \mathcal{G}$, a C^2 coordinate system linearlizing g near p,g) defines a C^2 diffeomorphism from the factor space $^s(p,g)/\mathcal{F}^{ss}(p,g)$ onto $\mathbb{R}$. Let $g_\#: W^s(p,g)/\mathcal{F}^{ss}(p,g) \circlearrowleft$ be the iffeomorphism induced by g.

emma 4. Let g, $\bar{g}$ be as in the Main Lemma. Let $_\#: W^s(p,g)/\mathcal{F}^{ss}(p,g) \to W^s(p,\bar{g})/\mathcal{F}^{ss}(p,\bar{g})$ be a homeomorphism that onjugates $g_\#$ and $\bar{g}_\#$. Then, there exists a homeomorphism $: \bigcup_{i=1}^{k} W^s(p_i,g) \cup W^s(p,g) \to \bigcup_{i=1}^{k} W^s(p_i,\bar{g}) \cup W^s(p,\bar{g})$ satisfying the ollowing properties:

1) $hg = \bar{g}h$,
2) h is compatible with the unstable foliations of p_i; that is, two points in a leaf of $\mathcal{F}^u(p_i,g)$ are sent into the same leaf f $\mathcal{F}^u(p_i,\bar{g})$,
3) h sends leaves of $\mathcal{F}^{ss}(p,g)$ onto leaves of $\mathcal{F}^{ss}(p,\bar{g})$ inducing $h_\#$ on the factor spaces,
4) h sends leaves of $\mathcal{F}^c(p,g)$ onto leaves of (a modified) center foliation $\mathcal{F}^c(p,\bar{g})$.

Proof. The proof goes by induction on the points p_i up to the point p. Since $p_1(g)$ is necessarily a source, h naturally sends $p_1(g)$ to $p_1(\bar{g})$. Now let A_i and $\bar{A}_i$ be fundamental domains for $W^s(p_i,g)$ and $W^s(p_i,\bar{g})$, $1 \le i \le k$. Let us suppose that h has been constructed up to p_j with the following properties:

1) h is differentiable along each leaf of $\mathcal{F}^u(p_i,g)$ off $W^s(p_i,g)$, $i < j$,
2) if $W^u(p_i,g) \cap W^s(p_\ell,g) \neq \emptyset$, then for each leaf F of $\mathcal{F}^u(p_i,g)$, in a neighborhood of $F \cap A_\ell$, h is C^1 near the inclusion map, $i < \ell < j$.

Let us construct h on $W^s(p_j,g)$. Suppose $W^u(p_{j-1},g) \cap \cap W^s(p_j,g) \neq \phi$, for otherwise we consider $W^u(p_{j-2},g)$. First we choose h as a diffeomorphism near the inclusion of neighborhood of $W^u(p_{j-1},g) \cap \partial_{ex} A_j$ in $W^s(p,g)$ onto a neighborhood of $W^u(p_{j-1},\bar{g}) \cap \partial_{ex} \bar{A}_j$ in $W^s(p,\bar{g})$. Near the interior boundary of A_j, h is defined by $hg = \bar{g}h$. Being C^1 near the inclusion map, this partial diffeomorphism can be extended to all of $W^u(p_{j-1},g) \cap \cap A_j$ (see [9]). To extend it to the leaves of $\mathfrak{F}^u(p_{j-1},g)$, we first observe that h being defined on $W^s(p_{j-1},g)$ induces a map from the space of leaves of $\mathfrak{F}^u(p_{j-1},g)$ onto the space of leaves of $\mathfrak{F}^u(p_{j-1},\bar{g})$: $F \in \mathfrak{F}^u(p_{j-1},g)$ is associated to $\bar{F} \in \mathfrak{F}^u(p_{j-1},\bar{g})$ if $h(F \cap W^s(p_{j-1},g)) = \bar{F} \cap W^s(p_{j-1},\bar{g})$. Thus the construction of h that we performed before for $W^u(p_{j-1},g)$ can be done along each leaf of $\mathfrak{F}^u(p_{j-1},g)$ near A_j in a parametrized way, the parameter space being a neighborhood of p_{j-1} in $W^s(p_{j-1},g)$. Then the continuity of h follows from the C^1 continuity of the leaves of the foliations $\mathfrak{F}^u(p_{j-1},g)$ and $\mathfrak{F}^u(p_{j-1},\bar{g})$. Now we consider p_{j-2}. If $W^u(p_{j-2},g) \cap W^s(p_{j-1},g) = \phi$, then $W^u(p_{j-1},g) \cap A_j$ is compact and disjoint from $W^u(p_{j-1})$ and we can repeat the previous construction. So let us suppose that $W^u(p_{j-2},g) \cap W^s(p_{j-2},g) \neq \phi$. Since $\mathfrak{F}^u(p_{j-2})$ and $\mathfrak{F}^u(p_{j-1})$ for g and $\bar{g}$ are compatible, h is already defined on the leaves of $\mathfrak{F}^u(p_{j-2})$ in a neighborhood V of $W^u(p_{j-1},g) \cap A_j$. On the other hand, $W^u(p_{j-2},g) \cap A_j$ off V is compact. So, we can repeat the previous argument extending h to the leaves of $\mathfrak{F}^u(p_{j-2},g)$ in A_j. The argument for all p_i, $i < j$, is similar. Thus we have constructed the conjugacy h on the stable manifold of all p_i up to p satisfying the properties mentioned above.

The construction of the conjugacy h on $W^s(p,g)$ requires special care. Roughly speaking, the idea is to define h as a "product" of two conjugacies: one is $h_{\#}$ defined in the space of ıe strong stable leaves and the other to be defined on the space of ıe weak stable leaves. However, since the center foliation may have everal leaves going through the periodic orbit p, we can not xpress the stable manifold as exactly the product of these two foliations. That is the reason we make a slight modification of one of ıe center foliations previously constructed, say for $W^s(p,\bar{g})$. he construction goes as follows. In $W^{ss}(p,g)$ we take a fundamental omain A_* whose exterior boundary $\partial_{ex} A_*$ is transverse to $W^u(p_i,g)$, ll $1 \le i \le k$. We now form a fundamental domain A for $W^s(p,g)$ hose exterior boundary $\partial_{ex} A$ is made of a "cylinder" obtained with eaves of the center foliation through $\partial_{ex} A_*$ and a pair of discs $_1$, D_2 contained in strong stable leaves. Of course, the interior oundary of A is given as $\partial_{in} A = g(\partial_{ex} A)$. Let $A_1 \subset D_1$ and $_2 \subset D_2$ be the annuli defined by the intersection of the center eaves through points of A with D_1 and D_2. Similar construction s performed for $W^s(p,\bar{g})$ with correspondent fundamental domains $\bar{A}_*$ nd $\bar{A}$, discs $\bar{D}_1$ and $\bar{D}_2$ and annuli $\bar{A}_1$ and $\bar{A}_2$. Now we contruct a conjugacy $h_*: A_* \to \bar{A}_*$ between g and $\bar{g}$ compatible with he unstable foliations $\mathcal{F}^u(p_i)$, $1 \le i \le k$, and the conjugacies reviously defined on $W^s(p_i)$. This can be done by induction on the oints p_i exactly as before since the $W^u(p_i)$ meet A_*, $\bar{A}_*$ transersally. The conjugacy h_* induces a homeomorphism $h_*: A_1 \cup A_2 \to \bar{A}_1 \cup \bar{A}_2$ in a natural way via projection through the center leaves. Je can then extend the homeomorphism h_* to all of $D_1 \cup D_2$, keepng it compatible with the unstable foliations $\mathcal{F}^u(p_i)$ and the conjugacies previously defined on $W^s(p_i)$, $1 \le i \le k$. Again this can be done as before because the unstable manifolds $W^u(p_i)$ are transverse to $D_1 \cup D_2$ and $\bar{D}_1 \cup \bar{D}_2$, respectively. At this point, we would like to define the conjugacy h as the "product" of $h_\#$ and h_* using the strong stable and the center foliations. To do this we have to modify the center foliation in $W^s(p,\bar{g})$. It is enough to do so in $\bar{A}$ in the region bounded by $\bar{D}_1-\bar{A}_1$ and $\bar{g}(\bar{D}_1)$ and in the region bounded by $\bar{D}_2-\bar{A}_2$ and $\bar{g}(\bar{D}_2)$. Let $\varphi: \bar{D}_1-A_1 \to \bar{g}(\bar{D}_1)$ be the homeomorphism defined by $\varphi(x) = y$, where $h_*^{-1}(x)$ and $g\, h_*^{-1}\, \bar{g}^{-1}(y)$ belong to the same leaf of $\mathfrak{F}^c(p,g)$. Let $\psi: \bar{D}_1-\bar{A}_1 \to \bar{g}(\bar{D}_1)$ be defined by $\psi(x) = z$, where x and z belong to the same leaf of $\mathfrak{F}^c(p,\bar{g})$. If $\psi = \varphi$, then no modification of $\mathfrak{F}^c(p,\bar{g})$ is necessary. So let us change $\mathfrak{F}^c(p,\bar{g})$ to get the second map to be equal to the first. Notice that $\varphi(x)$ and $\psi(x)$ belong to the same leaf of $\mathfrak{F}^u(p_i,\bar{g})$ for some $1 \le i \le k$. Moreover, $\psi\varphi^{-1}$ along each such a leaf is C^1 near the identity map. Let $\bar{X}$ be the vector field whose integral curves are the leaves of $\mathfrak{F}^c(p,\bar{g})$. Using the fact above and proceeding by induction on $\mathfrak{F}^u(p_i,\bar{g})$, $1 \le i \le k$, we can modify $\bar{X}$ near but off $\bar{g}(\bar{D}_1)$, so that the corresponding map $\bar{\psi}$ satisfies $\bar{\psi}\varphi^{-1} = 1$ on $\bar{g}(\bar{D}_1)$. Observe that the modification required for $\bar{X}$ along $W^u(p_k,\bar{g})$ is well known and we can perform it in a parametrized way along the leaves of $\mathfrak{F}^u(p_k,\bar{g})$. Using the fact that the foliations $\mathfrak{F}^u(p_i,\bar{g})$ are compatible, we proceed by downward induction on the indices $1 \le i \le k$. The procedure in the region bounded by $\bar{D}_2-A_2$ and $\bar{g}(\bar{D}_2)$ is similar. It is clear that the new center foliation, still denoted by $\mathfrak{F}^c(p,\bar{g})$, coincides with the previous one on $\partial\bar{A}$. Once it is defined on $\bar{A}$, it is defined on all of $W^s(p,\bar{g})$ simply through iterations by $\bar{g}$. It is now clear how to define the conjugacy $h: A \to \bar{A}$ using the maps $h_\#$, h_* and the foliations $\mathfrak{F}^{ss}(p)$ and $\mathfrak{F}^c(p)$ for g and $\bar{g}$. Again h is defined on all of $W^s(p,g)$ by the conjugacy equation $hg = \bar{g}h$. This completes the proof of the lemma.

<u>Remark</u>. In our case, $h_\#$ is differentiabe except at $W^{ss}(p,g)$. Correspondingly, the conjugacy h between g and $\bar{g}$ is differentiable along the leaves of $\mathfrak{F}^u(p_i)$, $1 \le i \le k$, except at $W^s(p_i,g)$ and $W^{ss}(p,g)$.

We need a parametrized version of a result of Kuiper [4].

emma 5. Let $\varphi: \mathbb{R}\times\mathbb{R}^n \to \mathbb{R}$ be of class C^2. Suppose that for each $\in$ R near 0 the function $\varphi_\mu: \mathbb{R}^n \to \mathbb{R}$, $\varphi_\mu(x) = \varphi(\mu,x)$ has a non-egenerate singularity at the point x_μ of index i and the curve $\mapsto (\mu,x_\mu)$ is C^1 and transversal to $\{0\}\times\mathbb{R}^n$ at 0. Then there is C^1 diffeomorphism ψ of a neighborhood of $(0,0) \in \mathbb{R}\times\mathbb{R}^n$ onto a eighborhood of $(0,x_o)$ such that

$$\psi(\mu,x) = (\mu,\psi_\mu(x))$$

$$\psi(\mu,0) = (\mu,x_\mu)$$

$$\varphi_\mu \circ \psi_\mu(x) = \sum_{j=1}^{i} - x_j^2 + \sum_{j=i+1}^{n} x_j^2 + \varphi_\mu(x_\mu)$$

orollary. Let $\varphi,\bar{\varphi}: \mathbb{R}\times\mathbb{R}^n \to \mathbb{R}$ be as in the above lemma. Let $: (\mathbb{R},\varphi_o(x_o)) \to (\mathbb{R},\bar{\varphi}_o(\bar{x}_o))$ be a germ of homeomorphism. Then there s a germ of homeomorphism $\psi: (\mathbb{R}\times\mathbb{R}^n, (0,x_o)) \to (\mathbb{R}\times\mathbb{R}^n, (0,\bar{x}_o))$ such hat

$$\psi(\mu,x) = (\mu,\psi_\mu(x))$$

$$\psi(\mu,x_\mu) = (\mu,\bar{x}_\mu)$$

$$\bar{\varphi}_\mu \circ \psi_\mu = \alpha \circ \varphi_\mu$$

urthermore, if α is C^1 then ψ is also C^1.

emma 6. Let $f_i: \mathbb{R}\times\mathbb{R}^n \to \mathbb{R}\times\mathbb{R}^n$ be linear isomorphisms, i=1,2, such that $_i(x_1,x_2) = (\mu_i x_1, A_i x_2)$ where μ_i is positive and A_i is a linear apping whose eigenvalues have norm smaller than one. Let Σ_i be a urve transversal to the planes $\{x_1\}\times\mathbb{R}^n$. Let $h: \Sigma_1 \to \Sigma_2$ be a omeomorphism such that $h(\Sigma_1 \cap \{0\}\times\mathbb{R}^n) = \Sigma_2 \cap \{0\}\times\mathbb{R}^n$. If $\lim_{x_1\to 0} \frac{|h^1(x_1,x_2)|}{|x_1|^\alpha}$, where $\alpha = \frac{\log \mu_2}{\log \mu_1}$, exists and is positive then extends to a homeomorphism of the closure of $\bigcup_{n=0}^{\infty} f_1^{-n}(\Sigma_1)$ to the closure of $\bigcup_{n=0}^{\infty} f_2^{-n}(\Sigma_2)$ which conjugates f_1 and f_2.

Proof. If $h^1(x_1,x_2) > 0$ for $(x_1,x_2) \in \Sigma_1$ and $x_1 > 0$ we define

$$h(x_1,0) = \begin{cases} (cx_1^{\alpha},0) & \text{if } x_1 \geq 0 \\ (-c|x_1|^{\alpha},0) & \text{if } x_1 < 0 \end{cases}$$

If $z \in f_1^{-n}(\Sigma_1)$ we define $h(z) = f_2^{-n}\, h\, f_1^n(z)$. It is easy to see that h is a homeomorphism and conjugates f_1 and f_2.

Proof of the Main Lemma

Let $g, \bar{g} \in \mathcal{G}$ be as in the statement of the Main Lemma. We are going to construct a conjugacy h between g and $\bar{g}$.

1st step. As in Lemma 4 we construct a homeomorphism $h\colon \bigcup W^s(p_i,g) \to \bigcup W^s(p_i,\bar{g})$ conjugating g and $\bar{g}$ and compatible with the unstable foliation of the periodic orbits p_i's. Similarly we construct a homeomorphism $h\colon \bigcup W^u(q_j,g) \to \bigcup W^u(q_j,\bar{g})$ conjugating g and $\bar{g}$ and compatible with the stable foliations of the periodic orbits q_j's.

2nd step. We define a homeomorphism $h_{\#}\colon W^s(p,g)/\mathcal{F}^{ss}(p,g) \to W^s(p,\bar{g})/\mathcal{F}^{ss}(p,\bar{g})$ whose expression, in the coordinate system induced on the factor spaces by the C^2 coordinate systems linearizing g and $\bar{g}$ near p, is linear in the logarithmic scale. More precisely we define $h_{\#}\colon R \to R$ by

$$h_{\#}(x) = \begin{cases} x^{\alpha} & \text{if } x \geq 0 \\ -|x|^{\alpha} & \text{if } x < 0 \end{cases} \qquad \text{where} \qquad \alpha = \frac{\log|\lambda(\bar{g})|}{\log|\lambda(g)|}.$$

Using Lemma 4 we can extend the homeomorphism constructed in step one to $W^s(p,g)$ so that h is compatible with the unstable foliations of p_i's, preserves the strong stable foliation of $W^s(p,g)$ and induces $h_{\#}$ on the space of leaves. Furthermore h sends leaves of the center foliation $\mathcal{F}^c(p,g)$ onto leaves of a center foliation $\mathcal{F}^c(p,\bar{g})$. Next we construct a center-unstable foliation $\mathcal{F}^{cu}(p,\bar{g})$ whose leaves intersects $W^s(p,\bar{g})$ in leaves of $\mathcal{F}^c(p,\bar{g})$. Since h preserves the center foliations it induces a homeomorphisms h_* from the space of leaves of $\mathcal{F}^{cu}(p,g)$ onto the space of leaves of $\mathcal{F}^{cu}(p,\bar{g})$. We will extend h to a conjugacy in a neighborhood of p preserving the enter-unstable foliations and inducing h_* on the spaces of leaves. efore that we need to construct an auxiliary homeomorphism which will llow us to define a homeomorphism $h^u_{\#}$ on the space of leaves of the trong unstable foliation of q.

nd step. Let $W^{cu}(p,g)$ be the center-unstable manifold whose image y the C^2 coordinate system linearizing g near p is the sum of he eigenspaces of $dg(p)$ associate to the expanding eigenvalues and he weakest contracting eigenvalue. Let $\mathfrak{F}^u(p,g)$ be the C^2 in-ariant foliation on $W^{cu}(p,g)$ whose leaves, in the above coordinate ystem, are the affine spaces parallel to the eigenspace of the ex-anding eigenvalues. We notice that $W^{cu}(p,g)$ is not necessarily a eaf of $\mathfrak{F}^{cu}(p,g)$ but it is tangent along $W^u(p,g)$ to the leaves of $^{cu}(p,g)$. Similarly we consider $W^{cs}(q,g)$ with a C^2 foliation $^s(q,g)$ and the corresponding objects for $\bar{g}$.

et $z \in W^u(p,g) \cap W^s(q,g)$ be a point of non-transversal intersection nd V a small neighborhood of z in $W^{cu}(p,g) \cap W^{cs}(q,g)$. Notice hat the complement of $\bigcup_{n=-\infty}^{\infty} g^n(V)$ in its closure is the union of he center manifolds $W^c(p,g) = W^{cu}(p,g) \cap W^s(p,g)$ and $W^c(q,g) =$ $W^{cs}(q,g) \cap W^u(q,g)$. Let $\rho: V \to W^c(p,g)$ be the projection along he leaves of $\mathfrak{F}^u(p,g)$. Let $F^s(w)$ be the leaf of $\mathfrak{F}^s(q,g)$ through he point $w \in W^c(q,g)$ and $\rho_w: F^s(w) \cap V \to W^c(p,g)$ be the res-riction of ρ. Since z is a point of quasi-transversal inter-section ρ_w is a C^2 function which has a unique nondegenerate singularity at z_w $(z_q=z)$ and the set $\Sigma = \{z_w: w \in W^c(q,g)\}$ is a C^1 curve which is transversal to $F(w) \cap V$ in V. Hence, the com-plement of $\bigcup_{n=-\infty}^{\infty} g^n(\Sigma)$ in its closure is $W^c(p,g) \cup W^c(q,g)$. Similarly we consider for $\bar{g}$ the function $\bar{\rho}: \bar{V} \to W^c(p,\bar{g})$ and the curve $\bar{\Sigma}$. Here $\bar{V}$ is again a neighborhood of $\bar{z}$ (the point of nontransversal intersection of $W^u(p,\bar{g}) \cap W^s(q,\bar{g})$ near z) in $W^{cu}(p,\bar{g}) \cap W^{cs}(q,\bar{g})$ and $\bar{\Sigma}$ is the set of singularities of $\bar{\rho}_{\bar{w}}$.

Since each leaf of $\mathcal{F}^{ss}(p,g)$ intersects $W^c(p,g)$ in a unique point, $h_\#$ induces a homeomorphism $\varphi_{-\infty}: W^c(p,g) \to W^c(p,\bar{g})$ which is linear in the logarithmic scale as before. Let $\varphi_0: \Sigma \to \bar{\Sigma}$ be the homeomorphism defined by $\bar{\rho}\varphi_0 = \varphi_{-\infty}\rho$. We observe that $\varphi_{-\infty}$ is differentiable off p and it is homeotopic to a C^1 diffeomorphism. Therefore φ_0 is also homeotopic to a diffeomorphism.

Clearly φ_0 extends to a homeomorphism φ_- of the closure of $\bigcup_{n=0}^{\infty} g^{-n}(\Sigma)$ conjugating g and $\bar{g}$ and $\varphi_- = \varphi_{-\infty}$ on $W^c(p,g)$.

Since $\dfrac{\log|\mu(\bar{g})|}{\log|\mu(g)|} = \dfrac{\log|\lambda(\bar{g})|}{\log|\lambda(g)|}$ we can, using Lemma 6, extend φ_0 to a homeomorphism φ of the closure of $\bigcup_{n=-\infty}^{\infty} g^n(\Sigma)$ conjugating g and $\bar{g}$. The restriction of φ to $W^c(q,g)$ is differentiable off q and induces a homeomorphism $h^u_\#: W^u(q,g)/\mathcal{F}^{uu}(q,g) \to W^u(q,\bar{g})/\mathcal{F}^{uu}(q,\bar{g})$.

From the corollary of Lemma 5 we can extend φ_0 to a homeomorphism $\tilde{\varphi}_0$ on V preserving the foliations $\mathcal{F}^u(p,g)$ and $\mathcal{F}^s(q,g)$. Since φ_0 is not differentiable at z, $\tilde{\varphi}_0$ is not a diffeomorphism but it is homeotopic to a diffeomorphism. It is easy to see that $\tilde{\varphi}_0$ extends to a homeomorphism $\tilde{\varphi}$ of the closure of $\bigcup_{n=-\infty}^{\infty} g^n(V)$ conjugating g and $\bar{g}$.

<u>4th step</u>. Using Lemma 4 we can extend h to a homeomorphism from $W^u(q,g)$ into $W^u(q,\bar{g})$ which conjugates g and $\bar{g}$, is compatible with the stable foliations of q_j's, preserves the strong unstable foliations and induces $h^u_\#$ on the space of leaves. Furthermore h sends leaves of the center foliation $\mathcal{F}^c(q,g)$ onto leaves of a center foliation $\mathcal{F}^c(q,\bar{g})$. Next we construct a center-stable foliation $\mathcal{F}^{cs}(q,\bar{g})$ whose leaves intersect $W^s(q,\bar{g})$ in the leaves of the center foliation. Since h preserves the leaves of the center foliations it induces a homeomorphism h^u_* on the space of leaves of the center-stable foliations.

<u>5th step</u>. Let U be a small neighborhood of z so that $U \cap W^{cs}(q,g) \cap W^{cu}(p,g) = V$ is contained in the domain of the homeo-

orphism $\tilde{\varphi}_o$ constructed in step 3. If U is small enough we can con- truct a C^2 retraction $\pi: U \to V$ such that the fibers of π are trans- ersal to the leaves of the foliation $\mathfrak{F}^{cu}(p,g) \cap \mathfrak{F}^{cs}(q,g)$; this is a fo- iation (with singularities) whose leaves are the intersection of the eaves of $\mathfrak{F}^{cu}$ with the leaves of $\mathfrak{F}^{cs}$. We demand the intersection of ach fiber of π with $W^{cs}(q,g)$ to be contained in a leaf of $\mathfrak{F}^{s}(q,g)$ and he intersection of each fiber with $W^{cu}(p,g)$ to be contained in a leaf f $\mathfrak{F}^{u}(p,g)$. The homeomorphisms h_* on the space of leaves of $^{cu}(p,g)$, h^u_* on the space of leaves $\mathfrak{F}^{cs}(q,g)$ define a homeomor- hism $\hat{h}$ on the space of leaves of $\mathfrak{F}^{cu} \cap \mathfrak{F}^{cs}$. Also, φ_o induces a omeomorphism $\tilde{h}$ on the space of fibers of π. Since the fibers of are transversal to the leaves of $\mathfrak{F}^{cu} \cap \mathfrak{F}^{cs}$ and they intersect n points, there is a unique homeomorphism $h: U \to \bar{U}$ which induces and $\tilde{h}$ on the corresponding spaces of leaves. Since $\tilde{\varphi}_o$ is sotopic to a diffeomorphism and π is C^2 we can extend h to a igger neighborhood in such way that the restriction of h to a eaf of $\mathfrak{F}^{cs} \cap \mathfrak{F}^{cu}$ is a diffeomorphism off U. Next we extend h o $\bigcup_{n=-\infty}^{\infty} g^n(U)$ by the equation $hg = \bar{g}h$. We claim that h is con- inuous on $W^u(q,g)$ and on $W^s(p,g)$. To prove this we have to con- truct two fibration on U which are preserved by h. Each leaf f $\mathfrak{F}^{cs}(q,g)$ intersects a fiber of π in a submanifold having the ame dimension as $W^{ss}(p,g)$. Hence this submanifold intersects $^{cu}(p,g)$ transversally at a unique point. This defines a fibration $^{ss}: U \to W^{cu}(p,g) \cap U$ which is compatible with π and with $\mathfrak{F}^{cs}(q,g)$. Similarly there is a unique fibration $\pi^{uu}: U \to U \cap W^{cs}(q,g)$ which is compatible with π and $\mathfrak{F}^{cu}(p,g)$. Let $\bar{\pi}^{ss}: \bar{U} \to W^{cu}(p,\bar{g}) \cap \bar{U}$ and $\bar{\pi}^{uu}: \bar{U} \to \bar{U} \cap W^{cs}(q,\bar{g})$ be the corresponding fibrations for $\bar{g}$. Clearly h maps fibers of π^{ss} (resp. π^{uu}) onto fibers of $\bar{\pi}^{ss}$ (resp. $\bar{\pi}^{uu}$) and therefore defines homeomorphism $\psi^u: U \cap W^{cu}(p,g) \to$ $\to \bar{U} \cap W^{cu}(p,\bar{g})$ and $\psi^s: U \cap W^{cs}(q,g) \to \bar{U} \cap W^{cs}(q,\bar{g})$ satisfying $\psi^s \pi^{ss} = \bar{\pi}^{ss} h$ and $\psi^u \pi^{uu} = \bar{\pi}^{uu} h$. Clearly ψ^u and ψ^s extends $\tilde{\varphi}_o$,

ψ^s preserves the foliation $\mathfrak{F}^s$ and ψ^u preserves the foliation $\mathfrak{F}^u$. This implies that ψ^u extends to a conjugacy on the closure of $\bigcup_{n=0}^{\infty} g^{-n}(U \cap W^{cu}(p,g))$ with $\psi^u = \varphi_{-\infty}$ at $W^c(p,g)$ and ψ^s extends to a conjugacy on the closure of $\bigcup_{n=0}^{\infty} g^n(U \cap W^{cs}(q,g))$ with $\psi^s = \varphi_\infty$ at $W^c(q,g)$. Iterating the fibers of π^s by g^{-n} we get a fibration on $\bigcup_{n=0}^{\infty} g^{-n} U$ which will converge to the strong stable foliation at p. Since $\psi^s \pi^{ss} = \bar{\pi}^{ss} h$, ψ^s extends to the closure of $\bigcup_{n=0}^{\infty} g^{-n}(U \cap W^{cu}(p,g))$ and since h preserves $\mathfrak{F}^{cu}$ it follows that h extends continuously to the closure of $\bigcup_{n=0}^{\infty} g^{-n}(U)$. Similarly we conclude that h extends continuously to the closure of $\bigcup_{n=0}^{\infty} g^n(U)$.

<u>6th step</u>. Here we extend h to a conjugacy in a neighborhood of p. First we extend π^{ss} to an invariant fibration over a neighborhood, in $W^{cu}(p,g)$ of a fundamental domain of $W^u(p,g)$ in such a way that π^{ss} is compatible with the stable foliations of the points q_j's. This can be obtained by induction on the behaviour as in the construction of the compatible system of stable foliations. Iterating the fibers of π^{ss} by g^{-n} we get a fibration over a neighborhood of p in $W^{cu}(p,g)$ whose fibers over points of $W^c(p,g)$ coincide with the leaf of the strong stable foliation through the same point. Suppose we have extended ψ^u to a conjugacy $\psi^u\colon W^{cu}(p,g) \to W^{cu}(p,\bar{g})$ which is compatible with the stable foliations of the periodic orbits q_j's. Then ψ^u defines a homeomorphism on the space of fibers of π^{ss} which together with the homeomorphism on the space of leaves of $\mathfrak{F}^{cu}$ we have constructed before define a unique homeomorphism on a neighborhood of p which conjugates g and $\bar{g}$ and extends h. So it remains to construct the extension of ψ^u.

First we extend ψ^u to a conjugacy on a fundamental domain of $W^u(p,g)$ compatible with the mappings already defined on the spaces of leaves of $\mathfrak{F}^s(q_j,g)$, $j = 1,\ldots,\ell$. As in Lemma 4 this can be done so that ψ^u is differentiable restricted to each leaf of any of these oliations. Notice that by construction ψ^u is already compatible with the leaves of $\mathcal{F}^{cs}(q,g)$ and the leaves of $\mathcal{F}^u(p,g)$. As in Lemma 3, we can now construct a fibration on a neighborhood of the fundamental domain of $W^u(p,g)$ in $W^{cu}(p,g)$ such that the fibers have dimension one, are transversal to the leaves of $\mathcal{F}^u(p,g)$ in $W^{cu}(p,g)$ and are compatible with the foliations of the periodic points q_j, $j = 1,\ldots,\ell$. A similar construction is performed in $W^{cu}(p,\bar{g})$. By preserving these fibrations and $\mathcal{F}^u(p,g)$, $\mathcal{F}^u(p,\bar{g})$ the extension of ψ^u is defined on all of $W^{cu}(p,g)$ onto $W^{cu}(p,\bar{g})$ and it is a conjugacy between g and $\bar{g}$. This completes the construction of the conjugacy h on a neighborhood of (p,g).

7th step. We now perform the extension of h to a conjugacy on a neighborhood of q. As in the step 6 we first extend π^{uu} to an invariant fibration over a neighborhood of q in $W^{cs}(q,g)$ compatible with the unstable foliations of the periodic orbits p_i's. The fibers of π^{uu} through points of $W^c(q,g)$ are leaves of the strong unstable foliation. Next we extend ψ^s to a conjugacy $\psi^s: W^{cs}(q,g) \to W^{cs}(q,\bar{g})$ as in (6). The difficulty here is that we can not expect that ψ^s will send leaves of $\mathcal{F}^s(q,g)$ onto leaves of $\mathcal{F}^s(q,\bar{g})$. So to define the extension of ψ^s we have to construct a continuous invariant fibration $\pi^s: W^{cs}(q,\bar{g}) \to W^c(q,\bar{g})$ such that the fibers of π^s are differentiable off $W^c(q,\bar{g})$ and coincide with the leaves of $\mathcal{F}^s(q,\bar{g})$ in U, and $\psi^s: \bigcup_{n=0}^{\infty} g^n(U \cap W^{cs}) \to \bigcup_{n=0}^{\infty} \bar{g}^{-n}(\bar{U} \cap W^{cs})$ induces a homeomorphism from the space of leaves of $\mathcal{F}^s(q,g)$ to the space of fibers of π^s as in [5]. Then we extend ψ^s to a conjugacy on $W^s(q,g)$ and to $W^{cs}(q,g)$ as before. This defines a homeomorphism from the space of fibers of π^{uu} onto the space of $\bar{\pi}^{uu}$. Using this and the homeomorphism h_*^u we define h on a neighborhood of q.

8th step. Here we conclude the construction of the conjugacy h between g and $\bar{g}$. So far h is defined on the spaces of leaves of $\mathcal{F}^u(p_i)$, $1 \le i \le k$, on the spaces of leaves of $\mathcal{F}^s(q_j)$, $1 \le j \le \ell$, and on a neighborhood N of p and q and thus on $B = \bigcup_n g^n(N)$, $n \in Z$. We proceed by induction constructing the conjugacy on neighborhoods of the periodic orbits q_j. Let A, $\bar{A}$ be fundamental domains for $W^s(q_1,g)$ and $W^s(q_1,\bar{g})$, $\bar{N} = h(N)$ and $\bar{B} = h(B)$. First we extend $h: B \to \bar{B}$ to $h: A \to \bar{A}$ conjugating g and $\bar{g}$ and being compatible with the unstable foliations of p, q and p_i, $1 \le i \le k$. This can be done as in Lemma 4 because h is differentiable restricted to the leaves of these foliations near the boundary of B. The conjugacy h is then defined on all of $W^s(q_1,g)$ by the equation $hg = \bar{g}h$. Over A we also construct an unstable foliation $\mathcal{F}^u(q_1,g)$ defined on A-B compatible with the unstable foliations of p, q and p_i, $1 \le i \le k$. Again this is possible because h restricted to the leaves of these foliations is differentiable near the boundary of B. Now $\mathcal{F}^u(q_1)$ is defined on a full neighborhood of q_1 off B by simply considering the iterates by g of the leaves through points of A and the leaf $W^u(q_1,g)$. Similarly for $\bar{g}$ and $q_1(\bar{g})$. Using the complementary foliations $\mathcal{F}^s(q_1)$ and $\mathcal{F}^u(q_1)$ for g and $\bar{g}$ and the conjugacy already defined on their spaces of leaves, we obtain h defined on a full neighborhood of q_1. The induction procedure for the remaining q_j is similar. In this way we have constructed a conjugacy h on the stable manifolds of all periodic orbits and thus on all of M. It is clearly continuous at the stable manifolds of p, q and q_j for $1 \le j \le k$. That h is continuous at the stable manifolds of p_i, $1 \le i \le k$, follows immediately from the fact that it preserves the unstable foliations of these periodic points. The Main Lemma is proved.

References

[1] C. Camacho, N. Kuiper and J. Palis: The topology of holomorphic flows with singularity, Publications Mathématiques, Inst. des Hautes Études Scientifiques, nº 48 (1978), p. 5-28. See also C.R. Acad. Sc. Paris, t. 282 (1976), p. 959-961.

[2] Yu. S. Il'yashenko: Remarks on the topology of singular points of analytic differential equations in the complex domain and Ladis' theorem, Funct. Analysis and its Applications vol. 11, nº 2 (1977), p. 105-113.

[3] M. Hirsch, C. Pugh and M. Shub: Invariant manifolds, Lecture Notes in Math., Springer-Verlag, 583 (1977).

[4] N. Kuiper: C^r functions near non-degenerate critical points, preprint Univ. of Warwick.

[5] W. de Melo: Moduli of stability of two-dimensional diffeomorphisms, to appear in Topology.

[6] S. Newhouse and J. Palis: Bifurcations of Morse-Smale dynamical systems, Dynamical Systems (ed. M. Peixoto), (1973), p. 303-366.

[7] S. Newhouse and J. Palis: Cycles and bifurcation theory, Astérisque 31 (1976), p. 43-140.

[8] S. Newhouse, J. Palis and F. Takens: Stable families of diffeomorphisms, IMPA preprint and to appear.

[9] R. Palais: Local triviality of the restriction map for embeddings, Comm. Math. Helvet. 34 (1960), p. 305-312.

[10] J. Palis: A differentiable invariant of topological conjugacies and moduli of stability, Astérisque 51 (1978), p. 335-346.

[11] J. Palis: Moduli of stability and bifurcation theory, Proc. Int. Congress of Math. Helsinki (1978), to appear.

[12] J. Palis: On Morse-Smale dynamical systems, Topology 16 (1977), p. 335-345.

[13] J. Palis and S. Smale: Structural stability theorems, Proc. Symp. Pure Math. A.M.S. 14 (1970), p. 223-232.

[14] S. Smale: The Ω-stability theorem, Proc. Symp. Pure Math. A.M.S. 14 (1970), p. 289-297.

[15] S. Sternberg: On the structure of local homeomorphism of euclidean n-spaces II, Amer. Journ. of Math. 80 (1958), p. 623-631.

[16] F. Takens: Partially hyperbolic fixed points, Topology 10 (1971), p. 133-147.

Instituto de Matemática Pura e Aplicada
Rua Luis de Camões, 68
Rio de Janeiro, RJ, Brasil

UNCOUNTABLY MANY DISTINCT TOPOLOGICALLY HYPERBOLIC EQUILIBRIA IN $\mathbb{R}^4$

Dean A. Neumann*

§1. Introduction.

We consider an isolated equilibrium, say 0, of a smooth flow ϕ on $\mathbb{R}^d$. We say that ϕ is *topologically hyperbolic* at 0 if it has the same rough features as the hyperbolic equilibrium of the *standard example* $\phi_{m,n}$ generated by the equations: $\dot{x} = -x$, $\dot{y} = y$ ($x \in \mathbb{R}^m$, $y \in \mathbb{R}^n$, $m + n = d$); viz., there is a neighborhood $B = D^m \times D^n$ of 0 with $\{0\}$ as maximal ϕ-invariant set in B and satisfying

(1) orbits of ϕ *enter* B on $\partial D^m \times D^n$, and *exit* B on $D^m \times \partial D^n$ (they then "bounce off" externally on $\partial D^m \times \partial D^n$);

(2) the stable (unstable) manifold of 0 in B is $D^m \times \{0\}$ (respectively $\{0\} \times D^n$).

(A more complete definition is given in the paper [5] of Wilson in this volume.)

It was conjectured by Coleman [1] that at such an equilibrium ϕ is locally topologically equivalent to $\phi_{m,n}$. This conjecture was more precisely reformulated by Wilson [4] in terms of isolating blocks. It is now known that the conjecture is true when $m = 1$ or $n = 1$, and that there is a C^∞ counterexample in the case $m = n = 2$[3]. A number of interesting questions remain unanswered, among them the possibility, of classifying in some reasonable way the topologically hyperbolic equilibria

*This research was supported in part by NSF grant MCS78-04371.

in a given dimension. The main result of the present paper is relevant to this question:

THEOREM. *There is an uncountable collection of pairwise inequivalent topologically hyperbolic equilibria in the case* $m = n = 2$.

This author wishes to thank Wes Wilson for a number of helpful discussions concerning the Coleman conjecture.

§2. Preliminaries and Notation

We first establish some of the necessary notation in the special case $m = n = 2$; the reader is referred to [3] for more detail.

If $B = D^2 \times D^2 = \{(x,y,z,w) \in \mathbb{R}^4 \mid x^2 + y^2 \leq 1, z^2 + w^2 \leq 1\}$ is an isolating block for the topologically hyperbolic equilibrium 0 of ϕ, then ϕ induces a *Poincaré map*

$$h: \partial D^2 \times (D^2 \setminus 0) \longrightarrow (D^2 \setminus 0) \times \partial D^2$$

that maps each entry point in ∂B to the exit point of the same orbit. We denote $\partial D^2 \times D^2$ by V', a *solid torus* with *centerline* $C' = \partial D^2 \times 0$; analogously set $V = D^2 \times \partial D^2$, $C = 0 \times \partial D^2$. Note that both $V \setminus C$, $V' \setminus C'$ are homeomorphic with $S^1 \times S^1 \times (0,1]$. It is convenient to use coordinates that exhibit this product stucture; thus we reparametrize by homeomorphisms $F, F': S^1 \times S^1 \times (0,1] \to V \setminus C, V' \setminus C'$ given by

$$F(e^{i\alpha}, e^{i\beta}, c) = (c \cos \alpha, c \sin \alpha, \cos \beta, \sin \beta),$$

$$F'(e^{i\alpha}, e^{i\beta}, c) = (\cos \alpha, \sin \alpha, c \cos \beta, c \sin \beta);$$

see Figure 1. Then $V\backslash C, V'\backslash C'$ are foliated by the level sets of the coordinate functions α, β, c. The subset of V' given by $\{\alpha = \alpha_0\}$ has closure D_{α_0}, a *meridional disc* spanning V'; also, we denote the single point of $D_{\alpha_0} \cap C$ by α_0. The corresponding

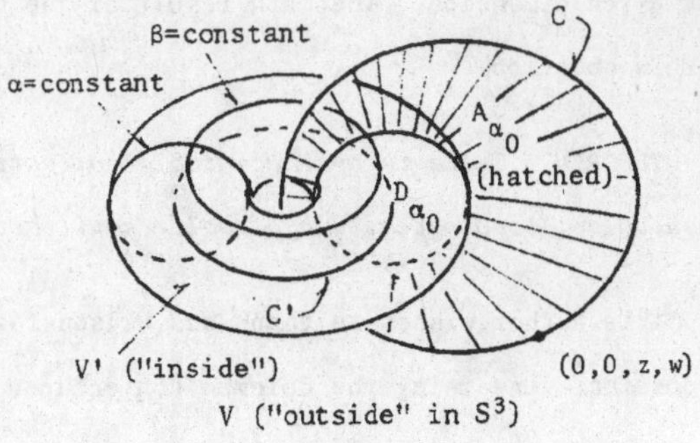

Figure 1

subset $\{\alpha = \alpha_0\}$ of V is a *half open annulus* A_{α_0}, with closure the annulus $A_{\alpha_0} \cup C$. The sets $D_{\beta_0} \subseteq V$, $A_{\beta_0} \subseteq V'$ are defined analogously. The sets given by $\{c = c_0\}$ are boundary parallel tori in both V and V', denoted $T_{c_0} \subseteq V$, $T'_{c_0} \subseteq V'$.

The remainder of this section is devoted to several preliminary observations that will facilitate the description of the examples below.

We first note that the local topological type of ϕ is completely determined by that of its Poincaré map h; we need only the obvious half of this assertion, namely that an equivalence $\tilde{\chi}: B \to B$ of ϕ with ϕ' restricts to a homeomorphism ("conjugacy" of the corresponding Poincaré maps) $\chi: \partial B \to \partial B$ satisfying

$$h' \circ \chi\Big|_{V'\backslash C'} = \chi\Big|_{V\backslash C} \circ h$$

(We consider equivalences defined on all of B for simplicity; the extension of the argument to the general case is straightforward.)

Thus we can restrict our attention to the construction of a collection of pairwise nonconjugate homeomorphisms $h: V'\backslash C' \to V\backslash C$.

Of course we can only use homeomorphisms that are realizable as Poincaré maps of topologically hyperbolic equilibria. Fortunately any homeomorphism $h: V'\backslash C' \to V\backslash C$ that is the identity on $T_1' = \partial V'$ and maps each T_c' to the corresponding T_c is realizable in this way; moreover in this case the corresponding flow can be taken as smooth as h. This assertion is proved in [5]. All our examples will be of this type.

It is easy to check that, in terms of the coordinates (α,β,c) defined above, the Poincaré map $h_{2,2}$ of the standard example $\phi_{2,2}$ is just the "identity". Thus $h_{2,2}$ maps each *punctured disc* $D_\alpha\backslash\alpha$ onto the corresponding annulus A_α, by a "reflection" in ∂D_α (cf. Figure 1). An arbitrary Poincaré map h restricts to the identity on $\partial V'$, hence h maps each $D_\alpha\backslash\alpha$ onto a half-open annulus A_α^h that has the same boundary curve in ∂V as A_α. If we have $h(T_c') = T_c$ $(0 < c \leq 1)$, then A_α^h is isotopic in $V\backslash C$ to A_α. We call any such annulus in $V\backslash C$ *longitudinal*. We note that the closure in V of a longitudinal annulus need not be a manifold at points of C. For a given Poincaré map h the associated A_α^h will be referred to as the *characteristic annuli* of h. These form the basis of our approach to equivalence, which is formalized in the following necessary condition:

ASSERTION. *If ϕ,ϕ' are locally equivalent topologically hyperbolic equilibria, then the characteristic annuli of the corresponding Poincaré maps h,h' satisfy the following condition: For any α_0*

there is an α, *a neighborhood* $N \subseteq V$ *of* C, *and a homeomorphism* χ *of* V *such that* $\chi(A^h_{\alpha_0})$ *does not intersect* $A^{h'}_\alpha$ *in* N.

For suppose this condition is not satisfied, but that there is a homeomorphism $\chi\colon \partial B \to \partial B$ with $h'\chi\big|_{V'\setminus C'} = \chi\big|_{V\setminus C} h$. Since $\chi(D_{\alpha_0})$ is a (meridional) disc in V', $\chi(D_{\alpha_0}\setminus\alpha_0)$ has one limit point on C'. On the other hand we have

$$\chi(D_{\alpha_0}\setminus\alpha_0) = h'^{-1}\chi h(D_{\alpha_0}\setminus\alpha_0) = h'^{-1}\chi(A^h_{\alpha_0}).$$

Since, by assumption, $\chi(A^h_{\alpha_0})$ meets *every* $A^{h'}_\alpha$ arbitrarily close to C, it follows that $\chi(D_{\alpha_0}\setminus\alpha_0)$ meets *every* D_α arbitrarily close to C'; i.e., that $\chi(D_{\alpha_0}\setminus\alpha_0)$ contains all of C' in its closure. This contradiction establishes the assertion.

§3. Construction of Examples

We describe the Poincaré map h of a typical example, and this is done by specifying the characteristic annuli A^h_α of h (cf. §2). We first describe a single such annulus $A^h = A^h_{\alpha_0}$. A^h is obtained by a sequence of modifications of the standard annulus $A = A_{\alpha_0}$. For a sequence $\{c_i\}$ monotonically decreasing to zero, the ith modification will change A only on the interior of the $T^2 \times I$ given by

$$U_i = \bigcup\{T_c \mid c_{i+1} \le c \le c_i\}.$$

The ith modification itself consists of pushing a subdisc of A out along an oriented simple closed curve γ_i to form a long tube and so the resulting disc winds at least twice along γ_i. The curve γ_i lies on one of the tori T_c, $c \in (c_{i+1}, c_i)$. Except for the choice of the level c, there are just two possible simple closed curves, the γ^0, γ^1 defined by Figure 2 (i.e., we have $\pi_1(\gamma_i) = \gamma^0$ or γ^1, where $\pi_1 : S^1 \times S^1 \times (0,1] \to S^1 \times S^1$ is the natural projection), so there are just two types of modification possible, say type 0 or 1. The result of a modification of type 0 is illustrated in Figure 3.

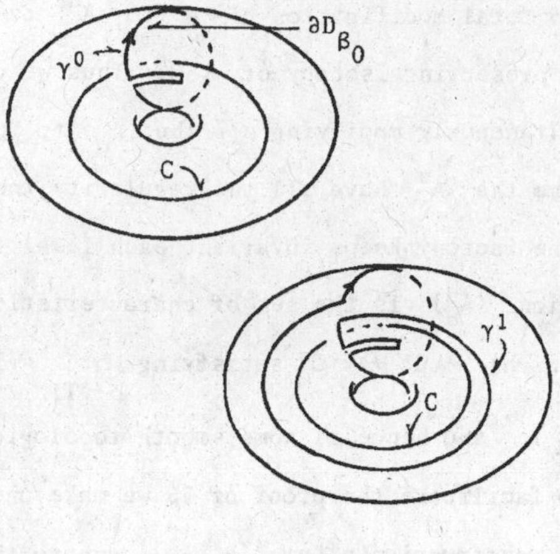

Figure 2

The subdisc E_i of A is replaced by the subdisc E_i^h in A^h, with $\partial E_i^h = \partial E_i$. Thus

$$A^h = A \setminus (\bigcup_{i \geq 1} E_i) . \cup . (\bigcup_{i \geq 1} E_i^h).$$

Note that the innermost disc on the modified annulus extends all the way through the "fold" in γ_i, so that E_i^h passes completely around the fold three times.

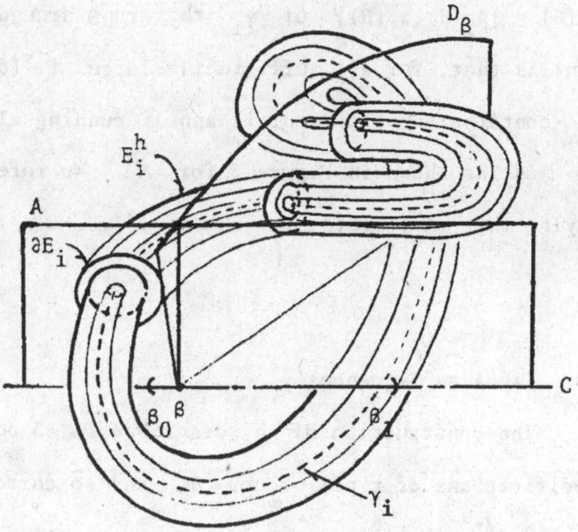

Figure 3

The total modification of A to A^h can be realized by a smooth c-level preserving isotopy of $V \setminus C$. Thus we can regard this isotopy as simultaneously modifying *all* the A_α to "parallel" annuli A_α^h. We assume the A_α^h have all the regularity indicated in Figure 3. Since the isotopy keeps invariant each level T_c, the resulting collection $\{A_\alpha^h\}$ is the set of characteristic annuli of a diffeomorphism $h: V' \setminus C' \to V \setminus C$ satisfying $h\big|_{T_1'} = \mathrm{Id}$ and $h(T_c') = T_c$ for $0 < c \leq 1$, and hence of some smooth topologically hyperbolic equilibrium.

To facilitate the proof of §5 we make one additional restriction in this construction. Let $[\beta_0, \beta_1]$ denote the closed subarc of C to which the folds on the γ_i converge; here C is parametrized by $\beta \in S^1$ as in §2. For $\beta \in (\beta_0, \beta_1)$ the meridional disc D_β meets γ_i in two points $p_i(\beta), q_i(\beta)$ that are joined by an oriented subarc $\sigma_i(\beta) = [p_i(\beta), q_i(\beta)]$ of γ_i that runs around the fold. Our requirement is that, for all sufficiently large i (depending on β) *every* A_α^h contains three concentric annuli running along $\sigma_i(\beta)$, and meeting D_β just as shown in Figure 3 for A^h. We refer to this later by saying that A_α^h meets D_β *canonically*.

§4. Lemma on Sequences

The construction of h described in §3 consists of a sequence of modifications of types 0 or 1, and so corresponds to a sequence $(s_i) \in 2^{\mathbf{Z}+}$. We will prove in the next section that flows ϕ, ϕ', corresponding to sequences $(s_i), (s_i')$ respectively, are equivalent only if some terminal segment of (s_i) matches some terminal segment of (s'_i). The existence of the desired uncountable collection is then a consequence of the following lemma.

DEFINITION. Two sequences $(s_i),(t_i) \in 2^{\mathbb{Z}^+}$ are *essentially different* if no terminal segment $(s_n,s_{n+1},\ldots)$ of (s_i) matches *any* terminal segment $(t_m,t_{m+1},\ldots)$ of (t_i).

LEMMA. *There is an uncountable collection* $\mathcal{H} \subseteq 2^{\mathbb{Z}^+}$ *of pairwise essentially different sequences.*

Proof: If we exclude the set $R \subseteq 2^{\mathbb{Z}^+}$ of sequences that terminate with repeated 1's, there is a one-one correspondence $[0,1) \leftrightarrow 2^{\mathbb{Z}^+} \setminus R$ given by binary expansion. Note that if $(s_i),(t_i) \in 2^{\mathbb{Z}^+}$ fail to be essentially different, then the corresponding real numbers $s = \sum_1^\infty s_i 2^{-i}$, $t = \sum_1^\infty t_i 2^{-i}$, and 1 are linearly dependent over the rationals $\mathbb{Q}$. Thus we take a Hamel basis $\mathcal{H}'$ of $\mathbb{R}$ over $\mathbb{Q}$, with $\mathcal{H}' \subseteq [0,1]$ and $\mathcal{H}' \cap \mathbb{Q} = \{1\}$, and let $\mathcal{H}$ be the corresponding subset of $2^{\mathbb{Z}^+}$.

§5. Proof of Inequivalence

Now suppose we have two Poincaré maps $h,h'\colon V'\setminus C' \to V\setminus C$, each constructed as above, and corresponding to the sequences $(s_i),(s'_i)$ respectively. We assume that h,h' correspond to equivalent equilibria and must show that then $(s_i),(s'_i)$ cannot be essentially different.

We use the necessary condition of §2; thus we assume that there is a homeomorphism $\chi: V \to V$ with $\chi(A^{h'})$ disjoint from one of the characteristic annuli of h. We denote this one A^h; the essential property of A^h will be that, for $\beta \in (\beta_0,\beta_1)$, it meet D_β canonically (cf. end of §3), and since all the A^h_α have this property it does not matter which one we are forced to use.

The idea of the proof is that, as χ is defined on C, it cannot produce much variation in the longitudinal (β) coordinate near C. Thus, because of the longitudinal fold in the γ_i', χ cannot unwrap the tubes $E_i^{h'}$, at least for sufficiently large i. It follows that, in order that $\chi(A^{h'})$ miss A^h, the $\chi(E_i^{h'})$ must be "nested" with the E_i^h. This nesting will define the desired correspondence between (s_i) and (s_i'). We attempt to give all the important steps of the proof without going into excessive detail.

For certain "general position" arguments it is convenient to work in the PL category. By [2, Theorem 1, p. 253] we may assume that $\chi(A^{h'})$ is PL without disturbing $\chi(A^{h'}) \cap A^h = \emptyset$. Analogously we may assume that $\chi(A^{h'})$ is locally in general position (transverse) with some fixed D_β whenever necessary.

Choose $\beta_-,\beta_+ \in (\beta_0,\beta_1)$ with $\beta_- < \beta_+$, and choose $\beta \in C$ so that $\chi(\beta) \in (\beta_-,\beta_+)$ (it will turn out that $\beta \in (\beta_0,\beta_1)$). Define $V_\varepsilon = C \cup \bigcup_{c \leq \varepsilon} T_c$. Choose $\varepsilon > 0$ so that:

(1) $\chi(D_\beta) \cap V_\varepsilon$ is between D_{β_-} and D_{β_+}, and

(2) for any i with $c_i \leq \varepsilon$, E_i^h meets D_{β_-}, D_{β_+} canonically.

For convenience assume $\varepsilon = c_{i_0}$ for some index i_0. Then for $i \geq i_0$ we have the situation illustrated in Figure 4.

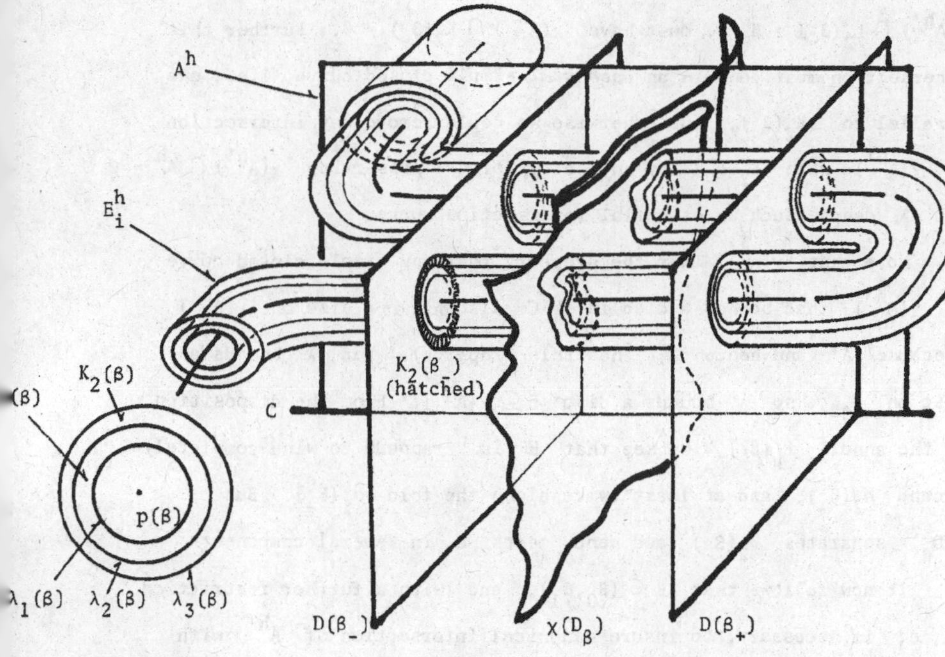

Figure 4

For $\beta \in (\beta_0, \beta_1)$ and E_i^h meeting D_β canonically, let $p(\beta)$ be the first intersection of γ_i with D_β, and let $\lambda_1(\beta), \lambda_2(\beta), \lambda_3(\beta)$ be the corresponding intersections of E_i^h with D_β (cf. Figure 4). Let $K_1(\beta), K_2(\beta)$ be the annuli on D_β bounded by the $\lambda_\mu(\beta)$, so $\partial K_\mu(\beta) = \lambda_\mu(\beta) \cup \lambda_{\mu+1}(\beta)$ $(\mu = 1,2)$; and let $L_1(\beta), L_2(\beta)$ be the corresponding annuli on E_i^h: $\partial L_\mu(\beta) = \lambda_\mu(\beta) \cup \lambda_{\mu+1}(\beta)$ $(\mu = 1,2)$. Let $W_1(\beta), W_2(\beta)$ denote the solid tori bounded by these annuli: $\partial W_\mu(\beta) = K_\mu(\beta) \cup L_\mu(\beta)$ $(\mu = 1,2)$. (Here, although these sets depend on i, it is convenient to suppress the subscript i in the notation.)

Now every longitudinal annulus in V, and $\chi(A^{h'})$ in particular, must intersect γ_i. Since γ_i is in the interior of $W_2(\beta_-)$, and

$\chi(A^{h'}) \cap L_2(\beta_-) = \emptyset$ we must have $\chi(A^{h'}) \cap K_2(\beta_-) \neq \emptyset$. Further this intersection must contain an *essential* simple closed curve, i.e., one parallel to $\partial K_2(\beta_-)$, for otherwise we could remove any intersection of $\chi(A^{h'})$ with $K_2(\beta_-)$ by an isotopy of V preserving $\chi(A^{h'}) \cap A^h = \emptyset$. Let λ denote such an essential intersection curve.

Note that $\chi(A^{h'})$ has the property that any simple closed curve on $\chi(A^{h'})$ that bounds a disc in $V \setminus C$ also bounds a disc on $\chi(A^{h'})$ (because A' and hence $A^{h'}$ has this property). But λ bounds a disc on D_{β_-}, so λ bounds a disc $E \subseteq \chi(A^{h'})$. From the disposition of the annuli $L_\mu(\beta_-)$, we see that E is "trapped" to wind completely around $W_2(\beta_-)$, and at least twice along the fold $\sigma_i(\beta_-)$. But $\chi(D_\beta)$ separates $W_2(\beta_-)$ and hence meets E in several components.

It now follows that $\beta \in (\beta_0, \beta_1)$, and (with a further restriction on ε, if necessary, to insure canonical intersection of $A^{h'}$ with D_β in $\chi^{-1}(V_\varepsilon)$) that these intersections of E with $\chi(D_\beta)$ must contain the curves $\chi(\lambda_1'(\beta)), \chi(\lambda_2'(\beta))$ corresponding to some $E_j^{h'} \subseteq A^{h'}$. Thus the annulus $\chi(K_1'(\beta))$ lies in the interior of $W_2(\beta_-)$ and therefore $\chi(W_1'(\beta)) \subseteq W_2(\beta_-)$. This is the desired nesting; note that $\chi(\gamma_j') \subseteq W_2(\beta_-)$ so that the types of the modifications $E_j^{h'}, E_i^h$ are both 0 or both 1, i.e., $s_j' = s_i$.

The nesting $\chi(W_1'(\beta)) \subseteq W_2(\beta_-)$ defines a correspondence $\zeta\colon j \mapsto i$. It is clear from the preceding that ζ is a well-defined function whose range contains a terminal segment of $\mathbb{Z}_+$, and that $s_{\zeta(j)} = s_j'$. That ζ is one-to-one may be seen as follows: note that $\zeta(j) = i$ if and only if $\chi(E_j^{h'})$ has an essential intersection $\chi(\lambda)$ with $K_1(\beta_-)$. If also $\chi(E_{j'}^{h'})$ meets $K_1(\beta_-)$ in an essential curve $\chi(\lambda')$, then both λ, λ' (appropriately oriented) would have the same linking number with γ_j' (as their images do with $\chi(\gamma_j')$). But this is impossible as $\lambda \subseteq E_j^{h'}$, $\lambda' \subseteq E_{j'}^{h'}$. Also, the domain of ζ contains some terminal segment of $\mathbb{Z}_+$. For, by a previous argument, applied now to χ^{-1} in place of χ, the annulus $\chi(K_2'(\beta)) \subseteq \chi(D_\beta)$ must contain an essential intersection with A^h, hence with some E_i^h. It is then not difficult to show that $\chi(W_1'(\beta))$ is contained in the corresponding $W_2(\beta)$ as desired.

Finally we show that, while ζ may not be monotone, it cannot reverse the order of two modifications of different type. For example, suppose γ_j' is "inside" $\gamma_{j'}'$ (i.e., $c_j < c_{j'}$), γ_j' is of type 0, $\gamma_{j'}'$ is of type 1, and $\zeta(j) = i$, $\zeta(j') = i'$. Then γ_j' bounds a disc in V that misses $\gamma_{j'}'$. It follows that γ_i bounds a disc in V that misses $\gamma_{i'}$ and hence that γ_i is inside $\gamma_{i'}$ also.

Thus ζ provides a matching of some terminal segment of (s_i') with a terminal segment of (s_i), as was to be shown.

REMARK. We describe briefly an example to indicate that the correspondence ζ defined above may not be monotone. Suppose in the construction of the A^h that both γ_i and γ_{i+1} are of type 0. Let $U \subseteq V$ be a small solid torus that contains γ_i and γ_{i+1}, but misses all the other γ_j's, situated as illustrated in Figure 5. There is a homeomorphism of U that interchanges γ_i and γ_{i+1}, and is the identity on the boundary of U, hence extends (by the identity) to a homeomorphism χ of all of V. The characteristic annuli A^h can be

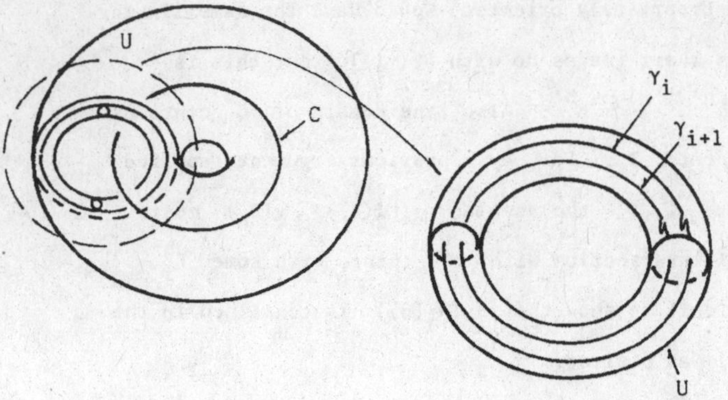

Figure 5

chosen to be invariant under χ. Then χ defines a self-equivalence of the corresponding topologically hyperbolic equilibrium that interchanges γ_i and γ_{i+1}.

REFERENCES

[1] C. Coleman, Hyperbolic stationary points, *Reports of the Fifth International Conference on Nonlinear Oscillations*, Vol. 2 (Qualitative methods), Kiev, 1970.

[2] E. E. Moise, *Geometric Topology in Dimensions 2 and 3*, Springer-Verlag, New York, 1977.

[3] D. A. Neumann, Topologically hyperbolic equilibria in dynamical systems, to appear in *J. Differential Equations*.

[4] F. W. Wilson, A reformulation of Coleman's conjecture concerning the local conjugacy of topologically hyperbolic singular points, *Structure of Attractors in Dynamical Systems* (Lecture Notes in Mathematics, Vol. 668), Springer-Verlag, New York, 1978.

[5] __________, Coleman's conjecture relating to topological hyperbolicity, these proceedings.

Bowling Green State University
Bowling Green, Ohio 43403

DYNAMICAL PROPERTIES OF CERTAIN NON-COMMUTATIVE SKEW-PRODUCTS

S. E. Newhouse

1. In this note we outline a general structure theory for certain homeo-orphisms of certain compact metric spaces. Detailed proofs of the results de-cribed here will appear elsewhere. Some of this work was done together with ai-Sang Young.

The homeomorphisms we shall discuss have enough "hyperbolicity" to be losely related to certain Axiom A diffeomorphisms. Roughly speaking, if certain small" invariant subsets are neglected, our homeomorphisms become topologically onjugate to those Axiom A diffeomorphisms on their non-wandering sets.

Our homeomorphisms arise naturally as non-wandering set restrictions of ertain diffeomorphisms considered by Abraham and Smale [1], [4, p. 40], Shub 10], and Hirsch, Pugh, Shub [7]. In this connection our results provide the irst detailed description of the orbit structures of elements in open sets of on-Axiom A diffeomorphisms.

Before stating the general theorem, let us consider an example of a homeo-orphism which illustrates our main results.

Let $\Sigma_2 = \{0,1\}^{\mathbb{Z}}$ be the set of bi-infinite sequences of 0's and 1's ith the compact open topology and $\phi : \Sigma_2 \to \Sigma_2$ be the 2-shift: $(\underline{a})(i) = \underline{a}(i+1)$ for $\underline{a} \in \Sigma_2$, $i \in \mathbb{Z}$. Let $T^2 = \mathbb{R}^2/\mathbb{Z}^2$ be the 2-torus, and et $L : T^2 \to T^2$ be a linear Anosov diffeomorphism with stable foliation F^s. et g be any diffeomorphism of T^2 which is homotopic to L and preserves ^{s} (i.e. $g(F^s_x) = F^s_{g(x)}$ where F^s_y is the leaf of F^s through y). In paricular,

L could be the linear automorphism induced by the matrix $\begin{pmatrix} 2 & 1 \\ 1 & 1 \end{pmatrix}$ and g could be a DA diffeomorphism as described in [12]. We assume that g normally expands F^s in the sense of Hirsch, Pugh, and Shub [7]. This means there is a continuous line bundle E^u over T^2 such that

(a) $T_y g(E^u_y) = E^u_{gy}$ for $y \in T^2$

and

(b) $$\inf_{\substack{|v|=1 \\ v\in E^u_y \\ y\in T^2}} ||T_y g(v)|| > \sup_{\substack{|v|=1 \\ v\in T_y F^s_y \\ y\in T^2}} ||T_y g(v)||.$$

Let $f : \Sigma_2 \times T^2 \to \Sigma_2 \times T^2$ be the map $f(\underline{a},y) = (\phi\underline{a}, f_{\underline{a}}(y))$ where

$$f_{\underline{a}}(y) = \begin{cases} Ly & \text{if } \underline{a}(0) = 0 \\ gy & \text{if } \underline{a}(0) = 1 \end{cases}.$$

One may think of f as the set of all random compositions of L and g. If $\pi : \Sigma_2 \times T^2 \to \Sigma_2$ is projection onto the first factor, then $\pi f = \phi\pi$. Let μ_ϕ be the Bernoulli $\left(\frac{1}{2},\frac{1}{2}\right)$ measure on Σ_2.

For a map ψ, let $M(\psi)$ denote the set of ψ-invariant probability measures, and let $h(\psi)$ be the topological entropy of ψ. If $\nu \in M(\psi)$, let $h_\nu(\psi)$ be the measure theoretic entropy of ψ relative to ν. It is well-known that $h(\psi) = \sup_{\nu\in M(\psi)} h_\nu(\psi)$ (see [5, p. 131]). We say $\nu \in M(\psi)$ is a maximal measure if $h_\nu(\psi) = h(\psi)$. If $\pi : X \to Y$ is a map and ν is a measure on X, we let $\pi_*\nu$ be the induced measure on Y.

<u>Theorem 1</u>. Suppose $f : \Sigma_2 \times T^2 \to \Sigma_2 \times T^2$ is as above. Then,

(a) $h(f) = h(\phi) + h(L)$

(b) f has maximal measures

(c) any maximal measure ν for f is such that $\pi_*\nu = \mu_\phi$.

Let m be the normalized Haar measure on T^2. Then $\mu_\phi \times m$ is the ique maximal measure of $\phi \times L : \Sigma_2 \times T^2 \to \Sigma_2 \times T^2$. For $x \in \Sigma_2$, write $= f_{\phi^{n-1}x} \circ f_{\phi^{n-2}x} \circ \ldots \circ f_x$.

eorem 2. In addition to the assumptions of theorem 1, suppose $B \subset \Sigma_2$ is a invariant set such that for each $x \in B$,

$$\limsup_{n\to\infty} \frac{1}{n} \log(\sup_{y\in\pi^{-1}x} ||T_y f_x^n | T_y F_y^s||) < 0 .$$

en, $f|\pi^{-1}(B)$ is topologically conjugate to $\phi \times L|\pi^{-1}(B)$. If $\mu_\phi(B) = 1$, en f has a unique maximal measure ν_f and the pair (f,ν_f) and $\times L, \mu_\phi \times m)$ are measure theoretically conjugate.

marks. 1. The inequality hypothesis of theorem 2 implies that for $x \in \Sigma_2$, f_x^n contracts the leaves of F^s for large n.

2. We do not know if the existence of a set B with $\mu_\phi(B) = 1$ as in theorem 2 is necessary for f to have a unique maximal measure. Perhaps any f satisfying the assumptions of theorem 1 already has a unique maximal measure

3. The map f is a non-commutative skew-product (i.e. the fiber maps do not commute). Unique maximal measures were found for certain commutative skew-products in [9].

4. Since the system $(\phi \times L, \mu_\phi \times m)$ is measure theoretically conjugate to a Bernoulli shift, so is (f,ν_f).

2. In this section we outline the proof of theorem 1.

First we recall some definitions and preliminary results. Let $f : X \to X$ a homeomorphism of the compact metric space X with metric d. Let $n > 0$ a positive integer, let $\delta > 0$ be a positive real number, and let K be a ompact subset of X. A set $E \subset K$ is an (n,δ,K)-separated set if for $\neq y$ in E there is a $j \in [0,n)$ such that $d(f^j x, f^j y) > \delta$. Let

$r(n,\delta,K,f)$ be the maximal cardinality of an (n,δ,K)-separated set. Let $h(K,f) = \lim_{\delta\to 0} \limsup_{n\to\infty} \frac{1}{n} \log r(n,\delta,K,f)$, and let $h(f) = h(X,f)$. The number $h(K,f)$ is called the topological entropy of the pair (K,f), and the number $h(f)$ is the topological entropy of f. The variational principle $h(f) = \sup_{\nu\in M(f)} h_\nu(f)$ has been generalized by Ledrappier and Walters as follows [8].

Let $X = \Lambda \times Y$, let $\pi : \Lambda \times Y \to \Lambda$ be projection on the first factor, and let $\phi : \Lambda \to \Lambda$, $f : \Lambda \times Y \to \Lambda \times Y$ be homeomorphisms such that $\pi f = \phi\pi$. Then for $\mu \in M(\phi)$,

$$\sup_{\substack{\nu\in M(f)\\ \pi_*\nu=\mu}} h_\nu(f) = h_\mu(\phi) + \int_\Lambda h(\pi^{-1}x,f)d\mu \quad .$$

Now we can outline the proof of theorem 1.

Statement (b) of theorem 1 follows from the fact that f is h-expansive [2]. That is, for some $\varepsilon > 0$, $h(W^s(x,\varepsilon),f) = 0$ for all $x \in \Sigma_2 \times T^2$ where

$$W^s(x,\varepsilon) = \{y \in \Sigma_2 \times T^2 : d(f^n x, f^n y) \leq \varepsilon \text{ for } n \geq 0\} .$$

It is not hard to see that for $\varepsilon > 0$ small and $x = (x_1, x_2) \in \Sigma_2 \times T^2$, $W^s(x, \varepsilon) = C \times I_\varepsilon$ where $C \subset \Sigma_2$ and I_ε is an interval in $F^s_{x_2}$ containing x_2. Because $W^s(x, \varepsilon)$ has this form, one can show that the maximal cardinality of an $(n,\delta,W^s(x,\varepsilon))$-separated set grows as a polynomial of degree one in n. Hence, $h(W^s(x,\varepsilon),f) = 0$.

For statements (a) and (c), one proves

(*) For each $x \in \Sigma_2$, $h(\pi^{-1}x,f) = h(L)$.

Assuming (*), one has by the above mentioned Ledrappier-Walters result that for any $\mu \in M(\phi)$,

$$\sup_{\substack{\nu\in m(f)\\ \pi_*\nu=\mu}} h_\nu(f) = h_\mu(\phi) + \int h(\pi^{-1}x,f)du$$
$$= h_\mu(\phi) + h(L) \quad .$$

This gives $h(f) = h(\phi) + h(L)$ which is (a). Statement (c) also follows since for $\mu \in M(\phi)$ and $\mu \neq \mu_\phi$ one has $h_\mu(\phi) < h_{\mu_\phi}(\phi)$.

We now briefly describe how (*) is proved. For notational convenience, if $s_1(n,\varepsilon)$ and $s_2(n,\varepsilon)$ are functions of (n,ε), let us write $s_1(n,\varepsilon) \sim s_2(n,\varepsilon)$ if

$$\lim_{\varepsilon\to 0} \limsup_{n\to\infty} \frac{1}{n} \log s_1(n,\varepsilon) = \lim_{\varepsilon\to 0} \limsup_{n\to\infty} \frac{1}{n} \log s_2(n,\varepsilon) \quad .$$

To prove (*), we fix $x \in \Sigma_2$, $n > 0$ and $\varepsilon > 0$ small. We wish to compute $r(n,\varepsilon,\pi^{-1}x,f)$. We first choose $N = N(\varepsilon, x) > 0$ and a periodic point z of ϕ of period $n + N$ such that $\tilde{d}(\phi^j x,\phi^j z)$ is very small for $j \in [0,n)$ where $\tilde{d}$ is the metric on Σ_2. With some work one shows $N(\varepsilon,x)$ can be chosen independent of x and $r(n,\varepsilon,\pi^{-1}x,f) \sim r(n+N,\varepsilon,\pi^{-1}z,f)$. Replacing $n + N$ by n, it suffices to compute $r(n,\varepsilon,\pi^{-1}z,f)$ where $\phi^n(z) = z$.

Let $T_z^2 = \pi^{-1}z$. Now f_z^n maps T_z^2 to itself and preserves F^s on T_z^2. Also, since f_z^n is homotopic to L^n on T_z^2, there is a point $y \in T_z^2$ such that $f_z^n(y) = y$. Since f_z^n normally expands F^s on T_z^2, there is a strong unstable manifold $W^{uu}(y)$ through y invariant by f_z^n. Let I_z be a small open interval about y in $W^{uu}(y)$. Then f_z^n (closure I_z) $\supset I_z$ and the local picture near y is as in figure 1.

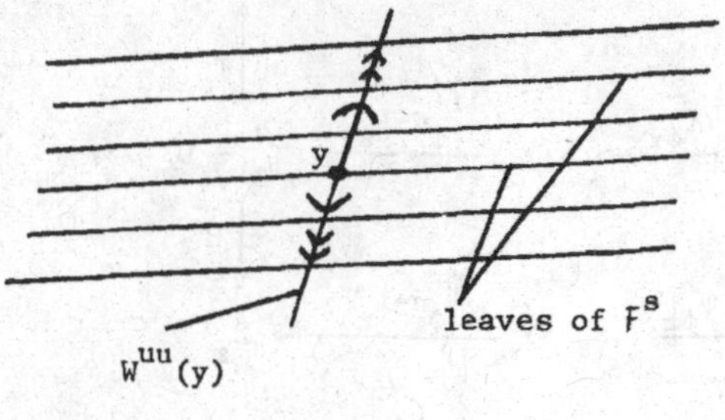

Figure 1

Cut T_z^2 open along I_z to give a new torus $\tilde{T}_z^2$, an open disk $\tilde{D}_z \subset \tilde{T}_z^2$, and a map $\tilde{\pi} : \tilde{T}_z^2 - \tilde{D}_z \to T_z^2$ which is 1-to-1 off the boundary of $\tilde{D}_z$ and 2-to-1 on $\partial\tilde{D}_z$ except at two points as in Figure 2.

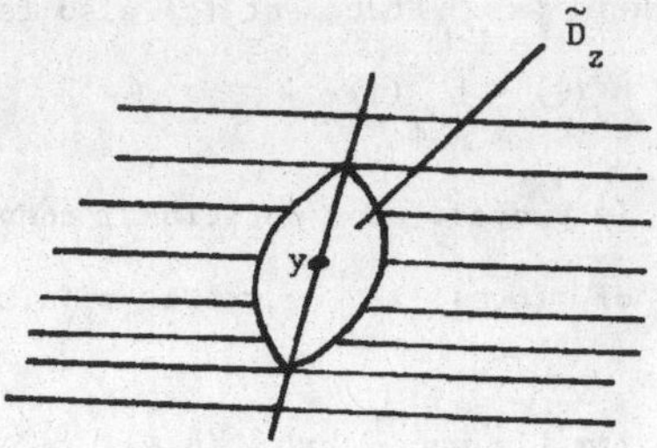

Figure 2

Then $\tilde{T}_z^2 - \tilde{D}$ inherits a one-dimensional foliation $\tilde{F}^s(z)$ such that $\tilde{\pi}\,\tilde{F}^s(z) = F^s$. Also, each component of $\tilde{F}^s(z)$ is a closed interval. If we form the quotient space Σ_z obtained by identifying points in a single component of $\tilde{F}^s(z)$, then Σ_z can be given the structure of a smooth branched 1-manifold in the sense of Williams [11]. Let $\overline{\pi}_z : \tilde{T}_z^2 \to \Sigma_z$ be the quotient map. Proceed similarly on $T^2_{\phi^j z}$ for $j \in [0,n)$ using the interval $I_{\phi^j z} = f_{\phi^{j-1} z} \circ \dots \circ f_z(I_z)$. Thus, we get tori $\tilde{T}^2_{\phi^j z}$, branched manifolds $\Sigma_{\phi^j z}$, and maps $\tilde{\pi}_j$, $\overline{\pi}_j$, $\tilde{f}_{\phi^j z}$, $\psi_{\phi^j z}$ so that the following diagram commutes

$$\begin{array}{ccccccc}
T_z^2 & \xrightarrow{f_z} & T_{\phi z}^2 & \xrightarrow{f_{\phi z}} & \cdots & \xrightarrow{f_{\phi^{n-1} z}} & T_z^2 \\
\uparrow{\scriptstyle \tilde{\pi}_z} & & \uparrow{\scriptstyle \tilde{\pi}_{\phi z}} & & & & \uparrow{\scriptstyle \tilde{\pi}_z} \\
\tilde{T}_z^2 - \tilde{D}_z & \xrightarrow{\tilde{f}_z} & \tilde{T}_{\phi z}^2 - \tilde{D}_{\phi z} & \xrightarrow{\tilde{f}_{\phi z}} & \cdots & \xrightarrow{\tilde{f}_{\phi^{n-1} z}} & \tilde{T}_z^2 - \tilde{D}_z \\
\downarrow{\scriptstyle \overline{\pi}_z} & & \downarrow{\scriptstyle \overline{\pi}_{\phi z}} & & & & \downarrow{\scriptstyle \overline{\pi}_z} \\
\Sigma_z & \xrightarrow{\psi_z} & \Sigma_{\phi z} & \xrightarrow{\psi_{\phi z}} & \cdots & \xrightarrow{\psi_{\phi^{n-1} z}} & \Sigma_z
\end{array}$$

Let us define $\tilde{f}_z^j = \tilde{f}_{\phi^{j-1}z} \circ \dots \circ \tilde{f}_z$ and $\psi_z^j = \psi_{\phi^{j-1}z} \circ \dots \circ \psi_z$. A set $E \subset \tilde{T}_z^2 - \tilde{D}_z$ is (n,ε)-separated if $x \neq y$ in E implies there is a $j \in [0,n)$ such that $d(f_z^j x, f_z^j y) > \varepsilon$. Set $r(n,\varepsilon,\tilde{T}_z - \tilde{D}_z,\tilde{f})$ to be the maximal cardinality of an (n,ε)-separated set in $\tilde{T}_z - \tilde{D}_z$. Similarly, define $r(n,\varepsilon,\Sigma_z,\psi)$. In the above diagram each $\tilde{\pi}_{\phi^j z}$ is at most 2-to-1 and each $\bar{\pi}_{\phi^j z}$ has its pre-images of points consisting of closed intervals. From this, it follows that $r(n,\varepsilon,\pi^{-1}z,f) \sim r(n,\varepsilon,\tilde{T}_z^2 - \tilde{D}_z,\tilde{f}) \sim r(n,\varepsilon,\Sigma_z,\psi)$. So the estimate of $r(n,\varepsilon,\pi^{-1}z,f)$ is reduced to $r(n,\varepsilon,\Sigma_z,\psi)$. Now each $\psi_{\phi^j z}$ is an expanding map from the branched 1-manifold $\Sigma_{\phi^{j-1}z}$ to $\Sigma_{\phi^j z}$. Using this, one can prove that $r(n,\varepsilon,\Sigma_z,\psi) \sim M_n(\psi)$ where $M_n(\psi)$ is the number of fixed points of ψ_z^n. By the Lefschetz trace formula, $M_n(\psi) = \text{trace } \psi_*^n - 1$ where $\psi_*^n \circ H_1(\Sigma_z,\mathbb{R}) \to H_1(\Sigma_z,\mathbb{R})$ is the map induced by ψ_z^n on the first homology group of Σ_z with real coefficients. The maps $\tilde{\pi}_z$ and $\bar{\pi}_z$ induce isomorphisms on the first real homology groups, so

$$\text{trace } \psi_*^n = \text{trace } \tilde{f}_{z*}^n = \text{trace } f_{z*}^n .$$

Since each $f_{\phi^j z}$ is homotopic to L, we get that f_z^n is homotopic to L^n, so

$$\text{trace } f_{z*}^n = \text{trace } L_*^n .$$

Moreover, it is known that

$$h(L) = \lim_{n\to\infty} \frac{1}{n} \log(\text{trace } L_*^n - 2) .$$

Thus,

$$\lim_{n\to\infty} \frac{1}{n} \log M_n(\psi) = h(L) .$$

This implies that

$$h(\pi^{-1}x,f) = h(\pi^{-1}z,f) = h(L)$$

as required.

3. Here we outline the proof of theorem 2.

Let 0 denote the origin in $\mathbb{R}^2$, and let $\tilde{\pi} : (\mathbb{R}^2, 0) \to (T^2, \bar{0})$ denote the canonical projection with $\tilde{\pi}(0) = \bar{0}$. Let $J : \Sigma_2 \times T^2 \to \Sigma_2 \times T^2$ be the affine bundle map over $\phi : \Sigma_2 \to \Sigma_2$ defined by $J(\underline{a}, y) = (\phi\underline{a}, Ly + f_{\underline{a}}(\bar{0}))$. Since Σ_2 is zero-dimensional, the map $\underline{a} \longmapsto f_{\underline{a}}(\bar{0})$ from Σ_2 to T^2 lifts to a map $\underline{a} \longmapsto \tilde{f}_{\underline{a}}(0)$ from Σ_2 to $\mathbb{R}^2$. Also, the affine bundle map J lifts to an affine bundle map $\tilde{J} : \Sigma_2 \times \mathbb{R}^2 \to \Sigma_2 \times \mathbb{R}^2$ where $\tilde{J}(\underline{a}, y) = (\phi\underline{a}, \tilde{L}(y) + \tilde{f}_{\underline{a}}(0))$ and $\tilde{L} : \mathbb{R}^2 \to \mathbb{R}^2$ is the lift of $L : T^2 \to T^2$. Moreover, the map $f : \Sigma_2 \times T^2 \to \Sigma_2 \times T^2$ lifts to a map $\tilde{f} : \Sigma_2 \times \mathbb{R}^2 \to \Sigma_2 \times \mathbb{R}^2$ so that $(\mathrm{id} \times \tilde{\pi}) \circ \tilde{f} = f \circ (\mathrm{id} \times \tilde{\pi})$ and $\tilde{f}(\underline{a}, 0) = (\phi\underline{a}, \tilde{f}_{\underline{a}}(0))$.

The conjugacy between $f|\pi^{-1}(B)$ and $\phi \times L|\pi^{-1}(B)$ is obtained by getting conjugac[y] between $f|\pi^{-1}(B)$ and $J|\pi^{-1}(B)$ and then between J and $\phi \times L$ on $\Sigma_2 \times T^2$. For the con[ju]gacy between $f|\pi^{-1}(B)$ and $J|\pi^{-1}(B)$, one uses parametrized versions of Franks argumen[t] in [6] after showing that $f|\pi^{-1}(B)$ has an invariant section. To get this section, o[ne] uses the hyperbolicity of $\tilde{L}$ to produce a continuous bundle map $H : \Sigma_2 \times \mathbb{R}^2 \to \Sigma_2 \times \mathbb{R}^2$ covering the identity map of Σ_2 such that $H\tilde{f} = \tilde{L}H$. Let F^u be the strong unstable foliation of F^s on $\Sigma_2 \times T^2$ for f. Let $\tilde{F}^s$ and $\tilde{F}^u$ be the lifts of those foliatio[ns] to $\Sigma_2 \times \mathbb{R}^2$. From the inequality estimate in the hypothesis in theorem 2 and method[s] as in Franks one can prove that for $x \in B$ and any $y, z \in \mathbb{R}^2$, $\tilde{F}^s_{(x,y)} \cap \tilde{F}^u_{(x,z)}$ is a unique point. From this it follows that $H|[\pi \circ (\mathrm{id} \times \tilde{\pi})]^{-1}B$ is a homeomorphism. Next one shows that $\tilde{J} : \Sigma_2 \times \mathbb{R}^2 \to \Sigma_2 \times \mathbb{R}^2$ has an invariant section, say $\tilde{\psi}$. This is a map $\tilde{\psi} : \Sigma_2 \to \mathbb{R}^2$ such that $\tilde{J}(\underline{a}, \tilde{\psi}(\underline{a})) = (\phi\underline{a}, \tilde{\psi}(\phi\underline{a}))$. This is done as follows.

Let $\mathbb{R}^2 = E^s \times E^u$ where E^s and E^u are the invariant contracting and expanding subspaces of $\tilde{L}$. Write $\tilde{\psi}(\underline{a}) = (\tilde{\psi}^s(\underline{a}), \tilde{\psi}^u(\underline{a})) \in E^s \times E^u$. The equation $\tilde{J}(\underline{a}, \tilde{\psi}(\underline{a})) = (\phi\underline{a}, \tilde{\psi}(\phi\underline{a}))$ determines equations for $\tilde{\psi}^s$ and $\tilde{\psi}^u$ which can be solved using the contraction map theorem.

Let $B_1 = [\pi \circ (\mathrm{id} \times \tilde{\pi})]^{-1}B$. Since $\tilde{J}$ has an invariant section and is topologically conjugate to $\tilde{f}$ on B_1, it follows that $\tilde{f}|B_1$ has an invariant section. This implies that $f|\pi^{-1}B$ has an invariant section, and then repeating the parametrized version of Franks arguments above one can show that H actually induces a conjugacy between f and J on $\pi^{-1}B$.

The final step in proving that $f|\pi^{-1}B$ is topologically conjugate to $\phi \times L|\pi^{-1}B$ involves showing that there is an affine conjugacy A from J to $\phi \times L$ on $\Sigma_2 \times T^2$. The section $\tilde{\psi}$ determines a section $\psi = \tilde{\pi} \circ \tilde{\psi} : \Sigma_2 \to T^2$ which is invariant under J. Then the map $A(\underline{a},y) = (\underline{a}, y - \psi(\underline{a}))$ is the required affine conjugacy from J to $\phi \times L$.

Let us now prove the other parts of theorem 2. Let $H : \pi^{-1}B \to \pi^{-1}B$ be a topological conjugacy from $f|\pi^{-1}B$ to $\phi \times L|\pi^{-1}B$ and let ν be any maximal measure for f. Then $\pi_*\nu = \mu_\phi$. If $\mu_\phi(B) = 1$, then $\nu(\pi^{-1}B) = 1$. Thus, $H_*\nu \in M(\phi \times L)$ and $h_{H_*\nu}(\phi \times L) = h(f)$. Since $h_\nu(f) = h(f) = h(\phi \times L)$, we get that $H_*\nu = \mu_\phi \times m$, or $\nu = H_*^{-1}(\mu_\phi \times m)$. This proves the uniqueness of ν and shows that H is a measure theoretic conjugacy from (f,ν) to $(\phi \times L, \mu_\phi \times m)$.

4. We now proceed to state more general versions of our results. The proofs follow the same general scheme described in sections 2 and 3.

Let $\phi : \Lambda \to \Lambda$ be a homeomorphism of a compact metric space Λ satisfying expansiveness and specification (see [3] for definitions). For example, ϕ might be a topologically mixing subshift of finite type or a topologically mixing mapping on a hyperbolic basic set of an Axiom A diffeomorphism. Bowen proved in [3] that ϕ has a unique measure of maximal entropy, say μ_ϕ. Let T^n be the n-dimensional torus, and let $L : T^n \to T^n$ be a linear Anosov diffeomorphism such that for some (or any) $x \in T^n$ the stable manifold of x is one-dimensional. Let $\mathrm{Diff}^1 T^n$ be the set of C^1 diffeomorphisms of T^n with the uniform C^1 topology. Let $0 \in T^n$ be the identity element of the group T^n, and let $F : \Lambda \to \mathrm{Diff}^1 T^n$ be a continuous map such that for each $x \in \Lambda$, $F(x)$ is homotopic to L and the map $x \longmapsto F(x)(0)$ from Λ to T^n is homotopic to the constant map $x \longmapsto 0$. Suppose that, for each $x \in \Lambda$, there is a one dimensional C^1 lamination $F(x)$ on T^n satisfying the following conditions.

(1) $F(x)$ is minimal (i.e. every leaf is dense in T^n)

(2) $F(x)F(x) = F(\phi x)$

(3) $F(x)$ normally expands $F(x)$ in the sense of Hirsch, Pugh, and Shub [7]

(4) $\bigcup_{x\in\Lambda} F(x)$ is a continuous lamination of $\Lambda \times T^n$.

To say that $F(x)$ is a one-dimensional C^1 lamination [7] of T^n means that $F(x)$ is a collection of C^1 immersed lines whose tangents vary continuously. To say that $\bigcup_{x\in\Lambda} F(x)$ is a continuous lamination of $\Lambda \times T^n$ means that if $x \in \Lambda$, $y \in T^n$, $F(x)_y$ is the leaf of $F(x)$ through y and $T_yF(x)_y$ is its tangent space at y, then $(x,y) \longmapsto T_yF(x)_y$ is a continuous line field on $\Lambda \times T^n$. To say that $F(x)$ normally expands $F(x)$ means there is a continuous codimension one subbundle $E^u(x)$ of T^n such that

(a) $T_yF(x)(E^u(x)_y) = E^u(\phi x)_{F(x)(y)}$

and

(b) $$\inf_{\substack{|v|=1 \\ v\in E^u(x) \\ x\in\Lambda, y\in T^n}} ||T_yF(x)(v)|| > \sup_{\substack{|v|=1 \\ v\in T_yF(x) \\ x\in\Lambda, y\in T^n}} ||T_yF(x)(v)|| .$$

Let $\pi : \Lambda \times T^n \to \Lambda$ be projection on the first factor.

<u>Theorem 1'</u>. Suppose $f : \Lambda \times T^n \to \Lambda \times T^n$ is the homeomorphism defined by $f(x,y) = (\phi x, F(x)(y))$ and that the above conditions hold. Then,

(a) $h(f) = h(\phi) + h(L)$

(b) f has maximal measures

(c) any maximal measure ν of f satisfies $\pi_*\nu = \mu_\phi$.

For $x \in \Lambda$ and $n > 0$, write $F^n(x) = F(\phi^{n-1}x) \circ \ldots \circ F(x)$. Let m be Haar measure on T^n.

Theorem 2'. In addition to the assumptions of theorem 1', suppose $B \subset \Lambda$ is a ϕ-invariant set such that for each $x \in B$,

$$\lim_{n\to\infty} \sup \frac{1}{n} \log(\sup_{y\in\pi^{-1}x} ||T_y F^n(x)|T_y F(x)_y|| < 0 \ .$$

Then, $f|\pi^{-1}B$ is topologically conjugate to $\phi \times L|\pi^{-1}B$. If $\mu_\phi(B) = 1$, then f has a unique maximal measure ν_f and the pairs (f,ν_f) and $(\phi \times L,\mu_\phi \times m)$ are measure theoretically conjugate.

References

1. R. Abraham and S. Smale, Nongenericity of Ω-stability, Proc. Symp. Pure Math., 14, Amer. Math Soc., Providence, RI, 1970, 5-8.

2. R. Bowen, Entropy-expansive maps, Trans. Amer. Math. Soc. 164 (1972), 323-333.

3. ________, Some systems with unique equilibrium states, Math. Syst. Theory 8 (1974), 193-202.

4. ________, On Axiom A diffeomorphisms, Regional Conference Series in Math. 35, Amer. Math. Soc., Providence, RI, 1978.

5. M. Denker, C. Grillenberger, and K. Sigmund, Ergodic Theory on Compact Spaces, Lecture Notes in Math. 527, Springer-Verlag, NY, 1976.

6. J. Franks, Anosov diffeomorphisms on tori, Trans. AMS 145 (1969), 117-124.

7. M. Hirsch, C. Pugh, and M. Shub, Invariant manifolds, Lecture Notes in Math. 583, Springer-Verlag, NY, 1977.

8. F. Ledrappier and P. Walters, A relativized variation principle, Bull. London Math. Soc., to appear.

9. B. Marcus and S. Newhouse, Measures of maximal entropy for a class of skew-products, Lecture Notes in Math 729, Springer-Verlag, NY, 1979, 105-126.

10. M. Shub, Symposium on differential equations and dynamical systems, Math. Institute, Univ. of Warwick, Coventry, England, 1969, p. 35.

11. R. Williams, One-dimensional non-wandering sets, Topology 6 (1967), 473-487.

12. __________, The "DA" maps of Smale and structural stability, Proc. Symp. Pure Math. 14, Amer. Math. Soc., Providence, RI, 1970, 329-334.

University of North Carolina
Chapel Hill, N.C. 27514

A Note on Explosive Flows

Zbigniew Nitecki

Let M be an open manifold, and denote by $W^r(M)$ the set of C^r flows on M whose non-wandering set is empty (the completely unstable flows). Takens and White [5] conjectured that this set is contained in the closure of its interior with respect to the strong C^r topology. When M is the complement of two simple circles in a 3-manifold and $r \geq 1$, a counterexample to this conjecture was constructed in [1]. On the other hand, when M is an open surface of finite genus, the conjecture has been shown true [2] for $r \geq 0$. These results leave open the conjecture for surfaces of infinite genus and for certain higher-dimensional manifolds, including $\mathbb{R}^n$, $n \geq 3$. The present note shows how to adapt the example of [1] to all higher-dimensional manifolds. I would like to thank Alan Dankner and Clark Robinson for helpful conversations concerning this example.

<u>Theorem</u>: If M is an open manifold and dim $M \geq 3$, there exist completely unstable C^r flows on M which cannot be C^r-perturbed into the interior of $W^r(M)$, for any $r \geq 1$.

The reader is referred to [1] for a more detailed discussion of the construction, terminology and notation.

<u>Proof of theorem when</u> dim $M = 3$:

By the arguments in [1], the crux of the problem is to construct the required example on a solid torus $D^2 \times S^1$ with finitely many points deleted. As in [1], our construction starts with a modification of the parallel flow $\dot{x} = 0$, $\dot{t} = 1$ on a flow box $P = D^2 \times [-1,1]$. We introduce two periodic saddles, σ_- and σ_+, with a cylinder of saddle connections consisting of one component of $W^u(\sigma_-)$ which coincides with a component of $W^s(\sigma_+)$. The other component of $W^u(\sigma_-)$ forms a cylinder Σ_- whose boundary consists of σ_- and a periodic sink s_-; similarly Σ_+ is a cylinder bounded by σ_+ and a periodic source s_+, joined by a component of $W^s(\sigma_+)$. Both components of $W^s(\sigma_-)$ enter P via its inset $\partial^- P$, while both components of $W^u(\sigma_+)$ leave P via its outset $\partial^+ P$. Finally, we require that the vector field in P satisfy the <u>mirror property</u>: $\dot{x}(x,-t) = -\dot{x}(x,+t)$, $\dot{t}(x,-t) = +\dot{t}(x,+t)$. This insures that any orbit entering $\partial^- P$ at $(p,-1)$ either stays in P forever or else leaves $\partial^+ P$ at $(p,+1)$ - that is, the Poincaré map from $\partial^- P$ to $\partial^+ P$, where defined, is the identity. The configuration of orbits and invariant manifolds in P is sketched in Fig. 1: it is identical to the "plug" in [1] and, originally, in [4].

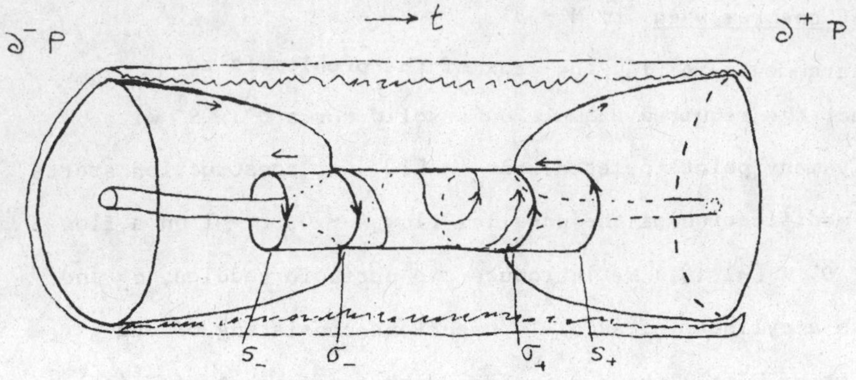

Figure 1

Now, we let $\sigma_\pm$ each bound a disc $\delta_\pm$ in P, on which we slow the vector field down to zero. The discs $\Delta_\pm$ (Fig. 2) consisting of $\delta_\pm$ and the closed cylinders $\Sigma_\pm$ are invariant sets for the flow; we remove them from P.

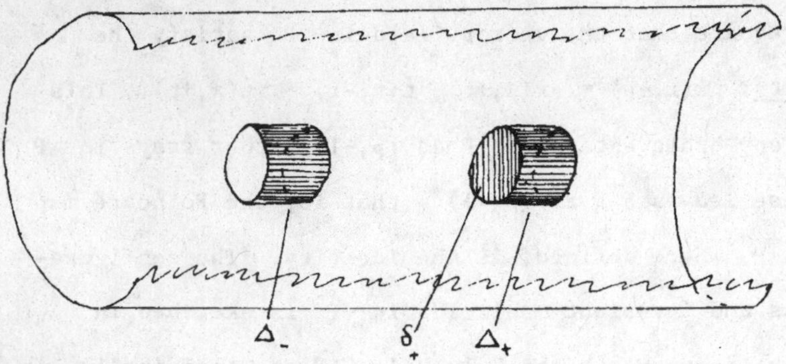

Figure 2

By contracting $\Delta_\pm$ to points $p_\pm$, we see, first, that $P\backslash[\Delta_+ \cup \Delta_-]$ is diffeomorphic to $P\backslash[p_+ \cup p_-]$, and second, that we can regard our flow as having two (degenerate) fixed points $p_{\pm}$ in place of the invariant discs $\Delta_{\pm}$. We note that any orbit entering $\partial^{-}P$ interior to the outer sheet of $W^{s}(\sigma_{-})$ limits on Δ_{-}, so that the stable set of p_{-} becomes a solid cone with vertex at p_{-}; similarly, the unstable set $W^{u}(p_{+})$ is a solid cone with vertex at p_{+}, while $W^{u}(p_{-}) = W^{s}(p_{+})$ is a solid ball with $p_{\pm}$ on its boundary (see Fig. 3). We denote by C_{+} the circle by which

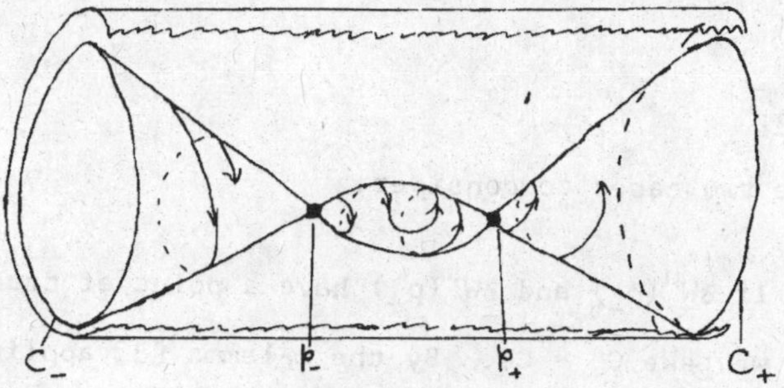

Figure 3

the boundary cone of the unstable set of p_{+} leaves $\partial^{+}P$:

$$C_{+} = \partial W^{u}(p_{+}) \cap \partial^{+}P$$

and similarly,

$$C_{-} = \partial W^{s}(p_{-}) \cap \partial^{-}P.$$

Note that the strong C^{1} topology on $P \setminus [p_{+} \cup p_{-}]$ forces the 1-jet of any perturbation to vanish at $p_{\pm}$. Thus, despite the degeneracy of the fixed points $p_{\pm}$, any strong C^{1}-perturbation of our flow will still have $p_{\pm}$ fixed with $W^{u}(p_{\pm})$ and $W^{s}(p_{\pm})$ emanating from $p_{\pm}$ as solid cones near the original ones. (The best way to see this is by means of Fig. 1).

Our example is based on the following

<u>Lemma</u>: Suppose a strong C^1 perturbation of the flow on P described above has $\partial W^u(p_-)$ transverse to $\partial W^s(p_+)$. Then there exist circles $\tilde{C}_\pm$ uniformly near $C_\pm$, such that any transversal to $\tilde{C}_-$ includes all of $\tilde{C}_+$ in the C^1-closure of its forward orbit by the perturbed flow.

<u>Proof of Lemma</u>:

There are two cases to consider:

<u>Case 1</u>: If $\partial W^u(p_-)$ and $\partial W^s(p_+)$ have a point of transverse intersection, we take $\tilde{C}_\pm = C_\pm$. By the λ-lemma [3] applied to σ_- in Fig. 1, a transversal at $C_- \subset W^s(\sigma_-)$ accumulates in forward time on all of $W^u(\sigma_-) = \partial W^u(p_-)$ in the C^1 sense. Thus, it crosses $\partial W^s(p_+) = W^s(\sigma_+)$ transversally, and the λ-lemma at σ_+ gives the conclusion of the lemma.

<u>Case 2</u>: If $\partial W^u(p_-)$ and $\partial W^s(p_+)$ are disjoint, then one is interior to the other. Assume $\partial W^u(p_-)$ is exterior to $\partial W^s(p_+)$, so that $\partial W^s(p_+)$ and hence all of $W^s(p_+)$ is interior to $W^u(p_-)$. Then $\partial W^u(p_-) = W^u(\sigma_-)$ closely follows $W^s(\sigma_+)$ and then $W^u(\sigma_+)$ on the outside, eventually crossing ∂^+P in a circle $\tilde{C}_+$ C^1 near (and slightly outside) C_+ (Fig. 4). We then take $\tilde{C}_- = C_-$. This time, a single application of the λ-lemma (the first half of the argument in Case 1) gives the desired conclusion.

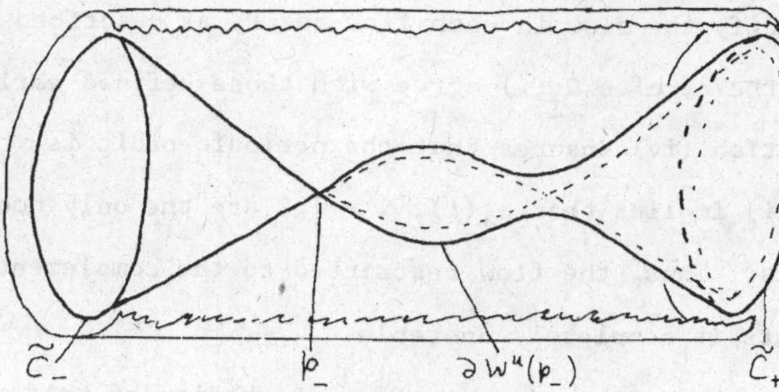

Figure 4

To construct our example on the solid torus $D^2 \times S^1$, we start with a flow having a single periodic sink, $0 \times S^1$, whose basin of attraction is the solid torus. Pick two flowboxes, P_1 and P_2, with P_2 "downstream" from P_1. Pick circles $C_{\pm}(i) = C(i) \times \{\pm 1\} \subset \partial^{\pm} P_i$, and align the flow boxes and circles so that (Fig. 5):

i) As $C_+(1)$ flows to $\partial^- P_2$, it crosses $C_-(2)$ transversally.

ii) When $C_+(2)$ returns to $\partial^- P_1$, it crosses $C_-(1)$ transversally.

iii) When $C_+(i)$ returns to $\partial^- P_i$, it is interior to the disc bounded by $C_-(i)$.

iv) The periodic sink $\{0\} \times S^1$ crosses $\partial^- P_i$ interior to $C_-(i)$.

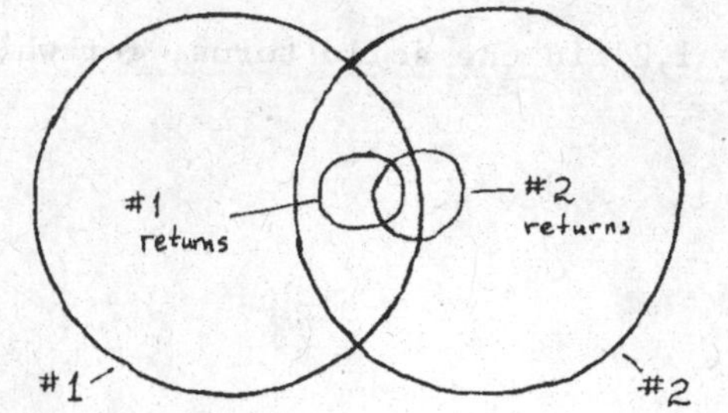

Figure 5

Now, we modify the flow in each flow box P_i as described above, so that the circles $C_{\pm}(i)$ agree with those defined earlier. Note that condition (iv) insures that the periodic orbit is broken, and (iii) implies that $p_{\pm}(i)$, $i = 1,2$ are the only non-wandering points. Thus, the flow restricted to the complement of these four points is completely unstable.

On the other hand, consider any C^1-perturbation of this completely unstable flow for which the hypotheses of the lemma hold in each "plug" P_i, $i = 1,2$. There are circles $\tilde{C}_{\pm}(i)$ near $C_{\pm}(i)$ such that a transversal at $\tilde{C}_-(i)$ piles up on all of $\tilde{C}_+(i)$. Starting from $\tilde{C}_-(1)$, such a transversal piles up on $\tilde{C}_+(1)$; but this transversally intersects $\tilde{C}_-(2)$, and so piles up on all of $\tilde{C}_-(1)$. In particular, if we start with a transversal at one of the points of intersection described in (ii), then this transversal returns arbitrarily near this point: in other words, any neighborhood of this point intersects itself in forward time, and the point in question is non-wandering.

Since the transversality of $\partial W^u(p_-)$ and $\partial W^s(p_+)$ is C^r-generic for any r, we see that the example is a flow on the complement of $p_{\pm}(i)$ $(i = 1,2)$ in the solid torus, for which a

C^r-generic C^r perturbation has nonwandering points. Thus we have a flow in W^r which is not in the closure of int W^r. Using the construction outlined in [1], we can now embed this solid torus in a flow on any open 3-manifold M for which points outside the solid torus wander. We then join each of the four points $p_{\pm}(i)$, $i = 1,2$ to an end of M by simple, disjoint arcs, first slowing down the flow to make them invariant. If we now delete these arcs from M, we obtain a manifold diffeomorphic to M. Although the argument using transversals might be affected by this deletion, the conclusion that neighborhoods of certain points intersect themselves in forward time is not affected. Thus, the theorem is proven when dim M = 3.

Proof of theorem when dim $M \geq 3$:

We note that the comment in [1] concerning the extension of the example to higher dimensions is incorrect: the perturbation arguments break down. However, we can create an n-dimensional version ($n > 3$) of our example on the solid torus by taking its product with a strong contraction on the (n-3)-disc ($\dot{x} = -Ax$, A a large scalar). This creates a flow on $D^2 \times S^1 \times D^{n-3}$ in which all unstable sets retain their dimension from the original example, while all stable sets increase dimension by $n - 3$ (and hence retain their codimension). From this consideration, it is clear that the transversality conditions (i), (ii) in our construction remain; similarly, conditions (iii) and (iv) are now statements about the interior of the open set $W^s(p_-(i)) \cap \partial^- P_i$ in $\partial^- P_i$.

The lemma needs to be modified so that the "circle" $\tilde{C}_- = S^1 \times \{-1\}$ of dimension 1 in $\partial^- P$ becomes a generalized cylinder $\tilde{C}_- = S^1 \times \{-1\} \times D^{n-3}$ of codimension one in $\partial^- P = D^2 \times \{-1\} \times D^{n-3}$. The argument for this lemma carries over, except for Case 2, where we need to analyze more carefully the ways in which $\partial W^u(p_-)$ can miss $\partial W^s(p_+)$. The easiest way to do this is by means of normal hyperbolicity theory [6,7], which allows us to invoke the persistence of normally hyperbolic invariant sets and simply repeat the three-dimensional arguments of the lemma inside a (3-dimensional) solid torus near the original one, and invariant under the perturbed flow. The technical problem with this is that our solid torus is not invariant, but merely flows into itself. While this can be handled via an appropriate modification of theorems for "overflowing invariant" sets as stated, for example, in [6], a technical device suggested by Alan Dankner allows us to deduce our conclusions directly from the results for (strictly) invariant manifolds.

We consider the flow on a neighborhood of our solid torus, which we think of as the product of a slightly larger solid torus with an (n-3)-disc. Now, we extend the (unperturbed) flow in the (3-dimensional) solid torus to a flow on the three-dimensional sphere, so that there is a single new (repelling) periodic orbit, linking the torus, such that every other orbit enters the solid torus at some (forward) time. We then consider the product of this flow on S^3 with a strong contraction on D^{n-3}.

Any C^1-perturbation of our original flow can be regarded as the restriction of a perturbation of our enlarged flow; the 3-sphere in the original flow is normally hyperbolic, so the perturbed flow has an invariant 3-sphere near S^3, and on this 3-sphere the new flow is C^1-near the original flow on S^3. Moreover, since the original solid torus formed part of a filtration for the original flow on S^3, there will be a new nearby solid torus with similar properties for the perturbation.

Thus, we can restrict our attention to the behavior inside this new solid torus, and the proof of the lemma, verbatim, gives the desired conclusion.

Again, we can use a device similar to the three-dimensional one in [1] to place the flow in $D^2 \times S^1 \times D^{n-3}$ inside a flow on any n-manifold M, with no nonwandering points elsewhere in M. Combining these elements, we obtain the required example on any open manifold M, $\dim M \geq 3$.

REFERENCES

1. Z. Nitecki, On the topology of the set of completely unstable flows. Trans. AMS 252 (1979) 147-162.
2. ______, Recurrent structure of completely unstable flows on surfaces of finite genus. Preprint, IHES
3. J. Palis, On Morse-Smale diffeomorphisms. Topology 8 (1969) 385-405.
4. C. Pugh, R. Walker, and F. Wilson, On Morse-Smale approximations--a counterexample. J.Diff. Eqns. 23 (1977) 173-182.
5. F. Takens and W. White, Vector fields with no nonwandering points. Am. J.Math. 98 (1976) 415-425.

6. N. Fenichel, Persistence and smoothness of invariant manifolds for flows. Indiana Univ. Math. J. 21 (1971) p. 205, thm. 1.

7. M. Hirsch, C. Pugh, and M. Shub, Invariant Manifolds. Lect. Notes in Math. 583 (Springer-Verlag, 1977).

Tufts University
Medford, MA 02155

INTERTWINING INVARIANT MANIFOLDS AND THE LORENZ ATTRACTOR

By Carles Perelló

The purpose of this paper is to show how the limitations imposed by some invariant two - manifolds, give an explanation to the existence of a complicated attractor in the Lorenz system.

To find out how the invariant manifolds are embedded in R^3 we have used a desk computer with a plotter and a simple Runge-Kutta method. The way we claim the manifolds are embedded is the simplest way to be in accordance with the numerical data. Our system is given by $x' = -10x+10y$, $y' = rx-y-xz$, $z' = -8z/3+xy$ and we are going to see what is the situation for different values of r in the range [0,50]. This system and related questions have been considered in [1], [2] , [3] , [4] and [5] .

For $r<1$ the origin is a global attractor and it bifurcates for $r = 1$ to three rest points: the origin O, and A and B which are attractors with real eigenvalues (we shall call all the relevant elements with some symbol independent of the parameter r , and keep their identity through coninuity). The origin has a one-dimensional unstable manifold with the two non-trivial orbits α tending to A and β tending to B , and the two dimensional stable manifold γ. At some value of r two of the eigenvalues of A (and symetrically of B) become conjugate complex, but the conditions are essentially the same, with γ properly embedded and dividing the space in the two basins of attraction of A and B .

For $r = r_1$, close to 14 (see [6]), we have α and β becoming homoclinic, and hence belonging to γ .

In this case γ is no longer a proper embedding of a plane and it comes to a butt with itself along two invariant curves which intersect at 0. In any point of this curves different from 0, γ looks locally like $\{x = 0\} \cup \{z = 0, \ x > 0\}$ in $\mathbb{R}^3$ (we name it a "T point"). In figure 1 we despict the situation. Observe that we still have two basins of attraction, one towards A and the others towards B.

The bifurcation at $r = r_1$ makes things very different: γ rolls up into two intertwining scrolls, and inside each of them a tube appears, which is the stable manifold corresponding to a periodic orbit which bifurcates out of the homoclinic orbit. All the orbits "inside" the tubes tend either to A or to B . The remaining orbits not lying in any asymptotic manifolds, are condemned to wander between the scrolls, and keep bouncing between the two unstable manifolds (δ and ε) of the periodic orbits around A and B . They tend to a complicated attractor. As r keeps increasing, the periodic orbits shrink to A and B, and they disappear, together with the associated tubes , for a value of $r = r_2$, close to 24.74 (a subcritical Hopf bifurcation). Under these conditions A and B are no more attractors, and all orbits not lying on an asymptotic manifold (of O , A or B), have the bouncing and unscrolling behavior above mentioned. Figure two tries to give an idea of the situation before the tubes disappear.

It may happen that for some values of r, the orbits α and β are in γ , i.e., they are homoclinic. In this case there is, as in the case $r = r_1$, a pair of invariant curves through 0 to which γ tends to a butt. One of the orbits in each of these curves winds out of the corresponding scroll. We do not intend to make a drawing of the situation but we remark that the bouncing and unscrolling behavior is also valid in this case.

In his paper [3], Guckenheimer shows for a system which has the basic features of the Lorenz one, that this homoclinic behavior of α and β is obtained through arbitrarily small perturbations of the non-homoclinic one, and viceversa. He shows also that in this case the attractor is of a different topological type to the one corresponding to α and β non-homoclinic.

For larger values of r the system undergoes more changes (see [6]), but we do not go into them.

By using the scrolls and the unstable manifolds of A and B which act as the bouncing barriers, it is easy to convince oneself of the complexity of the limit behavior of orbits. For instance, one may try to follow the strip obtained by saturating the segment DC of the invariant manifold of A . Taking into account that the strip cannot cross itself we see how it winds around A or B , piling up in a complicated way. In figure 3 we show the order in which the few first turns of the strip appear.

It seems that looking for invariant manifolds in other systems the way we do, may shed some light on the nature of complicated limit behavior in 3 dimensions, and it may turn out that such structures, i.e. intertwining manifolds and hence complicated attractors are abundant among dynamical systems.

REFERENCES

1]. Lorenz, E.N., "Deterministic nonperiodic flow", J.of the atmospheric sciences, 20 (1963),130-141.

2]. Ruelle, D., "The Lorenz Attractor and the problem of Turbulence" In L.N. 565, Springer-Verlag, 1976, (146-158)

[3]. Guckenheimer, J., "A strange, strange attractor". In Marsden,J. E., McCracken,M.,"Hopfbifurcation and its applications", Spri -- ger-Verlag, 1976, (368-381)

[4]. William, R.F.,"The structure of Lorenz attractors". In L.N. 615 Springer-Verlag, 1977, (94.112).

[5]. Marsden, J.E. "Attempts to relate the Na ier-Stokes equations to turbulence". In L.N. 615, Springer-Verlag, 1977,(1-22).

6]. Henon, M., Pomeau,Y.,"Two strange attractors with a simple struc ture". In L.N. 565, Springer-Verlag, 1976.(29-68).

Universitat Autònoma de Barcelona
Barcelona, Spain

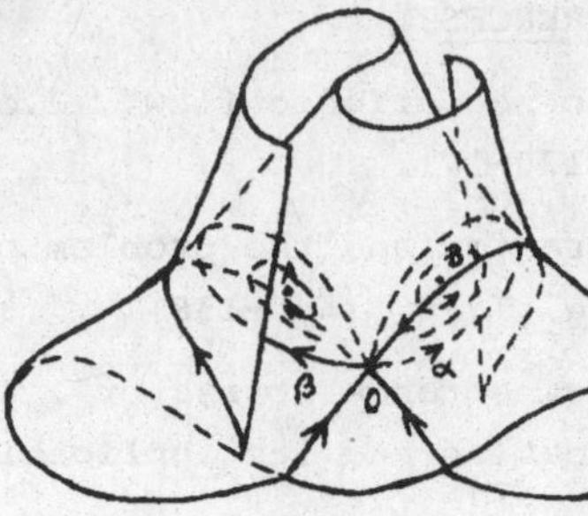

fig. 1

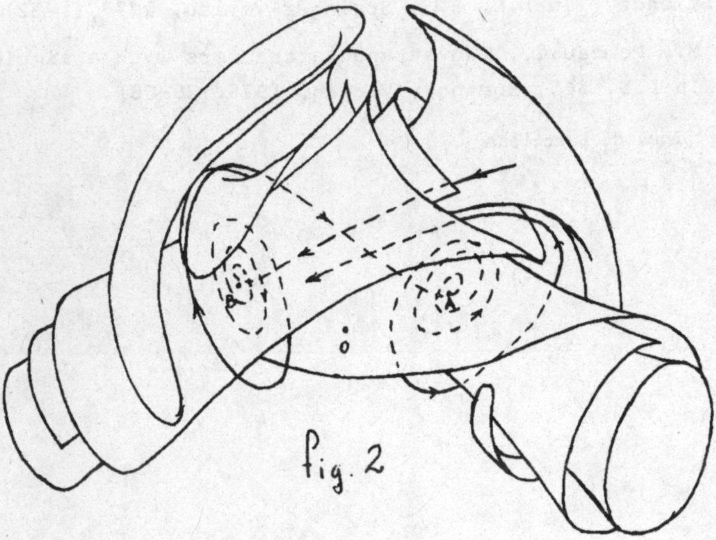

fig. 2

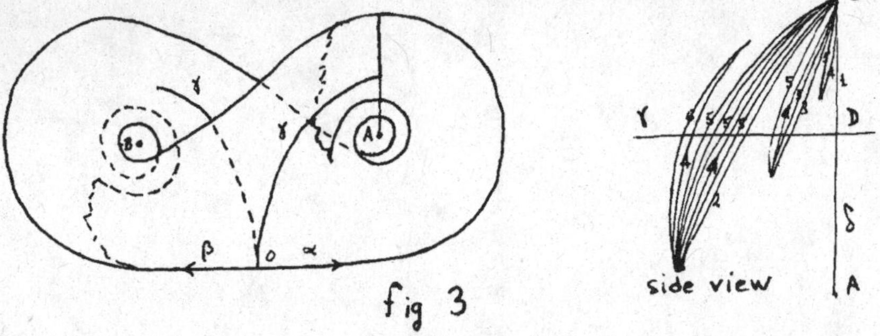

fig 3

COUNTING COMPATIBLE BOUNDARY CONDITIONS

Everett Pitcher

Abstract

The index theorem of M. Morse, which counts focal points weighted by multiplicity in positive regular problems, and the extension by the author, which counts focal points weighted by a signature in problems not required to be positive regular, are further extended to the counting of compatible sets of general self-adjoint boundary conditions. Circumstances under which the weight is the multiplicity are exhibited.

1. The differential equations.

The system of differential equations to be investigated is the general self-adjoint system of 2m first order equations in Hamiltonian form. In matrix notation it is

$$L(y) \equiv \mathcal{J} y' + \mathcal{A} y = 0 \tag{1.1}$$

where

$$\mathcal{J} = \begin{pmatrix} 0 & -1 \\ 1 & 0 \end{pmatrix} \qquad \mathcal{A} = \begin{pmatrix} C & -A^* \\ -A & -B \end{pmatrix} \tag{1.2}$$

when represented in m x m blocks. The entries in A, B, C will be supposed continuous and B and C are symmetric. Notations ' and * denote derivative and transpose. The equations are also written

$$(1.3') \qquad L_1(u,v) \equiv -v' + Cu - A^*v = 0$$

$$y = \begin{pmatrix} u \\ v \end{pmatrix}$$

$$(1.3'') \qquad L_2(u,v) \equiv u' - Au - Bv = 0$$

The second order self-adjoint system of m equations

$$(1.4) \qquad (R\eta' + Q\eta)' - (Q^*\eta' + P\eta) = 0,$$

in which R, Q, P are m x m matrices of continuous functions with R and P symmetric and R non-singular, reduces to the form (1.1) with the substitution

$$(1.5) \qquad u = \eta \qquad v = R\eta' + Q\eta.$$

Then

$$(1.6) \qquad A = -R^{-1}Q \qquad B = R^{-1} \qquad C = P - Q^*R^{-1}Q.$$

If B in (1.2) is non-singular, the transformation is reversible. On the other hand, (1.4) presents substantial difficulties when R is singular because the existence theorem may fail while (1.1) merely demands care when B is singular.

A convenient reference for the formulation is [R, Ch. VII, Sec. 2].

2. The boundary conditions.

One form of the general self-adjoint boundary conditions at two distinct points a^1 and t is

$$(2.1') \qquad u(a^1) = c^1\theta \qquad u(t) = c^2\theta$$

$$(2.1'') \qquad b\theta + c^{2*}v(t) - c^{1*}v(a^1) = 0,$$

in which c^1 and c^2 are m x r matrices of constants with $0 \leq r \leq 2m$ such that $(c^{1*}\ c^{2*})$ has rank r, θ is a column of r parameters, and b is an r x r symmetric matrix of constants. Equivalent forms and specialized forms will be described as needed. See [M, Ch.IV, Sec.2].

Self-adjoint boundary conditions at a single point are

$$(2.2') \qquad u(a^1) = c^1\theta$$

$$(2.2'') \qquad b^1\theta - c^{1*}v(a^1) = 0,$$

where c^1 is $m \times r$ of rank r, with $0 \leqq r \leqq m$, θ is a column of r parameters, and b^1 is $r \times r$ and symmetric. The solutions of the differential equation (1.1) satisfying the boundary conditions (2.2) form an m- dimensional vector space $\mathcal{F}$ called a *conjugate family*. They constitute a maximal isotropic subspace in the space of all solutions equipped with the metric $\langle y,\bar{y}\rangle = \bar{y}^*\mathcal{J}\, y$ and any maximal isotropic subspace can be described by a set of conditions of the form (2.2) at any preassigned point a^1.

Conditions (2.2) together with conditions

$$(2.3) \qquad u(t) = 0 \qquad\qquad t > a^1$$

are a set of self-adjoint boundary conditions (2.1) for which the differential equation (1.1) has a non-trivial solution if and only if t is a *focal point* of the conjugate family $\mathcal{F}$. The *multiplicity* of the focal point is the dimension of the space of such solutions.

If B is positive definite, the index theorem of M. Morse counts the number of focal points on an interval, weighted by multiplicity. It says that the number of focal points of $\mathcal{F}$ on the open interval (a^1,a^2) is equal to the index of the quadratic function

$$(2.4) \qquad J(y) = \theta^*v\theta + \int_{a^1}^{a^2} (u^*cu + v^*Bv)dx$$

evaluated on an appropriate space of functions, say vector functions y which are piecewise solutions of the differential equation (1.3") such that u is continuous and $u(a^1) = c^1\theta$ and $u(a^2) = c^2\theta$. The function v may have jump discontinuities. The *index* of a quadratic function is the least upper bound of the dimension of subspaces on which the function is negative definite. See [M, Ch.III, Th.6.2].

If B is not positive definite, the index as described above will in general be infinite. However, it is possible that focal points can be counted by appropriate modifications of the method. This is done in the author's papers [P2] and [P3].

In this paper the method is extended to general boundary conditions as follows. A set of boundary conditions (2.1) for the differential equation (1.1) is *compatible* if the system has a non-trivial solution and t is called a *compatible point*. The *multiplicity* of a compatible point is the dimension of the set of such solutions. Compatible points will be counted with a weight equal to a *signature* to be defined. Conditions will be noted under which the count can be effected with weight equal to multiplicity.

A simple example of the principal Theorem 3.1 is the following. Suppose that p is a continuous function and that w is a solution of the equation

$$w'' - pw = 0 \tag{2.5}$$

such that w' and p have no zeros in common. Then the number of zeros of w on an interval (a^1,a^2) differs by at most 1 from the count of zeros of w' on the same interval, where the weight of a zero of w' is +1 when p is negative and -1 when p is positive. With a more complicated statement, the hypothesis about common zeros may be relaxed. An *ad hoc* proof of the statement in the example is readily devised.

3. Adjusted index and signature.

Two conditions are imposed. The first is

Hypothesis N. The differential equation (1.1) is identically normal.

This means that the only solution of

$$v' + A^*v = 0 \qquad\qquad Bv = 0 \tag{3.1}$$

on any interval is $v = 0$ and is equivalent to the statement that in any solution y of (1.1), v is uniquely determined by u. See [R, Ch.VII, Sec.3] and [B, p.219].

The second condition is

Hypothesis D. The differential equation (1.1) is uniformly locally disconjugate.

This means that corresponding to an interval $[a^1,a^2]$ there is a number $\varepsilon > 0$ such that there is no non-trivial solution y for which $u(e) = 0 = u(f)$, $a^1 - \varepsilon < e < f < a^2 + \varepsilon$, and $f - e < \varepsilon$. Uniform local disconjugacy was introduced in [P2].

If B is non-singular, Hypotheses N and D hold. See [P3] for a more detailed discussion of Hypotheses N and D.

The adjusted index is defined by adapting the broken extremal technique of Morse. See [M, Ch.III]. Let

$$a^1 = a_0 < a_1 < \ldots < a_N < a_{N+1} = a^2 \tag{3.2}$$

be a partition with $a_{i+1} - a_i < \varepsilon$. Let

$$Z = (\theta, z^1, \ldots, z^N) \tag{3.3}$$

where z^i is a column of m rows. Let

$$Q(Z;t) = J(y) \tag{3.4}$$

where y is the unique solution of (1.1) on each interval (a_i,a_{i+1}) such that u satisfied the conditions.

$$u(a^1) = c^1\theta \qquad u(a_i) = z^i, \quad i = 1,\ldots,N \qquad u(t) = c^2\theta. \tag{3.5}$$

The quadratic function Q is singular if and only if t is compatible and its *nullity* $\nu(t)$ as a quadratic function is equal to the multiplicity of t as a compatible point. Its index $i^N(t)$ is independent of the precise location of the vertices a_i but in general depends on N.

The *type form* $\tau(z;x_1,x_2,x_3)$ is defined as the special case of Q with end conditions $u(x_1) = 0$, $u(x_3) = 0$ and one intermediate vertex (x_2,z) and with $a^1 - \varepsilon < x_1 < x_3 < a^2 + \varepsilon$ and $x_3 - x_1 < \varepsilon$. It is non-singular by virtue of Hypothesis D. Its index ρ is the *type* of the differential equation (1.1). If B is non-singular, the type of the equation is equal to the index of z^*Bz.

The *adjusted index* $\hat{i}(t)$ is defined by

$$\hat{i}(t) = i^N(t) - N\rho \tag{3.6}$$

and is independent of N. See P3, Sec.II .

The *signature* of an isolated compatible point t is

$$\sigma(t) = \hat{i}(t-) - \hat{i}(t+), \tag{3.7}$$

whence it follows that if compatible points are isolated then

$$\hat{i}(a^2) - \hat{i}(a^1+) = \Sigma -\sigma(t) \qquad a^1 < t < a^2. \tag{3.8}$$

This statement acquires force when conditions for isolated compatible points and a separate calculation of σ are introduced.

Let S(t) denote the space of solutions of the differential equation (1.1) with boundary conditions (2.1) and let $\nu(t)$ denote its dimension, namely the multiplicity of t as a compatible point. Define an *auxiliary form* by

$$q^t = Q_t(Z;t) \mid S(t). \tag{3.9}$$

It is a quadratic form on a space of dimension $\nu(t)$. Let σ_0 denote its signature. The following theorem is an instance of [P1, Th.3], where cases with q^t singular are also handled.

Theorem 3.1. If the auxiliary form q^t is non-singular for each t, then the compatible points are isolated and are counted with signature as weight by the formula

(3.10) $\hat{i}(a^2) - \hat{i}(a^1+) = \Sigma -\sigma_0(t)$.

In this theorem, the quadratic form in 0 variables is non-singular with signature 0.

If q^t is negative definite, then $\sigma_0(t) = -\nu(t)$.

Corollary 3.1. If the auxiliary form q^t is negative definite for each t, then the compatible points are isolated and are counted with multiplicity as weight by the formula

(3.11) $\hat{i}(a^2) - \hat{i}(a^1+) = \Sigma\nu(t)$.

If q^t is positive definite, the right hand side is $-\Sigma\nu(t)$.

4. Calculations with the auxiliary form.

There are two lemmas that facilitate the application of Corollary 3.1.

Lemma 4.1.

(4.1) $$Q_t(Z;t) = (u^*(t,t) \quad v^*(t,t))\, \mathcal{A}(t) \begin{pmatrix} u(t,t) \\ v(t,t) \end{pmatrix}$$

Observe that $Q_t(Z;t)$ is a quadratic form in the variables z^N and θ, for $v(t,t)$ depends on z^N.

Let

(4.2) $$F(t) = \int_{a^N}^{t} (u^*Cu + v^*Bv)dx$$

where u and v are components of the solution of (1.1) with

(4.3) $u(a_N,t) = z^N \qquad u(t,t) = c^2\theta.$

Thus

$$(4.4)\qquad Q_t(Z;t) = F'(t) = u^*(t,t)C(t)u(t,t) + v^*(t,t)B(t)v(t,t) + 2\int_{a_N}^{t} [u_t^*(x,t)C(x)u(x,t) + v^*(x,t)B(x)v(x,t)]dx.$$

Replace Cu by its value $v_x + A^*v$ from (1.3') and integrate the term $u_t^*v_x$ by parts. The integral becomes

$$(4.5)\qquad 2u_t^*(x,t)v(x,t)\Big|_{x=a_N}^{x=t} + 2\int_{a_N}^{t} [u_t(x,t)A^*(x) + v_t^*(x,t)B(x) - u_{xt}^*(x,t)]v(x,t)dx.$$

One notes that $u_t(a_N,t) = 0$ and that $u_t(t,t) + u_x(t,t) = 0$ from (4.3). Moreover, u, v satisfy (1.3") and, on differentiating with respect to t, one sees that the square bracket in the last integral is 0. Thus

$$(4.6)\qquad Q_t(Z;t) = u^*(t,t)C(t)u(t,t) + v^*(t,t)B(t)v(t,t) - 2u_x^*(t,t)v(t,t).$$

On replacing u_x by its value from (1.3") one obtains (4.1).

Lemma 4.2. If R is non-singular then

$$(4.7)\qquad \begin{pmatrix} 1 & Q^* \\ 0 & R \end{pmatrix} \mathcal{A} \begin{pmatrix} 1 & 0 \\ Q & R \end{pmatrix} = \begin{pmatrix} P & 0 \\ 0 & -R \end{pmatrix}$$

This is a direct calculation from (1.6) and shows that $\mathcal{A}$ and diag (P -R) are similar.

5. Weighting by multiplicity.

In order to use Corollary 3.1, it is convenient to have criteria that q^t is definite. The vocabulary of the negative definite case will be used.

First, from Lemma 4.1 it is seen that q^t is negative definite if $\mathcal{A}(t)$ is negative definite, without reference to the boundary conditions. From Lemma 4.2, this is seen to be the case for the differential equation (1.4) if R is positive definite and Q is negative definite.

Second, in (4.1) the second relation in (4.3) holds. That is

$$(5.1)\quad Q_t(Z;t) = (\theta^* v^*(t,t))\, W(t) \begin{pmatrix} \theta \\ v(t,t) \end{pmatrix},$$

where

$$(5.2)\quad W(t) = \begin{pmatrix} c^{2*}C(t)c^2 & -c^{2*}A^*(t) \\ -A(t)c^2 & -B(t) \end{pmatrix}.$$

Whereas $\mathcal{A}$ is a matrix of order 2m, the matrix W is of order m + r. If W(t) is negative definite on an interval, then so is q^t.

Third, suppose that the boundary conditions at a^1 and t are separated, that is, have the form

$$(5.3)\quad u(a^1) = c^1\theta^1 \qquad u(t) = c^2\theta^2$$

$$b^1\theta^1 - c^{1*}v(a^1) = 0 \qquad b^2\theta^2 - c^{2*}v(t) = 0$$

where θ^1 [resp. θ^2] is a column of r^1 [r^2] parameters with $0 \leqq r \leqq m$ [$0 \leqq r^2 \leqq m$]. The conditions at t can also be written in the form

$$(5.4)\quad p^2u(t) = q^2v(t)$$

where p^2 and q^2 are m x m matrices of constants, $(p^2\; q^2)$ has rank m, and p^2q^{2*} is symmetric. Equally well, the conditions at t have the form

$$(5.5)\quad u(t) = q^{2*}\phi \qquad v(t) = p^{2*}\phi$$

where ϕ is a column of m parameters. See [M, Ch.IV, Sec.2]. Then q^t is obtained by evaluating $\phi^* V(t)\phi$ on a subspace, where

$$(5.6) \quad V(t) = (q^2 \; p^2)\, \mathcal{A}(t) \begin{pmatrix} q^{2*} \\ p^{2*} \end{pmatrix}$$

$$= q^2 C(t) q^{2*} - p^2 A(t) q^{2*} - q^2 A^*(t) p^2 - p^2 B(t) p^{2*}.$$

If V(t) is negative definite on an interval, so is q^t.

References

[B] Bliss, G. A., Lectures on the Calculus of Variations, University of Chicago Press, 1946.

[M] Morse, M., The calculus of variations in the large, Am. Math. Soc. Coll. Publ. XVIII, 1934.

[P1] Pitcher, E., The variation in index of a quadratic function depending on a parameter, Bull. Am. Math. Soc., 65 (1959), 355-357.

[P2] Pitcher, E., Conjugate points without the condition of Legendre, Optimal Control and Differential Equations, A. B. Schwarzkopf, Walter G. Kelley, and Stanley B. Eliason, Editors, 223-244, Academic Press, 1978.

[P3] Pitcher, E., Counting focal points and characteristic roots, Bull. Inst. of Math. Acad. Sin., 6 (1978), 389-413.

[R] Reid, W.T., Ordinary differential equations, Wiley, 1971.

Lehigh University

STABLE MANIFOLDS FOR MAPS*)

DAVID RUELLE
IHES, 91 Bures-sur-Yvette,
FRANCE

MICHAEL SHUB
QUEENS COLLEGE OF
THE CITY UNIVERSITY OF NEW YORK
FLUSHING, NEW YORK

Here we present a stable manifold theorem for non-invertible differentiable maps of finite dimensional manifolds. There is a long history of stable manifold theorems for hyperbolic fixed points and sets, see for instance [1]. More recently Pesin [3] has proven theorems of a general nature which rely on measure theoretic techniques. Pesin's results have been extended in [5]. The results described in the present paper were arrived at by the two authors along different paths. The first author starting from a treatment of differentiable maps in Hilbert space [6] specializes to the finite dimensional case while the second starting from seminar notes by Fahti, Herman and Yoccoz applies graph transform as in [1].

We say that a map is of class $C^{r,\theta}$ if its r-th derivative is Holder continuous of exponent θ (Lipschitz if $\theta = 1$). Similarly for manifolds. In what follows class $\mathcal{C}$ will mean class $C^{r,\theta}$ with integer $r \geq 1$ and $\theta \in (0,1]$, or class C^r with $r \geq 2$, or class C^∞, or class C^ω (real analytic), or (complex) holomorphic. [Class $\mathcal{C}^{-1}$ will be respectively $C^{r-1,\theta}$, C^{r-1}, C^∞, C^ω, or holomorphic].

Throughout what follows, M will be a locally compact $\mathcal{C}$-manifold and $f: M \to M$ a $\mathcal{C}$-map such that fM is relatively compact in M. (In particular, if $fM = M$, then M is a compact manifold). We introduce the inverse limit.

$\tilde{M} = \{(x_n)_{n\geq 0} : x_n \in M$ and $fx_{n+1} = x_n\}$ and define $\tilde{\pi}(x_n) = x_0$, $\tilde{f}(x_n) = (y_n)$ where $y_n = x_{n+1}$ for $n \geq 0$. Notice that $\tilde{M}$ is compact, $\tilde{\pi}$ is continuous $\tilde{M} \to M$ with image $\bigcap_{n\geq 0} f^n M$, and $\tilde{f}$ is a homeomorphism of $\tilde{M}$. Furthermore $f\tilde{\pi} = \tilde{\pi}\tilde{f}^{-1}$.

We state in (1), (2), (3) below some (easy) consequences of the multiplicative ergodic theorems**). Our main results are the stable and unstable manifold

---

*) This work has been supported by NSF Research Grant

**) See Oseledec [2], Raghunathan [4].

theorems in (4), (5). It is likely that these results extend to general local fields (the multiplicative ergodic theorem does, see [4]). We have however not checked the ultrametric case.

(1) There is a Borel set $\Gamma \subset M$ such that $f\Gamma \subset \Gamma$, and $\rho(\Gamma) = 1$ for every f-invariant probability measure ρ. If $x \in \Gamma$, there are an integer $s \in [0,m]$, reals $\mu^{(1)} > \dots > \mu^{(s)}$, and spaces $T_xM = V_x^{(1)} \supset \dots \supset V_x^{(s)} \supset V_x^{(s+1)} \supset \{0\}$ such that

$$\lim_{n\to\infty} \frac{1}{n} \log \|Tf^n(x)u\| = \mu^{(r)} \quad \text{if} \quad u \in V_x^{(r)} \setminus V_x^{(r+1)}$$

for $r = 1,\dots,s$, and

$$\lim_{n\to\infty} \frac{1}{n} \log \|Tf^n(x)u\| = -\infty \quad \text{if} \quad u \in V_x^{(s+1)} .$$

The functions $x \to s$, $\mu^{(1)},\dots,\mu^{(s)}$, $V_x^{(1)},\dots,V_x^{(s)}$ are Borel and $x \to s$, $\mu^{(1)},\dots,\mu^{(s)}$, $\dim V_x^{(1)},\dots,\dim V_x^{(s)}$ are f-invariant.

(2) Similarly there is a Borel set $\tilde{\Gamma} \subset \tilde{M}$ such that $\tilde{f}\tilde{\Gamma} \subset \tilde{\Gamma}$ and $\tilde{\rho}(\tilde{\Gamma}) = 1$ for every $\tilde{f}$-invariant probability measure $\tilde{\rho}$. If $\tilde{x} = (x_n) \in \tilde{\Gamma}$, there are $s \in [0,m]$, $\mu^{(1)} > \dots > \mu^{(s)}$ and $\{0\} = \tilde{V}_{\tilde{x}}^{(0)} \subset \tilde{V}_{\tilde{x}}^{(1)} \subset \dots \subset \tilde{V}_{\tilde{x}}^{(s)} \subset T_{x_0}M$ such that if $(u_n)_{n\geq 0}$ satisfies $u_n \in T_{x_n}M$ and $Tf(x_{n+1})u_{n+1} = u_n$ and

$$\lim_{n\to\infty} \frac{1}{n} \log \|u_n\| < +\infty$$

then $u_0 \in \tilde{V}_{\tilde{x}}^{(s)}$. Conversely, for every $u_0 \in \tilde{V}_{\tilde{x}}^{(s)}$ there is such a sequence (u_n), it is unique and

$$\lim_{n\to\infty} \frac{1}{n} \log \|u_n\| = -\mu^{(r)} \quad \text{if} \quad u_0 \in \tilde{V}_{\tilde{x}}^{(r)} \setminus \tilde{V}_{\tilde{x}}^{(r-1)}$$

for $r = 1,\dots,s$.

(3) The map $\tilde{\pi}$ sends the $\tilde{f}$-invariant probability measures on $\tilde{M}$ <u>onto</u> the f-invariant probability measures on M. Almost everywhere with respect to every $\tilde{f}$-invariant probability measure $\tilde{\rho}$, the quantities $s \circ \tilde{\pi}$, $\mu^{(r)} \circ \pi$, $\dim V_{\tilde{\pi}(.)}^{(r+1)}$ occurring in (1) are equal to $s, \mu^{(r)}$, $m - \dim \tilde{V}_{(.)}^{(r)}$ in (2). This justifies the confusion in notation for s and $\mu^{(r)}$.

(4) Local stable manifolds

Let Θ, λ, r be f-invariant Borel functions on Γ with $\Theta > 0$, $\lambda < 0$, r integer $\in [0,s]$, and

$$\mu^{(r+1)} < \lambda < \mu^{(r)}$$

where $\mu^{(0)} = +\infty$, $\mu^{(s+1)} = -\infty$). Replacing possibly Γ by a smaller set retaining the properties of (1) one may construct Borel functions $\beta > \alpha > 0$ on Γ with the following properties.

(a) If $x \in \Gamma$ the set $W_x^\lambda = \{y \in M: d(x,y) \le \alpha(x)$ and $d(f^n x, f^n y) \le \beta(x) e^{n\lambda(x)}$ for all $n > 0\}$ is contained in Γ and is a C-submanifold of the ball $\{y \in M: d(x,y) \le \alpha(x)\}$. For each $y \in W_x^\lambda$, we have $T_y W_x^\lambda = V_y^{(r+1)}$. More generally, for every $t \in [0,s]$, the function $y \to V_y^{(t+1)}$ is of class C^{-1} on W_x^λ.

(b) If $y,z \in W_x^\lambda$, then

$$d(f^n y, f^n z) \le \gamma(x)\, d(y,z)\, e^{n\lambda(x)}.$$

(c) If $x \in \Gamma$, then $\alpha(f^n x)$, $\beta(f^n x)$ decrease less fast with n than the exponential $e^{-n\Theta}$.

The manifolds W_x^λ do not in general depend continuously on x, but the construction implies measurability properties on which we shall not elaborate here.

(5) Local unstable manifolds

Let Θ, μ, r be $\tilde{f}$-invariant Borel functions on $\tilde{\Gamma}$ with $\Theta > 0$, $\mu > 0$, r integer $\in [0,s]$, and

$$\mu^{(r+1)} < \mu < \mu^{(r)}$$

(where $\mu^{(0)} = +\infty$, $\mu^{(s+1)} = -\infty$). Replacing possibly $\tilde{\Gamma}$ by a smaller set retaining the properties of (2), one may construct Borel functions $\tilde{\beta} > \tilde{\alpha} > 0$ and $\tilde{\gamma} > 1$ on $\tilde{\Gamma}$ with the following properties.

(a) If $\tilde{x} = (x_n) \in \tilde{\Gamma}$ the set

$$\tilde{W}_{\tilde{x}}^\mu = \{\tilde{y} = (y_n) \in \tilde{M} : d(x_0,y_0) \le \tilde{\alpha}(\tilde{x}) \text{ and } d(x_n,y_n) \le \tilde{\beta}(\tilde{x}) e^{-n\mu(\tilde{x})}$$

for all $n > 0\}$ is contained in $\tilde{\Gamma}$; the map $\tilde{\pi}$ restricted to $\tilde{W}_{\tilde{x}}^\mu$ is injective and $\tilde{\pi}\tilde{W}_{\tilde{x}}^\mu$ is a C-submanifold of the ball $\{y \in M: d(x_0,y) \le \tilde{\alpha}(\tilde{x})\}$. For each

$\tilde{y} = (y_n) \in \tilde{W}^{\mu}_{\tilde{x}}$, we have $T_{y_0} \tilde{\pi} \tilde{W}^{\mu}_{\tilde{x}} = \tilde{V}^{(r)}_{\tilde{y}}$. More generally, for every $t \in [0,s]$, the function $y \mapsto \tilde{V}^{(t)}_{\tilde{\pi}^{-1} y}$ is of class C^{-1} on $\tilde{\pi} \tilde{W}^{\mu}_{\tilde{x}}$.

(b) If $(y_n), (z_n) \in \tilde{W}^{\mu}_{\tilde{x}}$, then

$$d(y_n, z_n) \le \tilde{\gamma}(\tilde{x})\, d(x_0, y_0) e^{-n\mu(x)} .$$

(c) If $\tilde{x} \in \tilde{\Gamma}$, then $\tilde{\alpha}(\tilde{f}^b \tilde{x})$, $\tilde{\beta}(\tilde{f}^n \tilde{x})$ decrease less fast with n than the exponential $e^{-n\Theta}$.

(6) Global stable and unstable manifolds exist under obvious transversality conditions (for instance, if $T_x f$ is a linear isomorphism), Under these conditions they are immersed submanifolds.

(7) The results described above for maps apply immediately to flows, via a time T map.

REFERENCES

[1] M. Hirsch, C. Pugh and M. Shub, Invariant manifolds, Lecture Notes in Math. no. 583, Springer, Berlin, 1977.

[2] V.I. Oseledec, Multiplicative ergodic theorem, Lyapunov characteristic numbers for dynamical systems. Trudy Moskov., Mat. Obsc. 19, 179-210 (1968). English transl. Trans. Moscow Math. Soc., 19, 197-221 (1968).

[3] Ya. B. Pesin, Invariant manifold families which correspond to non-vanishing characteristic exponents. Izv. Akad. Nauk SSSR, Ser. Mat. 40 no. 6, 1332-1379. (1976), English transl. Math. USSSR izv. 10, no. 6, 1261-1305 (1976).

[4] M.S. Raghunathan, A proof of Oseledec multiplicative ergodic theorem, Israel J. Math. To appear.

[5] D. Ruelle, Ergodic theory of differentiable dynamical systems. I.H.E.S. Publications Mathematiques. To appear.

[6] D. Ruelle, Invariant manifolds for flows in Hilbert space, to appear.

SINGULAR POINTS OF PLANAR VECTOR FIELDS

by

Stephen Schecter and Michael F. Singer*

Suppose the origin is an isolated singular point of an analytic vector field a $\mathbb{R}^2$. Write

$$\begin{aligned} \dot{x} &= X_d(x,y) + X_{d+1}(x,y) + \ldots \\ \dot{y} &= Y_d(x,y) + Y_{d+1}(x,y) + \ldots \end{aligned} \tag{1}$$

ach $X_i(x,y)$ (resp. $Y_i(x,y)$) is a homogeneous polynomial of degree i, the terms of egree i in the infinite series expansion of $\dot{x}$ (resp. $\dot{y}$) about the origin. The nteger $d \geq 1$ is called the <u>degree</u> of the singularity (we assume at least one of $_d(x,y)$, $Y_d(x,y)$ is not identically zero).

In a small enough neighborhood of the origin one of three things happens:

(i) All solution curves of (1) are closed: (0,0) is a <u>center</u>.

(ii) All solution curves of (1) spiral toward (or away from) (0,0): (0,0) s a <u>focus</u>.

(iii) Near (0,0) there is a finite number of <u>elliptic</u>, <u>hyperbolic</u>, and arabolic sectors (Figure 1).

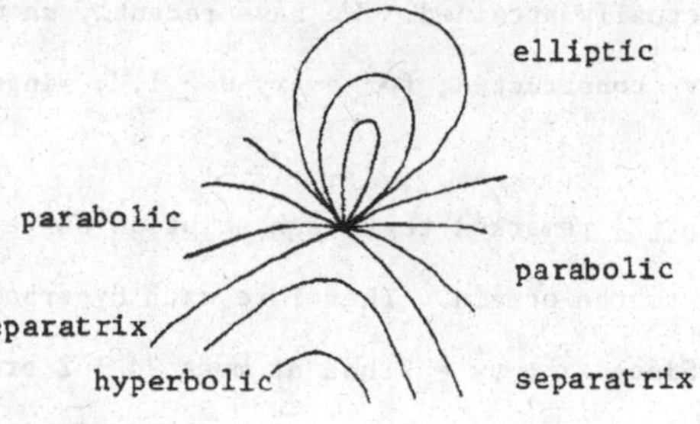

Figure 1

) Research of both authors partially supported by NSF grant MCS-7902524.

The question that interests us is: In the third case, at a singular point of degree d, what local topological pictures are possible? This is equivalent to asking how many different sectors of the various types there can be and how they can be arranged. (See [1], Chapter 8.) Our purpose here is to describe some progress toward answering this question. Our work is largely inspired by Bendixson's beautiful paper on differential equations in the plane [2] to which we will frequently refer. The best-known results of this paper are probably the Poincaré-Bendixson Theorem and the formula that gives the index of a singular point in terms of the numbers of elliptic and hyperbolic sectors; but there is much else there.

A separatrix is a solution curve of (1) that bounds a hyperbolic sector (Figure 1). Let

e = number of elliptic sectors

h = number of hyperbolic sectors

s = number of separatrices

Let us first briefly discuss e. Bendixson [2] noticed that each solution curve in an elliptic sector has at least one point where the tangent vector is vertical. Thus each elliptic sector contains a branch of $\dot{x} = 0$. Since $\dot{x} = 0$ has at most 2d branches at the origin (assuming $X_d(x,y) \not\equiv 0$, which can always be arranged by a linear change of coordinates), Bendixson concluded that $e \leq 2d$. Bendixson is silent on whether this bound is actually attained. We have recently shown [5] that in fact $e \leq 2d - 1$, and we have constructed, for every $d \geq 1$, a singularity of degree d with $e = 2d - 1$.

Regarding h, Bendixson [2] remarked that each solution curve in a hyperbolic sector has a point closest to the origin. Therefore each hyperbolic sector contains a branch of $x\dot{x} + y\dot{y} = 0$. Since $x\dot{x} + y\dot{y} = 0$ has at most $2d + 2$ branches at the origin (assuming $xX_d(x,y) + yY_d(x,y) \not\equiv 0$, which can be arranged by a linear change of coordinates), Bendixson concluded that $h \leq 2d + 2$. In this case, as Bendixson knew, the bound is sharp for every $d \geq 1$. We will see examples shortly.

It follows that $s \leq 4d + 4$. However, a singularity of degree d with $s = 4d + 4$ would have to have $2d + 2$ hyperbolic sectors and parabolic or elliptic sectors etween every pair of hyperbolic sectors: one readily doubts that this is possible. n fact we have the following result:

heorem 1. $s \leq \begin{cases} 4 & \text{if } d = 1. \\ 6 & \text{if } d = 2. \\ 4d-4 & \text{if } d \geq 3. \end{cases}$ These bounds are sharp.

To prove this result one uses the blowing-up construction for vector fields irst presented by Bendixson in [2]. In the remainder of the paper we will review his construction, use it to describe some interesting examples, and sketch the roof of Theorem 1. Details are in [4]. In an appendix, we describe how to onstruct a polynomial singularity having any desired configuration of elliptic, yperbolic, and parabolic sectors.

Blowing-up is used to analyze a degenerate singularity by replacing it by less egenerate singularities. We will blow up algebraically, analagous to the way lgebraic geometers blow up singularities of curves, rather than by polar oordinates, which is perhaps more usual in papers on differential equations. Our eason is that computations are easier when one blows up algebraically.

Consider the map η_1: $x\lambda$-plane $\to$ xy-plane given by $\eta_1(x,\lambda) = (x,x\lambda)$. The -axis collapses to $(0,0)$; the pencil of lines $y = \lambda_0 x$ through $(0,0) \in xy$-plane all lines through $(0,0)$ except the vertical) pulls back to the family of parallel ines $\lambda = \lambda_0$. The half-plane $x > 0$ (resp. $x < 0$) of the $x\lambda$-plane goes diffeo-orphically onto the corresponding half-plane of the xy-plane (but note that $x < 0$ oes upside-down).

Pull back (1) via η_1. We get

$$\begin{aligned} \dot{x} &= x^d X_d(1,\lambda) + x^{d+1} X_{d+1}(1,\lambda) + \ldots \\ \dot{\lambda} &= x^{d-1}[Y_d(1,\lambda) - \lambda X_d(1,\lambda)] + x^d[Y_{d+1}(1,\lambda) - \lambda X_{d+1}(1,\lambda)] + \ldots \end{aligned} \tag{2}$$

priori we should regard (2) as defined only on $\{(x,\lambda) : x \neq 0\}$, since on the -axis $D\eta_1$ is singular. However, since $d \geq 1$, (2) extends analytically to the -axis, so we regard (2) as defined on the entire $x\lambda$-plane. If $d > 1$ (2) is identically zero on the λ-axis. We distinguish two cases, depending on whether $Y_d(1,\lambda) - X_d(1,\lambda)$ is identically zero. Equivalently, the two cases depend on whether the homogeneous form $xY_d(x,y) - yX_d(x,y)$ is identically zero.

1. Type I singularities. $xY_d(x,y) - yX_d(x,y) \not\equiv 0$. Divide (2) by x^{d-1}. We get

$$\begin{aligned} \dot{x} &= xX_d(1,\lambda) + x^2X_{d+1}(1,\lambda) + \dots \\ \dot{\lambda} &= Y_d(1,\lambda) - \lambda X_d(1,\lambda) + x[Y_{d+1}(1,\lambda) - \lambda X_{d+1}(1,\lambda)] + \dots \end{aligned} \tag{3}$$

Off the λ-axis, (2) and (3) have the same solution curves. Since $\dot{x} = 0$ when $x = 0$, the λ-axis is invariant under (3). Singularities of (3) on the λ-axis occur at solutions of $Y_d(1,\lambda) - \lambda X_d(1,\lambda) = 0$, a polynomial of degree at most $d + 1$ that is not identically zero. It follows that (3) has at most $d + 1$ singularities on the λ-axis, and the sum of their degrees is also at most $d + 1$.

Example 1. For every $d \geq 2$ there exist Type I singularities of degree d that when blown up yield $d + 1$ degree 1 singularities on the λ-axis, all saddles (Figure 2a)[1]. From the blown-up picture we can deduce the picture of the original

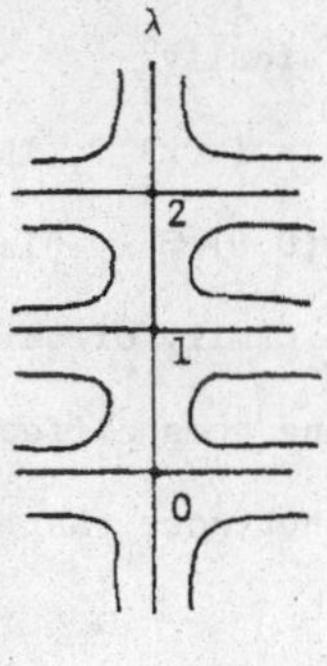

Figure 2a (d = 3)

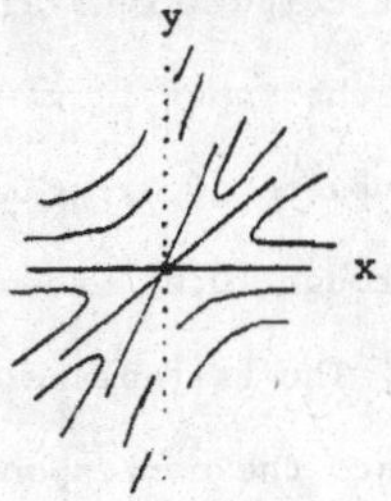

Figure 2b

singularity except near the vertical direction (Figure 2b), about which blowing up via η_1 (horizontal blowing up) tells us nothing.

---

[1] For example

$$\dot{x} = F(x,y) + x\,\frac{\partial F}{\partial y}(x,y)$$

$$\dot{y} = y\,\frac{\partial F}{\partial y}(x,y)$$

where $F(x,y) = \prod_{j=1}^{d}(y - (d + j + 1)x)$. The singularities on the λ-axis are at $\lambda = 0,1,\dots,d$.

To complete the picture we use vertical blowing up: let

: μy-plane $\to$ xy-plane be $\eta_2(\mu,y) = (\mu y, y)$. η_2 collapses the μ-axis to (0,0); ıe pencil of lines $x = \mu_o y$ through $(0,0) \in xy$-plane (all lines through (0,0) :cept the horizontal) pulls back to the family of parallel lines $\mu = \mu_o$. We ıll back (1) via η_2 and, in the case of a Type I singularity, divide by y^{d-1} and ınd up with a vector field leaving the μ-axis invariant We care only about a ›ssible singularity at $(0,0) \in \mu y$-plane, which corresponds to the vertical .rection. Such a singularity occurs if x divides $xY_d - yX_d$. In Example 1 there ; no singularity at $(0,0) \in \mu y$-plane (Figure 3a). Therefore the original vector

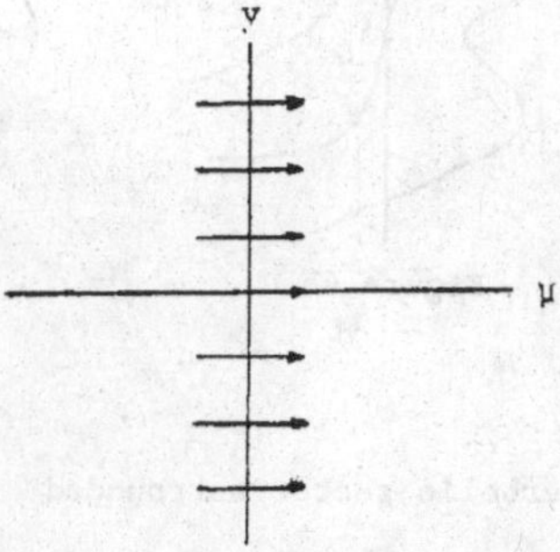

Figure 3a

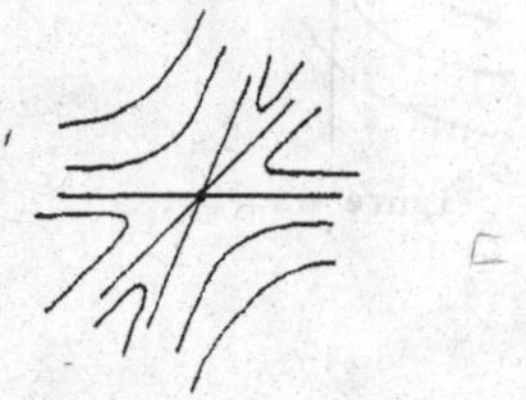

Figure 3b

ield is transverse to the y-axis near (0,0), so we can complete the picture Figure 3b). Thus Example 1 has $h = 2d + 2$, the maximum, and $s = 2d + 2$.

If we regard the set of singularities produced by blowing up as (1) all ingularities on the λ-axis and (2) the singularity at the origin of the μy-plane, f there is one, then at most $d + 1$ singularities in all are produced (which xplains why in Example 1 there could not be a singularity at $(0,0) \in \mu y$-plane), ıd the sum of their degrees is at most $d + 1$.

2. Type II singularities. $xY_d(x,y) - yX_d(x,y) \equiv 0$. We must have

$_d(x,y) = yQ_{d-1}(x,y)$ and $X_d(x,y) = xQ_{d-1}(x,y)$, $Q_{d-1}(x,y) \not\equiv 0$. We can divide (2) by ı x^{d-1} (since the x^{d-1} term of $\dot\lambda$ vanishes), obtaining

$$\begin{aligned}\dot{x} &= Q_{d-1}(1,\lambda) + xX_{d+1}(1,\lambda) + \ldots \\ \dot{\lambda} &= Y_{d+1}(1,\lambda) - \lambda X_{d+1}(1,\lambda) + x[\ ] + \ldots\end{aligned} \tag{4}$$

(4) is transverse to the λ-axis except at solutions of $Q_{d-1}(1,\lambda) = 0$.

(i) If $Q_{d-1}(1,\lambda_o) = 0$ but $Y_{d+1}(1,\lambda_o) - \lambda_o X_{d+1}(1,\lambda_o) \neq 0$, $(0,\lambda_o)$ is a point of <u>tangency</u> of (4) to the λ-axis. We have one of the following pictures:

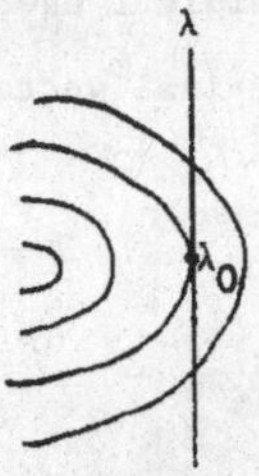

Figure 4a

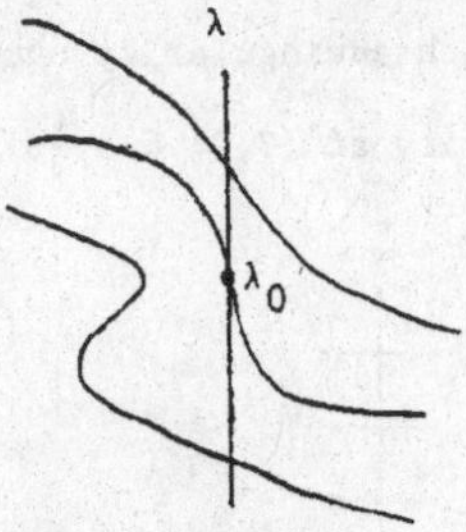

Figure 4b

When Figure 4a occurs, the original singularity has a hyperbolic sector surrounded by parabolic sectors and an elliptic sector opposite.

(ii) If $Q_{d-1}(1,\lambda_o) = 0$ and $Y_{d+1}(1,\lambda_o) - \lambda_o X_{d+1}(1,\lambda_o) = 0$, $(0,\lambda_o)$ is a singularity of (4).

Since $Q_{d-1}(1,\lambda)$ is a nonzero polynomial of degree at most $d - 1$, the number of tangencies of (4) to the λ-axis plus the sum of the degrees of singularities of (4) on the λ-axis is at most $d - 1$. If we add in a possible singularity or tangency at $(0,0) \in \mu y$-plane resulting from vertical blowing up, the bound still holds.

<u>Example 2</u>. For every $d \geq 2$ there exist Type II singularities of degree d that when blown up yield $d - 1$ degree 1 singularities on the λ-axis, all saddles (Figure 5)[2].

---

[2]For example,

$$\begin{aligned}\dot{x} &= xQ_{d-1}(x,y) - xyQ_{d-1}(x,y) \\ \dot{y} &= yQ_{d-1}(x,y) + x^2Q_{d-1}(x,y) + x^4 \frac{\partial}{\partial y} Q_{d-1}(x,y)\end{aligned}$$

where $Q_{d-1}(x,y) = \prod_{j=1}^{d-1} (y - jx)$.

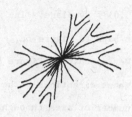

Figure 5 (d = 4)

ıch singularities have h = 2d - 2 and s = 4d - 4.

It may happen that blowing up once does not suffice to reduce the original ingularity to singularities one understands. Then one must blow up again. We llustrate this process by an example. Suppose we must analyze the singularity

$$\begin{aligned} \dot{x} &= 2x^2y + 2y^3 - x^4 \\ \dot{y} &= 2x^3y \end{aligned} \tag{5}$$

t turns out that this is the degenerate saddle pictured in Figure 6a. Since x oes not divide $xY_3 - yX_3$ we need only blow up horizontally. We get a single

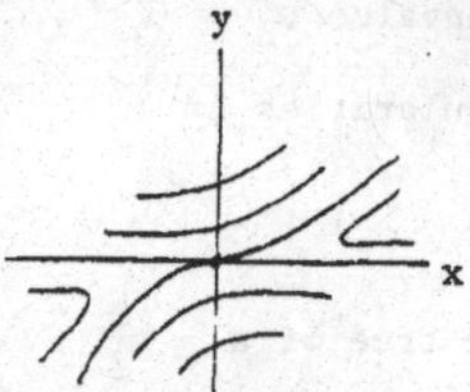

Figure 6a

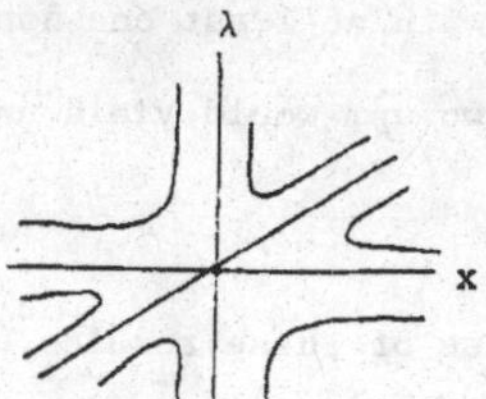

Figure 6b

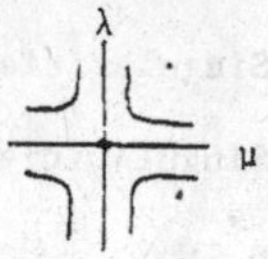

Figure 6c

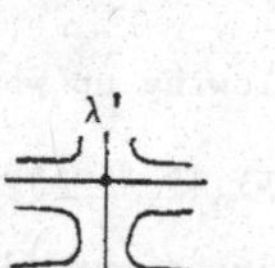

Figure 6d

egree 2 singularity at (0,0) ∈ xλ-plane (Figure 6b). Blowing this up vertically roduces a degree 1 saddle at (0,0) ∈ μλ-plane (Figure 6c); blowing up horizontally roduces two degree 1 saddles at (0,0) and (0,1) ∈ xλ'-plane (Figure 6d). Now we re done: from Figure 6c,d we could reconstruct Figure 6b and then Figure 6a.

Notice that in Figure 6b the two separatrices that lie along the λ-axis disappear when we pass to Figure 6a, since the λ-axis collapses to $(0,0) \in$ xy-plane. In Figure 6c all four separatrices disappear by the time we get back to Figure 6a, since both axes ultimately collapse to $(0,0) \in$ xy-plane. We say that in Figure 6b $(0,0)$ has one ghost direction; in Figure 6c $(0,0)$ has two ghost directions. The number of ghost directions at a singularity is simply the number of axes through it that ultimately collapse into the original singularity.

To keep track of successive blow-ups it is useful to define the tree $\mathcal{T}$ of a singularity. Start with a vertex representing the original singularity. Connect it to vertices that represent the singularities produced when we blow it up. If we wish to blow up one of the new singularities, connect its vertex to some new ones, and so on. The tree of (5) looks like:

When do we stop? Bendixson [2] gave the first proof of the following result, under assumptions to be explained in a moment.

Theorem 2. (Bendixson, Dumortier). Repeated blowing up eventually reduces a singularity to:

1. Tangencies (that follow a Type II singularity).
2. Singularities of degree 1 with at least one nonzero eigenvalue[3].
3. Singularities that, if blown up, would yield no new singularities or tangencies[4].

We stop blowing up when we get to one of these results. Thus the tree of a singularity is finite. Terminal vertices are those we do not blow up further.

Bendixson proved Theorem 2 for analytic singularities (1) with $\dot{x}$ and $\dot{y}$ relatively prime in the ring of real analytic functions. His proof also works for C^∞ singularities such that the Taylor series of $\dot{x}$ and $\dot{y}$ about the singularity are relatively prime in the ring of real power series. Dumortier [3] has extended

[3] Such singularity is a node, focus, center, saddle or saddle-node.

[4] Such a singularity is a node, focus or center. For Type I singularities this happens when $xY_d - yX_d$ has no real linear factors; for Type II singularities it happens when Q_{d-1} has no real linear factors.

heorem 2 to C^∞ singularities satisfying a Łojasiewicz condition. It follows from umortier's result, for example, that blowing up terminates for the singularity

$$\dot{x} = x^2 + y^2$$
$$\dot{y} = x^2 + y^2$$

ut one could not conclude this from Bendixson's result. It is Dumortier's result hat allows us to consider any isolated analytic singularity; of course, we could ave been even more general.

Now we sketch the proof of Theorem 1.

Each separatrix of (1) corresponds to a separatrix of a terminal saddle or addle-node in the tree of (1), or to half of a solution curve through a terminal angency in the tree of (1). However, not every separatrix of a terminal saddle or addle-node corresponds to a separatrix of (1): there is, to begin with, the ifficulty of ghost directions.

Define a function S: vertex set of $T \to$ nonnegative integers as follows. Let be a vertex of T; if V represents a singularity, let deg(V) = degree of that ingularity.

If V represents a tangency, let S(V) = 1.

If V is the initial vertex of T or if V represents a singularity that follows Type II singularity, let S(V) = deg(V) + 1.

If V represents a singularity that follows a Type I singularity, let (V) = deg(V) + 1 - number of ghost directions (always 1 or 2).

To see the significance of this definition, let us look at certain terminal ertices. At the tangency of Figure 4a, S = 1 and the number of separatrices ontributed is 2. At the terminal degree 1 singularities of Figure 5, S = 2 and he number of separatrices contributed by each singularity is 4. At the ingularities of Figure 6d, S = 1 and the number of separatrices contributed by ach singularity is 2. At the singularity of Figure 6c, S = 0 and the number of eparatrices contributed is 0. In fact we have:

Lemma 1. If V is a terminal vertex of T, the number of separatrices contributed by V is at most 2S(V).

Moreover,

Lemma 2. Let V represent a Type I singularity that blows up to produce k singularities represented by $V_1,\ldots,V_k$. Then $\sum_{i=1}^{k} S(V_i) \leq S(V)$.

Besides the fact that $\Sigma\deg(V_i) \leq \deg(V) + 1$, the proof of Lemma 2 requires only that one look at the number of ghost directions.

Let $S(T) = \Sigma S(V)$, where the sum is over all terminal vertices V of T. Using Lemmas 1 and 2 and the fact that S (initial vertex) = d + 1, we have:

Proposition 1. If all vertices of T represent Type I singularities, then $S(T) \leq d + 1$, so $s \leq 2d + 2$.

When we blow up a Type II singularity, however, S can increase, as Example 2 shows. To get a handle on how much S can increase, we use a second function P : vertex set of $T \to$ nonnegative integers. P is defined as follows:

If V represents a tangency, let P(V) = 0.

If V represents a singularity, let P(V) = deg(V) - 1.

Define $P(T)$ analogously to $S(T)$.

It turns out that when we blow up a Type II singularity, any increase in S is offset by a decrease in P, i.e., S + P does not increase[5]. For example, let V represent a Type II singularity with deg(V) = m. Then S(V) = m - 1, m, or m + 1 and P(V) = m - 1, so $S(V) + P(V) \geq 2m - 2$. Suppose V blows up to produce k singularities $V_1,\ldots,V_k$ and no tangencies. We have said that $\Sigma\deg(V_i) \leq m - 1$, so $\Sigma S(V_i) \leq m - 1 + k$ and $\Sigma P(V_i) \leq m - 1 - k$. Therefore $\Sigma[S(V_i) + P(V_i)] \leq 2m - 2 \leq S(V) + P(V)$.

---

[5] S stands for "separatrices" and P stands for "potential" (to cause trouble). When trouble occurs, i.e., when S increases, the potential for trouble declines, so the total trouble that can occur is finite.

Unfortunately, it is possible for S + P to increase when we blow up a Type I ingularity. This happens when a Type I singularity of degree m blows up to yield single singularity of degree m + 1. It turns out that such an increase in S + P s temporary: as one continues to blow up, eventually S + P goes back down.

If V represents the initial singularity, S(V) + P(V) = 2d, so we have:

roposition 2. $S(T) \leq S(T) + P(T) \leq 2d$, so $s \leq 4d$.

Proposition 2 is an improvement over the bound $s \leq 4d + 4$ mentioned earlier. nen d = 1, the bound of Proposition 2 is the best possible.

The next result says that if $d \geq 2$, S + P must actually decrease at some point n the course of blowing up:

roposition 3. If $d \geq 2$, then $S(T) \leq 2d - 1$, so $s \leq 4d - 2$.

When d = 2, the bound of Proposition 3 is the best possible (consider xample 1).

Here is a sketch of the proof of Proposition 3:

1. If T has no Type II singularities, then by Proposition 1 $(T) \leq d + 1 \leq 2d - 1$ (since $d \geq 2$).

2. If T contains a Type II singularity V with deg(V) = m and S(V) = m or + 1, then S + P decreases when we blow up V.

3. If T contains a Type II singularity V with deg(V) = m and S(V) = m - 1, + P need not decrease when we blow up V. However, it turns out that S + P must ave decreased earlier in order that such a singularity be present in the tree.

Of course a careful proof must show that these decreases in S + P do not imply offset temporary increases in S + P of the sort mentioned earlier.

The bound on S in Proposition 3 is the best possible, but not the bound on s. or each $d \geq 3$ there exist singularities with $S(T) = 2d - 1$, but they do not have d - 2 separatrices. Here is the simplest class of examples: For each $d \geq 3$ there xists a Type I singularity of degree d that blows up to yield one degree 1 saddle and one degree d Type II singularity. The latter then blows up to yield d - 1 degree 1 saddles (Figure 7)[6]. We have $S(T) = 1 + 2(d - 1) = 2d - 1$, and there appear

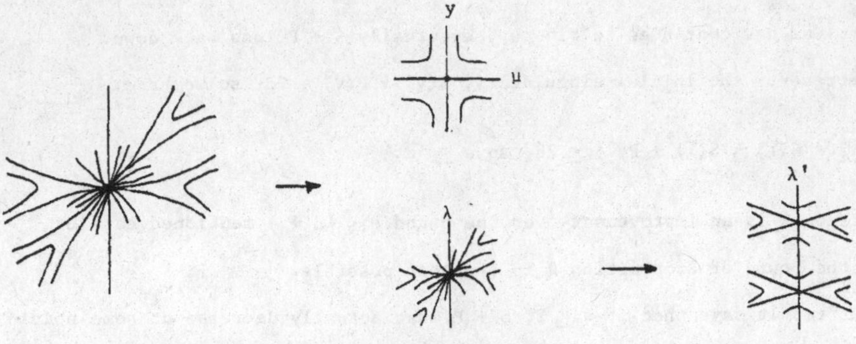

Figure 7 (d = 3)

at first glance to be 4d - 2 separatrices. However, the two non-ghost separatrices of the first saddle do not correspond to separatrices of our original singularity: one sees from Figure 7 that they correspond to curves that lie in the middle of parabolic sectors. Thus $s = 4d - 4$, not $4d - 2$.

The remainder of the proof consists in showing that when $d \geq 3$ and $s = 2d - 1$, the singularity must be very similar to the example just described.

<u>Appendix</u>. Constructing Polynomial Singular Points of Prescribed Type.

Consider an isolated singular point q of a C^1 planar vector field Y such that some solution curve of Y approaches q with a limiting direction. Our goal is to construct a polynomial vector field X with a singularity at (0,0) that is <u>topologically equivalent</u> to that of Y at q. By this it is most convenient to

---

[6]For example

$$\dot{x} = xy^{d-1} + \sum_{i=2}^{d} a_i x^{2i-1} y^{d-i}$$

$$\dot{y} = 2y^d + \sum_{i=2}^{d} a_i x^{2i-2} y^{d+1-i} + \sum_{i=2}^{d} b_i x^{2i+2} y^{d-i}$$

where $\lambda^{d-1} + \sum_{i=2}^{d} a_i \lambda^{d-i} = \prod_{j=1}^{d-1} (\lambda - j)$ and $\sum_{i=2}^{d} b_i \lambda^{d-i} = \frac{d}{d\lambda} \prod_{j=1}^{d-1} (\lambda - j)$

o mean the following: there are neighborhoods U of (0,0) and V of q and an rientation preserving homeomorphism $\phi : U \to V$ that takes solution curves of X o those of Y. ϕ need not preserve the parameterization of solution curves. learly, however, ϕ either preserves the time direction of every solution curve or everses the time direction of every solution curve.

The topological equivalence class of Y at q is determined by the arrangement, n order, of elliptic, hyperbolic, and parabolic sectors about q. Every elliptic ector has a parabolic sector on each side of it. Thus to describe Y at q we eed only list, in counterclockwise order, the hyperbolic sectors, the elliptic ectors, and those parabolic sectors bounded on each side by a hyperbolic sector. e do this by means of a finite <u>sector sequence</u> $S_1, \ldots S_n$ of symbols from the set E,H,P}. A sector sequence must satisfy:

(i) $e \equiv h \bmod 2$

(ii) If $S_i = P$, then $S_{i-1} = S_{i+1} = H$.

ere e = number of E's in $\{S_i\}$, h = number of H's in $\{S_i\}$ (and, for future use, = number of P's in $\{S_i\}$). The first condition is required by Bendixson's formula or the index of a singularity: since $i = \frac{1}{2}(e - h + 2)$ is an integer, $e \equiv h \bmod 2$. n the second condition, the subscripts are mod n. Any sector sequence determines topological equivalence class of singularities, and two sector sequences etermine the same class iff they are identical after a cyclic rearrangement. The ector sequence of the singularity pictured in Figure 8a is HE or EH.

To construct models, we will use vector fields X having a Type I singularity t (0,0) such that $xY_d - yX_d$ has no repeated factors and x does not divide $Y_d - yX_d$. When we blow up such a singularity horizontally, the resulting vector ield $\overline{X}$ has only saddles, nodes, and saddle-nodes on the λ-axis. (At each ingularity $(0,\lambda)$ of $\overline{X}$, the λ-axis is an eigendirection for $D\overline{X}$ with nonzero igenvalue.) At each singularity $(0,\lambda)$ of $\overline{X}$ select one solution curve of $\overline{X}$ in > 0 and one in $x < 0$ asymptotic to $(0,\lambda)$. These curves and the λ-axis divide a eighborhood of the λ-axis into <u>strips</u>, each called elliptic, hyperbolic or arabolic according to whether the corresponding region in the xy-plane is elliptic, hyperbolic or parabolic. See Figure 8b, where, for example, strip 2 is hyperbolic. Notice the upper left and lower right strips are identified, as are

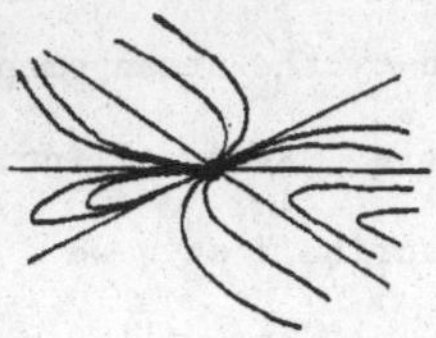

Figure 8a

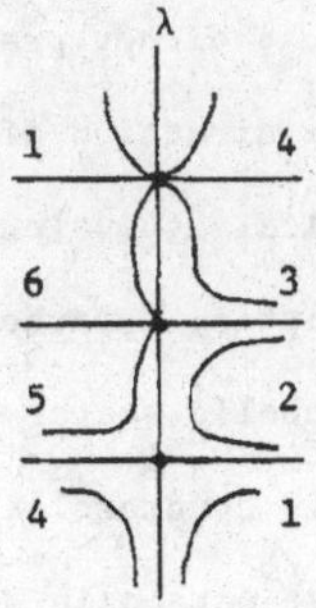

Figure 8b

the lower left and upper right strips, since the assumption that x does not divide $xY_d - yX_d$ implies that the identified strips correspond to parts of a single sector of X at (0,0). Thus if $\overline{X}$ has m singularities on the λ-axis, there are exactly 2m strips. Label the strips from 1 to 2m in the order shown. Associate with the singularity a <u>strip sequence</u> of 2m symbols from the set {E,H,P}; each symbol indicates whether the corresponding strip is elliptic, hyperbolic, or parabolic. In the example of Figure 8 the stip sequence is PHPPPE.

Let Y at q have <u>sector</u> sequence $\{S_i\}$. Construct a new sequence $\{S'_i\}$ as follows: (i) If $S_i = E$ and $S_{i+1} = H$, or $S_i = H$ and $S_{i+1} = E$, add a P after S_i. (The subscripts are to be interpreted mod n.) (ii) If $S_i = P$, add another P after it. Thus if

$$\{S_i\} = \text{EEHPHHE}$$

then

$$\{S'_i\} = \text{EEPHPPHHPE}.$$

We will return to this example shortly. Let t denote the number of strings of E's in $\{S_i\}$. These strings are mod n: thus in our example t = 1. Proviso (i) has us add 2t P's to $\{S_i\}$; proviso (ii) has us add p P's to $\{S_i\}$. Thus the length of

$S'_i\}$ is $L = e' + h' + p' = e + h + (2t + 2p)$, which is even since $e \equiv h \bmod 2$. he sequence $\{S'_i\}$ has the other important property that no E and H are adjacent.

Now we show how to construct a Type I polynomial singularity X of degree $= \frac{1}{2}(L + 2) = \frac{1}{2}(e + h + 2t + 2p + 2)$ such that $xY_d - yX_d$ has no repeated linear actors, x does not divide $xY_d - yX_d$, and the strip sequence of X is S'_i. Then X t (0,0) is topologically equivalent to Y at q.

First we sketch the flow of $\overline{X}$, the horizontal blow-up of X. $\overline{X}$ has singuarities on the λ-axis at $\lambda = 1,\ldots,d - 1$. Along the λ-axis, $\overline{X}$ points in the egative direction for $\lambda > d - 1$ and reverses direction at $\lambda = 1,\ldots,d - 1$. We ust now decide, for each singularity of $\overline{X}$, whether the singularity is to be a node, addle, or saddle-node. Our decisions will make the strip sequence of X be $\{S'_i\}$. t a singularity $(0,i)$ we draw the line $\lambda = i$. The half-line $\lambda = i$, $x > 0$ (resp. $= i$, $x < 0$) will separate two strips to the right of the λ-axis (resp. left of he λ-axis). We label these strips E, H, or P so as to have the strip sequence $'_i$; then we label each half-line S (saddle) or N (node) according to the two trips that it separates:

One strip E, one strip P: label N.

One strip H, one strip P: label S.

Both strips E: label N.

Both strips H: label S.

Both strips P: label N.

he label S on, say, the half-line $\lambda = i$, $x > 0$ indicates that this half-line eparates two hyperbolic sectors of $(0,i)$. The label N on the same half-line ould indicate that $(0,i)$ has a neighborhood U such that $U \cap \{(x,\lambda) : x > 0\}$ is parabolic sector of $(0,i)$. In our example we have $d = 6$ and the diagram in igure 9a. This diagram becomes the phase portrait of Figure 9b.

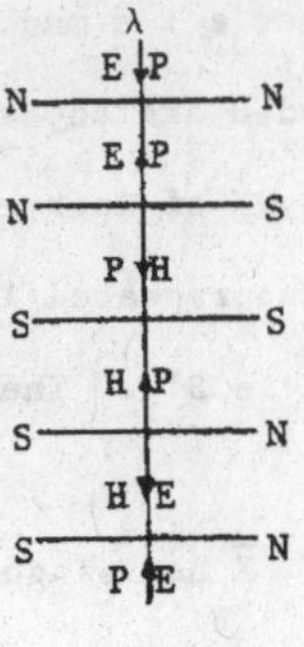

Figure 9a

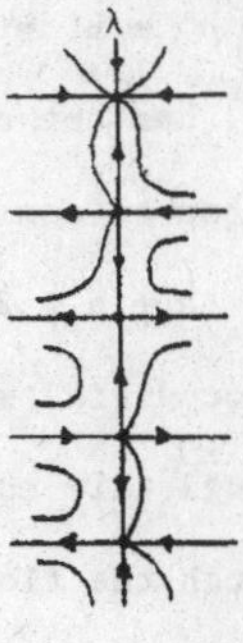

Figure 9b

Thus, as Figure 9 illustrates, we have determined for each singularity (0,i) whether the x-direction is to be expanding, contracting, or a center manifold.

We now specify a polynomial vector field X such that the flow of $\overline{X}$ looks topologically like the flow we have sketched (although the lines $\lambda = i$ will not be invariant). X will be of the form

$$\dot{x} = X_d(x,y) + X_{d+1}(x,y)$$
$$\dot{y} = Y_d(x,y)$$

with each X_i, Y_i homogeneous of degree i. Then $\overline{X}$ will be

$$\dot{x} = xX_d(1,\lambda) + x^2X_{d+1}(1,\lambda)$$
$$\dot{\lambda} = Y_d(1,\lambda) - \lambda X_d(1,\lambda) - x\lambda X_{d+1}(1,\lambda)$$

Choose $X_d(x,y)$ so that for each $i = 1,\ldots,d-1$,

(i) $X_d(1,i) > 0$ if $\overline{X}(0,i)$ is to be expanding in the x-direction;

(ii) $X_d(1,i) < 0$ if $\overline{X}(0,i)$ is to be contracting in the x-direction;

(iii) $X_d(1,i) = 0$ if $\overline{X}(0,i)$ is to have the x-direction as a center manifold.

In addition,

(iv) $X_d(1,\lambda) = \lambda^d$ + lower order terms.

Next choose $X_{d+1}(x,y)$ so that at the saddle-nodes the flow along the center manifold is in the desired direction. Finally choose $Y_d(x,y)$ so that

$$Y_d(1,\lambda) - \lambda X_d(1,\lambda) = -(\lambda^2 + 1) \prod_{i=1}^{d-1} (\lambda - i).$$

his can be done since $X_d(1,\lambda) = \lambda^d$ + lower order terms. Then the flow of $\overline{X}$ near he λ-axis is as desired.

References

1] A. A. Andronov, et. al, Qualitative Theory of Second Order Dynamic Systems, John Wiley and Sons, New York, 1973.

2] I. Bendixson, "Sur les courbes définies par des équations différentielles," Acta Math., 24(1901), 1-88.

3] F. Dumortier, "Singularities of vector fields on the plane," J. of Diff. Eq., 23(1977), 53-106.

4] S. Schecter and M. F. Singer, "Separatrices at Singular Points of Planar Vector Fields," preprint.

5] ________________, "Elliptic Sectors at Singular Points of Planar Vector Fields," preprint.

Department of Mathematics
North Carolina State University
Raleigh, North Carolina 27650

Gradient Vectorfields Near Degenerate Singularities

by

Douglas Shafer

Let U be an open neighborhood of $0 \in R^n$ and let $V:U \to R$, $V(0) = 0$, be an at least C^2 function with an isolated singularity at 0. For any Riemannian metric g the vectorfield $X = \text{grad}_g V$ has an isolated critical point at 0. The question treated here is: How can the topological equivalence class (topological type) of X change as V is held fixed and g varies over all Riemannian metrics?

If the singularity of V is non-degenerate then the critical point of X is hyperbolic, and a simple proof gives invariance. Again, it is easy to see that if X has an attractor or repeller at 0 then the topological type is the same for all Riemannian metrics. Thus we restrict attention to a degenerate singularity of V.

This question was first raised in [6, p. 229], and an example of a potential with two inequivalent gradient vectorfields was given in [6, p. 231]. See also [3, p. 105]. This paper treats the case $n = 2$. A fuller exposition and complete proofs will appear in [7]. For terminology see [4].

The singularity of V is isolated if and only if its first partials vanish together only at (0,0). Under this condition $V^{-1}(0) - \{0\}$ is a finite even number of arcs, and because X is a gradient there is a one-to-one correspondence between these arcs and hyperbolic sectors of X.

<u>Lemma</u> ([1]) If X and Y are at least C^1 vectorfields in the plane with the same configuration of elliptic, hyperbolic, and parabolic sectors in a neighborhood of an isolated singularity, then they are topologically equivalent in a neighborhood of the singularity.

As a gradient X can have no elliptic sectors, so by the lemma and the one-to-one correspondence above the only possible change in the topological type of X with change in g is "opening" or "closing" of parabolic sectors. (See Figure 1.) Thus there are only finitely many equivalence classes for a particular potential V.

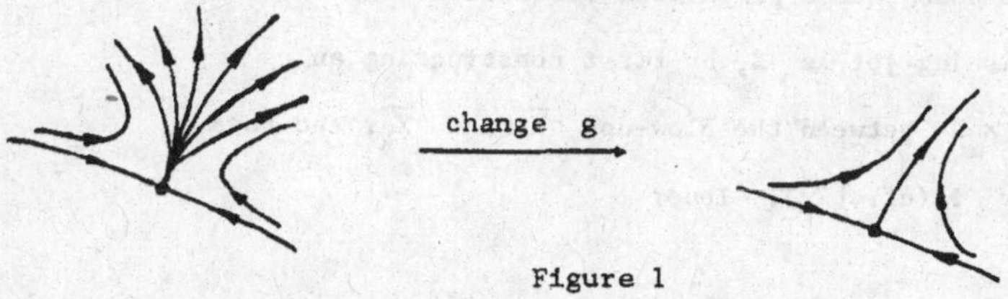

Figure 1

Parabolic sectors may be located by means of "polar blowing-up" ([2], [8]). Essentially we write X down in polar coordinates and divide through successively by r. More precisely, we obtain a vectorfield $\overline{X}$ on $S^1 \times R$ such that $\overline{X}$-orbits near $S^1 \times \{0\}$ are carried onto X-orbits near (0,0) by $\Phi: S^1 \times R \to R^2: (\phi, r) \mapsto (r\cos\phi, r\sin\phi)$. For example, when $V = \frac{1}{3}(x^3 + y^3)$ and g is the standard Riemannian metric, $X = x^2\partial/\partial x + y^2\partial/\partial y$ and $\overline{X}(\phi, r) = \cos\phi\sin\phi(\sin\phi - \cos\phi)\partial/\partial\phi + (\cos^3\phi + \sin^3\phi)r\partial/\partial r$. (see Figure 2.)

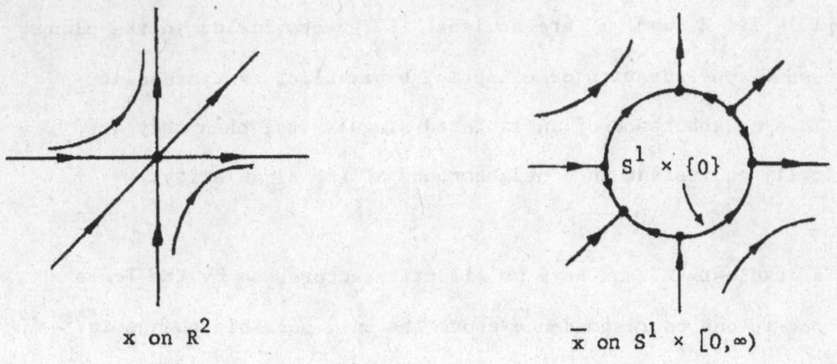

Figure 2

In addition, under certain conditions on X_k, as in Theorem 1 below, the singularities of $\overline{X}$ on $S^1 \times \{0\}$ are less degenerate than that of X at $(0,0)$, so in fact we can construct an equivalence in R^2 between X and the homogeneous polynomial vectorfield X_k with the same first non-vanishing jet as X, by first constructing an equivalence in $S^1 \times R$ between the blow-ups $\overline{X}$ and $\overline{X}_k$, and then carrying it down by Φ (cf. [2]). Thus:

<u>Theorem 1</u>. If $V = U + W$, U homogeneous of degree $k > 1$ (the first non-zero jet), W at least C^{k+2}, and $U_x(p) = U_y(p) = 0$ $\Leftrightarrow p = (0,0)$, then $X = \text{grad}_g V$ is topologically equivalent to $X_k = \text{grad}_{g(0,0)} U$. That is, X is topologically determined by its first non-vanishing jet, which depends only on U and the constant terms of the Riemannian metric.

The condition on U implies that the coefficients of $\partial/\partial\phi$ and $r\partial/\partial r$ in $\overline{X}$ cannot vanish simultaneously on $S^1 \times \{0\}$, which gives normal hyperbolicity. The phase portrait of X can change at Riemannian metric g only if singularities of $\overline{X}$ on $S^1 \times \{0\}$ are not hyperbolic, which means the tangential eigenvalue is 0. A computation of this condition yields a (constructible) polynomial $P_k(x,y)$, homogeneous of degree $2k$ and depending only on the coefficients of U, such that this non-hyperbolicity can occur only if $P_k(x_0, y_0) = 0$ for some $(x_0, y_0) \in R^2$.

Using the blowing-up construction and Theorem 1 we obtain:

Theorem 2A. Under the conditions in Theorem 1:

(a) When $k = 2$, there are five (non-empty) equivalence classes: (i)/(ii) source/sink, (iii) saddle, and (iv)/(v) two hyperbolic and two negative/positive parabolic sectors. (Classes (i)-(iii) are non-degenerate.) The topological type of X is same for all Riemannian metrics.

(b) When $k = 3$, there are three (non-empty) equivalence classes: (i) two hyperbolic sectors, (ii) two hyperbolic and two parabolic sectors, and (iii) six hyperbolic sectors. The topological type of X is invariant under changes in the Riemannian metric iff $P_3(x,y) < 0$ $\forall (x,y) \in R^2$ iff $R[U_x, U_y] < 0$ iff X has six hyperbolic sectors. (Here $R[U_x, U_y]$ is a number, the resultant of U_x and U_y ; see [5].)

(c) When $k = 4$, there are six (non-empty) equivalence classes: (i)/(ii) attractor/repeller, (iii) eight hyperbolic sectors, (iv)/(v) two hyperbolic and two negative/positive parabolic sectors, and (vi) four hyperbolic sectors. X is invariant iff either $(0,0)$ is an attractor/repeller or $P_4(x,y) < 0$ $\forall (x,y) \in R^2$.

(d) When $k \geq 5$, X is invariant only if X has an attractor/repeller at $(0,0)$ or $P_k(x,y) < 0$ for all $(x,y) \in R^2$.

Having Theorem 1 before us, we may take a more geometrical approach to the problem. Since $X_k = \text{grad}_{g(0,0)}U$ is homogeneous, its sector-separating null solutions are straight lines through $(0,0)$ along which X_k is radially directed. Thus to locate parabolic sectors we count the number of such lines between adjacent lines of zeros of U (the latter being in one-to-one correspondence with hyperbolic sectors of X_k). But since X_k is $g(0,0)$-perpendicular to level curves of U, it is radially directed along lines of tangency of level curves of U and multiples of the $g(0,0)$-unit sphere, i.e., the unit sphere in the norm induced by $g(0,0)$. This explains geometrically why a hyperbolic critical point in R^2 cannot change even when not a source or sink. We may assume $U = xy$, so between axes level curves are concave-outward everywhere. No matter how the metric is varied, the $g(0,0)$- sphere, an ellipse, can only be tangent to level curves along two lines. For the same reason, if U is a cubic with three lines of zeros, there can be geometrically no change, since non-zero level curves are concave-outward everywhere, while the cubic $U = \frac{1}{3}(x^3 + y^3)$ has level curves as shown below, so the unit sphere in the usual Riemannian metric gives three lines of tangency--hence a pair of parabolic sectors, but a $g(0,0)$-unit sphere that is a narrow ellipse gives only one--hence no parabolic sectors. See Figure 3.

Thus a necessary condition for change in the equivalence class of X with change in g is that level curves of U have positive curvature (be concave-inward) at some point (i.e., along some line). This condition is sufficient only in low degree; otherwise it is possible that level

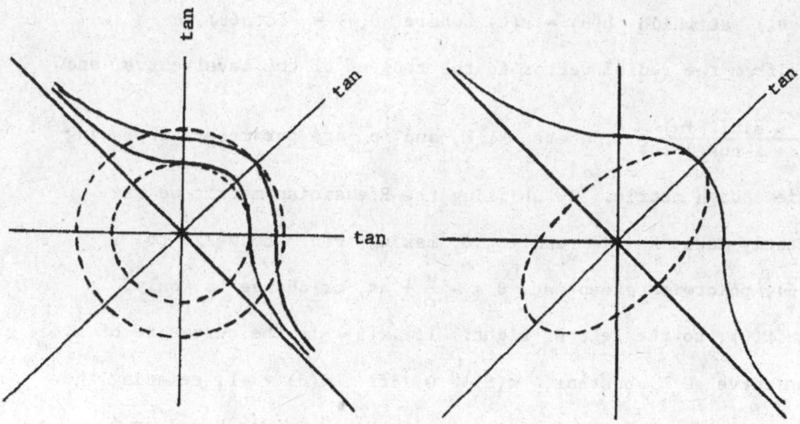

usual metric: 3 lines of tangency $\quad \begin{pmatrix} 3 & 1 \\ 1 & 3 \end{pmatrix}^{-1}$: 1 line of tangency

(dotted curves are multiples of unit spheres)

Figure 3

curves can be so complicated that a parabolic sector may never close for any Riemannian metric. We have:

Theorem 3. In the conditions of Theorem 1, the topological type of X changes at some Riemannian metric only if U has a line of zeros and level curves of U have positive curvature at some point. This condition is sufficient if $\deg U = 3$ or 4 (or 5, when there is more than one line of zeros).

The technique of proof is as follows. If $\ell_1 : \theta = \theta_1$ and $\ell_2 : \theta = \theta_2$ are adjacent lines of zeros of U, X_k is radially directed at values of

$\theta \in (\theta_1, \theta_2)$ at which $D(\theta) = P(\theta)$, where $D(\theta) = -\cot\alpha(\theta)$, α = the angle from the radial vector to the tangent of the level curve, and

$P(\theta) = \dfrac{R\sin(2\theta+u)}{K - R\cos(2\theta+u)}$, where R, K, and u are parameters depending on the Riemannian metric. By shifting the Riemannian metric we may independently make $K - R$ tend to 0, making $P(\theta)$ converge to $\cot(\theta + \frac{u}{2})$, pointwise except at $\theta = -\frac{u}{2} + n\pi$, or change u only, shifting $P(\theta)$ to the left or right. If $\kappa(\theta)$ is the curvature of the plane curve $U \equiv$ constant, $\kappa(\theta) > 0$ iff $\alpha'(\theta) > -1$, relating the shape of D to that of the level curve of U. For deg U = 3 or 4, $D(\theta)$ is relatively simple, and we may determine its crossing of $P(\theta)$, for various Riemannian metrics, to count parabolic sectors.

When the phase portraits of the vectorfields in Theorem 2A(c) are examined, we can use Theorem 3 to complete the characterization of change in topological equivalence class of X when deg $U \leq 4$:

<u>Theorem 2B</u>. In Theorem 2A(c), classes (i), (ii), and (iii) are invariant. X may or may not change from class (vi), depending on U, while there always exist Riemannian metrics changing X from (iv) to (vi) and from (v) to (vi), but never from (iv) to (v) or conversely.

The same question in R^n, $n \geq 3$, seems to be much more difficult. Simple generalizations of the lemma above do not hold: in R^3 there are inequivalent vectorfields X and Y with geometrically the same phase portraits (the configurations of "generalized" sectors are the same). While the blow-up of X, when restricted to $S^{n-1} \times \{0\}$ is a gradient, even when n = 3 there may be saddle connections or even curves of critical points. The analogue of Theorem 1 has been established only in simple cases.

References

[1] A. A. Andronov, et al, Qualitative Theory of Second Order Dynamic Systems, John Wiley and Sons, New York, 1973.

[2] F. Dumortier, Singularities of Vectorfields in the Plane, J. Differential Equations 23 (1977) 53-106.

[3] John Guckenheimer, Bifurcation and Catastrophe, in Dynamical Systems (M. M. Peixoto, ed.), Academic Press, New York, 1973.

[4] Philip Hartman, Ordinary Differential Equations, John Wiley and Sons, New York, 1964.

[5] F. S. Macaulay, The Algebraic Theory of Modular Systems, Cambridge University Press, London, 1916.

[6] Manifolds-Amsterdam 1970 (N. Kuiper, ed.), Lecture Notes in Mathematics 197, Springer-Verlag, New York, 1971: "Problems Concerning Manifolds," Problem D 4c of R. Thom, p. 229; "A Solution: An example as requested in the problem of R. Thom, D 4c," by F. Takens, p. 231.

[7] D. Shafer, Topological Equivalence of Gradient Vectorfields, to appear in Trans. Amer. Math. Soc.

[8] F. Takens, Singularities of Vectorfields, I.H.E.S. Pub. Math. 43 (1974) 47-100.

Department of Mathematics
University of North Carolina at Charlotte
Charlotte, North Carolina 28223

INVARIANT CURVES NEAR PARABOLIC POINTS AND REGIONS OF STABILITY

Carles Simó

Abstract.- In this paper we consider analytic area preserving mappings (APM) near a fixed point. If the fixed point is parabolic a criterion is given which ensures the existence of invariant curves and, hence, stability. An heuristic discussion of the limit of the stability region completes the paper.

§1. Introduction.- It is well known that stability is easily decided near a fixed point of an APM, $T : R^2 \to R^2$, if the fixed point (that we take as the origin) is hyperbolic or elliptic (with some nondegeneracy conditions). We deal here with the parabolic case with nondiagonal linear part :

$$T\begin{pmatrix}x\\y\end{pmatrix} = \begin{pmatrix}x_1\\y_1\end{pmatrix} = \begin{pmatrix}x + f(x,y)\\x+y+g(x,y)\end{pmatrix} \qquad (1.1),$$

where f,g begin with terms of second order. From now on we only consider parabolic points with linear part given by the matrix $\begin{pmatrix}1&0\\1&1\end{pmatrix}$.

An instability criterion for (1.1) was given by Levi-Civita [5]. Let a_{22} be the coefficient of y^2 in $f(x,y)$. If $a_{22} \neq 0$, T is unstable at the origin.

However in [3] Chirikov and Izraelev present a mapping given by

$$\begin{pmatrix}x_1\\y_1\end{pmatrix} = \begin{pmatrix}x - y^3\\x+y-y^3\end{pmatrix} \qquad (1.2),$$

which has the origin as parabolic fixed point. Some simulation displays an stability region near the origin. For instance, we can produce 50,000 iterates of points in the x axis. Up to x = .525 the points appear to be stable. After that value they are unstable with some exceptions that are related to elliptic periodic points (mainly 38/5-periodic at x = .5614913442, 44/6-periodic at .5893141038 and at .6118599570, 22/3-periodic at .6006277462, etc.).

It is apparent that in (1.2) the Levi-Civita criterion does not apply. We search for a criterion which characterizes the mappings for which a parabolic point of the type (1.1) is stable. First we need some normal form that is presented in the next section.

A related question is the stability of solutions of certain second order finite difference equations. Let δ be the central difference operator. Then the equation $\delta^2 z_n - \phi(z_n) = 0$, with ϕ beginning with terms of second order is easily converted to

$$\begin{aligned}x_{n+1} &= x_n + \phi(x_n + y_n)\\ y_{n+1} &= x_n + y_n\end{aligned} \qquad (1.3),$$

of the form (1.1) setting $y_n = z_n$. An equivalent form is

$$(x_n, y_n) \to (x_n + \phi(y_n), x_n + y_n + \phi(y_n)) = (x_n + \phi(y_n), y_n + x_{n+1}).$$

That this is not far from the general (1.1) equation is seen from the normal form.

§2. A normal form near parabolic points.- With a change of variables (1.1) can be reduced to a similar transformation with the function g suppressed [5]. Unfortunately the change is not an APM. We see that a mapping like (1.3) is area reserving for every function ϕ . For polynomial APM of prime degree Engel [4] howed that a reduction to the form (1.3) is always possible.

Using standard techniques (see, f.i. [6] ,§23) we obtain the following emma.- Let T be an APM of the form (1.1). For every integer $n \geq 2$ there is a olynomial change of variables C such that the transformed mapping $T^* = C^{-1}TC$ s of the form

$$T^*\begin{pmatrix}x\\y\end{pmatrix} = \begin{pmatrix}x + F_n(x+y) + 0_{n+1}\\ x + y + 0_{n+1}\end{pmatrix} \qquad (2.1),$$

here F_n is a polynomial with terms of degrees 2 to n and 0_k means an analytic unction beginning with terms of order k .

roof : We set

$$\begin{aligned} x &= \xi + \phi(\xi,\eta)\\ y &= \eta + \psi(\xi,\eta)\end{aligned} \qquad (2.2),$$

s the equations giving C . We impose, formally, that T^* be given by the relations $,\eta) \to (\xi + \overline{f}(\xi,\eta) , \xi + \eta)$. From (1.1), (2.2) and the desired form we obtain the quations

$$\begin{aligned} \phi(\xi+\overline{f}(\xi,\eta) , \xi+\eta) &= f(\xi+\phi(\xi,\eta) , \eta+\psi(\xi,\eta)) + \phi(\xi,\eta) - \overline{f}(\xi,\eta)\\ \psi(\xi+\overline{f}(\xi,\eta) , \xi+\eta) &= g(\xi+\phi(\xi,\eta) , \eta+\psi(\xi,\eta)) + \phi(\xi,\eta) + \psi(\xi,\eta)\end{aligned} \qquad (2.3).$$

e add to (2.3) one more condition

$$\phi_\xi + \psi_\eta + \phi_\xi\psi_\eta - \phi_\eta\psi_\xi = 0 \qquad (2.4),$$

mposing that (2.2) be formally area preserving.

Let

$$\phi = \sum_{k\geq 2}\sum_{j=0}^{k}\phi_{k,j}\xi^{k-j}\eta^j \quad ; \quad \psi = \sum_{k\geq 2}\sum_{j=0}^{k}\psi_{k,j}\xi^{k-j}\eta^j \quad ; \quad \overline{f} = \sum_{k\geq 2}\sum_{j=0}^{k}\overline{f}_{k,j}\xi^{k-j}\eta^j \ .$$

nserting in (2.3), (2.4) we get a linear system for the coefficients of the omogeneous part of order k and where the independent terms are functions of f,g nd of the terms of lower order. From (2.4) and the second of equations (2.3) we ave equations like

$$(k-j)\phi_{k,j} + (j+1)\psi_{k,j+1} = \alpha \quad , \quad j = 0 \text{ to } k-1$$

$$\sum_{i=j+1}^{k}\binom{i}{j}\psi_{k,i} - \phi_{k,j} = \beta \quad , \quad j = 0 \text{ to } k-1 \qquad (2.5),$$

$$\phi_{k,k} = \gamma$$

here α, β, γ are known quantities. System (2.5) is invertible. We remark that $\phi_{k,0}$ does not enter in (2.5) and remains arbitrary. The first of equations (2.3) llows us to obtain $\overline{f}_{k,j}$ via

$$\sum_{i=j+1}^{k}\binom{i}{j}\phi_{k,j} + \overline{f}_{k,j} = \delta \quad , \quad j = 0 \text{ to } k \quad , \quad \delta \text{ known.}$$

We perform the previous stepd from $k = 2$ to $k = n$. The transformed mapping is iven by

$$\xi_1 = \xi + \sum_{k=2}^{n} \overline{f}_k(\xi,\eta) + 0_{n+1}$$
$$\eta_1 = \xi + \eta \qquad\qquad + 0_{n+1} \qquad (2.6).$$

Let z, z_1, ζ, ζ_1 be, respectively, the vectors $(x,y)^T$, $(x_1,y_1)^T$, $(\xi,\eta)^T$, $(\xi_1,\eta_1)^T$. Some relations hold for the Jacobians :

$$|D_z z_1| = 1 \quad ; \quad |D_\zeta z| = 1 + 0_n \quad ; \quad |D_{\zeta_1} z_1| = 1 + 0_n .$$

Then $|D_\zeta \zeta_1| = 1 + 0_n$, but from (2.6) we have

$$|D_\zeta \zeta_1| = 1 + \sum_{k=2}^{n} ((\overline{f}_k)_\xi - (\overline{f}_k)_\eta) + 0_n ,$$

and hence $(\overline{f}_k)_\xi = (\overline{f}_k)_\eta$, $k = 2$ to n, which ends the proof of the Lemma.

§3.Existence of invariant curves.- The following theorem characterizes the parabolic points of the studied type for which there exist invariant curves.
Theorem.- Let be $F_n(u) = a_m u^m + 0_{m+1}$. The mapping T^* given by (2.1) has invariant curves surrounding the origin if and only if m is odd and a_m is negative.
Proof : An easy modification of the Levi-Civita argument proves the instability if m is even or if being m odd, a_m is positive. Then we can restrict ourselves to the case $m = 2k - 1$, $a_m = -1$ (with a suitable scaling).

We introduce a hamiltonian flow in the plane through the differential equations

$$\frac{dx}{dt} = -H_y , \quad \frac{dy}{dt} = H_x \quad ; \quad H(x,y) = \frac{x^2}{2} + \frac{xy^{2k-1}}{2} + \frac{y^{2k}}{2k} \qquad (3.1).$$

The system (3.1) has all the orbits closed near the origin. The orbit given by the curve $H(x,y) = \frac{y_0^{2k}}{2k}$ has a period $\tau = O(y_0^{-k+1})$. We introduce polar coordinates in the plane through

$$\begin{pmatrix} x \\ y \end{pmatrix} \to \begin{pmatrix} r = (2kH(x,y))^{1/2k} \\ \alpha = 2\pi t/\tau \end{pmatrix} \qquad (3.2),$$

where t is the time elapsed on the orbit of (3.1) from $(0,y_0)$ to (x,y) .

The unit time mapping associated to (3.1) can be written

$$(r,\alpha) \to (r, \alpha + cr^{k-1}) \qquad (3.3),$$

where c is a function of r only which is $O(1)$ when $r \to 0$. This map is a twist that mimics the initial mapping (2.1). Let (r_1,α_1) be the image of (r,α) under (2.1) in the variables given by (3.2). Using a Newton polygon argument to analyze dominant terms one readily checks that $r_1 = r + O(r^k)$ and $\alpha_1 = \alpha + cr^{k-1} + O(r^k)$, so we have a perturbation of (3.3).(Details at the Appendix). The same method gives $2\pi/c = 4\sqrt{k} \int_0^1 (1 - s^{2k})^{-1/2} ds + O(r^{k-1})$.

As the perturbations in r, α are small compared to cr^{k-1}, a standard argument (see [6] , §34) ends the proof using the twist theorem.

§4.Examples.- As a Corollary of the Theorem we have that parabolic points with negative odd dominant nonlinearity are stable. We remark that when we only consider he linear part, mapping (1.1) is unstable. The form of the invariant curves near he origin is approximated by $H(x,y) = \text{constant}$. The number of iterations to omplete a revolution on the invariant curve whichs cuts the y axis at r is $0(r^{-k+1})$. he rotation number tends to zero when the curve approaches the origin.

The Theorem can be applied to the APM of §1 : $(x,y) \to (x - (x+y)^3, x+y)$. ven more : the point (1,0) is a 6-periodic parabolic point for this APM. The inear part at (1,0) is given by $\begin{pmatrix}1 & 0\\ 6 & 1\end{pmatrix}^T$ and the reduction of the Lemma annihilates he quadratic terms and produces a negative cubic term in F_n . This originates some table island containing the point (1,0).

For quadratic APM there are only two cases of parabolic points [7]. One of hem, given by $(x,y) \to (x+(x+y)^2, x+y)$ is unstable. The other one is $(x,y) \to (-x-(x+y)^2, -x-y)$. Squaring the mapping and scaling x we obtain the form 1.1), and the application of the Lemma produces an F_n with negative cubic dominant erm. Hence we get an stable point.

5. Stability region.- Once the existence of invariant curves has been established he natural question is how large is the region of stability. In this section we resent some computations and an heuristic approach.

We consider again the model mapping (1.2). Some eriodic hyperbolic points on the x axis are given n Table 1 with the greatest eigenvalue. Figure 1 is qualitative picture of the invariant manifolds ssociated to the hyperbolic points of Table 1.

Period	x	λ
9	.44531289	1.3573
17/2	.49730852	1.4008
25/3	.51704025	1.7816
33/4	.52571513	3.2818
49/6	.53108962	47.866
8	.53228493	5.5153
23/3	.55127186	8.3681
15/2	.57461099	4.5457
29/4	.61826715	26.836
7	.63401098	5.0030

Table 1

Mapping (1.2) has periodic points in the x axis nd in the $y = x/2$ line. Approaching the origin on he x axis starting at the point (1,0) we encounter 6-periodic parabolic point and after that, points f increasing period (7, 8, 9, 10, ...). Points of eriod $k \equiv 2 \pmod 4$ are elliptic. The remaining ones yperbolic. Approaching the origin on the $y = x/2$ ine starting at the point (.80178 , .40089), we ncounter a 6-periodic hyperbolic point. Points of ncreasing period k are elliptic if $k \equiv 0 \pmod 4$ and yperbolic otherwise. Points of fraccionary period are easily detected between the ndicated ones.

Hyperbolic points of even period produce homoclinic points. The ones with odd eriod in $y = 0$ give rise to heteroclinic points with the ones of the same period on the line $y = x/2$.

In fig.1 we see that heteroclinic points connect the invariant manifolds of points of period 49, 8, 23, 15, 29, 7 . The connection between points of periods 15 and 29 is not exactly seen. They connect through some auxiliary point, f.i. the hyperbolic satellite point of period 44 near the 22 elliptic periodic point.

The seven periodic point in $y = x/2$ produces an heteroclinic point with the six periodic point on the same line. This one quickly connects with infinity. We have obtained a transition chain in the sense of [1]. A similar mechanism for the escaping of points of APM has been proposed by Bartlett [2].

We remark that the manifolds of the periodic points of periods 33, 25, 9 (see fig.1) do not produce heteroclinic points between them. The fact that the mechanism for sending out of the region of stability a point is broken seems to agree with the previous estimates of the stability region. In section 1 we mention that the limit of stability is around .525 . Point 33-periodic is located at .5257 .

As a conclusion we conjecture that the limit of the region of stability coincides with the apparition of heteroclinic points associated to hyperbolic periodic points of different period.

§6. Open questions.- In order to locate the limit of stability according to the previous conjecture we list some questions :

a) Predict the location of higher periodic points near the parabolic (or, in general, elliptic) stable fixed point. The period n must be high but the distance

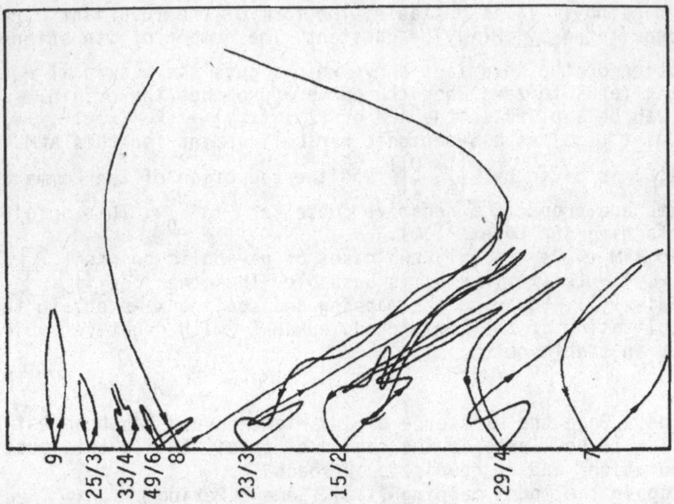

Figure 1

to the fixed point must remain finite.

b) Estimate the differential of T^n at the periodic points of a). This gives the direction of the tangent to the invariant manifolds.

c) Estimate the second differential of T^n at the periodic points of a). This is an indication of the bending of the invariant manifolds. A bounding of this bending can provide a proof of the existence of "micro" heteroclinic points. The name micro comes from the fact that the related hyperbolic points are extremely close.

An analytical answer to the preceding questions can allow us to predict the region of stability without numerical simulation of the mapping.

References.

[1] Arnold,V.I.,Avez,A.: Ergodic problems of classical mechanics. Benjamin (1968).

[2] Bartlett,J.: Global stability of area preserving mappings, in Long-time predictions in dynamics, ed.V.Szebehely and B.D.Tapley. Reidel (1976),pp.99-110.

[3] Chirikov,B.D.,Izraelev,F.M.: Some numerical experiments with a nonlinear mapping : stochastic component, in Transformations ponctuelles et leurs applications, Colloque international du C.N.R.S. n°229 (1973), pp.409-428.

[4] Engel,W.: Ganze Cremona-Transformationen von Primzahlgrad in der Ebene, Math. Annalen 136, 319-325 (1958).

[5] Levi-Civita,T.: Sopra alcuni criteri di instabilità, Annali di Matematica, Ser.III, 5, 221-307 (1901).

[6] Siegel,C.L.,Moser,J.K.: Lectures on Celestial Mechanics. Springer (1971).

[7] Simó,C.: Una nota sobre las aplicaciones cuadráticas que conservan área, in Actas V Congreso de Matemáticos de Expresión Latina, Palma 1977, pp.361-369.

Universitat de Barcelona
Barcelona, Spain

ppendix.- We give here the details of the proof which involve some computation.

From (3.1) we have that for small positive values of h, the curves $H(x,y) = h$ are lmost coincident with the curves $x^2/2 + y^{2k}/2k = h$. Hence, $x = O(r^k)$, $y = O(r)$ (see (3.2) nd recall $k \geqq 2$). Again from (3.1) we get by derivation

$$\dot{x} = -y^{2k-1} - \frac{2k-1}{2} x y^{2k-2} \quad , \quad \dot{y} = x + \frac{1}{2} y^{2k-1} \quad ,$$

$$\ddot{x} = -(2k-1)xy^{2k-2} + O(r^{4k-3}) \quad , \quad \ddot{y} = -y^{2k-1} + O(r^{4k-3}) \quad ,$$

$$\dddot{y} = -(2k-1)xy^{2k-2} + O(r^{4k-3}) \; .$$

ll the other derivatives are $O(r^{4k-3})$. Therefore, the unit time mapping is given by

$$\overline{x} = x - y^{2k-1} - (2k-1)xy^{2k-2} + O(r^{4k-3}),$$

$$\overline{y} = y + x - \frac{2k-1}{6} x y^{2k-2} + O(r^{4k-3}),$$

nd the initial APM is expressed, using the Lemma, as

$$x_1 = x - (x+y)^{2k-1} + \sum_{j=2k}^{4k-4} a_j (x+y)^j + O(r^{4k-3}),$$

$$y_1 = x + y \qquad\qquad + O(r^{4k-3}).$$

The dominant terms in the difference are

$$x_1 - \overline{x} = O(y^{2k}, xy^{2k-1}, r^{4k-3}),$$

$$y_1 - \overline{y} = O(xy^{2k-2}, r^{4k-3}).$$

We rewrite $\overline{x}$, $\overline{y}$ using (3.2) and (3.3) getting $\overline{r} = r$, $\overline{\alpha} = \alpha + cr^{k-1}$. Looking for the ifferences $r_1 - \overline{r}$, $\alpha_1 - \overline{\alpha}$ we have

$$\frac{r_1^{2k}}{2k} = H(x_1, y_1) = \frac{x^2 + O(r^{3k-1})}{2} + \frac{O(r^{3k-1})}{2} + \frac{y^{2k} + O(r^{3k-1})}{2k} = \frac{r^{2k}}{2k} + O(r^{3k-1}) \; .$$

ence $r_1 = r + O(r^k)$. This value measures the perturbation of the twist given by the nit time map, transversal to the flow of (3.1). In order to obtain estimates for $_1 - \overline{\alpha}$ (i.e., the perturbation tangent to the flow) we project $\Delta = (x_1 - \overline{x}, y_1 - \overline{y})^T$ on he velocity vector $v = (\dot{x}, \dot{y})^T$. Let Δ_v be the projection. We easily obtain

$$\Delta_v = \Delta \cdot v / \|v\| = O(y^{4k-1}, x^2 y^{2k-2}, r^{5k-3}) / \|v\| \; .$$

The elapsed time for that change in α be produced is

$$t_\Delta = \Delta_v / \|v\| = O(y^{4k-1}, x^2 y^{2k-2}, r^{5k-3}) / (x^2 + xy^{2k-1} + \frac{5}{4} y^{4k-2} + O(r^{5k-3})).$$

e recall that from (3.1), (3.2) we have $\frac{x^2}{2} + \frac{xy^{2k-1}}{2} + \frac{y^{2k}}{2k} = \frac{r^{2k}}{2k}$. To discuss the ounds of t_Δ we consider different regions in the plane.

If $|y| \overset{\Delta}{=} |x|^\beta$ for some $\beta \geqq 1/k$, there is an $r_0(\beta)$ such that for r in $(0, r_0(\beta))$ we ave $r^k/(2\sqrt{k}) < |x| < 3r^k/(2\sqrt{k})$ and, therefore, $y = O(r)$. This gives the estimate $_\Delta = O(r^{4k-2})/r^{2k} = O(r)$.

If $|x| = |y|^\beta$ for some $\beta \geqq k$, there is an $r_0'(\beta)$ such that for r in $(0, r_0'(\beta))$ we ve $\frac{r}{2} < |y| < 3\frac{r}{2}$. First we suppose $\beta \leqq (2k+1)/2$. Then $t_\Delta = O(r^{2\beta+2k-2})/r^2 = O(r)$. In he opposite case, $\beta > (2k+1)/2$, we get $t_\Delta = O(r^{4k-1})/r^{4k-2} = O(r)$.

As the angular velocity is cr^{k-1} we have $\alpha_1 - \overline{\alpha} = O(r^k)$. This ends the estimates f the perturbation.

We want to remark that the same method can be applied to other interesting cases. It is known that an elliptic point with linear part which has as eigenvalue a fourth root of the unity is not forcely stable nor the invariant curves around the fixed point (if they exist) look like circles. Take the easiest nontrivial case : a quadratic APM. In suitable coordinates we have $T : (x,y) \rightarrow (x^2-y,x)$. It is known from numerical experiments (see, f.i., [7]) that near the origin it seems to be invariant curves which look like a latin cross, whose arms have a width of the order of the square of its length.

Let $f = T^4$. We compute

$$f\begin{pmatrix}x\\y\end{pmatrix} = \begin{pmatrix}x+2x^2y-x^4-4xy^3+2y^5+8x^3y^2+0_6\\ y-2xy^2+y^4+4x^3y-2x^5-4x^2y^3+0_6\end{pmatrix},$$

getting a parabolic point with linear part $\begin{pmatrix}1&0\\0&1\end{pmatrix}$. Note that if we only take into account the third order terms,the origin is unstable under f. Instead, we introduce a hamiltonian flow as in (3.1) with $H(x,y) = x^2y^2-xy^4-x^4y+(x^6+y^6)/3$, whose level curves $H = h$ have the desired latin cross shape if h is a small enough positive number.

Let $r(x,y)$ be defined by $r^{12} = H$. Then the period of the orbit through (x,y) is $O(-\ln r/r^6)$. The mapping f can be seen as a perturbation of the unit time map under the flow of the given hamiltonian The required estimates are made as before and we get the existence of invariant curves. Extensions to other maps near a parabolic fixed point with diagonal linear part can be made introducing hamiltonians whose dominant terms (using Newton's polygon) define simple closed curves around the origin in every neighborhood of the fixed point.

Motion under the influence of a strong constraining force.

by

Floris Takens

1. Introduction and statement of the results.

Classical conservative mechanical systems with a finite number of degrees of freedom are discribed mathematically by 2nd order differential equations of the form

$$\ddot{x} = -(\text{grad } V)(x)$$

with $x \in \mathbb{R}^n$ and $V : \mathbb{R}^n \to \mathbb{R}$ a smooth function. $\mathbb{R}^n$ has the usual Euclidean structure, and in this paper we shall assume that V is bounded below. In some generalizations, the configuration space $\mathbb{R}^n$ is replaced by a Riemannian manifold, in which case $\ddot{x}$ is replaced by the covariant acceleration, i.e., the covariant derivative of the tangent vector $\dot{x}$ along $x(t)$. These equations typically represent the motion of one or more masspoints under the influence of mutual or external potentials.

A mechanical system with a constraint can be mathematically described as a pair, consisting of an equation of the above form together with a submanifold $M \subset \mathbb{R}^n$. The trajectories of such a constrained system are curves $x(t)$ in $\mathbb{R}^n$ such that $x(t) \in M$ for all $t \in \mathbb{R}$ and

$$\ddot{x}(t) + (\text{grad } V)\ (x(t)) \perp T_{x(t)}(M).$$

Note that for each $x(0) \in M$ and $\dot{x}(0) \in T_{x(0)}(M)$ there is exactly one such trajectory. It should also be observed that these trajectories are exactly those curves on M for which the covariant acceleration (with respect to the Riemannian metric in M which is induced from the Euclidean structure in $\mathbb{R}^n$) equals $-\text{grad}\,(V|M)$. So if we introduce a constraint $M \subset \mathbb{R}^n$ to a mechanival system, we obtain a new mechanical system with configuration space M.

The physical explanation, in case our system is made up of mass-points, is the following: $M \subset \mathbb{R}^n$ represents the set of all possible positions, or configurations, of the mass-points; such a restriction might be the consequence of inelastic connections between the various mass-points. From this it seems desirable to obtain the equations for the trajectories of the constrained system by first introducing "elastic constraints" and then considering the limit for more and more inelastic constraints. To make this somewhat more precise we say that an elastic constraint to a submanifold $M \subset \mathbb{R}^n$ is described by a smooth function $W : \mathbb{R}^n \to \mathbb{R}$ such that $W|M = 0$ and $W > 0$ on $\mathbb{R}^n - M$. The corresponding trajectories are solutions of

$$\ddot{x} = -(\text{grad}(V + W))\,(x). \qquad {**}$$

This equation expresses the fact that it takes extra energy not to be in M. The function W is of course not determined by M: in practical cases this function should be derived from the actual way the constraints are imposed. Intuitively one sees that the bigger W is, the more inelastic the constraint is.

In order to compare these two descriptions of constraints, namely the mathematical description, leading to *, and the physical description leading to **, we would like to relate the solutions of * to those of ** when W is "very big". In order to make this "very big" precise, we keep W fixed but multiply it with $\lambda \in \mathbb{R}$ and take the limit for $\lambda \to \infty$. This leads to the equation

$$\ddot{x} = -(\text{grad}(V + \lambda . W))\,(x). \qquad {*\lambda}.$$

From now on we also assume that in each point of M, the 2nd order derivative of W, normal to the tangent space of M, is positive definite.

We want to compare solutions of * with curves which can be obtained as limits of solutions of $*\lambda$ with $\lambda \to \infty$. We say that a curve $x(t)$ is a limiting solution of $*\lambda$ for $\lambda \to \infty$, or a solution of $*\,\infty$, if there is a sequence of solutions $x_{\lambda_i}(t)$ of $*\lambda_i$ with uniformly bounded energy and converging to $x(t)$. This energy is defined as usual: if $x_\lambda(t)$ is a solution of $*\lambda$, then its energy $E_\lambda(t)$ is defined by $E_\lambda(t) = \frac{1}{2}(\dot{x}_\lambda(t)^2 + (V + \lambda W)(x_\lambda(t))$.

It is an important fact that for any solution $x_\lambda(t)$ of $*\lambda$, the energy is independent of the time t.

Since V is bounded below and W is non-negative, $|\dot{x}_\lambda|$ is uniformly bounded for solutions with bounded energy. This means that, in order to obtain the limiting solutions of $*\,\infty$ we may assume that all the immages of our sequence $x_{\lambda_i}(t)$ of solutions of $*\lambda_i$, for t in some bounded interval $I \subset \mathbb{R}$, are contained is a fixed compact set $K \subset \mathbb{R}^n$. By restricting to this compact set we have for solutions $x_{\lambda_i}(t)$, of $*\lambda_i$, with uniformly bounded energy, that

$$\rho(x_{\lambda_i}(t), M) = O(\lambda_i^{-\frac{1}{2}}),$$

where ρ denotes the Euclidean distance. Hence the limiting solutions will be in M. Since $|\dot{x}_\lambda|$ is uniformly bounded, they will be Lipschitz.

For λ big it is usefull to "decompose" a solution $x_\lambda(t)$ of $*\lambda$ into
1^e a motion along M, obtained by projecting $x_\lambda(t)$ to M;
2^e a motion perpendicular to M.
This goes as follows. Let $e_1(x), \ldots, e_k(x)$ be an orthonormal basis of the normal space $T_x(M)^\perp$, depending smoothly on $x \in M$. Then every point $\overline{x}$ near M can be uniquely written as

$$\overline{x} = \overline{x}_M + \sum_{i=1}^{k} \overline{x}_i . e_i(\overline{x}_M)$$

with $\overline{x}_M \in M$ and $\overline{x}_i$ small. Applying this to a solution $x_\lambda(t)$ of $*\lambda$ we get a decomposition

$$x_\lambda(t) = x_{\lambda,M}(t) + \sum_{i=1}^{k} x_{\lambda,i}(t).e_i(x_{\lambda,M}(t)).$$

We refer to $x_{\lambda,M}(t)$ as the motion along M and to $x_{\lambda,i}(t)$ as the normal motion with respect to the basis $e_1,\ldots,e_k$. From [1.] and the calculations in this paper, see section 2, we conclude that $|\ddot{x}_{\lambda,M}(t)|$ is uniformly bounded for a family of solutions $x_\lambda(t)$ of $*\lambda$ with uniformly bounded energy.

In order to describe in detail the relation between solutions of $*$ and $*\infty$, we first restrict to the case where $\dim(M) = n-1$. Here there is only one normal vector $e_1(x)$, also denoted by e or $e(x)$. $\sigma : M \to \mathbb{R}$ assigns to each $x \in M$ the second order derivative of W in the direction $e(x)$. For a solution $x_\lambda(t) = x_{\lambda,M}(t) + x_{\lambda,1}(t).e$ of $*\lambda$, we define the <u>normal energy</u> as

$$E_{N,\lambda}(t) = \tfrac{1}{2}.(\dot{x}_{\lambda,1}(t))^2 + \tfrac{1}{2}.\lambda.\sigma(x_{\lambda,M}(t)).(x_{\lambda,1}(t))^2.$$

Note that this normal energy plus the kinetic and potential energy of the motion along M equals the energy, modulo terms which tend to zero for $\lambda \to \infty$.

If $x_{\lambda_i}(t)$ is a sequence of solutions of $*\lambda_i$, $\lambda_i \to \infty$, converging to $x(t)$, then we may assume that also the normal energies $E_{N,\lambda_i}(t)$ of these solutions converge, say to $E(t)$ (if necessary by taking a subsequence). According to [1.] the set of all limiting solutions is just the set of all solutions of:

$$E(t) = C.(\sigma(x(t)))^{\frac{1}{2}}$$
$$\frac{D}{dt}\dot{x}(t) = -(\mathrm{grad}(V + C.\sigma^{\frac{1}{2}}))x(t) \qquad (1.)$$

where $\frac{D}{dt}$ denotes the covariant derivative along $x(t)$, and where C is a non-negative number which is constant for each solution. One may say that C depends on the "initial condition": if we consider a limit of solutions x_{λ_i} with $x_{\lambda_i}(0) = x(0) \in M$ and $\dot{x}_{\lambda_i}(0) = \dot{x}(0)$ fixed and with $\dot{x}(0)$ at some angle with $T_{x(0)}(M)$ then, if this angle is big we have more normal motion and the constant C will be big.

We can now also give a description of the normal motion $x_{\lambda,1}$ for λ big: $x_{\lambda,1}$ describes a "harmonic ascillation" with slowly variing frequency and amplitude. The frequency being

$$\approx \frac{1}{2\pi}\sqrt{\lambda.\sigma(x_{\lambda,M}(t))}$$

and the amplitude being

$$\approx \frac{2.C}{\lambda.\sqrt{\sigma(x_{\lambda,M}(t))}}$$

Next we describe a more complicated situation. Assume M has co-dimension k in $\mathbb{R}^n$ and assume that there is a smooth orthonormal basis $e_1(x),\dots, e_k(x)$, $x \in M$ for the normal bundle of M such that

$$d^2_{e_i(x)\, e_j(x)} W = \delta_{ij}\sigma_i(x)$$

for certain functions $\sigma_i : M \to \mathbb{R}$. In this case we say that $d^2_N W$, the second order normal derivative of W along M, is diagonalizable. For a solution

$x_\lambda(t) = x_{\lambda,M}(t) + \sum_{i=1}^{k} x_{\lambda,i}(t).e_i$ we define the ith normal energy as

$$E_{\lambda,i}(t) = \tfrac{1}{2}.(\dot{x}_{\lambda,i}(t))^2 + \tfrac{1}{2}.\sigma_i(x_{\lambda,M}(t)).(x_{\lambda,i}(t))^2.$$

Again from [1.] we know that any limiting solution x(t) is the limit of a sequence of solutions $x_{\lambda_k}(t)$ of $*\lambda_k$, $\lambda_k \to \infty$, such that also the normal energies converge, say to $E_i(t)$, and that these limiting solutions and their normal energies satisfy:

$$E_i(t) = C_i(t).(\sigma_i(x(t)))^{\frac{1}{2}},$$

$$\frac{D}{dt}(\dot{x}(t)) = -(\mathrm{grad}(V + \Sigma C_i(t).\sigma_i^{\frac{1}{2}}))(x(t)), \tag{2.}$$

if for some $\bar{t}$, $C_1(\bar{t}) = .. = C_k(\bar{t}) = 0$ then they are zero for all t.

$C_i(t)$ now depends continuously on t. In this paper we prove

Theorem 1. If, in the above situation, $\sigma_i(x) \neq \sigma_j(x)$ for all $x \in M$, $i \neq j$ and if $\pm\sigma_i(x)^{\frac{1}{2}} \pm \sigma_j(x)^{\frac{1}{2}} \pm \sigma_k(x)^{\frac{1}{2}} \neq 0$ for all $x \in M$ and i,j,k,

then the set of all limiting solutions is exactly the set of all solutions of (2) with $C_i(t)$ independent of t.

This theorem will be proved in the sections 2 and 3. If the assumptions on σ_i in the above theorem hold we say that there is no strong resonance (in the normal frequences).

Also if these conditions of no strong resonance or even of $d^2_N W$ being diagonalizable are violated, the "total normal energy $\Sigma E_i(t)$)" of the limiting solution depends continuous on the time t : it is the difference of the total energy and the energy of the motion along M.

The results on limiting solutions discussed so far show that under the following conditions these limiting solutions are just the solutions of *:
- if all σ_i are constant or
- if the "initial velocity" is tangent to the constraining submanifold, i.e., if all C_i are zero.

Even if neither of these conditions are satisfied these limiting solutions are completely described by (2) (with C_i constant) if $d^2_N W$ is diagonalizable and if there are no strong resonances.

We now proceed to show that if $d^2_N(W)$ is not diagonalizable, there is no "deterministic" equation any more for the limiting solutions. In fact we shall analyze the situation where dim (M) = 2, k = 2 (hence n = 4) and where there is an orthonormal basis $\tilde{e}_i(x_1,x_2)$ of normal vectors such that

$$d^2_{\tilde{e}_i,\tilde{e}_j} W = \begin{pmatrix} a(x_1,x_2)+x_1 & x_2 \\ x_2 & a(x_1,x_2)-x_1 \end{pmatrix};$$

x_1,x_2 are coordinates in M and we concentrate on the situation near $(x_1,x_2)=(0,0)$.

Before going into the details of this example, we explain why it may be considered as the simplest example where $d^2_N W$ is not diagonalizable. First we remark that for any basis $\tilde{e}_1,..,\tilde{e}_k$ of normal vectors, which we shall assume to be orthonormal,

$$d^2_{\tilde{e}_i\,\tilde{e}_j} W(x) = (d^2_N(W))(x)$$

is a symmetric $k \times k$ matrix which depends smoothly on $x \in M$. If all the eigenvalues of this matrix are different, we can locally diagonalize. The set of symmetric matrices, having at least one eigenvalue with multiplicity geater than one, has codimension two in the set of all symmetric matrices. The proof of this fact is straight foreward; see also [2.]. So the lowest dimension of M, in which we generically may expect that $d^2_N W$ is not locally diagonalizable, is two. Also M must have at least co-dimension two. So in our example the dimensions are as low as possible. Also our example occures in an "open set of constraining data" in the following sense: if we perturb M,W to M',W' so that M' is near M and $d^2_N W'$ near $d^2_N W$ then there are coordinates x_1',x_2' on W', and an orthonormal basis $\tilde{e}_1',\tilde{e}_2'$ of normal vectors of M' such that

$$d^2_{\tilde{e}_i',\tilde{e}_j'} W' = \begin{pmatrix} a'(x_1',x_2')+x_1' & x_2' \\ x_2' & a'(x_1',x_2')-x_1' \end{pmatrix},$$

where a' is a function near a.

From now on we shall restrict our attention to a small neighbourhood U of the point $(x_1,x_2) = (0,0)$. On $U \setminus (0,0)$ we can diagonalize $d^2_N W$ up to sign: this gives a basis $\pm \bar{e}_1$, $\pm \bar{e}_2$ such that $d^2_{\bar{e}_i \bar{e}_j} W = \delta_{ij}\sigma_i$. From a simple calculation we see that we may take

$\sigma_1 = a + \sqrt{x_1^2 + x_2^2}$ and $\sigma_2 = a - \sqrt{x_1^2 + x_2^2}$. To give an explicit formula for $\bar{e}_i$ we use polar coordinates r, ϕ : $x_1 = r\cos\phi$, $x_2 = r\sin\phi$. Then we may take

$$\bar{e}_1 = \cos\tfrac{\phi}{2}.\tilde{e}_1 + \sin\tfrac{\phi}{2}.\tilde{e}_2$$

$$\bar{e}_2 = -\sin\tfrac{\phi}{2}.\tilde{e}_1 + \cos\tfrac{\phi}{2}.\tilde{e}_2.$$

To see that these are eigenvectors of $d^2_N W$, we write

$$\begin{pmatrix} a+x_1 & x_2 \\ x_2 & a-x_1 \end{pmatrix} = \begin{pmatrix} a & o \\ o & a \end{pmatrix} + r.\begin{pmatrix} \cos\phi & \sin\phi \\ \sin\phi & -\cos\phi \end{pmatrix} =$$

$$\begin{pmatrix} a & o \\ o & -a \end{pmatrix} + r.\begin{pmatrix} \cos^2\frac{\phi}{2} - \sin^2\frac{\phi}{2} & 2.\sin\frac{\phi}{2}.\cos\frac{\phi}{2} \\ 2.\sin\frac{\phi}{2}.\cos\frac{\phi}{2} & \sin^2\frac{\phi}{2} - \cos^2\frac{\phi}{2} \end{pmatrix}.$$

The occurence of $\frac{\phi}{2}$ in the definition of $\bar{e}_i$ means that these vectors are only defined up to sign.

If $x_{\lambda_i}(t)$ is a sequence of solutions of $*\lambda_i$ and $x_{\lambda_i}(t) \in U$ for all $t \in I$, $\lambda_i \to \infty$, then, for i,t so that $x_{\lambda_i}(t) \neq (0,0)$, the normal energies are defined as before, but now using the basis $\pm\bar{e}_1$, $\pm\bar{e}_2$.

<u>Proposition 2</u>. Let $x_{\lambda_i}(t)$ be a sequence as above such that

$0 \in \text{int}(I)$; $x_{\lambda_i}(0) = (0,0)$;

$\dot{x}_{\lambda_i}(0)$ converges to a non-zero vector;

$x_{\lambda_i}(t)$ converges to $x(t)$ for $t \in I$.

Then we may assume, taking if necessarily a subsequence, that $E_{\lambda_i,1}(t)$ and $E_{\lambda_i,2}(t)$ converge for $t \neq 0$ to $E_1(t)$ and $E_2(t)$.

In this case $x(t)$, $E_1(t)$, $E_2(t)$ satisfy:

$$\left.\begin{aligned}
E_1(t) &= \begin{cases} C_1.\sigma_1^{\frac{1}{2}}(x(t)) & \text{for } t<0; \\ C_2.\sigma_1^{\frac{1}{2}}(x(t)) & \text{for } t>0; \end{cases} \\
E_2(t) &= \begin{cases} C_2.\sigma_2^{\frac{1}{2}}(x(t)) & \text{for } t<0; \\ C_1.\sigma_2^{\frac{1}{2}}(x(t)) & \text{for } t>0; \end{cases} \\
\frac{D}{dt}\dot{x}(t) &= \begin{cases} -(\mathrm{grad}(V + C_1.\sigma_1^{\frac{1}{2}} + C_2.\sigma_2^{\frac{1}{2}}))x(t) & \text{for } t<0, \\ -(\mathrm{grad}(V + C_2.\sigma_1^{\frac{1}{2}} + C_1.\sigma_2^{\frac{1}{2}}))x(t) & \text{for } t>0; \end{cases} \\
&\dot{x}(t) \text{ is continuous.}
\end{aligned}\right\} \quad (3)$$

We shall prove this proposition in section 4. Here we continue to derive from this proposition our main theorem. First however we need to give some clarification of the content of the present proposition. For a correct interpretation of the equation for $\frac{D}{dt}\dot{x}(t)$ we would like to have an existence and uniqueness theorem for solutions of

$$\frac{D}{dt}\dot{x}(t) = -(\mathrm{grad}(V+\Sigma))\, x(t),$$

where Σ is a function having in $(0,0)$ a point of non-differentiability like $\sigma_1^{\frac{1}{2}}$ or $\sigma_2^{\frac{1}{2}}$. Such a proposition will be given in section 4 for initial conditions of the form $x(0) = (0,0)$, $\dot{x}(0) \neq 0$; in case $\dot{x}(0) = 0$ there is in general no uniqueness of solutions.

Since $|\ddot{x}_{\lambda,M}(t)|$ is uniformly bounded, $x(t)$ will be a C^1 curve going with positive velocity through $(0,0)$ at time $t = 0$. For such curves we have $\lim_{t \nearrow 0} \pm \bar{e}_1(x(t)) = \lim_{t \searrow 0} \pm \bar{e}_2(x(t))$ and vice-versa; this follows from the above expressions for $\bar{e}_i$. This explains why we may expect that for $t = 0$ the values of E_1 and E_2 interchange.

Next, we want to show from the last proposition that there are also limiting solutions where the normal energies do not completely interchange, i.e., such that instead of (3) we have:

$$\left.\begin{aligned} E_1(t) &= \begin{cases} C_1\sigma_1^{\frac{1}{2}} & \text{for } t<0 \\ C_1'\sigma_1^{\frac{1}{2}} & \text{for } t>0 \end{cases} \\ E_2(t) &= \begin{cases} C_2\sigma_2^{\frac{1}{2}} & \text{for } t<0 \\ C_2'\sigma_2^{\frac{1}{2}} & \text{for } t>0 \end{cases} \\ &\text{with } C_1+C_2=C_1'+C_2' \text{ for any } C_1' \text{ between } C_1 \text{ and } C_2 \end{aligned}\right\} \quad (3'.)$$

Note that $C_1+C_2=C_1'+C_2'$ follows from the continuity of the normal energy.

To prove this we consider a two-parameter family of curves $x_{\lambda_i,\mu}(t)$ such that $x_{\lambda_i,\mu}$ is a solution of $*\lambda_i$, $x_{\lambda_i,0}(0)=(0,0)$, $\dot{x}_{\lambda_i,0}(0)$ tends to a non-zero limit, the initial conditions $x_{\lambda_i,\mu}(-\bar{t})$, $\dot{x}_{\lambda_i,\mu}(-\bar{t})$,

$C_{j,\lambda_i,\mu}(-\bar{t})=C_{j,\lambda_i}(-\bar{t})$ depend continuously on μ and have a limit for $i\to\infty$; $C_{j,\lambda_i,\mu}(-\bar{t})$ determines the limiting energy by

$$E_j(-\bar{t})=C_{j,\lambda_i,\mu}(-\bar{t}).\sigma(x_{\lambda_i,\mu}(-\bar{t}))^{\frac{1}{2}}.$$

This means that there are limiting solutions $x_\mu(t)=\lim\limits_{i\to\infty} x_{\lambda_i,\mu}(t)$. We also assume that, for $\mu\neq 0$, $x_{\lambda_i,\mu}(t)\neq(0,0)$ for $|t|\leq\bar{t}$.

From all the previous work we know that

1e $C_{j,\lambda_i,\mu}(\bar{t})$ is well defined for μ near zero; it is continuous in μ;

2e for $\mu=0$, $\lim\limits_{i\to\infty} C_{j,\lambda_i,o}(\bar{t})=C_{j'}=\lim\limits_{i\to\infty} C_{j',\lambda_i}(-\bar{t})$, where $j'=2$ if $j=1$ and $j'=1$ if $j=2$; for $\mu\neq 0$, $\lim\limits_{i\to\infty} C_{j,\lambda_i,\mu}(\bar{t})=C_j$.

Hence for any C_1' between C_1 and C_2 there is a function $\mu(i)$, with $\lim\limits_{i\to\infty}\mu(i)=0$, such that $\lim\limits_{i\to\infty} C_{1,i,\mu(i)}(\bar{t})=C_1'$. Hence there is a limiting solution as announced.

If in the above reasoning we invert time then we see that for C_1', C_2' fixed and any C_1 between C_1' and C_2' there is a corresponding limiting solution with $C_2 = C_1' + C_2' - C_1$. This proves our main.

Theorem 3. Let $\ddot{x} = -(\text{grad } V)(x)$, $V : \mathbb{R}^4 \to \mathbb{R}$, and $x \in \mathbb{R}^4$, describe a mechanical system. Let $M \subset \mathbb{R}^4$ be a 2-dimensional submanifold and $W : \mathbb{R}^4 \to \mathbb{R}$ a function which is zero on M, positive on $\mathbb{R}^4 - M$ and with positive definite normal derivative $d^2_N(W)$ with eigenvalues $\sigma_1 \geq \sigma_2$. Let $q \in M$ be a point in M where $\sigma_1 = \sigma_2$, but where $d^2_N(W)$ is still "generic" as described before. Then the limiting solutions of

$$\ddot{x} = -\text{ grad }(V + \lambda W)(x)$$

for $\lambda \to \infty$ near q and with non-zero velocity are exactly those curves in M for which:

$$\frac{D}{dt}\dot{x}(t) = -(\text{grad}(V + \Sigma\, C_i(t).\sigma_i^{\frac{1}{2}}))x(t),$$

with $C_i(t) \geq 0$ locally constant if $x(t) \neq q$; for $x(t_0) = q$, $C_1(t)$, $C_2(t)$ may have any discontinuity at $t = t_0$ provided $C_1 + C_2$ remains constant.

Remark 4. The behaviour of limiting solution in a point q as above which "arrive" with zero velocity at q, as well as the behaviour near more complicated non-diagonalizable points, which may be generic for higher dimensional M, is completely unknown.

Acknowledgement. This research started with an attempt to clarify some provoking remarks in [3; § 17, A and § 21.A]. The references [1.] and [4.] were kindly pointed out to me by V. Arnold.

2. The equations of motion in local coordinates.

Let M, W and V be as in the first section. We shall take local coordinates $x_1,\dots, x_m, y_1,\dots, y_s$ in a neighbourhood of a point in M so that $M = \{y_1 = \dots = y_s = 0\}$, so that curves $t \mapsto (x_1,\dots, x_m, t.y_1,\dots, t.y_s)$ are straight lines, and so that $(\frac{\partial}{\partial y_1},\dots,\frac{\partial}{\partial y_s})|_M$ is an orthonormal basis of the normal bundle of M. In these coordinates the kinetic energy, or the Riemannian metric, has the form

$$T = \tfrac{1}{2}\Sigma\dot{\lambda}_i^{\,2} + \tfrac{1}{2}\Sigma g_{\alpha\beta}(x,y)\dot{x}_\alpha\dot{x}_\beta + \Sigma h_{\alpha i}(x,y)\dot{x}_\alpha\dot{y}_i$$

where $h(x,0) = 0$; in the summations, the Greek indices have to be summed over $1,\dots, m$ and the Latin indices over $1,\dots, s$. This summation convention will be used throughout this section. We assume $d^2_N(W)$ to be diagonalizable, i.e., we assume that for a smooth orthonormal basis $e^1(x),\dots, e^s(x)$ of normal vectors, $W = \tfrac{1}{2}\Sigma <y, e^i>^2.\sigma_i(x) + \tfrac{1}{3}\Sigma B_{ijk}(x).<y,e^i>.<y,e^j>.<y,e^k> + O(|y|^4)$; we assume B_{ijk} to be symmetric in i,j,k.

Our Lagrangian L or $L_\lambda = T - V - \lambda W$ determines the equations of motion: $\frac{\partial L}{\partial x_\alpha} = \frac{d}{dt}(\frac{\partial L}{\partial \dot{x}_\alpha})$ and $\frac{\partial L}{\partial y_i} = \frac{d}{dt}(\frac{\partial L}{\partial \dot{y}_i})$. Writing this in local coordinates, using ∂_i, ∂_α instead of $\frac{\partial}{\partial y_i}$ or $\frac{\partial}{\partial x_\alpha}$, we obtain:

$$\frac{\partial L}{\partial \dot{x}_\alpha} = \frac{\partial T}{\partial \dot{x}_\alpha} = \sum_\beta g_{\alpha\beta}.\dot{x}_\beta + \sum_i h_{\alpha i}.\dot{y}_i$$

$$\frac{d}{dt}(\frac{\partial L}{\partial \dot{x}_\alpha})\ \Sigma\,\partial_\gamma g_{\alpha\beta}.\dot{x}_\gamma.\dot{x}_\beta + \Sigma\,\partial_i g_{\alpha\beta}.\dot{y}_i.\dot{x}_\beta + \Sigma g_{\alpha\beta}.\ddot{x}_\beta + \Sigma\,\partial_\gamma h_{\alpha i}.\dot{x}_\gamma.\dot{y}_i + \Sigma\,\partial_j h_{\alpha i}.\dot{y}_j.\dot{y}_i$$

$$+ \Sigma h_{\alpha i}.\ddot{y}_i.$$

$$\frac{\partial L}{\partial x_\alpha} = \tfrac{1}{2}\Sigma\,\partial_\alpha g_{\beta\gamma}.\dot{x}_\beta.\dot{x}_\gamma + \Sigma\,\partial_\alpha h_{\beta i}.\dot{x}_\beta.\dot{y}_i - \partial_\alpha V\ -\lambda.[\Sigma\tfrac{1}{2}\partial_\alpha\sigma_i.<y,e^i>^2$$

$$+\Sigma\sigma_i.<y,e^i>.<y,\partial_\alpha e^i> + \tfrac{1}{3}\Sigma\,\partial_\alpha B_{ijk}.<y,e^i>.<y,e^j>.<y,e^k>$$

$$+\Sigma B_{ijk}.\langle y,e^i\rangle.\langle y,e^j\rangle.\langle y,\partial_\alpha e^k\rangle + O(|y|^4)].$$

$$\frac{\partial L}{\partial \dot y_i} = \frac{\partial T}{\partial \dot y_i} = \dot y_i + \Sigma h_{\alpha i}.\dot x_\alpha.$$

$$\frac{d}{dt}\left(\frac{\partial L}{\partial \dot y_i}\right) = \ddot y_i + \Sigma \partial_\beta h_{\alpha i}.\dot x_\beta.\dot y_i + \Sigma \partial_j h_{\alpha i}.\dot y_j.\dot y_i + \Sigma h_{\alpha i}.\ddot x_\alpha.$$

$$\frac{\partial L}{\partial y_i} = \tfrac{1}{2}\Sigma \partial_i g_{\alpha\beta}.\dot x_\alpha.\dot x_\beta + \Sigma \partial_i h_{\alpha j}.\dot x_\alpha.\dot y_j - \partial_i V - \lambda.[\Sigma \sigma_j.\langle y,e^j\rangle\, e^j_i$$

$$+\Sigma B_{ljk}.\langle y,e^l\rangle.\langle y,e^j\rangle e^k_i + O(|y|^3)],$$

where $e_1{}^k,\ldots,e_s{}^k$ are the components of e^k in the $y_1,\ldots,y_s$ coordinates.

We consider now solutions $\gamma_\lambda(t) = (x_\lambda(t), y_\lambda(t))$ of these equations of motion with $t\in I$, I a compact interval, so that <u>the energy</u> is uniformly bounded and, so that $\cup\, \mathrm{Im}(\gamma_\lambda)$ is contained in a compact part of the domain of definition of our coordinates. In this situation we have the following <u>estimates</u>:

I. $|y_\lambda(t)| = O(\lambda^{-\frac{1}{2}})$ and hence $|h_{\alpha i}(x_\lambda(t), y_\lambda(t))|$, $|\partial_\beta h_{\alpha i}(x_\lambda(t), y_\lambda(t))|$ etc. are all $O(\lambda^{-\frac{1}{2}})$;

II. $|\dot y_\lambda(t)| = O(1)$, i.e., is uniformly bounded for $\lambda\to\infty$;

III. $|\dot x_\lambda(t)| = O(1)$;

IV. $|\ddot x_\lambda(t)| = O(1)$;

V. $|\ddot y_\lambda(t)| = O(\lambda^{\frac{1}{2}})$.

In order to prove IV and V (the others are trivial aryway) we notice that the equations of motion ca be written as

$$\begin{pmatrix} g & h \\ h^T & \mathrm{id}\end{pmatrix}\begin{pmatrix}\ddot x \\ \ddot y\end{pmatrix} = \begin{pmatrix} O(1) \\ O(\lambda^{\frac{1}{2}})\end{pmatrix}$$

with $|h| = O(\lambda^{-\frac{1}{2}})$ and both g and g^{-1} bounded.

We have:

$$\begin{pmatrix} g^{-1} & -g^{-1}h \\ -h^Tg^{-1} & id \end{pmatrix}\begin{pmatrix} g & h \\ h^T & id \end{pmatrix} = Id + O(\lambda^{-1})$$

and hence

$$\begin{pmatrix} \ddot{x} \\ \ddot{y} \end{pmatrix} = (Id + O(\lambda^{-1})).\begin{pmatrix} g^{-1} & -g^{-1}h \\ -h^Tg^{-1} & id \end{pmatrix}\begin{pmatrix} O(1) \\ O(\lambda^{\frac{1}{2}}) \end{pmatrix} = \begin{pmatrix} O(1) \\ O(\lambda^{\frac{1}{2}}) \end{pmatrix}.$$

Next we define the <u>normal</u> <u>energies:</u>

$E_i(x,y,\dot{y}) = \frac{1}{2}.<e^i,\dot{y}>^2 + \frac{1}{2}.\lambda.\sigma_i(x).<e^i,y>^2$. The rate of change $\frac{dE_i}{dt}$ of the ith normal energy along a solution is given by

$$\frac{dE_i}{dt} = <e^i,\dot{y}>.<\dot{e}^i,\dot{y}> + <e^i,\dot{y}>.<e^i,\ddot{y}> +$$
$$+ \tfrac{1}{2}.\lambda.\dot{\sigma}_i.<e^i,y>^2 + \lambda.\sigma_i.<e^i,y>.<\dot{e}^i,y> + \lambda.\sigma_i.<e^i,y><e^i,\dot{y}^i>,$$

where $\dot{e}^i$ and $\dot{\sigma}_i$ denotes the directional derivative of e^i,σ_i in the direction $\dot{x}$. We want to show that $\frac{dE_i}{dt}$ is uniformly bounded for solutions γ_λ, $\lambda \to \infty$, with uniformly bounded energy (and hence satisfying the above estimates I,...,V). The only terms in $\frac{dE_i}{dt}$ which may not be uniformly bounded are

$$<e^i,\dot{y}>.<e^i,\ddot{y}> + \lambda.\sigma_i.<e^i,y>.<e^i,\dot{y}>.$$

From the equations of motion we see that

$$\ddot{y} = -\lambda.\Sigma\sigma_j.<y,e^j>.e^j + O(1);$$

from this we conclude that $\frac{dE_i}{dt} = O(1)$.

In section 4, we will have to use the fact that $\frac{dE_i}{dt}$ is bounded also in case $g_{\alpha\beta}$ is <u>positive</u> <u>semidefinite</u>: this happens if we use polar coordinates on M. In this case the above proof does not work : $\dot{x}$, and hence $\dot{e}^i$, $\dot{\sigma}_i$ etc., may not be bounded. It follows however from a simple analysis of the above prove that in this case we still have:

Lemma 5. If in the above situation $g_{\alpha\beta}$ is positive semi-definite, then, if we restrict to solutions $\gamma_\lambda = (x_\lambda, y_\lambda)$ with

$|\dot{x}_\lambda|$ and $|\sum_\alpha h_{\alpha i}(x_\lambda, y_\lambda).(\ddot{x}_\lambda)_\alpha|$ uniformly bounded, then also $\frac{dE}{dt}$ is uniformly bounded for these solutions.

In the rest of this section we assume that $g_{\alpha\beta}$ is positve definite. We also make the following simplifying assumption : $e^i = (0,..,0,1,0,..,0)$, i.e., $e^i j = \delta^i{}_j$. We now have

$$\frac{dE_i}{dt} = \dot{y}_i.\ddot{y}_i + \tfrac{1}{2}.\lambda.\partial_\alpha\sigma_i.\dot{x}_\alpha.y_i{}^2 + \lambda.\sigma_i.y_i.\dot{y}_i + O(\lambda^{-\frac{1}{2}}).$$

In this we subsitute the value of $\ddot{y}$, obtained from the equations of motion. We obtain:

$$\frac{dE_i}{dt} = -\dot{y}_i.\sum_{\alpha\beta}\partial_\beta h_{\alpha i}\dot{x}_\beta\dot{x}_\alpha \quad \text{(1.)}$$

$$-\dot{y}_i.\Sigma\,\partial_j h_{\alpha i}.\dot{y}_j.\dot{x}_\alpha \quad \text{(2.)}$$

$$-\dot{y}_i.\Sigma\, h_{\alpha i}.\ddot{x}_\alpha \quad \text{(3.)}$$

$$+\tfrac{1}{2}.\dot{y}_i.\Sigma\,\partial_i g_{\alpha\beta}.\dot{x}_\alpha.\dot{x}_\beta \quad \text{(4.)}$$

$$+\dot{y}_i.\Sigma\,\partial_i h_{\alpha j}.\dot{x}_\alpha.\dot{y}_j \quad \text{(5.)}$$

$$-\dot{y}_i.\frac{\partial V}{\partial y_i} \quad \text{(6.)}$$

$$-\lambda.\dot{y}_i.\sigma_i.y_i \quad \text{(7.)}$$

$$-\lambda.\dot{y}_i.\sum_{l.j} B_{ijl}.y_j.y_l \quad \text{(8.)}$$

$$+\tfrac{1}{2}.\lambda.\sum_\alpha(\partial_\alpha\sigma_i).\dot{x}_\alpha.y_i{}^2 \quad \text{(9.)}$$

$$+\lambda.\sigma_i.y_i.\dot{y}_i \quad \text{(10.)}$$

$$+O(\lambda^{-\frac{1}{2}}).$$

We are interested in the integral of this expression for $\frac{dE_i}{dt}$, say from t_1 to t_2, modulo terms which go to zero for $\lambda \to \infty$. To be more precise, we consider a sequence of solutions $\gamma_{\lambda_r} = (x_{\lambda_r}, y_{\lambda_r})$, $\lambda_r \to \infty$, such that each $x_{\lambda_r}(t)$ converges. By taking a subsequence, we may assume that x_{λ_r} and $\dot{x}_{\lambda_r}$ converge uniformly. We want to find the limit of $\int \frac{dE_i}{dt}$ for such a sequence of solutions.

We observe that (7) and (10) cancel and that (1) and (3) go uniformly to zero for $\lambda \to \infty$. As we shall see in the next section, we have for a solution $(x(t), y(t))$:

$$\lambda^{\frac{1}{2}}.y_i(t_0 + \lambda^{-\frac{1}{2}}\bar{t}) = \sqrt{\frac{E_i(t_0).2}{\sigma_i(x(t_0))}}\ (a.\sin((\sigma_i(x(t_0)))^{\frac{1}{2}}.\bar{t})$$

$$+ b.\cos((\sigma_i(x(t_0)))^{\frac{1}{2}}.\bar{t}) + O(\lambda^{-\frac{1}{2}}),\ a^2 + b^2 = 1.$$

For each $A > 0$ fixed, $O(\lambda^{-\frac{1}{2}})$ holds uniformly for all t_0, $|\bar{t}| \leq A$ and λ. There is a corresponding formula for $\dot{y}_i$. The meaning of this formula is the following. For $\lambda \to \infty$, y_i and $\dot{y}_i$ look more and more like harmonic oscillations with slowly variing amplitude and frequency. In fact, if we only consider a finite number of oscillations, the differences between y_i or $\dot{y}_i$ and the corresponding harmonic oscillators are of the order "amplitude times $\lambda^{-\frac{1}{2}}$".

This means for example that the contribution of (6) ($= -\dot{y}.\frac{\partial V}{\partial y_i}$) in $\int \frac{dE_i}{dt}$ goes to zero: $\frac{\partial V}{\partial y_i}$ goes uniformly to a continuous limit and $\dot{y}_i$ is a rapid oscillation. For the same reason the contribution of (4) goes to zero. In order ot deal with terms containing products of the form $\dot{y}_i\dot{y}_j$ etc., we observe that the product of two harmonic oscillations, with frequencies $\nu_1 + \nu_2$ is the same as the sum of two harmonic oscillations with frequencies $\nu_1 + \nu_2$ and $\nu_1 - \nu_2$. For $\lambda \to \infty$ these oscillations will become rapid provided $\nu_1 + \nu_2 \neq 0$ and $\nu_1 - \nu_2 \neq 0$. In this way we see that if all normal frequencies are different, i.e., if we have $\sigma_i(x) \neq \sigma_j(x)$ for all $x \in M$, $i \neq j$, the contributions in $\int \frac{dE_i}{dt}$ of (2.) (except for $-\dot{y}_i \sum_\alpha \partial_i h_{\alpha i}.\dot{y}_i.\dot{x}_\alpha$), (5.) (except for $+\dot{y}_i \sum_\alpha \partial_i h_{\alpha i}.\dot{x}_\alpha.\dot{y}_i$), (7.) and (10.) tend to zero; the above exceptions cancel. Applying the same method to a product of three oscillations (in (8.)) one has to assume that the functions

$$\pm\,\sigma_i^{\frac{1}{2}}(x)\ \pm\,\sigma_j^{\frac{1}{2}}(x)\ \pm\,\sigma_k^{\frac{1}{2}}(x)$$

are nowhere zero. So finally we are left with the term (9). Modulo terms of order $O(\lambda^{-\frac{1}{2}})$ the integral of this term equals the integral of $\dfrac{(\Sigma\,\partial_\alpha\sigma_i.\dot{x}_\alpha).E_i}{2.\sigma_i}$.

Observing that $\Sigma\,\partial_\alpha\sigma_i.\dot{x}_\alpha = \dfrac{d\sigma_i}{dt}$, we see that

$$\int\{\frac{dE_i}{dt} - \frac{d\sigma_i}{dt}\cdot\frac{E_i}{2.\sigma_i}\} \to 0$$

for $\lambda\to\infty$. Since σ_i converges uniformly to a positive continuous function, we also have

$$\int\sigma_i^{-\frac{1}{2}}.\frac{dE_i}{dt} - \tfrac{1}{2}.E_i.\sigma_i^{-\frac{3}{2}}.\frac{d\sigma}{dt} =$$

$$\int\frac{d}{dt}(\sigma_i^{-\frac{1}{2}}.E_i) \to 0$$

for $\lambda\to\infty$. This clearly implies that in the limit, $E_i.\sigma_i^{-\frac{1}{2}}$ is constant. This proves theorem 1.

3. The normal vibrations.

We consider again solutions $\gamma(t) = (x(t),y(t))$ of $*\lambda$ as in section 2. Without further mentioning we assume, as before, that our solutions belong to a converging sequence γ_{λ_r}. We are especially interested in the behaviour of $y(t)$ in a neighbourhood of some t_0. Consider the following transformation:

$$\bar{t} = \lambda^{\frac{1}{2}}.(t-t_0) \quad ; \quad t = t_0 + \lambda^{-\frac{1}{2}}\,\bar{t}\,;$$

$$\bar{y}_i = \lambda^{\frac{1}{2}}.y_i \quad ; \quad y_i = \lambda^{-\frac{1}{2}}\,\bar{y}_i\,;$$

so $\dfrac{d}{dt} = \lambda^{\frac{1}{2}}\,\dfrac{d}{d\bar{t}}$.

We now apply this transformation to the part "$\dfrac{d}{dt}(\dfrac{\partial L}{\partial\dot{y}_i}) = \dfrac{\partial L}{\partial y_i}$" of the equation of motion in the following way: the differentiations of y_i (but not of x_α) with respect to t will be transformed to differentiations with respect to $\overline{t}$; differentiations with respect ot t are denoted by $\dot{}$, differentiations with respect to $\overline{t}$ by $'$, so $\ddot{y}_i = \overline{y}_i''.\lambda^{-\frac{1}{2}}$. Also ∂_j will not be transformed : it continues to denote $\frac{\partial}{\partial y_j}$. In this way we obtain:

$$\lambda^{\frac{1}{2}}.\overline{y}_i'' + \Sigma\, \partial_\beta h_{\alpha i}.\dot{x}_\beta.\dot{x}_\alpha + \Sigma\, \partial_j h_{\alpha i}.\overline{y}_j'.\dot{x}_\alpha + \Sigma\, h_{\alpha i}\ddot{x}_\alpha =$$

$$\tfrac{1}{2}\Sigma\, \partial_i g_{\alpha\beta}.\dot{x}_\alpha.\dot{x}_\beta + \Sigma\, \partial_i h_{\alpha j}.\dot{x}_\alpha.\overline{y}_j' - \partial_i V - \lambda^{\frac{1}{2}}.\sigma_i.\overline{y}_i - \Sigma\, B_{l j i}.\overline{y}_l.\overline{y}_j + O(\lambda^{-\frac{1}{2}}).$$

This implies that:

$$\overline{y}_i''(\overline{t}) = -\sigma_i(x(t_0 + \lambda^{-\frac{1}{2}}\overline{t})).\overline{y}_i(\overline{t}) + O(\lambda^{-\frac{1}{2}}).$$

If we restrict in this case $\overline{t}$ to a fixed interval, say $|\overline{t}| \leq A$, then we even have:

$$\overline{y}_i''(\overline{t}) = -\sigma_i(x(t_0)).\overline{y}_i(\overline{t}) + O(\lambda^{-\frac{1}{2}}).$$

This means that the solutions are of the form

$$\overline{y}_i(\overline{t}) = \sqrt{\frac{E_i(t_0).2}{\sigma_i(x(t_0))}}\left\{ a.\sin(\sigma_i(x(t_0))^{\frac{1}{2}}.\overline{t}) + \right.$$

$$\left. b.\cos(\sigma_i(x(t_0))^{\frac{1}{2}}.\overline{t}) \right\} + O(\lambda^{-\frac{1}{2}}),$$

with $a^2 + b^2 = 1$. A similar fomula holds for $\overline{y}_i'$. This shows that $\overline{y}_i(t)$ is an almost harmonic oscillation as stated in section 2.

4. The equations of motion in polar coordinates.

In this section we want to show an existence and uniqueness theorem for solutions of

$$\ddot{x} = -(\text{grad } V)(x) \quad \text{...}(1.)$$

in a situation where V is no longer differentiable (but still continuous). We consider the following situation:

$x \in \mathbb{R}^2$;

V is smooth except in (0,0);
(-grad V), expressed in polar coordinates, is smooth, i.e., there is a smooth vectorfield $\tilde{X}$ on $\mathbb{R}^2$ such that $\Phi_*(\tilde{X}(r,\phi)) = (-\text{grad } V)(r \cos \phi, r \text{ sing } \phi)$, where

$\Phi(r,\phi) = (r \cos \phi, r \sin \phi)$.

We want to show:
Proposition 6. Under the above assumptions there is for each $0 \neq v \in T_0(\mathbb{R}^2)$ a unique solution γ of (1.) such that $\gamma(0) = (0,0)$ and $\dot{\gamma}(0) = v$. This solution is smooth.
Proof. Writing (1) in polar coordinates, we obtain:

$$\left.\begin{aligned} &\ddot{r} \cos \phi - 2 \dot{r}\dot{\phi} \sin \phi - r\ddot{\phi} \sin \phi - r\dot{\phi}^2 \cos \phi = \\ &\qquad - \cos \phi . \tilde{X}_r + r . \sin \phi . \tilde{X}_\phi \\ &\ddot{r} \sin \phi + 2 \dot{r}\dot{\phi} \cos \phi + r\ddot{\phi} \cos \phi - r\dot{\phi}^2 \sin \phi = \\ &\qquad - \sin \phi . \tilde{X}_r - r \cos \phi . \tilde{X}_\phi . \end{aligned}\right\} \quad (1')$$

This is equivalent with:

$$\left.\begin{aligned} &\ddot{r} - r\dot{\phi}^2 = -\tilde{X}_r \\ &r\ddot{\phi} + 2\dot{r}\dot{\phi} = -r\tilde{X}_\phi \end{aligned}\right\} \quad (1'')$$

Clearly any solution of (1), transformed to polar coordinates, will be a solution of (1'), at least outside (0,0), or $\{r = 0\}$. A solution $\gamma(t)$ of (1) which goes through the origin, say $\gamma(0) = (0,0)$ with positive velocity gives in polar coordinates a well defined curve for $t \neq 0$. Since grad V is bounded, $\ddot{\gamma}$ is bounded and hence the curve $r(t)$, $\phi(t)$, such that $\gamma(t) = (r(t) \cos \phi(t), r(t) \sin \phi(t))$, has a limit for $t \to 0$. Also $\dot{r}(t)$ and $\dot{\phi}(t)$ have a limit in that case. Hence we have to analyse solutions of (1'), or (1"), defined for $t > 0$, such that for $t \to 0$, $r(t) \to 0$, $\phi(t) \to \bar{\phi}$, $\dot{r}(t) \to \bar{\dot{r}}$ for given values of $\bar{\phi}$ and $\bar{\dot{r}} > 0$.

For this we use 1", transforming it to a system of first order differential equations in the independent variables $r,\phi,\dot{r},r\dot{\phi}$:

$$\frac{d}{dt}(r) = \dot{r}$$

$$\frac{d}{dt}(\dot{r}) = r\dot{\phi}^2 - \tilde{X}_r = \frac{(r\dot{\phi})^2}{r} - \tilde{X}_r$$

$$\frac{d}{dt}(\phi) = \dot{\phi} = \frac{(r\dot{\phi})}{r}$$

$$\frac{d}{dt}(r\dot{\phi}) = -\dot{r}\dot{\phi} - r\tilde{X}_\phi = -\frac{(r\dot{\phi}).\dot{r}}{r} - r\tilde{X}_\phi .$$

If we write this system of differential equations as a vectorfield and multiply it with r, we have:

$$r.\dot{r}\frac{\partial}{\partial r} + ((r\dot{\phi})^2 - r.\tilde{X}_r)\frac{\partial}{\partial \dot{r}} + (r\dot{\phi})\frac{\partial}{\partial \phi} + (-\dot{r}.(r\dot{\phi}) - r^2\tilde{X}_\phi)\frac{\partial}{\partial(r\dot{\phi})}$$

We have now to analyse integral curves of this vectorfield which tend to

$$r = 0,\ \phi = \bar{\phi},\ \dot{r} = \dot{\bar{r}} > 0,\ (r\dot{\phi}) = 0 \text{ for } t \to -\infty .$$

We note that $\{r = 0,\ (r\dot{\phi}) = 0\}$ consists of singularities of this vectorfield. In points of this surface of singularities where $\dot{r} \neq 0$, the above vectorfield is normally hyperbolic [5.] with one expanding and one contracting eigevalue. Hence each point $r = 0$, $r\dot{\phi} = 0$, ϕ, $\dot{r} > 0$ has a one-dimensional stable manifold. This stable manifold represents the required solution of (1".) and hence of (1.). The differentiability of this solution follows from the differentiability of the above vectorfield and the stable manifold theorem [5.].

Now we continue the proof of proposition 2. In the situation, described in the assumptions of this proposition, we take coordinates as in section 2 so that $\frac{\partial}{\partial y_i}\big|_{y=0} = \tilde{e}_i$, see section 1. Note that the second normal derivative of W can now note be diagonalized. We change the coordinates x_1, x_2 in M to r,ϕ:

$x_1 = r\cos\phi$, $x_2 = r\sin\phi$. As a consequence we obtain a situation, where $g_{\alpha\beta}$ is only positive semidefinite, but where the 2nd normal derivative can be locally diagnonalized near any point of $\{r = 0\}$. Observe that, if $\hat{e}_1(r,\phi)$, $\hat{e}_2(r,\phi)$ is a smooth basis which puts $d^2_N W$ in diagonal form then, if for $r < 0$ $\hat{e}_1$ corresponds to $\bar{e}_1$ (see section 1) then for $r > 0$, $\hat{e}_1$ corresponds to $\pm e_2$. Hence, in order to show that for a limiting solution the normal energies interchange it is enough ot show that they are continuous with respect ot the basis $\hat{e}_1$, $\hat{e}_2$ in the present context of polar coordinates.

This problem was considered in section 2, lemma 5. We have to show that the solutions which tend here to a limit satisfy (in polar coordinates):

- $|\dot{r}|$ and $|\dot{\phi}|$ are uniformly bounded;
- $|h_{ri}(r,\phi,y)\ddot{r} + h_{\phi i}(r,\phi,y)\ddot{\phi}|$ is uniformly bounded for $i = 1,2$.

The first condition is simple: it follows from the fact that we only consider solutions $\gamma_\lambda = (x_\lambda, y_\lambda)$ with $x_\lambda(0) = (0,0)$ and $\dot{x}_\lambda(0)$ converging to a non-zero limit, and from the fact that $|\ddot{x}|$ is uniformly bounded. To show that the second condition holds, we observe that, since $|\ddot{x}|$ is uniformly bounded, also $|\ddot{r}|$ and $|r.\ddot{\phi}|$ are uniformly bounded. On the other hand, since $\frac{\partial}{\partial\phi}|_{r=0}$ corresponds to the zero vector in the x,y coordinates,

$$h_{\phi i}(0,\phi,y) = 0 \text{ and } h_{\phi i}(r,\phi,y) = O(|r|).$$

From this second condition follows.

REFERENCES

1. H. Rubin, P. Ungar, Motion under a strong constraining force, Comm. Pure and Appl. Math. 10 (1957), 65-87.
2. V.I. Arnold, Lectures on bifurcations in versal families, Russian Math. Surveys 27 (1972),(5), 54-123.
3. V.I. Arnold, Les méthodes mathématiques de la mécanique classique, Editions MIR, Moscow, 1976.
4. D.G. Ebin, The motion of slightly compressible fluids viewed as a motion with strong constraining force, Annals of Math. 105 (1977), 141-200.
5. M.W. Hirsch, C.C. Pugh, M. Shub, Invariant manifolds, Lecture Notes in Math. Springer-Verlag 583 (1977).

Mathematisch Instituut
Rijksuniversiteit
Groningen
Nederland

I.M.P.A.
Rua L. de Camões 68
Rio de Janeiro
Brasil

Conjugacies of Topologically Hyperbolic Fixed Points: a Necessary Condition on Foliations.

Russell B. Walker

C. Coleman [1] conjectured that if the flow near a fixed point topologically mimics the flow near a differentiably hyperbolic fixed point, then the two are locally orbit-conjugate. Wilson [6] used a uniformity condition on the conjugating homeomorphisms to give an affirmative response in the cases that the stable or unstable dimension is one. D. Neumann [3] constructed a four-dimensional counterexample, the key to which is a leaf no conjugating homeomorphism can "unhook". More recently Neumann [4] has found uncountably many examples in $\mathbb{R}^4$ which are pairwise non-conjugate.

In this note, necessary conditions for the orbit-conjugacy of two "topologically hyperbolic" fixed points in $\mathbb{R}^{m+n}$ are stated in terms of the relative positions of key foliations (Theorem 1). Next a useful corollary is developed which describes how the conjugating homeomorphism must preserve the number of "fluctuations" of given arcs. This corollary is then applied to certain carefully chosen arcs to give a new proof of Neumann's original counterexample and to distinguish countably many "multiple" Neumann examples (Theorem 2).

Counterexamples in dimensions ≥ 5 are then constructed (Theorem 3). I would like to thank the referee who showed how my five-dimensional construction could be repeated in higher dimensions.

Notations and Definitions

φ and ψ are differentiable flows on $\mathbb{R}^{m+n}$ with $B_{m,n} = D^m \times D^n$, an isolating block of the origin. Such flows will be called $B_{m,n}$ - flows if the following are satisfied: 1) $b^+ = \partial D^m \times D^n$ is the ingressing set, 2) $b^- = D^m \times \partial D^n$, the egressing set, 3) $\tau = \partial D^m \times \partial D^n$, the tangentcy set, 4) $a^+ = \partial D^m \times \{0\} \subset b^+$, the points of b^+ which do not exit $B_{m,n}$ in positive time, and 5) $a^- = \{0\} \times \partial D^n \subset b^-$ which do not exit in negative time.

Coleman's conjecture is that all $B_{m,n}$ - flows are orbit-conjugate to $\varphi_{m,n}$, the standard example on $\mathbb{R}^{m+n}$:

$$\dot{x} = -x \qquad (x \in \mathbb{R}^m)$$

$$\dot{y} = y \qquad (y \in \mathbb{R}^n)$$

$b^+\backslash a^+ = \partial D^m \times (D^n\backslash 0) \approx S^{m-1} \times S^{n-1} \times (0,1]$ has (μ,ν,r)- coordinates while $b^-\backslash a^- = (D^m\backslash 0) \times \partial D^n \approx S^{m-1} \times (0,1] \times S^{n-1}$ has (μ,r,ν)- coordinates. φ and ψ induce Poincaré maps $\Phi,\Psi : b^+\backslash a^+ \to b^-\backslash a^-$. The standard example induces $\Phi_{m,n} : (\mu,\nu,r) \to (\mu,r,\nu)$.

Foliation Notations

$\{\tau_r^+\} = \{r = \text{const}\}$ foliates $b^+\backslash a^+$ where each $\tau_r^+ \approx S^{m-1} \times S^{n-1} \times \{r\}$, $(r > 0)$. Similarly, $\{\tau_r^- \approx S^{m-1} \times \{r\} \times S^{n-1} : r > 0\}$ foliates $b^-\backslash a^-$. $\Phi_{m,n}(\tau_r^+) = \tau_r^-$.

Let $L^+(\mu)$ be the disk, $\{\mu\} \times D^n$, in b^+; $L^+(\nu)$, the annulus, $\partial D^m \times \{\nu\} \times [0,1]$ in b^+; $L^-(\mu)$, the annulus, $\{\mu\} \times [0,1] \times \partial D^n$; and $L^-(\nu)$, the disk, $D^m \times \{\nu\}$.

$\mathcal{A}_r^+ = \{\ell_r^+(\mu) = \tau_r^+ \cap L^+(\mu)\}$ is a product foliation of τ_r^+. Similarly, $\mathcal{B}_r^+ = \{\ell_r^+(\nu) = \tau_r^+ \cap L^+(\nu)\}$ foliates τ_r^+ while $\mathcal{A}_r^- = \{\ell_r^-(\mu) = \tau_r^- \cap L^-(\mu)\}$ and $\mathcal{B}_r^- = \{\ell_r^-(\nu) = \tau_r^- \cap L^-(\nu)\}$ foliates τ_r^-. $\Phi_{m,n}(\mathcal{A}_r^+) = \mathcal{A}_r^-$ and $\Phi_{m,n}(\mathcal{B}_r^+) = \mathcal{B}_r^-$.

Let d_μ denote the S^{m-1}-metric inherited from $\mathbb{R}^{m+n}$. d_μ induces a pseudo-metric, d_μ, on b^+ and on b^-: $d_\mu((\mu,\nu,r),(\mu',\nu',r')) = d_\mu(\mu,\mu')$. Similarly for d_ν. Each $L^+(\mu_0)$ has a d_μ-neighborhood, $N^\mu_\varepsilon(L^+(\mu_0)) = \{L^+(\mu) : d_\mu(\mu,\mu_0) < \varepsilon\}$. Similarly in other cases.

If $h : B_{m,n} \to B_{m,n}$ is the conjugating homeomorphism, $h^+ = h|b^+$, $h^- = h|b^-$, $h^+_0 = h|a^+$, and $h^-_0 = h|a^-$. h^+_0(resp. h^-_0) is a function of $\mu \in a^+$ (resp. $\nu \in a^-$).

In the $B_{2,2}$-case, T^-_r and T^+_r are 2-torri while $\ell^+_r(\mu)$, $\ell^+_r(\nu)$, etc. are circles. The following depicts these foliations and their images under an r-preserving Poincaré map.

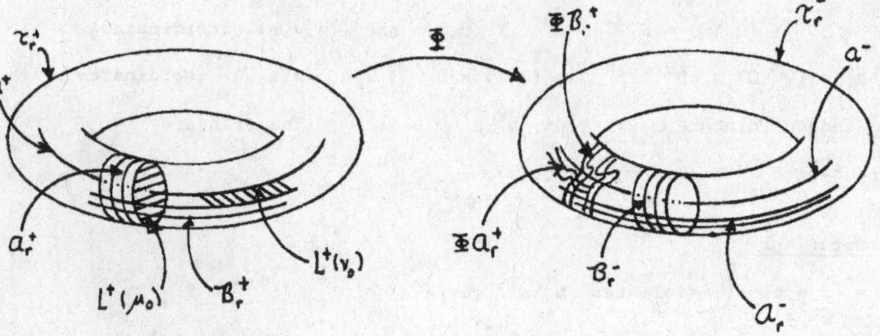

Figure 1.

<u>The Main Result</u>

<u>Theorem 1</u>: φ and ψ are $B_{m,n}$-flows. $h : B_{m,n} \to B_{m,n}$ carries orbits of φ to orbits of ψ. Then $\forall \varepsilon > 0$, $\exists r_\varepsilon > 0$ such that $\forall r < r_\varepsilon$,

1) $h^+\ell_r^+(\mu) \subset N_\varepsilon^\mu(L^+(h_0^+\mu))$,

2) $h^+\Phi^{-1}\ell_r^-(\nu) \subset \Psi^{-1}N_\varepsilon^\nu(L^-(h_0^-\nu))$,

3) $h^-\ell_r^-(\nu) \subset N_\varepsilon^\nu(L^-(h_0^-\nu)$, and

4) $h^-\Phi\ell_r^+(\mu) \subset \Psi N_\varepsilon^\mu(L^+(h_0^+\mu))$.

Remarks: 1) This is basically a simple concept requiring much notational wrappings. These restate that h^+ (resp. h^-) is continuous in the d_μ (resp. d_ν) pseudo-metric, and that these foliational properties can be uniformly pushed (or pulled) through the blocks via the Poincaré maps.

2) The main purposes in expressing these in foliational terms are to better facilitate geometric visualization of the nature of h and to establish a framework for the pursuit of sufficient conditions. It is the author's intent to use the generic theorems of [5] in this effort.

Proof: So $\Psi h^+ = h^-\Phi$.

(Showing 1): Assume not. Then $\exists\varepsilon_0 > 0$ such that for arbitrarily small $r > 0$, $\exists x \in \ell_r^+(\mu)$ such that $d_\mu(h^+(x),L^+(h_0^+\mu)) > \varepsilon_0$. (Here d_μ is the induced Hausdorff pseudo-metric). Or there exist $\{x_i \in \ell_{r_i}^+(\mu_i)\} \to \mu_0 \in a_+$ such that $d_\mu(h^+(x_i),L^+(h_0^+(\mu_i)) > \varepsilon_0$. Since $\mu_i \to \mu_0$, for i sufficiently large, $d_\mu(h^+(x_i),L^+(h_0^+(\mu_0)) > \varepsilon_0/2$. But $h_0^+\mu_0 \in L^+(h_0^+\mu_0)$, so for large i ,

$$d_\mu(h^+(x_i),h_0^+(\mu_0)) > \varepsilon_0/2$$

contradicting that h^+ is continuous.

(Showing 2,3, and 4): 3) follows similarly as 1) from the continuity of h^- and that $h_0^-\nu \in L^-(h_0^-\nu)$. Since $h^+\Phi^{-1} = \Psi^{-1}h^-$, 2) follows by taking Ψ^{-1} of both sides of 3) . Lastly, 4) follows similarly as 2) from 1) .

Fluctuations

Conditions 3) and 4) of Theorem 1 are now used in the $B_{2,2}$ - case to show that h^- "preserves" the "ν - fluctuations" of arcs in τ_r^- .

Consider $\tilde{\nu} \circ \Gamma_r : I \to \mathbb{R}$ where $\Gamma_r : I \to \tau_r^-$ is a closed arc and $\tilde{\nu} : b^- \to \mathbb{R}$, a lift of the circular coordinate function $\nu : b^- \to S^1$. For given $0 < \Delta < \pi$, $FL_\nu(\Gamma_r,\Delta)$ or the # of ν - fluctuations, and $\{s_0,s_1,\ldots\} \subset I$, a sequence of fluctuation points, are recursively defined:

$$s_0 = 0$$

$$s_1 = \min\{s \in I : |\tilde{\nu}\Gamma_r(s) - \tilde{\nu}\Gamma_r(s_0)| = \Delta\} \text{ if it exists; otherwise, } FL_\nu(\Gamma_r,\Delta) = 0.$$

If s_1 exists then define

$$\sigma = \operatorname{sign}(\tilde{\nu}\Gamma_r(s_1) - \tilde{\nu}\Gamma_r(s_0))$$

for $i > 1$, if s_{i-1} exists then define

$$s_i = \min\{s > s_{i-1} : \exists s_{i-1} \le \bar{s} \le s \text{ such that } \tilde{\nu}\Gamma_r(s) - \tilde{\nu}\Gamma_r(\bar{s}) = (-1)^{i-1}\sigma\Delta\}$$

if it exists; otherwise $FL_\nu(\Gamma_r,\Delta) = i - 1$. It is left to the reader to show that $FL_\nu(\Gamma_r,\Delta)$ is well-defined.

If $\tilde{\nu}\Gamma_r$ is monotonically increasing, s_1 exists but no s_2 exists because $\tilde{\nu}\Gamma_r$ cannot decrease. Thus $FL_\nu(\Gamma_r,\Delta) = 1$. Furthermore, if $\Gamma_r \cap \ell_r^-(\nu) \neq 0$, $\forall\nu$, $FL_\nu(\Gamma_r,\Delta) \ge 1$.

Now assume $h_0^- = Id_{a^-}$. If $\varepsilon < \frac{\Delta}{4}$, and r is small enough, then by Theorem 1(3) ,

$$|\tilde{\nu}h^-\Gamma_r(s) - \tilde{\nu}h^-\Gamma_r(\bar{s})| > \frac{\Delta}{2}$$

at each of the $i > 1$ stages in the above definition; no Δ - sized fluctuations may be compressed smaller than $\frac{\Delta}{2}$. Thus $FL_\nu(\Gamma_r,\Delta) \le FL_\nu(h^-\Gamma_r,\Delta/2)$.

When $h_0^- \neq Id_{a^-}$, intervals of length Δ in a^- can be compressed coniderably under h_0^-. But because a^- is compact, there is a nonzero lower ound on "compression",

$$C(\Delta) = \min(\{d_\nu(h_0^-(\nu), h_0^-(\nu+\Delta)) : \nu \in a^-\}, \Delta) .$$

f $\varepsilon < \frac{C(\Delta)}{2}$, $FL_\nu(\Gamma_r, \Delta) \le FL_\nu(h^-\Gamma_r, C(\Delta)/2)$. Formally,

Corollary 1: For all $0 < \Delta < \pi$, $\exists r_\Delta > 0$ such that $\forall\, 0 < r < r_\Delta$, and losed arcs $\Gamma_r : I \to \tau_r^-$,

$$FL_\nu(\Gamma_r, \Delta) \le FL_\nu(h^-\Gamma_r, C(\Delta)/2) .$$

imilarly for $\Gamma_r : I \to \tau_r^+$ and $h^+\Gamma_r$.

he Construction of Multiple Neumann Examples.

D. Neumann's examples [3], is of an r-preserving $B_{2,2}$-flow, φ, which specially "hooks" each $\ell_r^+(\mu_0)$. He shows that $\Phi\ell_r^+(\mu_0)$ meets each leaf of ℓ_r^- and that no conjugating homeomorphism can unhook $\Phi\ell_r^+(\mu_0)$ and thereby isrupt these intersections. Of course, no leaf of $\Phi_{2,2}\mathcal{A}_r^+$ meets every leaf of $\mathcal{A}_r^-$.

In the proof which follows, the assumption that $\varphi \sim \varphi_{2,2}$ leads to the xistence of an arc in τ_r^- having one ν-fluctuation which has a preimage under h^- having three ν-fluctuations, a contradiction.

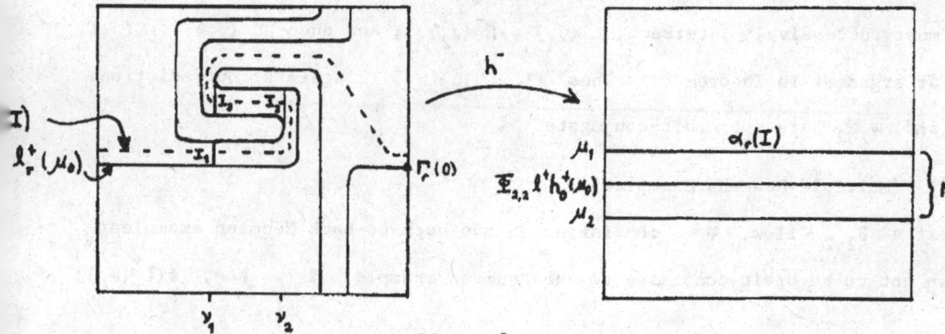

Figure 2.

Theorem 2 (Multiple Neumann Examples)

There exist countably many pairwise non-orbit-conjugate $B_{2,2}$ - flows.

Proof: (Part 1)

Let φ be an r - preserving $B_{2,2}$ - flow such that for small r , $\Phi\ell_r^+(\mu_0)$ is knotted as shown (Figure 2). An isotopy of $b^-\backslash a^-$ which simultaneously knots each $\Phi\ell_r^+(\mu_0)$ (r small) is easy to construct. Some involved differential topology is required to show that such an isotopy can be extended to all of $B_{2,2}$ [7] .

Assume φ and $\varphi_{2,2}$ are orbit-conjugate via h . So $\Phi_{2,2}h^+ = h^-\Phi$. Theorem 1(4) implies $\forall\varepsilon > 0$ and r sufficiently small, $h^-\Phi\ell_r^+(\mu_0) \subset \Phi_{2,2}N_\varepsilon^\mu L^+(h_0^+\mu) = N_\varepsilon^\mu L^-(h_0^+\mu) = N$. Assuming $\varepsilon < \pi$, $\partial(N \cap \tau_r^-) = \ell_r^-(\mu_1) \cup \ell_r^-(\mu_2)$. Let $\alpha_r : I \to \tau_r^-$ be such that $\alpha_r(I) = \ell_r^-(\mu_1)$ and $\widetilde{\nu}\alpha_r$ is monotonically increasing. Thus $\forall\ 0 < \Delta < \pi$, $FL_\nu(\alpha_r,\Delta) = 1$. Call $\Gamma_r = (h^-)^{-1}\alpha_r$. Since $\alpha_r(I) \cap h^-\Phi\ell_r^+(\mu_0) = \phi$, $\Gamma_r(I) \cap \Phi\ell_r^+(\mu_0) = \phi$.

First assume h^- is r - preserving . Simple toral know arguments imply that as $t \in I$ increases, $\Gamma_r(t)$ successively meets I_1, I_2 , and I_3 as shown. Also assume $\Gamma_r(0)$ is as shown. Assume I_1 and $I_3 \subset L^-(\nu_1)$ and $I_2 \subset L^-(\nu_2)$, $(\nu_1 \neq \nu_2)$. Then if $\Delta = |\nu_1 - \nu_2|$, $FL_\nu(\Gamma_r,\Delta) \geq 3$. By Corollary 1 , for r sufficiently small, $FL_\nu(\alpha_r, C(\Delta)/2) \geq 3$, a contradiction.

When h^- is not r - preserving, because Γ_r cannot intersect $\Phi L^+(\mu_0)$, $\Gamma_r(t)$ must successively intersect $L^-(\nu_1)$, $L^-(\nu_2)$, and then $L^-(\nu_1)$. (c.f. a similar argument in Theorem 3). Thus $FL_\nu(\Gamma_r,\Delta) \geq 3$, again a contradiction. So φ and $\varphi_{2,2}$ are not orbit-conjugate.

(Part 2) (Multiple Neumann Examples).

Next a $B_{2,2}$ - flow, φ , consisting of two back-to-back Neumann examples is shown not to be orbit-conjugate to the Neumann example, ψ . Now, $\Phi(\ell_r^+(\mu_0))$

s doubly hooked while $\Psi(\ell_r^+(\mu_1))$ is singly hooked. ν_1 and ν_2 are as they were in Part 1. Assume that $|\bar\nu_1 - \bar\nu_2| = |\bar\nu_3 - \bar\nu_4| = \Delta = |\nu_1 - \nu_2|$ as shown below:

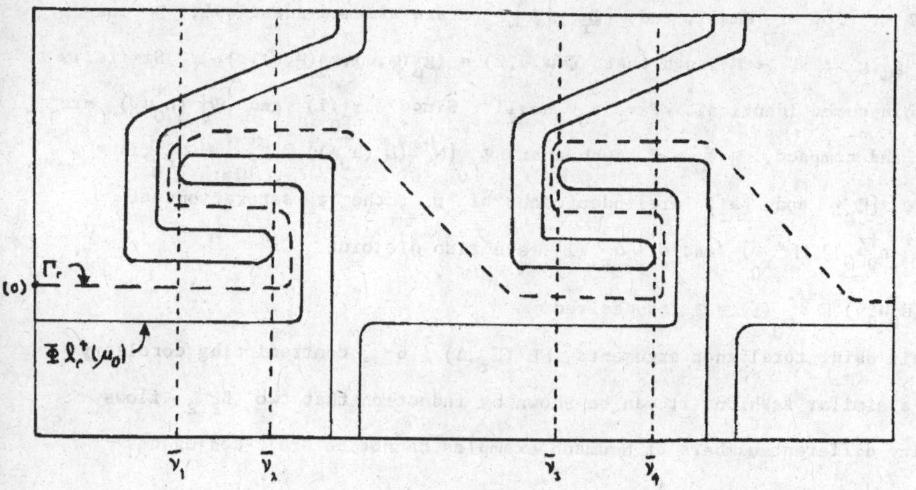

Figure 3.

Assume φ is orbit-conjugate to ψ via h . So $h^-\Phi = \Psi h^+$. In general, $h_0^+(\mu_0)$ is not near μ_1 . It cannot be argued that for small ε , $\forall r$ small enough that $h^-\Phi(\ell_r^+(\mu_0)) \subset \Psi N_\varepsilon^\mu(\ell_r^+\mu_1)$.

For all r , there exist closed arcs $\bar\alpha_r : I \to \tau_r^-$ such that $\bar\alpha_r(I) \cap \Psi(\ell_r^+(\mu_1)) = \emptyset$ and $FL_\nu(\bar\alpha_r,\Delta) = 3$ (e.g. Γ_r of figure 2). Do there exist similar arcs $\alpha_r(\mu)$ for each $\Psi(\ell_r^+(\mu))$? The Neumann flow, ψ , is assumed regular in the following sense: $\forall\mu$, $\Psi\ell_r^+(\mu)$ ν - fluctuates less than $\Psi\ell_r^+(\mu_1)$ (both considered as closed arcs). So the other leaves of $\Psi\mathcal{O}_r^+$ are "hooked" less than $\Psi\ell_r^+(\mu_1)$. In particular, each $\Psi\ell_r^+(h_0^+\mu_0)$ is hooked less. Thus $\forall r$ small, $\exists\ \alpha_r : I \to \tau_r^-$ such that $\alpha_r(I) \cap \Psi\ell_r^+(h_0^+\mu_0) = \emptyset$ and $FL_\nu(\alpha_r,\Delta) \le 3$ because such arcs α_r which miss $\Psi\ell_r^+(h_0^+\mu_0)$ need not

ν - fluctuate more often than $\tilde{\alpha}_r$. Let $\Gamma_r = (h^-)^{-1}\alpha_r$.

Next, it is shown that for r small enough, $h^- \Phi \mathcal{L}_r^+(\mu_0) \cap \alpha_r(I) = \emptyset$. Theorem 1(4) implies it suffices to produce $\varepsilon > 0$ such that $\Psi N_\varepsilon^\mu(L^+(h_0^+\mu_0)) \cap \alpha_r(I) = \emptyset$. For r small, the $\{\Psi_r = \Psi | \tau_r^-\}$ are assumed identical; $\exists$ continuous $g_\mu, g_\nu : \mathbb{R}^2 \to \mathbb{R}$ such that $\Psi(\mu,\nu,r) = (g_\mu(\mu,\nu), g_\nu(\mu,\nu), r)$. Similarly $\{\alpha_r\}$ are assumed identical. Fix r_0 small. Since $\alpha_{r_0}(I)$ and $\Psi_{r_0}(h_0^+\mu_0)$ are disjoint and compact, $\exists\ \varepsilon_0 > 0$ such that $\Psi_{r_0}[N_{\varepsilon_0}^\mu(L^+(h_0^+\mu)) \cap \tau_{r_0}^+] \cap \alpha_{r_0}(I) = \emptyset$. Since the $\{\Psi_r\}$ and $\{\alpha_r\}$ are independent of r , the r - saturations of $\Psi_r[N_{\varepsilon_0}^\mu(L^+(h_0^+\mu_0)) \cap \tau_{r_0}^+]$ and of $\alpha_{r_0}(I)$ are also disjoint. Or $\Psi N_{\varepsilon_0}^\mu(L^+(h_0^+\mu_0)) \cap \alpha_{r_0}(I) = \emptyset$ as desired.

Again using toral knot arguments, $FL_\nu(\Gamma_r,\Delta) \geq 6$, contradicting corollary 1 . In a similar fashion, it can be shown by induction that two $B_{2,2}$ - flows containing different numbers of Neumann examples cannot be orbit-conjugate.

Remark: These techniques cannot distinguish a double Neumann example from a $B_{2,2}$ - flow containing one Neumann example and one "up-side-down" Neumann example, back-to-back.

Higher Dimensional Examples

In this section a $B_{2,n}$ - flow $(n > 2)$ is constructed which is not conjugate to $\varphi_{2,n}$. Multiple non-conjugate examples can be constructed by gluing together a succession of these. The author especially thanks the referee for the clean form of this proof and its extension to the $n > 3$ cases.

Theorem 3

There exist $B_{2,n}$ - flows, φ , $(n > 2)$ which are not orbit-conjugate to $\varphi_{2,n}$.

roof: In this case, $b^-\backslash a^- \approx S^1 \times S^{n-1} \times (0,1]$ has coordinates $(\mu,(\nu^1,\nu^2),r)$ here $\nu^1 \in [-1,1]$ and $\nu^2 \in S^{n-2}$. Let $\bar{\Sigma},\bar{\Delta},\bar{G}$, and $\bar{I}_i \subset S^1 \times [-1,1]$ be as hown below (i = 1,2, or 3). $\bar{G}$ is the open shaded region.

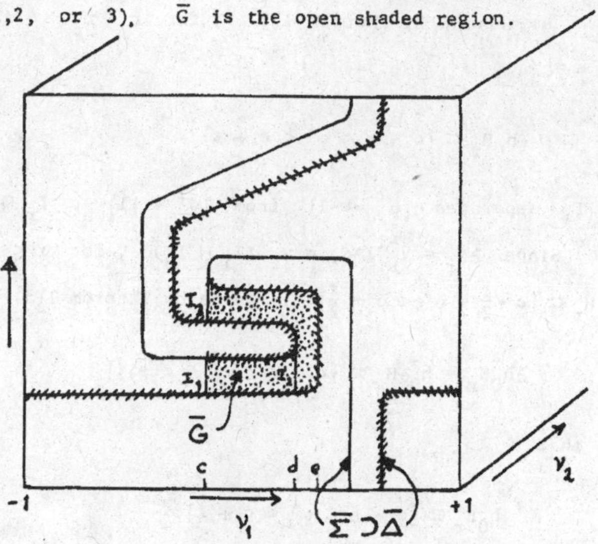

Figure 4.

Σ,Δ,G, and $I_i \subset b^-\backslash a^-$ are their (ν^2,r)-saturations. For example $\Sigma = \{(\mu,\nu^1,\nu^2,r) : (\mu,\nu^1) \in \bar{\Sigma}\}$. By [7], there exists an r and ν^2-preserving $B_{2,n}$-flow, φ, such that $\Phi(\ell_r^+(\mu_0)) = \Sigma \cap \tau_r^-$ (for all small r).

Assume φ and $\varphi_{2,n}$ are orbit-conjugate. So $h^-\Phi = \Phi_{2,n-2}h^+$. Define $\gamma_r : a^- \to b^-\backslash a^-$ by

$$\gamma_r(\nu^1,\nu^2) = (h^-)^{-1}(h_0^+(\mu_0)+\varepsilon,\nu^1,\nu^2,r)$$

where $\varepsilon > 0$ is small. Thus the image of γ_r is $(h^-)^{-1}\ell_r^-(h_0^+\mu_0+\varepsilon)$. $\pi : b^-\backslash a^- \to a^-$ is radial projection and $H_r = \gamma_r^{-1}G$.

By Theorem 1(4), for r smaller if necessary, $\Gamma_r = \gamma_r(a^-)$ does not meet Σ. (In Condition 4, replace Φ by $\Phi_{2,n}$ and Ψ by Φ. Then Γ_r

misses $\Phi N_\varepsilon^{\mu}(L^+(h_0^+\mu_0))$.)

Assume I_1 and $I_3 \subset \{\nu^1 = c\}$, that $I_2 \subset \{\nu^2 = d\}$, and that $\pi G = \{c < \nu^1 < e\}$, as shown. By Theorem 1(3) , for $r < r_\varepsilon$, $d_\nu(h_0^-, \pi\gamma_r) < \varepsilon$. Then since $\pi\gamma_r H_r = \pi G$

$$h_0^- H_r \subset \{c - \varepsilon < \nu^1 < c + \varepsilon\} \tag{1}$$

Because $\Delta \cup I_i$ separates $\{\nu^1 = -1\}$ from $\{\nu^1 = +1\}$, $\Gamma_r \cap I_i \neq \emptyset$, $i = 1,2,$ or 3 . Since $\partial H_r = \gamma_r^{-1}(\partial G) = \gamma_r^{-1}(I_1 \cup I_3)$, for $r < r_{(\varepsilon/2)}$ (Theorem 1(3)), $\partial H_r \subset \{c - \frac{\varepsilon}{2} < \nu^1 < c + \frac{\varepsilon}{2}\}$. Hence by Theorem 1(3) again,

$$\partial h_0^- H_r = h_0^- \partial H_r \subset \{c - \varepsilon < \nu^1 < c + \varepsilon\} \tag{2}$$

Next we show that

$$h_0^- H_r \subset \{c - \varepsilon < \nu^1 < c + \varepsilon\} \tag{3}$$

Assume $h_0^- H_r$ (which is open) meets $C = \{\nu^1 \geq c + \varepsilon\}$. (2) implies $\partial h_0^- H_r \cap C = \emptyset$. So since C is connected, $C \subset h_0^- H_r$ contradicting (1) . The other inequality in (3) follows from (1) .

Because $\Gamma_r \cap I_2 \neq \emptyset$, $h_0^- H_r$ meets $\{d - \varepsilon < \nu^1 < d + \varepsilon\}$ thus contradicting (3) . So φ and $\varphi_{2,n}$ are not orbit-conjugate.

Question. Do there exist $B_{3,3}$ - flows not orbit-conjugate to $\varphi_{3,3}$?

References

[1] C. Coleman, "Hyperbolic Stationary Points," Reports of the Fifth International Conference on Nonlinear Oscillations, Vol 2, Kiev (1970), 222-226.

[2] P. Hartman, "A Lemma in the Theory of Structural Stability of Differential Equations," Proc. Amer. Math. Soc. 11 (1960), 610-620.

3] D. Neumann, "Topological Hyperbolic Equilibria in Dynamical Systems," to appear.

4] __________, "Uncountably Many Distinct Topologically Hyperbolic Equilibria in $\mathbb{R}^4$," (these proceedings).

5] R. Walker, "Morse and Generic Contact between Foliations," Oct. 1979, Transactions of Am. Math. Soc.

6] W. Wilson, "A Uniform Continuity Condition which is Equivalent to Coleman's Conjecture," to appear.

7] __________, "Coleman's Conjecture on Topological Hyperbolicity," (These Proceedings).

niversity of Colorado
oulder, Colorado 80309

Coleman's Conjecture on Topological Hyperbolicity

F. Wesley Wilson

C. Coleman conjectured [1] that if Φ is a flow on a Euclidean space, which has a rest point at the origin, and which looks sufficiently like some flow near a differentiably hyperbolic rest point, then there is a local conjugacy between Φ and the differentiable model. The author [2] presented a tighter statement of this problem using isolating blocks, and found an analytic condition on the Poincaré Mapping which is equivalent to Coleman's Conjecture [3] . D. Neumann [4] has constructed an example of a flow on $\mathbb{R}^4$ which is topologically hyperbolic in this latter sense, but which fails to be locally conjugate to any differentiably hyperbolic flow; i.e., his example contradicts Coleman's conjecture.

In this survey, we shall give a precise statement of the original problem, a careful description of the known results, a procedure for building a large variety of smooth examples, and a list of several related problems.

<u>1. The Statement of the Problem</u>. We shall begin by reviewing the basic definitions concerning flows and isolating blocks, and then we shall give a precise statement of Coleman's Conjecture.

<u>Definition 1.1</u>. $\Phi : M \times \mathbb{R} \to M$ is a C^r <u>flow</u> $(r \geq 0)$ provided that

1. Φ is a <u>flow</u> on M ; i.e., Φ is a continuous mapping which satisfies $\Phi(x,0) = x$ and $\Phi(x,s+t) = \Phi(\Phi(x,s),t)$ for all x in M and s,t in $\mathbb{R}$.
2. M is a C^{r+1} manifold and Φ has a C^r tangent vector field defined by $\dot{\Phi}(x) = \frac{d}{dt}\Phi(x,t)\Big|_{t=0}$.

Note that in particular, a C^0 flow is more than just continuous; it is required to have a continuous tangent vector field. The set of C^r flows $(r \geq 0)$ is precisely the class of flows, for which the smoothing techniques of [5], [6], [7] can be applied.

<u>Definition 1.2.</u> A C^k <u>isolating-block-with-corners</u> for a C^k flow Φ on a C^k n-manifold M is a closed neighborhood in M with the following properties:

1. The boundary ∂B of B contains C^k $(n-1)$ - submanifolds b_+, b_- such that b_+ is the strict ingress set for Φ, b_- is the strict egress set for Φ, b_+ and b_- are differentiably transverse to the flow, and $b_+ \cup b_- = \partial B$.

2. $\tau = b_+ \cap b_-$ (tangency set) is a C^k $(n-2)$ - submanifold which has the property that for each $x \in \tau$, $\Phi(x,t)$ is exterior to B for all small non-zero values t.

3. The positively invariant set for $\Phi|B$ is A_+ and the negatively invariant set is A_-; their intersection $I = A_+ \cap A_-$ is the maximal closed invariant set in B.

4. These sets intersect the boundary of B in the sets

$$a_+ = \partial B \cap A_+ = b_+ \cap A_+$$

$$a_- = \partial B \cap A_- = b_- \cap A_-$$

structure is discussed in more detail in [6] and [8]. Much of the intui- can be gleaned from the following example, which is also of special importance our problem.

dard Example. Let $\Phi_{m,n}$ denote the flow on $\mathbb{R}^{m+n}$ which is generated by the tor fields

$$\dot{\Phi}_{m,n} \left| \begin{array}{l} \dot{x} = -x \\ \dot{y} = y \end{array} \right. \qquad (x,y) \in \mathbb{R}^m \times \mathbb{R}^n = \mathbb{R}^{m+n}$$

the product of unit disks $B = D^m \times D^n$ is an isolating block with

$$b_+ = \partial D^m \times D^n \qquad b_- = D^m \times \partial D^n$$

$$A_+ = D^m \times 0 \qquad A_- = 0 \times D^n$$

$$a_+ = \partial D^m \times 0 \qquad a_- = 0 \times \partial D^n$$

$$\tau = \partial D^m \times \partial D^n$$

inition 1.3. A C^1 flow Φ on $\mathbb{R}^{m+n}$ is differentiably hyperbolic of type (m,n) at the equilibrium point p if the Jacobian matrix J_p of Φ at p has m eigenvalues with negative real part and n eigenvalues with positive real part.

It is easy to show that if A is an $(m+n) \times (m+n)$ matrix which has m eigenvalues with negative real part and n eigenvalues with positive real part, then the flow generated by $\dot{x} = Ax$ is globally conjugate with the Standard Example $\Phi_{m,n}$. It therefore follows from Hartman's Theorem [9] that if Φ is differentiably hyperbolic of type (m,n) at p, then Φ is locally conjugate with $\Phi_{m,n}$. Consequently, up to local conjugacy, the Standard Examples are representative of all differentiably hyperbolic equilibria.

Definition 1.4. Let p be a rest point for the C^r flow Φ $(r \geq 0)$. Φ is topologically hyperbolic of type (m,n) at p if there is a C^1 system of coordinates (x,y) near p with $p \sim (0,0)$ and so that in these coordinates $B_\Phi = D^m \times D^n$ is an isolating block for Φ with

$$b_+ = \partial D^m \times D^n \qquad b_- = D^m \times \partial D^n$$
$$A_+ = D^m \times 0 \qquad A_- = 0 \times D^n .$$

Coleman's Conjecture. If Φ is topologically hyperbolic of type (m,n), then there is a local conjugacy $h : B_\Phi \to D^m \times D^n$ between Φ and the Standard Example $\Phi_{m,n}$.

2. The Poincaré Mapping of an Isolating Block.

Since A_+ and A_- are the respective positively invariant and negatively invariant sets of $\Phi|B$, it follows that each trajectory of $B - (A_+ \cup A_-)$ must leave B in positive and in negative time. Since the sets b_+ and b_- are transverse to the flow, it follows that the time-to-escape varies continuously with the point in $B - (A_+ \cup A_-)$. Indeed, a direct application of the Implicit Function Theorem yields the following precise statement.

Lemma 2.1. Let Φ be a C^r flow and let B be a C^r isolating-block-with-corners for Φ. Then there are C^r functions

$$t_{\pm} : B - A_{\pm} \to \mathbb{R}$$

which satisfy $\Phi[x,t_{\pm}(x)] \in b_{\pm}$.

In particular, since $(b_+ - a_+) \subset B - A_+$, it follows that there is a C^r mapping $P_\Phi : (b_+ - a_+) \to (b_- - a_-)$ defined by $P_\Phi(x) = \Phi[x,t_+(x)]$. Since the inverse for P_Φ is defined by $\Phi[y,t_-(y)]$, it follows that P_Φ is a C^r diffeomorphism. We call P_Φ the <u>Poincaré Mapping</u> of Φ and B .

<u>Example</u>. For the Standard Example $\Phi_{m,n}$ we have

$$(b_+ - a_+) = \partial D^m \times (D^n - 0) = S^{m-1} \times (0,1] \times S^{n-1} = (D^m - 0) \times \partial D^n = b_- - a_- .$$

Using coordinates $(\mu,\rho,\nu) \in S^{m-1} \times (0,1] \times S^{n-1}$, and integrating the differential equation from $\partial D^m \times D^n$ to $D^m \times \partial D^n$ (unit disks) we discover that

$$t_+(\mu,\rho,\nu) = -\ell n\, \rho$$

and

$$P_\Phi(\mu,\rho,\nu) = (\mu,\rho,\nu) .$$

<u>Theorem 2.2.</u> Let Φ and Ψ be C^0 flows on $\mathbb{R}^{m+n}$ which are topologically conjugate of type (m,n) at the origin with respect to the usual coordinates and isolating block $B = D^m \times D^n$. A necessary and sufficient condition for $\Phi|B$ and $\Psi|B$ to be conjugated by a homeomorphism $h : B \to B$ is that there be homeomorphisms $h_+ : b_+ \to b_+$ and $h_- : b_- \to b_-$ such that

$$\tilde{h}_+ = h_+|(b_+ - a_+) \qquad \tilde{h}_- = h_-|(b_- - a_-)$$

conjugate the Poincaré Mappings P_Φ and P_Ψ ; i.e., $\tilde{h}_- \cdot P_\Phi = P_\Psi \cdot \tilde{h}_+$.

The proof of this theorem is very similar to the proof given for [3 : Theorem 2.1] (in that case, Ψ was taken to be the Standard Example $\Phi_{m,n}$). We shall describe the essential steps of the proof to this theorem, but omit the verification of details which are exactly the same as in [3] . First of all, note that it is easy to verify that a conjugacy h between flows does induce a conjugacy between Poincaré mappings by $\tilde{h}_+ = h|(b_+ - a_+)$ and $\tilde{h}_- = h|(b_- - a_-)$. Therefore, the condition is necessary.

Suppose that homeomorphisms $h_+ : b_+ \to b_+$ and $h_- : b_- \to b_-$ are given so that $\tilde{h} \circ P_\Phi = P_\Psi \circ \tilde{h}_+$. Since

$$(b_+ - a_+) = S^{m-1} \times (0,1] \times S^{n-1} = \tau \times (0,1]$$

we can choose a product foliation $\mathcal{F}_+^\Phi$ for $(b_+ - a_+)$. Then there are foliations $\mathcal{F}_-^\Phi$ on $b_- - a_-$, $\mathcal{F}_+^\Psi$ on $b_+ - a_+$, and $\mathcal{F}_-^\Psi$ on $b_- - a_-$ induced by the homeomorphisms P_Φ, $\tilde{h}_+$, and $\tilde{h}_- \cdot P_\Phi$ respectively, and since $\tilde{h}_- \circ P_\Phi = P_\Psi \circ \tilde{h}_+$, the foliation $\mathcal{F}_-^\Psi$ is also the image of $\mathcal{F}_+^\Psi$ under P_Ψ. By [6 : Theorem 3.3] we can find monotone Lyapunov functions

$$V_\Phi : B \to [-1,1] \quad \text{for } \Phi|B$$

$$V_\Psi : B \to [-1,] \quad \text{for } \Psi|B$$

such that the level surfaces of V_Φ intersect ∂B in the foliations $\mathcal{F}_+^\Phi$ and $\mathcal{F}_-^\Phi$ and the level surfaces of V_Ψ intersect ∂B in the foliations $\mathcal{F}_+^\Psi$ and $\mathcal{F}_-^\Psi$. Using the full strength of Lemma 2.1, we can define homeomorphisms $g_\Phi : V_\Phi^{-1}(c) - A_- \to b_+$ and $g_\Psi : V_\Psi^{-1}(c) - A_- \to b_+$ by translating along the Φ and Ψ trajectories, respectively. Then the composition $g_\Psi^{-1} \circ h_+ \circ g_\Phi$ gives a homeomorphism of $V_\Phi^{-1}(c) - A_-$ onto $V_\Psi^{-1}(c) - A_+$ for each $-1 \le c \le 1$. The union of these homeomorphisms gives a homeomorphism of $B - A_-$ onto $B - A_-$, which carries Φ-trajectory segments to Ψ-trajectory segments. Similarly, we can push forward along to b_- and define a homeomorphism of $B - A_+$ onto $B - A_+$ which carries Φ-trajectory onto Ψ-trajectory segments. Since $\tilde{h}_- \circ P_\Phi = P_\Psi \circ \tilde{h}_+$, it follows that these homeomorphisms coincide on $B - (A_+ \cup A_-)$, and so their union h is a conjugacy on $B - (A_+ \cap A_-) = B - 0$. By the Invariance of Domain Theorem, the extension $h(0) = 0$ is a homeomorphism and since 0 is an invariant set, h is the desired local conjugacy.

<u>Corollary 2.3 (cf. [3 : Theorem 2.1])</u>. Let Φ be a C^r flow on $\mathbb{R}^{m+n}$ which is

opologically hyperbolic at the origin with respect to the usual coordinates and $= D^m \times D^n$. A necessary and sufficient condition for Φ to be locally conjugate o the Standard Example $\Phi_{m,n}$ is that there be homeomorphisms $\tilde{h}_+ : b_+ - a_+ \to b_+ - {}_+$ and $\tilde{h}_- : b_- - a_- \to b_- - a_-$ which have extensions to homeomorphisms $h_+ : b_+ \to {}_+$ and $h_- : b_- \to b_-$ and which factor the Poincaré Mapping P_Φ ; i.e., such that $_\Phi = \tilde{h}_- \circ \tilde{h}_+$.

roof. Apply the Theorem, and recall that for the Standard Example, the Poincaré apping in essentially the identity mapping (also the mapping $\tilde{h}_-$ in the Corollary s the inverse to the homeomorphism which is dèsignated by the same symbol in the tatement of the Theorem).

orollary 2.4. Coleman's Conjecture is true if $m = 1$ or $n = 1$.

roof. By the Invariance of Domain Theorem for any homeomorphism $h : D^n - 0 \to D^n$ 0 the extension $h(0) = 0$ defines a homeomorphism.

orollary 2.5. Let Φ, Ψ be flows which satisfy the hypotheses of Theorem 2.2. necessary and sufficient condition for $\Phi|B$ and $\Psi|B$ to be conjugated by a omeomorphism $h : B \to B$ is that there are homeomorphisms

$$\tilde{h}_+ : S^{m-1} \times (0,1] \times S^{n-1} \to S^{m-1} \times (0,1] \times S^{n-1}$$

$$\tilde{h}_- : S^{m-1} \times (0,1] \times S^{n-1} \to S^{m-1} \times (0,1] \times S^{n-1}$$

uch that $\tilde{h}_- \circ P_\Phi = P_\Psi \circ \tilde{h}_+$, such that $\tilde{h}_+$ and $\tilde{h}_+^{-1}$ are uniformly continuous in he $S^{n-1} \times D^n$ metric, and such that $\tilde{h}_-$ and $\tilde{h}_-^{-1}$ are uniformly continuous in he $D^m \times S^{n-1}$ metric.

roof. This corollary is merely a restatement of Theorem 2.2, using the characteri-ation of extendability in terms of uniform continuity, c.f. [4: Lemma 3.3 and 3.4].

Corollary 2.3 shows how we can approach the problem of building a counterexam-le to Coleman's Conjecture. We want to build a flow Φ on B such that the oincaré mapping $P_\Phi : b_+ - a_+ \to b_- - a_-$ has the property that for any homeomorphism

$h_+ : b_+ \to b_+$, the homeomorphism $\tilde{h}_- = P_\Phi \circ \tilde{h}_+^{-1}$ fails to have an extension to a homeomorphism $h_- : b_- \to b_-$. Note that there are two ways which this condition can fail:

1. $\tilde{h}_-$ may not have any continuous extension to a_- ,

2. $\tilde{h}_-$ has a continuous extension to a_- (in which case the extension is unique) and the extension may fail to be a homeomorphism, i.e. $\tilde{h}_-^{-1}$ may fail to have a continuous extension. Observe that $\tilde{h}_-^{-1} \circ \tilde{h}_- = \mathrm{id}$ always has an extension; so that if $\tilde{h}_-$ and $\tilde{h}_-^{-1}$ both have extensions, then both extensions are homeomorphisms. D. Neumann's counterexample to Coleman's conjecture provides a flow Φ so that the homeomorphism $\tilde{h}_-$ fails to have a continuous extension to a_- . Perhaps there is a less complicated example with the property that $\tilde{h}_-$ has an extension, but the extension fails to be a homeomorphism.

The Theorem provides a procedure for discussing the question of when any two topologically hyperbolic flows of type (m,n) are locally conjugate. Since we seek h_+ and h_- so that $\tilde{h}_- \circ P_\Phi = P_\Psi \circ \tilde{h}_+$, the condition which must be satisfied for Φ and Ψ to be inequivalent is that for any h_+ ,

$$\tilde{h}_- = P_\Psi \circ \tilde{h}_+ \circ P_\Phi^{-1}$$

fails to have an extension to a homeomorphism of b_- onto itself. Again, the condition can fail in two ways.

<u>3. Embedding Homeomorphisms into Topologically Hyperbolic Flows</u>. Since the conditions which we have just posed are stated in terms of Poincaré Mappings of flows, it would be useful to have a set procedure by which any homeomorphism P , from some fairly general class of homeomorphisms, can be realized as the Poincaré Mapping of some topologically hyperbolic flow. If the flow Φ is to be topologically hyperbolic of type (m,n) , then P_Φ must be a homeomorphism of $S^{m-1} \times (0,1] \times S^{n-1}$ and the restriction of P_Φ to $\tau = S^{m-1} \times \{1\} \times S^{n-1}$ must be the identity. We shall now sketch an argument that says that it is also necessary that P be

sotopic to the identity relative to τ : The trajectory segments of the Standard Example provide an analytic coordinate system for $B - (A_+ \cup A_-) \cup \tau$ as

$$(S^{m-1} \times (0,1) \times S^{n-1}) \times [0,1]$$

Let $\pi : B - (A_+ \cup A_-) \to (b_+ - a_+)$ be the projection onto the first component in this coordinate system. If it were the case that the Φ trajectories were always transverse to the hypersurfaces

$$L_t = (S^{m-1} \times (0,1] \times S^{n-1}) \times \{t\} .$$

Then the desired isotopy could be built by following the Φ - trajectories forward to L_t and the projecting back to $b_+ - a_+$ by π. This condition is satisfied in the following sense.

<u>Lemma 3.1.</u> Let Φ be a topologically hyperbolic flow of type (m,n) and let P_Φ be the Poincaré Mapping of Φ. Then there is a flow Ψ with the property that $P_\Psi = P_\Phi$ and such that the Ψ - trajectories are transverse to the levels $\{L_t\}_{t\in[0,1]}$.

We shall sketch the idea of a proof. Using the time parameterization of the Φ trajectories, we can represent $B - (A_+ \cup A_- \cup \tau)$ as $S^{m-1} \times (0,1) \times S^{n-1} \times [0,1]$ and so we obtain a foliation with leaves $\{\tilde{L}_t = ((S^{m-1} \times (0,1) \times S^{n-1}) \times \{\tau\})\}$. There is an isotopy of B which carries these leaves to the leaves of $\{L_t\}$ and this isotopy carries the Φ - trajectories onto trajectories of a flow Ψ which is transverse to the foliation $\{L_t\}$.

We have begged the question of differentiability in the above "proof". First of all, by our definition, all flows have tangent vector fields. Consequently, the homeomorphism of B must have enough differentiability to induce a tangent vector field on the new flow. Indeed, if we began with a C^r flow, then we would like to conclude with a C^r flow, putting further differentiability requirements on the homeomorphism of B. Finally, we also need that the flow Ψ be differentially transverse to the foliation $\{L_t\}$.

Supposing that all of this has been achieved, we note that since $\pi|L_t$ is an analytic diffeomorphism, then the isotopy which we have constructed has an additional property: it is differentiable in the time parameter at each (x,t) in its domain (the derivative is the π-image of component of the tangent vector of the flow along L_t).

Definition 3.1. The C^r diffeomorphisms $p,q : X \to X$ are strongly C^r isotopic provided that there is a C^r mapping $H : X \times I \to X$ such that

1. $H(x,t) = p(x)$ for t near 0 and
   $H(x,t) = q(x)$ for t near 1 ,
2. $H(\cdot,t) : X \to X$ is a C^r diffeomorphism for each $0 \le t \le 1$,
3. $\frac{d}{dt}H(x_0,t)\Big|_{t=t_0}$ exists and is a C^r function of x_0 and t_0 .

This last condition is precisely what is required for the associated flow isotopy

$$\chi(x,t) = (H(p^{-1}(x),t),t)$$

to be a C^r flow on $X \times I$. (This idea was first introduced by R. Thom [10] ; a nice use of this procedure is found in [11: pages 62-63]). Again, these analytic details require care and we shall leave them for future study if the need arises. We have called them to the attention of the reader, because they motivate the statement of the next theorem.

Theorem 3.2. Let $P : S^{m-1} \times (0,\varepsilon] \times S^{n-1} \to S^{m-1} \times (0,\varepsilon] \times S^{n-1}$ be a C^r diffeomorphism which is strongly C^r isotopic to the identity relative to $\tau_\varepsilon = S^{m-1} \times \{\varepsilon\} \times S^{n-1}$ $(r \ge 0)$. Then there is a C^r flow Φ on B , which coincides with the Standard Example $\Phi_{m,n}$ in a neighborhood of $A_+ \cup A_- \cup \tau$ and which has the property that

$$P_\Phi\Big|(S^{m-1} \times (0,\varepsilon] \times S^{n-1}) = P .$$

Proof. By the previous discussion, a strong isotopy $H : X \times I \to X$ induces a flow sotopy on $S^{m-1} \times (0,\varepsilon] \times S^{n-1} \times [0,1]$. We need to find an appropriate palce to mbed this flow in the Standard Example $\Phi_{m,n}$. Note that $V(x,y) = \|y\|^2 - \|x\|^2$ s a monotone Lyapunov function for $\Phi_{m,n}$, and so

$$L = \{(x,y) \mid \|x\| = \|y\|\}$$

s a cross section for $\Phi_{m,n}|B - (A_+ \cup A_-)$. Also, since the time to cross from $(\mu,\rho,\nu) \in b_+$ to $(\mu,\rho,\nu) \in b_-$ is $-\ell n(\rho)$, it follows that the trajectories with ime length greater than or equal to one are those with $\rho \le e^{-1}$. These trajecories cross L at $\|x\| = \|y\| \le e^{-\frac{1}{2}}$. Let $\tilde{L}_\varepsilon = \{(x,y) \mid \|x\| = \|y\| \le \varepsilon\}$ and let

$$S = \{\Phi_{m,n}(x,y,t) \mid \|x\| = \|y\| \le \varepsilon, \ |t| \le \tfrac{1}{2}\} .$$

f $\varepsilon \le e^{-\frac{1}{2}}$, then $S \subset B$ and S has a system of coordinates $\tilde{L}_\varepsilon \times [-\frac{1}{2},\frac{1}{2}] \approx \tilde{L}_\varepsilon \times$ $0,1]$. We define the desired flow Ψ to coincide with the flow isotopy of H n S and with $\Phi_{m,n}$ on $B - S$. Then Ψ is a C^r flow on $B - 0$. In order to et a C^r flow which has the desired properties, it is necessary to change the time arameterization of the trajectories of Ψ . Note that this kind of change has no ffect on the Poincaré Mapping. The easiest way to reparameterize a C^r flow is o integrate the product of a positive scaler function and the tangent vector field f the flow, along the trajectories of the flow. In this case, we use the vector ield $f(x) \cdot \dot{\Psi}(x)$ where $f(0) = 0$, $f(x) = 1$ for $\|x\| \ge \frac{1}{2}$, and f is so flat t the origin that all derivatives of $f\dot{\Phi}$, up to order r , vanish at the origin c.f. rough composition of mappings [12]). The result of integrating this vector ield along the trajectories of Ψ is a C^r flow which satisfies all of the reuirements of the theorem.

<u>eumann's Example</u>. By Corollary 2.4, the first possible counterexample to Coleman's onjecture is $m = 2$, $n = 2$. D. Neumann has given an example with $m = 2$ and $= 2$, which is inequivalent to the Standard Example $\Phi_{2,2}$. In this situation $_+ - a_+ = S^1 \times (0,1] \times S^1 = b_- - a_-$. However, $b_+ = S^1 \times D^2$ and $b_- = D^2 \times S^1$.

Using the convention that the homotopy generator for $S^1 \times S^1$ which bounds in the solid trees is called a meridian and the complementary homotopy generator is a longitude, we see that while the Poincaré Mapping for $\Phi_{2,2}$ is the identity, it does have the effect of interchanging the meridians and the longitudes. Let $T_\rho = S^1 \times \{\rho\} \times S^1$. Neumann's Poincaré Mapping carries each T_ρ to itself and begins with the identity mapping $(1 \geq \rho \geq e^{-1})$ and gradually distorts as ρ decreases until at $\rho = \frac{1}{4}$ the image of longitude circles is

The mapping on each T_ρ is the same for $0 < \rho \leq \frac{1}{4}$. Now it is clear that this example can be made C^∞ smooth and that it is C^∞-isotopic to the identity relative to $\tau = S^1 \times \{1\} \times S^1$ (note that for $r = \infty$, the notions of C^r isotopic and strongly C^r isotopic coincide). Hence by Theorem 3.4, there is a C^∞ flow whose Poincaré Mapping is the one prescribed by Neumann.

It is a fact that, when one constructs $\tilde{h}_- = P_\Phi \circ \tilde{h}_+$, for this homeomorphism, there is no extension of $\tilde{h}_-$ to a_- since radial lines from $b_- - a_-$ are mapped to infinite spirals by $P_\Phi \circ \tilde{h}_+$ for every choice of h_+. These details are included in [4], together with another approach to the smoothing question.

5. Related Results and Questions.

How many inequivalent examples are there? Can they be classified by some algebraic invariants; i.e., is there an obstruction to local conjugacy? First of all, we note that showing that two arbitrary examples are inequivalent is more difficult than showing that some example is inequivalent to a Standard Example since on the one hand we are dealing with an extension problem for

$$\tilde{h}_- = P_\Phi \cdot \tilde{h}_+$$

while on the other hand, we are dealing with an extension problem for

$$\tilde{h}_- = P_\Psi \cdot \tilde{h}_+ \circ P_\Phi^{-1}$$

(where h_+ must be allowed to vary over all homeomorphisms of b_+). In these proceedings [13], D. Neumann describes uncountably many inequivalent examples of topologically hyperbolic flows of type (2,2). Curiously, no simple way for extending his results to other dimensional situations (m,n) has been found.

A rather different procedure for proving that examples are not locally conjugate has been studied by R. Walker [14] (these proceedings). His method is weaker in that it cannot distinguish all of Neumann's new examples from each other, but it is stronger in that it does distinguish examples of type (m,2) and type (2,n) for arbitrary $m \geq 2$ and $n \geq 2$. These examples all involve a spiraling along the set $a_+ = S^1$ or $a_- = S^1$. For $m > 2$ and $n > 2$, both a_+ and a_- are simply connected, and some other kind of example will be necessary. However, to date, Coleman's Conjecture remains unresolved for $m > 2$ and $n > 2$.

Closely related (even for $m = 2$ and $n = 2$) is the question of whether or not there is an example where $\tilde{h}_-$ has an extension to a continuous mapping, but where this extension fails to be a homeomorphism. If such examples exist, then there are probably so many different local conjugacy classes that there is no hope of an algebraic classification.

References

1. C. Coleman, Hyperbolic Stationary Points, Reports of the Fifth International Congress on Nonlinear Oscillations, Vol. 2 (Qualitative Methods), Kiev (1970), 222-226.

2. F.W. Wilson, A reformulation of Coleman's conjecture concerning the local conjugacy of topologically hyperbolic singular points, Structure of Attractors in Dynamical Systems, (Lecture Notes in Mathematics, Vol. 668), Springer-Verlag, New York, 1978.

3. F.W. Wilson, A uniform continuity condition which is equivalent to Coleman's Conjecture, Jour. Diff. Equ., to appear.

4. D. Neumann, Topologically hyperbolic equilibria in dynamical systems, to appear.

5. F.W. Wilson, Smoothing derivatives of functions and applications, Trans. A.M.S.

6. F.W. Wilson, Special structure for C^r monotone Lyapunov functions, Jour. Diff. Equ., to appear.

7. F.W. Wilson and J.A. Yorke, Lyapunov functions and isolating blocks, Jour. Diff. Equ. 13(1973) 106-123.

8. C. Conley and R. Easton, Isolated invariant sets and isolating blocks, Trans. AMS 158(1971) 35-61.

9. P. Hartman, A lemma in the theory of structural stability of differential equations, Proc. AMS 11(1960), 610-620.

10. R. Thom, La classification des immersions, Seminar Bourbaki, 1957.

11. J. Milnor, Lectures on the h-Cobordism Theorem, Princeton Mathematical Notes, Princeton University Press, Princeton, 1965.

12. R. Abraham and J.Robbin, Transversal Mappings and Flows, Benjamin Press, New York, 1967.

13. D. Neumann, Uncountably many distinct topologically hyperbolic equilibria in $\mathbb{R}^4$, these proceedings.

14. R. Walker, Conjugacies of topologically hyperbolic fixed points: a necessary condition on foliations, these proceedings.

University of Colorado
Boulder, Colorado 80309

POPULATION DYNAMICS FROM GAME THEORY

E.C. Zeeman.

Introduction.

We study a class of cubic dynamical systems on a n-simplex. They arise in iology at both ends of the evolutionary scale, in models of animal behaviour and aolecular kinetics. The game theoretical aspects also suggest possible applications in ie social sciences.

Game theory was introduced into the study of animal behaviour by Maynard Smith nd Price [6, 7, 8] in order to explain the evolution of ritualised conflicts within a species, s for example when individuals compete for mates or territory. They defined the notion f an evolutionarily stable strategy (ESS) in a non-zero sum game. Each individual can lay one of n+1 strategies, and different points of the n-simplex Δ represents populations ith different proportions playing the various strategies. The pay-off represents fitness, r reproductive success, and an ESS is a point of Δ representing a population resistant) mutation, because mutants are less fit.

However, an ESS is a static concept, and so, following Taylor and Jonker [14], e introduce a dynamic into the game by assuming the hypothesis that the growth rate of aose playing each strategy is proportional to the advantage of that strategy. This gives flow on Δ whose flow lines represent the evolution of the population. In Section 1 we erify that if there is an ESS then it is an attractor of the flow, thereby sharpening a esult of [14; see also 4]. The converse is not true : an attractor may not necessarily e an ESS because locally the flow may spiral in elliptically towards the attractor (an ventuality that is not always covered by the notion of ESS due to the linearity of its efinition). We show there is also a global difference between an ESS and an attractor : an ESS lies in the interior of Δ then it must have the whole interior as its basin of ttraction and so there cannot be any other attractor, whereas if an attractor lies in the aterior of Δ then its basin can be smaller, and the game may admit other competing ttractors on the boundary. This is illustrated in Example 1, which gives a flow on a -simplex with a non-ESS attractor in the interior and an ESS attractor at a vertex, ividing Δ into two basins of attraction.

Meanwhile at the other end of the evolutionary scale studies by Eigen and chuster [1] of the evolution of macromolecules before the advent of life have led to xactly the same types of equation. The resulting dynamics have been studied by chuster, Sigmund, Wolff and Hofbauer [11, 12]. Here we are given n+1 chemicals, and ifferent points in Δ represent mixtures of them in different proportions. The dynamic epresents their enzymatic action upon each other, and an attractor represents a mixture aat remains stable because of mutual cooperation. For instance the example mentioned bove would represent a mixture of three chemicals, and if they happen to be added to the iixture in the right order, so that initial conditions fall into the basin of the interior ttractor, then the mixture will develop into a stable cooperative mixture of all three chemicals; but if they are added in the wrong order, so that the initial conditions fall into the other basin, then only one of the chemicals will survive and the other two will be excluded. Schuster and Sigmund have also applied the dynamics to animal behaviour in the battle of the sexes [13].

One of the main benefits of the dynamic approach is that it allows the notion of structural stability [9, 10, 15] to be introduced into game theory : a game is stable if sufficiently small perturbations of its pay-off matrix induce topologically equivalent flows. A property is called robust if it persists under perturbations. In Section 2 we study the fixed points, since they seem to be the most important feature determining the nature of the flows. For example a stable game can have at most one fixed point in the interior of each face of Δ. We show that an isolated fixed point is robust, and give a sufficient condition for there to be robustly no fixed points (and hence no periodic orbits) in the interior of Δ. These constraints limit the type of bifurcations that can occur in parametrised games : for instance elementary catastrophes [15] cannot occur, but we give examples to show that exchanges of stability can occur if an interior fixed point runs into another one on the boundary, and that Hopf bifurcations [5] are also possible.

In Section 3 we begin to tackle the classification problem, up to topological equivalence. We conjecture that stable classes are dense, and finite in number for each n. These conjectures are plausible because a game is determined by its pay-off matrix, and therefore the space of games on an n-simplex is the same as the space of real $(n+1)\times(n+1)$ matrices. For $n = 1$ it is easy to verify the conjectures, and show there are only 2 stable classes (up to flow reversal). For $n = 2$ we conjecture further, that a stable game is determined by its fixed points, and that there are therefore 19 stable classes (up to flow reversal) as illustrated in Figure 11. This conjecture is surprising because it implies that for $n = 2$ there are no periodic orbits in stable games, and therefore no generic Hopf bifurcations. In fact at the end of the paper we prove that all Hopf bifurcations on a 2-simplex are degenerate (thereby correcting a mistake in [14]), and the proof involves going some way towards proving the last conjecture. On the other hand such a conjecture would be false in higher dimensions, because when $n \geq 3$ generic Hopf bifurcations do occur, as is shown by Example 6, which is an elegant example due to Sigmund and his coworkers [11]. In higher dimensions the number of stable classes proliferates, but this is primarily due to the combinatorial possibilities of what can happen on the boundary of Δ, and if the flow is given on the boundary there seem to be relatively few stable extensions to the interior. For example if there are no fixed points in the interior we conjecture the extension is unique and gradient-like on the interior. If there is a fixed point then periodic orbits may also appear, but I do not know if strange attractors can occur.

In applications where perturbations are meaningful it is best to use stable models since they have robust properties. In another paper [16] we analyse the original game of Maynard Smith [6, 8] about animal conflicts, which gives a flow on a tetrahedron since there are 4 strategies involved. The retaliator is the best strategy, but it turns out to be only a weak attractor because the game is unstable. When the game is stabilised it becomes a proper attractor, but at the same time another competing attractor appears, surprisingly, which is a mixture of hawks and bullies, and which has biological implications for the evolution of pecking orders.

Section 1. ESS's and attractors.

Suppose competing individuals in a population can play one of n+1 strategies, abelled $i = 0,1,\ldots,n$. Let x_i denote the proportion of the population playing strategy i. hen $x = (x_0, x_1, \ldots, x_n) \in \Delta$, where Δ denotes he n-simplex in $\mathbb{R}^{n+1}$ given by $\Sigma x_i = 1$, $x_i \geq 0$. et $\mathring{\Delta}$ denote the interior of Δ given by $x_i > 0$, nd $\partial\Delta$ its boundary. Let $X_0, X_1, \ldots, X_n$ denote he vertices of Δ. We shall use x to denote mbiguously the population, the point in Δ, the ow matrix, and its transposed column matrix.

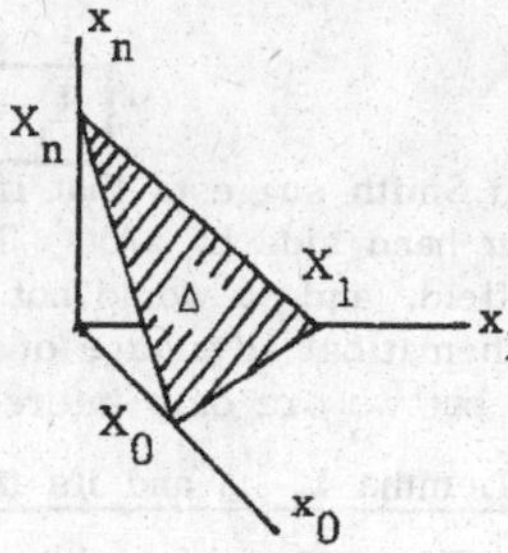

Figure 1.

The game is determined by the pay-off matrix $A = (a_{ij})$, which is a real (n+1)x(n+1) atrix. Pay-off means expected gain, and if an individual plays strategy i against another ndividual playing strategy j, then the pay-off to i is defined to be a_{ij}, while the pay-off j is a_{ji}. This is a non-zero sum game, and therefore A is not necessarily skew-ymmetric. If the population x is large the probability of an opponent playing j is x_j, nd therefore

$$\text{pay-off to } i \text{ against } x = \sum_j a_{ij} x_j = (Ax)_i ,$$
$$\text{pay-off to } x \text{ against } x = \sum_i x_i (Ax)_i = xAx .$$

two populations x, y play against each other

$$\text{pay-off to } x \text{ against } y = xAy.$$

Interpretation of the pay-off. There are three implicit assumptions : (i) Each ndividual plays a fixed pure strategy. If individuals were allowed to play mixed trategies then we should have to represent the population by a distribution on Δ rather an a point of Δ, and this leads to more complicated, but related, dynamics [see 2, 16]. lowever, in this paper we keep to pure strategies. (ii) Individuals breed true, in other ords if an individual plays strategy i so do his offspring. Of course this avoids the question f sex, but in applications to sex-related strategies, one can assume that the related sex reeds true. (iii) Pay-off is related to reproductive fitness, in other words the more ay-off the more offspring. In other applications the pay-off can represent rewards, eading to sociological adaptation rather than biological evolution.

Definition of evolutionarily stable strategy (ESS). Given $e \in \Delta$, call e an ESS of if, $\forall x \in \Delta - e$,

$$\text{either } xAe < eAe$$
$$\text{or } xAe = eAe \text{ and } xAx < eAx.$$

other words a mutant x strain will be less fit than e because it either loses out against or against itself. It will be convenient to write

$$fx = eAe - xAe, \qquad gx = eAx - xAx ,$$

o that the condition becomes $fx > 0$ or $fx = 0$ and $gx > 0$.

Definition of the dynamic. The main hypothesis is that the growth rate of those laying each strategy is proportional to the advantage of that strategy. By suitable choice of time scale we can make the factor of porportionality equal to 1.

$\therefore$ growth rate of x_i = (pay-off to i) - (pay-off to x)

$$\therefore \frac{\dot{x}_i}{x_i} = (Ax)_i - xAx$$

$$\therefore \boxed{\dot{x}_i = x_i[(Ax)_i - xAx]}$$

Maynard Smith suggests that if might be sometimes biologically more appropriate to divide the right-hand side by xAx. This would change the length but not the direction of the vector field, and so would not alter the phase portrait. The above dynamic does have the mathematical advantage of being polynomial, indeed cubic. The dynamic is defined on $\mathbb{R}^n$, but we are only interested in Δ.

Lemma 1. Δ and its faces are invariant.

Proof. The n-plane containing Δ given by $\Sigma x_i = 1$ is invariant because

$$(\Sigma x_i)^{\bullet} = \Sigma \dot{x}_i = xAx - (\Sigma x_i)xAx = 0 .$$

Similarly, given any q-dimensional face Γ then the q-plane containing Γ is invariant. Hence Δ and its faces are invariant.

Induced flow. Let φ_A denote the induced flow on Δ. Examples of such flows on a 2-simplex can be seen in Figure 11 below. The reverse flow is given by reversing the sign, $-\varphi_A = \varphi_{-A}$. If Γ is a face of Δ we write $\Gamma < \Delta$, and we shall use the symbol Γ to denote ambiguously both the subset of Δ and the subset of $\{0, 1, \dots, n\}$ corresponding to the vertices; thus $i \in \Gamma$ is an abbreviation for $X_i \in \Gamma$. If $A|\Gamma = \{a_{ij}; i, j \in \Gamma\}$ denotes the corresponding submatrix, then the induced flow on Γ satisfies $\varphi_A|\Gamma = \varphi_{A|\Gamma}$.

Attractors. For the most part we shall only need to consider point attractors. Recall the definition : a point is an attractor of the flow if it is the ω-limit of a neighbourhood, and the α-limit of only itself. Its basin of attraction is the (open) set of points of which it is the ω-limit. It is hyperbolic if its eigenvalues have negative real part.

Theorem 1. An ESS is an attractor, but not conversely. This result was first proved in [14] under the extra hypothesis that the ESS was regular, and giving the extra conclusion that the attractor was hyperbolic. Another proof is given in [4]. The Theorem shows that from the point of view of smooth dynamics an attractor is a more general notion than an ESS, and better characterisation of the resistance to mutation. Theorem 2 and Example 1 below show that there are also global differences between them.

Proof of Theorem 1. Suppose we are given an ESS e of A. We shall show that

$$V = \prod x_i^{e_i}$$

is a Lyapunov function for φ_A. In other words we shall prove there is a neighbourhood N of e such that

$$\left.\begin{array}{ll} (1) & \nabla V.(e-x) > 0 \\ (2) & \dot{V} > 0 \end{array}\right\} \quad \forall x \in N - e$$

By (1) V increases radially towards e, and so e is the maximum and there are no stationary points of V in N-e. By (2) all orbits inside a level curve of V tend to e, and so e is an attractor, as required. The proof of the two conditions is divided into two cases, according as to whether e lies in the interior or boundary of Δ.

<u>Proof of (1) when $e \in \mathring{\Delta}$.</u> Let $N = \mathring{\Delta}$. If $x \in \mathring{\Delta} - e$ then $V > 0$ and

$$V_i = \frac{\partial V}{\partial x_i} = V\frac{e_i}{x_i}$$

$\therefore \nabla V.(e-x) = \Sigma V_i(e_i - x_i) = V\Sigma\frac{e_i}{x_i}(e_i - x_i) = V\Sigma\frac{(e_i - x_i)^2}{x_i}$, since $\Sigma e_i = \Sigma x_i = 1$. $\therefore \nabla V.(e-x) > 0$, since $x \neq e$.

<u>Proof of (2) when $e \in \mathring{\Delta}$.</u> Recall that

$$fx = eAe - xAe, \quad gx = eAx - xAx.$$

Given $x \in \mathring{\Delta} - e$, and $t \in \mathbb{R}$, let $x_t = tx + (1-t)e$. Then $x_t \in \Delta$ for $|t|$ sufficiently small, since $e \in \mathring{\Delta}$.

$\therefore f(x_t) \geq 0$, since e an ESS. But $f(x_t) = tfx$.

$\therefore tfx \geq 0$ for $|t|$ sufficiently small. $\therefore fx = 0$.

$\therefore gx > 0$ since e an ESS.

$\therefore \dot{V} = \Sigma V_i \dot{x}_i = V\Sigma\frac{e_i}{x_i}x_i((Ax)_i - xAx) = Vgx > 0.$

This completes the proof of Theorem 1 for the case $e \in \mathring{\Delta}$. Notice that in this case, since $N = \mathring{\Delta}$, the basin of attraction of e contains $\mathring{\Delta}$. But the basin $\subset \mathring{\Delta}$, because $\partial\Delta$ is invariant. $\therefore$ the basin $= \mathring{\Delta}$.

<u>Proof of (1) when $e \in \partial\Delta$.</u> Suppose $e \in \mathring{\Gamma}$, $\Gamma < \Delta$.

Let $N_1 = \mathring{\Gamma} \cup \mathring{\Delta}$,

$G = \partial\Delta - \mathring{\Gamma} = \Delta - N_1$.

If $x \in N_1 - e$ then $x_i \neq 0$, $i \in \Gamma$.

$$\therefore V_i = \begin{cases} V\frac{e_i}{x_i}, & i \in \Gamma \\ 0, & i \notin \Gamma \end{cases}$$

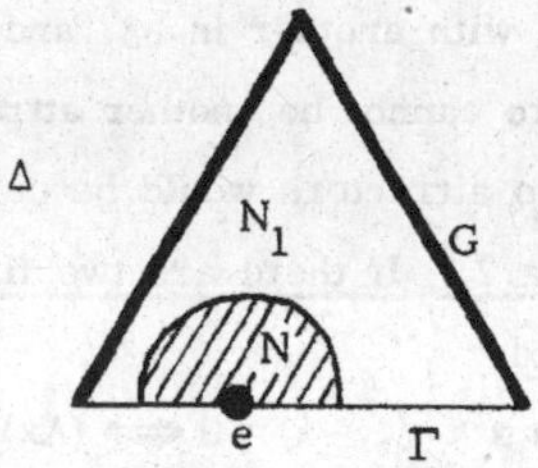

Figure 2.

$\therefore \nabla V.(e-x) = \sum_{i \in \Gamma} V\frac{e_i}{x_i}(e_i - x_i) = V\sum_{i \in \Gamma}\frac{(e_i - x_i)^2}{x_i} + V(1 - \sum_{i \in \Gamma} x_i) > 0$, because the first term > 0 and the second term ≥ 0. (Note that the proof given in [4] for this step does not work, and the proof given for the next step is incomplete).

<u>Proof of (2) when $e \in \partial\Delta$.</u> If $x \in N_1$ then $\dot{V} = \sum_{i \in \Gamma} V\frac{e_i}{x_i}x_i((Ax)_i - xAx) = Vgx$, since $e_i = 0$, $i \notin \Gamma$. Therefore we have to find a neighbourhood N of e in N_1 such that g is positive on N - e, but the problem this time is that f may not vanish on N. Let $G_0 = G \cap f^{-1}0$. (Notice $G_0 \supset \partial\Gamma$). Then $g > 0$ on G_0 by the ESS condition. $\therefore g > 0$ on an open neighbourhood G_1 of G_0 in G. Let $G_2 = G - G_1$. Then G_2 closed in G, and therefore compact. Since $f > 0$ on G_2, the function $\frac{g}{f}$ is defined and continuous on G_2, and therefore bounded since G_2 compact. Choose ε, $0 < \varepsilon < \frac{1}{2}$ such that $|\frac{g}{f}| < \frac{1}{2\varepsilon}$ on G_2. $\therefore \varepsilon|gx| < \frac{1}{2}fx$, $\forall x \in G_2$. Let N be the neighbourhood of e in N_1 given by

$$N = \{x_t = tx + (1-t)e;\ x \in G,\ 0 \leq t < \varepsilon\}.$$

Now

$$g(x_t) = t^2 gx + t(1-t)fx.$$

If $0 < t < \varepsilon$ and $x \in G_1$ then on the right-hand side the first term >0 and the second term ≥0. On the other hand if $x \in G_2$ then the second term >0, and the first term is smaller, because

$$|tgx| < \varepsilon|gx| < \tfrac{1}{2}fx, \text{ by above, } < (1-t)fx, \text{ since } t < \varepsilon < \tfrac{1}{2}.$$

Therefore in both cases $g > 0$. This completes the proof of Theorem 1 for the case $e \in \partial\Delta$. Finally the negative converse, that an attractor is not necessarily an ESS, is established by Example 1 below. A similar counterexample is given in [14], but ours has the extra subtlety of illustrating a global difference between the basins of attraction of an ESS and an attractor, as indicated by the following theorem.

Theorem 2. If an ESS lies in $\mathring{\Delta}$ then its basin of attraction is $\mathring{\Delta}$, and there are no other attractors. If an attractor lies in $\mathring{\Delta}$ then its basin may be smaller than $\mathring{\Delta}$, and there may be other attractors in $\partial\Delta$ (but not in $\mathring{\Delta}$).

Proof. We have already shown in the proof of Theorem 1 that an ESS in $\mathring{\Delta}$ has basin $\mathring{\Delta}$; therefore there cannot be another attractor in Δ otherwise its basin would have to be a non-empty open set in Δ disjoint from $\mathring{\Delta}$, which is impossible since $\mathring{\Delta}$ is dense in Δ. The second half of Theorem 2 is established by Example 1 below, which illustrates an attractor in $\mathring{\Delta}$ with another in $\partial\Delta$, and hence the basin of the former must be smaller than $\mathring{\Delta}$. There cannot be another attractor in $\mathring{\Delta}$, otherwise by Lemma 2 below the line joining the two attractors would be pointwise fixed, so neither would be an attractor.

Lemma 2. If there are two fixed points in $\mathring{\Delta}$ then the line joining them is pointwise fixed.

Proof. Given $x \in \mathring{\Delta}$, $\dot{x} = 0 \Longleftrightarrow (Ax)_i = xAx, \ \forall i$

$$\Longleftrightarrow (Ax)_i \text{ independent of } i, \text{ since } \Sigma x_i = 1 .$$

Given e, x fixed in $\mathring{\Delta}$, and $t \in \mathbb{R}$, then x_t is also fixed since

$$(Ax_t)_i = (A(tx+(1-t)e))_i = t(Ax)_i + (1-t)(Ae)_i$$

is independent of i. This completes the proof of Lemma 2.

Example 1. Non-ESS attractor.

$$A = \begin{pmatrix} 0 & 6 & -4 \\ -3 & 0 & 5 \\ -1 & 3 & 0 \end{pmatrix}$$

Figure 3.

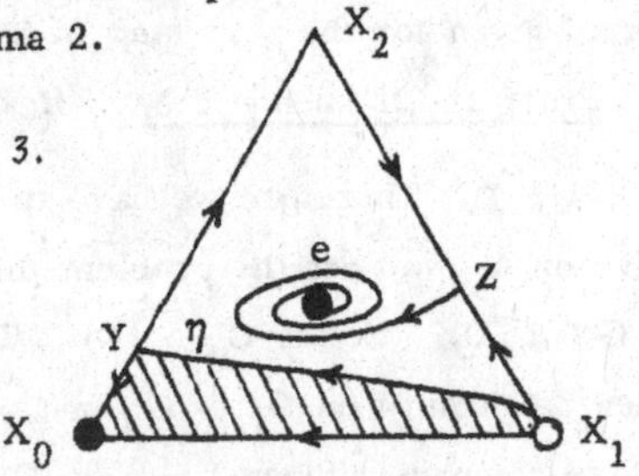

φ_A is a flow on the triangle $X_0X_1X_2$. There is an attractor at the barycentre $e = (\frac{1}{3}, \frac{1}{3}, \frac{1}{3})$ with eigenvalues $\frac{1}{3}(-1 \pm i\sqrt{2})$. However, e is not an ESS because $fX_0 = 0$ but $gX_0 = -\frac{4}{3}$. On the other hand X_0 is another attractor which is an ESS. The other fixed points are a repellor at X_1, and saddles at X_2, $Y = (\frac{4}{5}, 0, \frac{1}{5})$ and $Z = (0, \frac{5}{8}, \frac{3}{8})$. As visual notation for all he figures in this paper we use a solid dot for an attractor and an open dot for a epellor, and we always put in the insets and outsets of the saddles, as in Figure 3. Here insets and outsets are short for the usual more cumbersome terms "stable and nstable manifolds"). In the proof of Theorem 7 below we show that

$$V = x_0^4 x_1^5 x_2^{-10}(-4x_0 - 5x_1 + 10x_2)$$

s a global Lyapunov function for the flow in $\check{\Delta}$. Therefore the inset η of Y flows away rom the repellor X_1, and all other orbits in $\check{\Delta}$-e flow away from X_1 and towards one or ther of the two attractors, e and X_0. Hence η separates Δ into the basins of attractions f the two attractors, as illustrated in Figure 3, where the basin of X_0 is shown shaded. t also follows from the proof of Theorem 7 that this example is in fact stable. This xample completes the proof of Theorems 1 and 2.

Figure 3 illustrates qualitatively why an attractor need not be an ESS, and eveals exactly where the notion of ESS fails. The local reason that e is not an ESS is hat the orbits spiral in somewhat elliptically; therefore a mutant X_0-strain will initially ave a slight advantage over e, but it will also stimulate the growth of an X_2-strain that vill soon wipe out that advantage, and which will in turn be wiped out by an X_1-strain, and o on, as the orbit spirals in towards e. Meanwhile the global reason that e is not an ESS is that its basin is not the whole of $\check{\Delta}$.

In the application to chemical reactions, e represents cooperative behaviour, while X_0 represents exclusive behaviour. The fact that both types of behaviour occur in he same example shows that one cannot divide all stable systems into cooperative or xclusive, as might be suggested by the emphasis on this dichotomy in [11].

Section 2. Stability, fixed points and bifurcations.

Equivalence. Let M_{n+1} denote the space of games with n+1 strategies, which we identify with the space of real (n+1)x(n+1) matrices. Define $A, B \in M_{n+1}$ to be equivalent, written $A \sim B$, if there exists a face-preserving homeomorphism of Δ onto itself throwing φ_A-orbits onto φ_B-orbits. Here face-preserving means that each face is mapped onto another face, not necessarily onto itself.

Stability. Call A stable if it has a neighbourhood of equivalents in M_{n+1}. Note that this is a form of structural stability, with the proviso that we are confining ourselves to a special type of dynamical system, and to a restricted form of equivalence. A stable class is an equivalence class of stables. (Note that each stable class is open in M_{n+1}, but may have some unstable equivalents on its boundary, so the full equivalence class may be slightly larger than the stable class.)

Conjecture 1. Stables are dense in M_{n+1}.

Conjecture 2. For each n there are only a finite number of stable classes.

In other words we are suggesting that this is a well-behaved piece of mathematics. Although the dynamical systems involved are non-linear and possess some unexpected properties, nevertheless they appear to be qualitatively fairly simple, and there are so few of them that it seems plausible to try and classify them, at least in the lower dimensic When $n = 1$ it is easy to verify the conjectures are true (see Section 3 below). When n = : we go some way towards proving them (see Theorems 6, 7). For all n the limitations on the possible configurations of fixed points impose considerable constraints on the types of flows and bifurcations that can occur, and so we begin by examining the fixed points.

Theorem 3. A stable game has at most one fixed point in the interior of each face of Δ (including $\mathring{\Delta}$).

Before we prove Theorem 3 consider some examples. In Example 1 above there are 6 fixed points, one in the interior of each face except the edge X_0X_1. Figure 11 below illustrates all the different possible configurations of fixed points that can occur in stable games on a 2-simplex. The following example shows that for any n it is possible to have a stable game with exactly one fixed point inside every face. If a game is unstable there may be more than one fixed point - for instance $A = 0$ has every point fixed.

Example 2. Let I denote the identity matrix. Then φ_I has a fixed point at the barycentre of each face. The vertices are attractors, the barycentre e of Δ a repellor, and the rest are saddles.

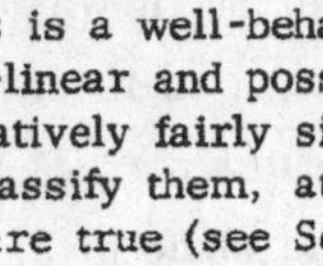

Figure 4.

Proof. Consider the reverse flow φ_{-I}. If $x \in \Delta - e$ then $fx = 0$ and $gx = |e-x|^2 > 0$. Therefore e is an ESS. Therefore by Theorem 1 e is an attractor, and by Theorem 2 ther are no other fixed points in $\mathring{\Delta}$. Hence e is a repellor for φ_I. Similarly there is a fixed point at the barycentre of each face, and no others. One can verify that this particular example is in fact a gradient flow, $\dot{x} = \nabla(\sigma_3 - \sigma_2^2)$, where σ_k is the kth symmetric function of the x_i's. Hence, by induction on the faces, it is structurally stable [9], and therefore stable.

Notation. Let u denote ambiguously the row vector $u = (1, 1, \ldots, 1)$ and its transposed olumn vector.

Proof of Theorem 3. Suppose $n \geq 1$, otherwise the result is trivial. Let $Q \subset M_{n+1}$, denote the set of matrices all of whose symmetric $q \times q$ minors are non-ero, for $1 \leq q \leq n+1$. Then Q is open dense in M_{n+1} being the complement of an lgebraic subset. Therefore any stable class meets Q. Therefore it suffices to prove the esult for games in Q, since the result is invariant under equivalence, and so let $A \in Q$. herefore A^{-1} exists since $\det A \neq 0$. If $x \in \mathring{\Delta}$ is a fixed point of φ_A then $(Ax)_i$ is ndependent of i, by the proof of Lemma 2.

$\therefore$ Ax = multiple of u. $\quad\therefore$ x = multiple of $A^{-1}u$.

ut the vector subspace $[A^{-1}u]$ of $\mathbb{R}^{n+1}$ generated by $A^{-1}u$ pierces Δ in at most one point, nd so x is unique. Therefore φ_A has at most one fixed point in $\mathring{\Delta}$. The same holds for ach face of Δ, using the fact that the corresponding minor is non-zero. This completes he proof of Theorem 3.

Robustness. A property of φ_A is called robust if it is preserved under erturbations; in other words the property is shared by φ_B for all B in a neighbourhood f A. Otherwise it is called transient. For example if A is stable then all topological roperties of φ_A are robust, and if A is unstable some property of φ_A is transient. But ve shall also consider robust properties of unstable games, as illustrated in the following heorem, which we need for both bifurcations (see the Corollary below) and applications 16].

Theorem 4. (i) Having an isolated fixed point in $\mathring{\Delta}$ is robust. (ii) If (adjA)u has oth positive and negative components then φ_A has no fixed points and no periodic orbits n $\mathring{\Delta}$, and this is robust.

Remarks : In part (i) it is necessary that the fixed point be isolated, otherwise consider he example $A = 0$; here every point is fixed but A has arbitrarily small perturbations vith no fixed points in $\mathring{\Delta}$. Nevertheless the result is surprising because isolated fixed oints are not robust amongst dynamical systems in general. For example consider the lynamic $\dot{y} = y^2$, $y \in \mathbb{R}$ (the fold catastrophe); here the origin $y = 0$ is an isolated fixed oint, but the perturbation $\dot{y} = y^2 + \varepsilon$, $\varepsilon > 0$, has none.

In part (ii) the hypothesis on (adjA)u is necessary because otherwise the absence of ixed points in $\mathring{\Delta}$ is not robust (for instance put $\varepsilon = 0$ in Example 3 below).

Proof of Theorem 4(i). Suppose φ_A has an isolated fixed point $e \in \mathring{\Delta}$. Notice this mplies no other fixed points in $\mathring{\Delta}$ by Lemma 2. There are three cases accordingly as o whether the rank, $r(A) = n+1, n$, or less.

Case 1 : $r(A) = n+1$. Here e is a multiple of $A^{-1}u$. Let $L_A = [A^{-1}u]$, the ector subspace of $\mathbb{R}^{n+1}$ generated by $A^{-1}u$. Then $e \in L_A \cap \mathring{\Delta}$. Therefore $L_A \neq 0$, and

L_A pierces $\mathring{\Delta}$ in e. Therefore if B is a sufficiently small perturbation of A, $L_B = [B^{-1}u] \neq$ and L_B pierces $\mathring{\Delta}$ in a point e_B near e. Hence e_B is the required unique fixed point of φ_B in $\mathring{\Delta}$.

<u>Case 2 : r(A) = n.</u> Choose $x \in \mathbb{R}^{n+1}$, $x \neq 0$, such that $Ax = 0$. If x is not a multiple of e, let $x_t = tx + (1-t)e$. For t sufficiently small $[x_t]$ pierces $\mathring{\Delta}$ in a point, λx_t say, $\neq$ e. Furthermore λx_t is fixed under φ_A since $(A\lambda x_t)_i = \lambda(1-t)eAe$, which is independent of i. Therefore e is not isolated, since $\lambda x_t \to e$ as $t \to 0$, a contradiction. Therefore x is a multiple of e. Therefore $Ae = 0$.

Suppose $(\mathrm{adj}A)u = 0$. Then the matrix obtained by replacing any one column of A by u has zero determinant. Since $r(A) = n$ there are n linearly independent columns, and so u is dependent upon them. Therefore there exists $y \in \mathbb{R}^{n+1}$, $y \neq 0$, such that $Ay = u$. Therefore y is not a multiple of e since $Ae = 0$. For small t let $y_t = ty + (1-t)e$, and let $\lambda y_t = [y_t] \cap \mathring{\Delta}$. Then λy_t is fixed under φ_A since $(A\lambda y_t)_i = \lambda t$, which is independent of i. Therefore again e is not isolated, a contradiction. Therefore $(\mathrm{adj}A)u \neq 0$.

Furthermore $(\mathrm{adj}A)u$ is a multiple of e because all columns of adjA are multiples of e, since $r(A) = n$ and $Ae = 0$. Let $L_A = [(\mathrm{adj}A)u]$. Then $L_A \neq 0$ and L_A pierces $\mathring{\Delta}$ in e. Therefore if B is a sufficiently small perturbation of A, then $L_B = [(\mathrm{adj}B)u] \neq 0$ and L_B pierces $\mathring{\Delta}$ in a point e_B near e. Furthermore e_B is fixed under φ_B since Be_B = multiple of $B(\mathrm{adj}B)u = (\det B)u$.

There remains to verify that e_B is isolated, and so suppose $x \in \mathring{\Delta}$ is any fixed point of φ_B. For sufficiently small perturbations, $r(B) \geq r(A) = n$. If $r(B) = n+1$ then x = multiple of $B^{-1}u$ = multiple of $(\mathrm{adj}B)u$, and so $x = e_B$. If $r(B) = n$, then

$$(xBx)(\mathrm{adj}B)u = (\mathrm{adj}B)(xBx)u = (\mathrm{adj}B)Bx = (\det B)x = 0.$$

$$\therefore xBx = 0, \text{ since } (\mathrm{adj}B)u \neq 0.$$

$$\therefore Bx = (xBx)u = 0 .$$

But Be_B = multiple of $(\det B)u = 0$.

$\therefore$ x = multiple of e_B, since $r(B) = n$.

$\therefore x = e_B$; so we have shown that e_B is the unique fixed point of φ_B in $\mathring{\Delta}$, and therefore isolated.

<u>Case 3 : rA < n.</u> Since the eigenspace of 0 has dimension ≥ 2, we can choose $x \in \mathbb{R}^{n+1}$, $x \neq$ multiple of e, such that $Ax = 0$. Then, as in case 2, this implies that e is not isolated, a contradiction.

<u>Proof of Theorem 4(ii).</u> Let $L_A = [(\mathrm{adj}A)u]$. Then $L_A \neq 0$ and L_A does not meet Δ, since by the hypothesis L_A meets the positive quadrant only in the origin. If B is a sufficiently small perturbation then $L_B = [(\mathrm{adj}B)u] \neq 0$ and L_B does not meet Δ, since Δ is compact. Also $rB \geq rA \geq n$, since $\mathrm{adj}A \neq 0$. Therefore by the arguments in Cases 1 and 2 above, any fixed point of φ_B in $\mathring{\Delta}$ must lie in $L_B \cap \mathring{\Delta}$, which is empty. Therefore

either φ_A nor φ_B has any fixed points in $\mathring{\Delta}$.

To show that no fixed points in $\mathring{\Delta}$ implies no periodic orbits in $\mathring{\Delta}$, we use an rgument of Sigmund et al. [11]. For suppose that was an orbit of period T. Let (t), $0 \le t \le T$ denote the flow round the orbit, and let

$$e = \int_0^T x dt, \quad \lambda = \int_0^T xAx dt.$$

hen $e \in \mathring{\Delta}$, since each $e_i > 0$, and

$$(Ae)_i - \lambda = \int((Ae)_i - xAx)dt = \int \frac{\dot{x}_i}{x_i} dt = [\log x_i]_0^T = 0$$

herefore $(Ae)_i = \lambda$, independent of i, and so e is fixed, a contradiction. This completes he proof of Theorem 4.

<u>Bifurcations.</u> We now examine the types of bifurcation that can occur in parametrised ames. First we use Theorem 4 to show that there are no elementary catastrophes, the ypical bifurcations of gradient systems [15]. Then we shall give some examples to show hat classical Hopf bifurcations [5] and exchanges of stability can occur.

<u>Corollary to Theorem 4.</u> <u>Elementary catastrophes cannot occur.</u>

<u>roof.</u> If an elementary catastrophe occured in $\mathring{\Delta}$ then some perturbation would have more han one isolated fixed point in $\mathring{\Delta}$, which is impossible by Lemma 2. If an elementary atastrophe occured in $\partial\Delta$, then some perturbation would contain a fold catastrophe, where he variation of a parameter causes two isolated fixed points to coalesce and disappear. Iow it is quite possible to make an isolated point in $\mathring{\Delta}$ run into another one in the boundary, n $\mathring{\Gamma}$ say, $\Gamma < \Delta$, so that at the critical parameter value they coalesce to form an isolated ixed point in $\mathring{\Gamma}$, but it is then impossible to make the latter disappear because it is robust by Theorem 4(i) applied to $\mathring{\Gamma}$. Therefore elementary catastrophes cannot occur.

<u>Example 3.</u> <u>Exchange of stabilities bifurcation.</u>

.et $A_\varepsilon = \begin{pmatrix} 0 & 1 \\ \varepsilon & 0 \end{pmatrix}$

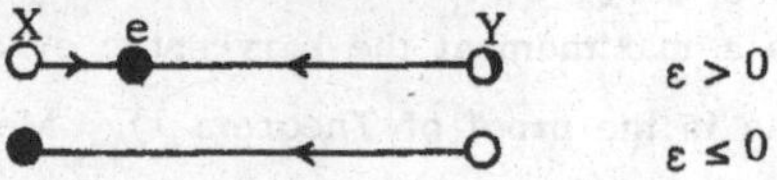

nd let φ_ε denote the induced flow. It is easy to verify there are two cases according to he sign of the parameter ε. If $\varepsilon > 0$ then φ_ε has an attractor at $e = \left(\frac{1}{1+\varepsilon}, \frac{\varepsilon}{1+\varepsilon}\right)$, and epellors at the two vertices of the 1-simplex. If $\varepsilon \le 0$ then φ_ε has an attractor at $X = (1,0)$ and a repellor at $Y = (0,1)$. Therefore A_0 is unstable at the critical parameter alue $\varepsilon = 0$. It is easy to verify A_ε is stable if $\varepsilon \neq 0$ (see Section 3 below). As $\varepsilon \to 0_+$ he attractor e runs into X and donates its attractiveness to X.

Mathematically the bifurcation is best understood by considering the induced flow on he line $\mathbb{R}$ containing Δ. If $\varepsilon < 0$ there is an additional repellor $e \in \mathbb{R}$ outside Δ.

Thus as the parameter passes through the critical value the fixed points e, X cross and xchange stabilities. Taking coordinates (x, y) the dynamic is given by

$$\dot{x} = x(y-(1+\varepsilon)xy), \quad \dot{y} = y(\varepsilon x-(1+\varepsilon)xy).$$

Putting $x = 1-y$, we can use y as a single variable for $\mathbb{R}$, with origin at X, and then the dynamic is equivalent to the single equation

$$\dot{y} = -y^2+y^3 + \varepsilon(y-2y^2+y^3).$$

Within the constraint imposed by the games this is indeed a versal unfolding of the germ $\dot{y} = -y^2+y^3$ at $y = 0$, since the constraint requires that X be kept fixed, but if we were to allow arbitrary perturbations on $\mathbb{R}$ then a versal unfolding would include an additional constant term, thereby giving a catastrophe surface with a fold curve through the origin. Then our constraint would be the same as taking the tangential section of this surface at the origin, thereby recovering the above unfolding as the classical exchange of stabilities bifurcation.

The following example shows the same phenomenon in one higher dimension. Here a saddle in a 2-simplex runs into, and exchanges stabilities with, an attractor on an edge. The details of proof are left to the reader (see also Figure 11).

$$A_\varepsilon = \begin{pmatrix} 0 & 2+\varepsilon & -1 \\ 2+\varepsilon & 0 & -1 \\ 1 & 1 & 0 \end{pmatrix}$$

Figure 5.

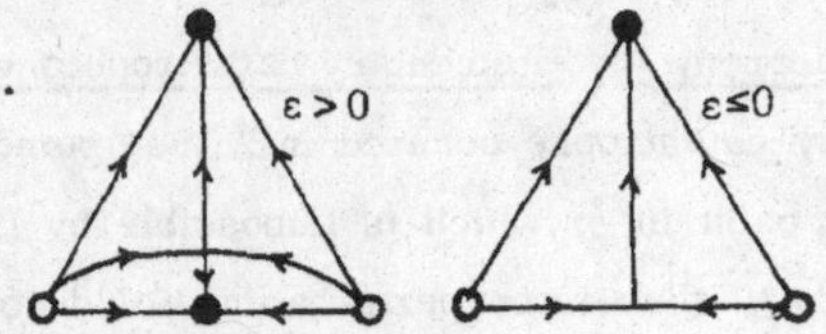

Example 4. The rock-scissors-paper game.

$$A = \begin{pmatrix} 0 & 1 & -1 \\ -1 & 0 & 1 \\ 1 & -1 & 0 \end{pmatrix}$$

Figure 6.

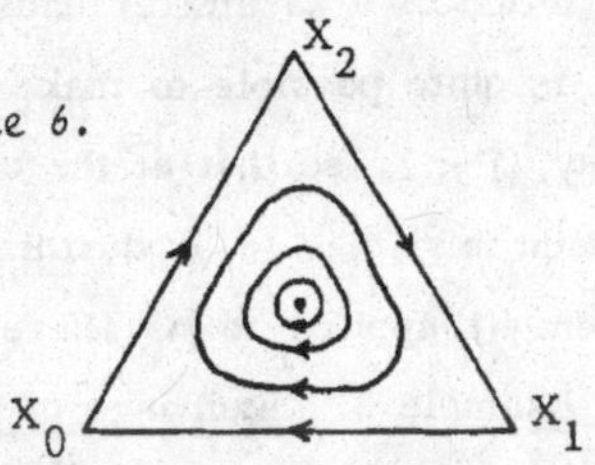

The associated dynamic is given by permuting cyclically

$$\dot{x}_0 = x_0(x_1-x_2).$$

Let $V = x_0x_1x_2$. Then V has a maximum at the barycentre e, and no other stationary points in $\mathring{\Delta}$ (by an argument as in the proof of Theorem 1). Meanwhile

$$\dot{V} = \Sigma\frac{V}{x_i}\dot{x}_i = (x_1-x_2) + (x_2-x_0) + (x_0-x_1) = 0 .$$

Therefore the orbits of φ_A in $\mathring{\Delta}$-e are the level curves of V, which are smooth simple closed curves surrounding e. The following perturbation shows that A is unstable.

Example 5. Degenerate Hopf bifurcation.

$$A_\varepsilon = \begin{pmatrix} 0 & 1+\varepsilon & -1 \\ -1 & 0 & 1+\varepsilon \\ 1+\varepsilon & -1 & 0 \end{pmatrix}$$

Figure 7.

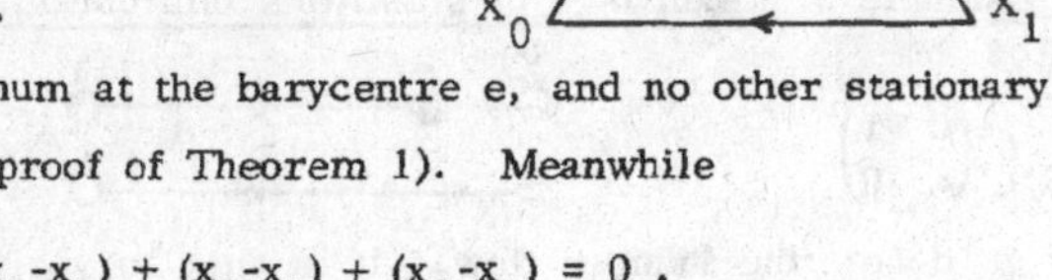

At the initial parameter value $\varepsilon = 0$ we have the previous example. When $\varepsilon \neq 0$ the same function V becomes a Lyapunov function for the flow, as we now show. The dynamic is given by permuting cyclically

$$\dot{x}_0 = x_0(x_1-x_2+\varepsilon(x_1-\sigma)), \text{ where } \sigma = x_0x_1+x_1x_2+x_2x_0 .$$

herefore $\dot{V} = \varepsilon(1-3\sigma)$. But σ has a maximum of $\frac{1}{3}$ at the barycentre e, and no other tationary points in Δ Therefore if $\varepsilon > 0$ then $\dot{V} > 0$ on $\mathring{\Delta} - e$, and so e is an *attractor vith basin of attraction $\mathring{\Delta}$. Similarly if $\varepsilon < 0$ then $\dot{V} < 0$ on $\mathring{\Delta} - e$, and so e is a repellor vith basin of repulsion $\mathring{\Delta}$. Therefore as the parameter passes through the critical value ne flow exhibits a Hopf bifurcation as the fixed point switches from attractor to repellor [5].

Notice that this is a "degenerate" Hopf bifurcation in the sense that all the cycles ccur at the critical value $\varepsilon = 0$, and so there are no small cycles before or after passing nrough the critical value. This type of Hopf bifurcation is called "degenerate" because it as codimension ∞ in the space of all 2-dimensional flows. However in our context it turns ut to be typical rather exceptional, because in Theorem 6 below we show that it has odimension 1, and in Theorem 7 that all Hopf bifurcations on a 2-simplex are of this autre. On the other hand if we raise the dimension by one then generic Hopf bifurcations o appear, as illustrated by the next example.

Example 6. Generic Hopf bifurcation.

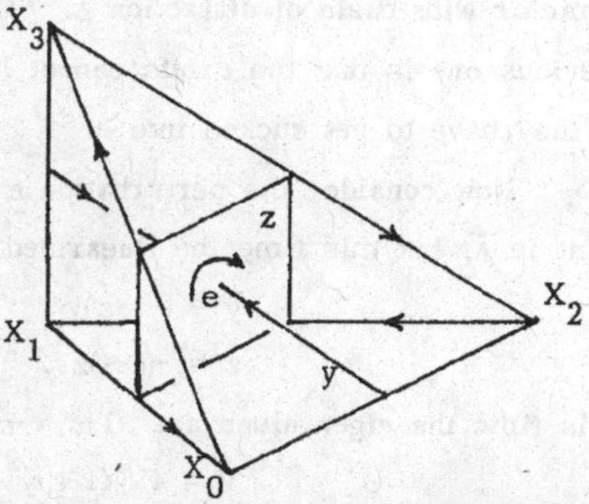

Figure 8.

$$A = 4\begin{pmatrix} 0 & 1 & \varepsilon & 0 \\ 0 & 0 & 1 & \varepsilon \\ \varepsilon & 0 & 0 & 1 \\ 1 & \varepsilon & 0 & 0 \end{pmatrix}$$

his example is due to Sigmund and his coworkers [11 part (ii)], and they have generalised to all $n \geq 3$. We first consider the critical case $\varepsilon = 0$, which they call the hypercycle, ince it represents a cycle of 4 chemicals each catalyzing the next. We shall show the arycentre e of the tetrahedron Δ is an attractor with basin $\mathring{\Delta}$. It is convenient to choose oordinates $(y, z) \in \mathbb{R} \times \mathbb{C}$, centred at e, given by

$$y = (x_0+x_2) - (x_1+x_3)$$
$$z = z_1+iz_2 = (x_0-x_2) + i(x_1-x_3)$$

here, for this example only, the notation i means $\sqrt{-1}$. The dynamic is given by ermuting cyclically

$$\dot{x}_0 = x_0(4x_1 - 1 + y^2) .$$

herefore in terms of y, z the dynamic can be rewritten

$$\dot{y} = -y+4z_1z_2+y^3$$
$$\dot{z} = -iz-(1-i)y\bar{z}+y^2z.$$

---

lternatively we could deduce this from Theorems 1 and 2, because e is an ESS, since $x = 0$ and $gx = \varepsilon(\frac{1}{3}-\sigma) > 0$ on $\Delta - e$. However this argument fails to generalise when we eed it for classification in Theorem 6 below.

The linear approximation at the fixed point is

$$\dot{y} = -y$$
$$\dot{z} = -iz \; .$$

Therefore the fixed point has eigenvalues -1, $\pm i$. Nevertheless e turns out to be an attractor, unlike the previous example. For consider the Lyapunov function $V = x_0 x_1 x_2 x_3$, which has a maximum at e and no other stationary points in $\mathring{\Delta}$. Then $\dot{V} = 4Vy^2$, and so $\dot{V} > 0$ on $\mathring{\Delta}$ except on the plane $y = 0$. If $y = 0$ and $z_1 z_2 \neq 0$ then $\dot{y} = 4z_1 z_2 \neq 0$, and so the orbit crosses this plane transversally. If $y = z_1 = 0$ and $z_2 \neq 0$ then $\dot{z}_1 = z_2 \neq 0$, and so the orbit crosses the z_2-axis transversally. Similarly orbits cross the z_1-axis transversally. Therefore V decreases strictly along all orbits in $\mathring{\Delta}$-e. Hence e is an attractor with basin of attraction $\mathring{\Delta}$. The subtlety of this example compared with the previous one is that the orbits cannot linger in the eigenspace of the eigenvalues $\pm i$, and so they have to get sucked into e.

Now consider the perturbation $\varepsilon \neq 0$. The barycentre e is again the unique fixed point in $\mathring{\Delta}$, but this time the linearised equations at e are :

$$\dot{y} = (-1+\varepsilon)y$$
$$\dot{z} = -(\varepsilon+i)z \; .$$

This time the eigenvalues are $-1+\varepsilon$, $-\varepsilon \pm i$, and

$$\dot{V} = 4V[(1-\varepsilon)y^2 + 2\varepsilon |z|^2] \; .$$

Hence if $0 < \varepsilon < 1$ then e is an attractor (indeed an ESS) with basin $\mathring{\Delta}$. On the other hand if $\varepsilon < 0$ then e is a 1-saddle. For small $\varepsilon < 0$ there must be an attracting small closed cycle near e by the Hopf bifurcation theorem [5], since there are no small cycles for $\varepsilon \geq 0$. This attracting cycle is shaped like the seam on a tennis ball, and as ε decreases it expands out to the cycle $X_3 X_2 X_1 X_0$ on the boundary.

Figure 9.

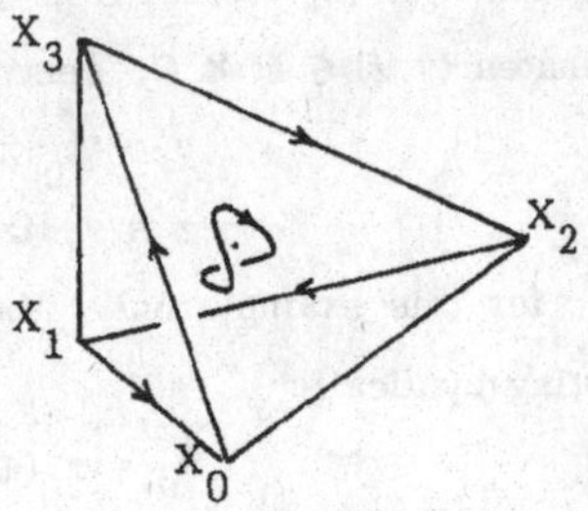

Chemically this example represents a mixture of 4 chemicals, and the Hopf bifurcation represents the continuous transition from a stable equilibrium into a little chemical clock - the precursor, perhaps, of the first biological clock? With only 3 chemicals this is impossible because by Theorem 7 below all Hopf bifurcations on a 2-simplex are degenerate as in the previous example, and so instead of getting a continuous transition from equilibrium to clock one would get a catastrophic breakdown of equilibrium, leading to the exclusion of two of the chemicals.

Section 3. Classification.

The eventual aim of classification is to list the equivalence classes, both the stable lasses and their bordering relations with those of higher codimension, and to describe he qualitative nature of the resulting flows, bifurcations and catastrophes. In particular he classification would involve giving criteria for two matrices to be equivalent, in other vords to induce topologically equivalent flows.

We begin very modestly in Lemma 3 by finding the condition for two matrices to nduce the same flow. For instance if a constant is added to a column of A then the flow is naltered. The interpretation of this in terms of game theory is as follows : if the ay-off to all strategies is increased equally then the relative advantage of each strategy is naltered, and so the evolution is the same. Therefore given any matrix we can, without ltering the flow, reduce its diagonal to zero by subtracting a suitable constant from each olumn. This simplifies the classification problem by reducing the dimension of the lassifying space; it also explains why we have chosen zero diagonal in all our examples.

Notation. Let $K_n \subset M_n$, be the set of n×n matrices all of whose columns are nultiples of u. Let $Z_n \subset M_n$, be the set of matrices with zero diagonal. Since $Z_n \cap K_n = 0$ we can write M_n as the direct sum or topological product

$$M_n = Z_n \times K_n .$$

et Z_n^+ denote the dense subset of Z_n consisting of matrices with zero diagonal and non-ero off-diagonal terms.

Lemma 3. Given $A, B \in M_{n+1}$ then $\varphi_A = \varphi_B \Longleftrightarrow A-B \in K_{n+1}$.

roof. Since $\dot{x}$ depends linearly upon A it suffices to prove $\varphi_A = 0$ if and only if $A \in K_n$.

$$\begin{aligned}
\varphi_A = 0 &\Longrightarrow \dot{x} = 0, \ \forall x \in \Delta \\
&\Longrightarrow (Ax)_i \text{ independent of } i, \ \forall i, x, \text{ such that } x_i \neq 0 \\
&\Longrightarrow a_{ii}t + a_{ij}(1-t) = a_{ji}t + a_{jj}(1-t), \ \forall i, j, t, \text{ such that } \\
&\qquad 0 < t < 1 \ (\text{putting } x_i = t, \ x_j = 1-t) \\
&\Longrightarrow a_{ij} = a_{jj}, \ \forall i, j \quad (\text{comparing coefficients}) \\
&\Longrightarrow A \in K_n .
\end{aligned}$$

onversely,
$$\begin{aligned}
A \in K_n &\Longrightarrow a_{ij} \text{ independent of } i, \ \forall i, j \\
&\Longrightarrow (Ax)_i \text{ independent of } i, \ \forall i, x \\
&\Longrightarrow x \text{ fixed}, \ \forall x, \text{ and so } \varphi_A = 0 .
\end{aligned}$$

Corollary. Every equivalence class in M_{n+1} is of the form $E \times K_{n+1}$, where E s an equivalence class of Z_{n+1}. Therefore stables are dense in M_{n+1} if and only if they re dense in Z_{n+1}, and to classify equivalence and stable classes in M_{n+1} it suffices to lassify them in Z_{n+1}.

Classification for n = 1. The corollary enables us to dispose of this case at once. ere Δ is a 1-simplex, and Z_2 consists of games of the form $A = \begin{pmatrix} 0 & a \\ b & 0 \end{pmatrix}$. By examining ne fixed points it is easy to verify there are 4 equivalence classes, as follows. In the rst two classes there is a fixed point $e = \left(\frac{a}{a+b}, \frac{b}{a+b}\right) \in \mathring{\Delta}$, which is an attractor in the rst class, and a repellor in the second. As usual, attractors are indicated by solid dots nd repellors by open dots. In class (iv) all points are fixed. Equivalences can be constructed by mapping fixed points to fixed points and extending piecewise linearly. If one of the variables changes sign while the other remains non-zero there is an exchange of stabilities bifurcation as in Example 3 above.

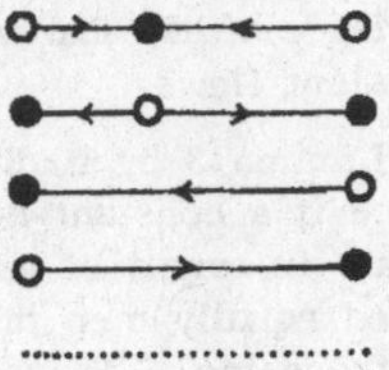

(i) $a, b > 0$

(ii) $a, b < 0$

(iii) $a \geq 0 \geq b$, not both zero

$a \leq 0 \leq b$, not both zero

(iv) $a = b = 0$.

Therefore A is stable $\Leftrightarrow A \in Z_2^+$. Therefore there are 3 stable classes (or 2 up to flow reversal since (i) is the reverse of (ii)), given by

(i) $a, b > 0$

(ii) $a, b < 0$

(iii) $a > 0 > b$ or $a < 0 < b$.

Lemma 4. $A \in Z_{n+1}$ and A stable $\Rightarrow A \in Z_{n+1}^+$.

Proof. Suppose not. Then $a_{ij} = 0$ for some $i \neq j$. Let Γ denote the edge ij. If $a_{ji} \neq 0$ then there are no fixed points in $\mathring{\Gamma}$, and a perturbation making a_{ij} the same sign as a_{ji} will introduce a fixed point in $\mathring{\Gamma}$, making an extra fixed point in the 1-skeleton of Δ, and hence an inequivalent flow. Therefore A is unstable. If $a_{ji} = 0$ then Γ is pointwise fixed, and a perturbation making a_{ij} non-zero will have no fixed point in $\mathring{\Gamma}$, making one fewer pointwise-fixed edge in the 1-skeleton, and hence an inequivalent flow. Therefore again A is unstable and the Lemma is proved.

Saddle points. Recall a fixed point is called hyperbolic if its eigenvalues have non-zero real part. It is called a saddle of index r, or more briefly an r-saddle, if the inset (= stable manifold) has dimension r and the outset (= unstable manifold) has dimension n - r. For instance an attractor is an n-saddle, and a repellor is a 0-saddle.

Lemma 5. If $A \in Z_{n+1}^+$ then all the vertices of Δ are hyperbolic. The index of X_j equals the number of negative terms in the jth column, and the inset, outset of X_j are open subsets of the faces $\{i; a_{ij} \leq 0\}$, $\{i; a_{ij} \geq 0\}$ respectively.

Proof. Taking x_i, $i \neq j$, as local coordinates at X_j, the linearization of the dynamic at X_j is $\dot{x}_i = a_{ij} x_i$, $i \neq j$. Hence the eigenvalues of X_j are a_{ij}, $i \neq j$, which are non-zero by the hypothesis $A \in Z_{n+1}^+$. Therefore X_j is hyperbolic with the required index. Since the faces of Δ are invariant X_j is an attractor, repellor of the induced flows on the two faces specified, and so its basins of attraction in them are open subsets of them, and these are the same as its inset, outset under φ_A.

Combinatorial equivalence. Given $A, B \in Z_{n+1}^+$ call them sign equivalent if corresponding off-diagonal elements have the same sign. Denote a sign class by the corresponding matrix of signs. Given a permutation σ of $\{0, 1, \ldots, n\}$ let σA denote the matrix obtained by permuting both rows and columns by σ. Call A, B combinatorially quivalent if $\sigma A, B$ are sign equivalent for some σ.

Lemma 6. Stable classes refine combinatorial classes.

roof. By Lemma 4 stable classes in Z_{n+1} are contained in Z^+_{n+1}, and so it suffices to how that $A \sim B$ implies A is combinatorially equivalent to B. Let h be a face-preserving omeomorphism of Δ inducing $A \sim B$. In particular h permutes the vertices, by a ermutation σ, say. If σ_* denotes the induced linear homeomorphism of Δ then σ_* gives n equivalence $A \sim \sigma A$. Therefore $h\sigma_*^{-1}$ gives an equivalence $\sigma A \sim B$, that fixes the ertices. But any equivalence maps insets to insets and outsets to outsets. Therefore σA s sign equivalent to B by Lemma 5, and so A is combinatorially equivalent to B, as equired.

Therefore the problem of classifying stable classes can be split into two, firstly he listing of combinatorial classes, and then the decomposition of them. It is a traightforward combinatorial task to list them, since each is characterised by the fixed oints in the 1-skeleton of Δ, although the list tends to get large as n increases. Meanwhile o decompose a combinatorial class it suffices to consider a single sign class (since all he other sign classes decompose isomorphically). In each sign class there seems to be elatively few equivalence classes, although to establish the actual decomposition in each ase appears to be a non-trivial problem.

Theorem 5. The number of combinatorial class (up to sign reversal) is as follows:-

n	1	2	3	...
number of classes	2	10	114	

The case n = 1 has been already done above; we shall prove the case n = 2 and leave n = 3 o the reader. Figure 10 illustrates the 10 cases for n = 2 by giving in each case an xample of the flow on the 1-skeleton. As abbreviated visual notation we only put an rrow on an edge if there is no fixed point in the interior of the edge, and otherwise ndicate the fixed point by a solid, open dot according as to whether it is an attractor, epellor for that edge, although of course when the flow is extended over the interior it nay in fact turn out to be a 1-saddle, depending upon the coefficients in A.

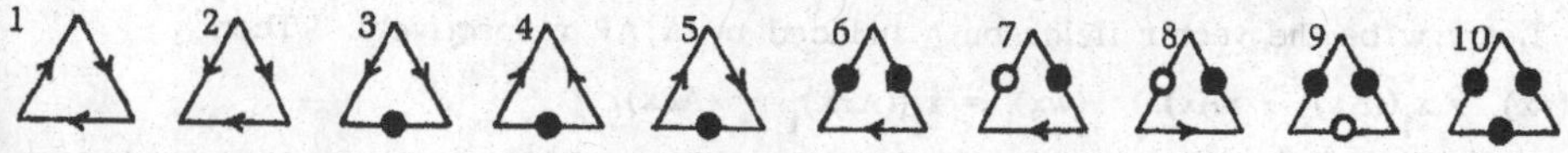

Figure 10.

roof. We compute the number N_r of classes having fixed points inside r edges by listing he inequivalent ways of putting arrows on the other edges. $N_0 = 2$ because the arrows can e cyclic or not, giving classes 1,2. $N_1 = 3$ because up to flow reversal we can choose he fixed point to be an attractor, and then the opposite vertex can be a repellor, attractor r saddle, giving classes 3,4,5. $N_2 = 3$ because if the two fixed points are similar the irection of the arrow does not matter, giving class 6, but if they differ it does, giving lasses 7,8. Finally $N_3 = 2$ because the three fixed points can be similar or not, giving lasses 9,10.

Now comes the more difficult business of decomposing combinatorial classes into table classes. We only attempt this for n = 2, because this dimension seems to have the following convenient property.

Conjecture 3. If $n = 2$ the fixed points determine the stable classes. The conjecture looks harmless, but is surprising because it implies there are no periodic orbits in the stable classes. This in turn implies that there are no generic Hopf bifurcations, but we prove this result separately in Theorem 7 below. (For $n \geq 3$ there are generic Hopf bifurcations by Example 6 above). The conjecture also implies the classification :

Corollary to Conjecture 3. If $n = 2$ there are 19 stable classes (up to flow reversal) as shown in Figure 11. Taylor and Jonker [14] and Schuster et al. [11, Part (iv)] have published computer drawings of some, but not all, of these 19 classes. We have arranged them in Figure 11 so that the first 5 classes are those with an attractor in the interior, the next 4 are those with a saddle, and the last 10 are those without a fixed point in the interior;of the latter the first 4 have one attractor on the boundary, the next 5 have two attractors, and the last has three attractors. In each class we have chosen a representative matrix such that, if there is a fixed point in the interior, it is the barycentre and, if not, the fixed points on the edges are at their barycentres. We have labelled each class by the combinatorial class containing it (see Figure 10), with a suffix if necessary, and a minus sign in those cases without a fixed point in the interior where the reverse flow has been chosen in order to maximise the number of attractors on the boundary. The three combinatorial classes 2, 3 and 8 are in fact equal to stables classes, but the other seven combinatorial classes each contain more than one stable class. In particular class 1 contains both the class shown and its reversal. It can be shown that the 19 cases are the only possible stable configurations of fixed points, and that these configurations are dense. In the last 14 cases it is easy to verify by Poincaré-Bendixson theory [3] that the fixed points determine the topology of the phase portrait, but in the first 5 cases this is not so obvious because it is necessary to prove the non-existence of periodic orbits surrounding the attractor. We prove this for class 1 in Theorem 6 below, but my proof for the other 4 classes is incomplete. Before we prove this we simplify the problem by showing how to move a fixed point in the interior to the barycentre.

Let P be a positive diagonal matrix $P = \begin{pmatrix} p_0 & & 0 \\ & \ddots & \\ 0 & & p_n \end{pmatrix}$, where $p_i > 0$, $i = 0, 1, \ldots, n$.

Let $p: \Delta \to \Delta$ be the induced projective map given by $(px)_i = \pi^{-1} p_i x_i$, where $\pi = \Sigma p_i x_i$.

Lemma 7. p induces an equivalence $AP \sim A$.

Proof. Let v, w be the vector fields on Δ induced by A, AP respectively. Then

$$(vx)_i = x_i((Ax)_i - xAx), \qquad (wx)_i = x_i((APx)_i - xAPx).$$

The derivative maps $((Dp)w)_i = \sum_j \frac{\partial}{\partial x_j}(px)_i w_j = \pi^{-1} p_i w_i - \pi^{-2} p_i x_i \Sigma_j p_j w_j$

$$\therefore ((Dp)wx)_i = \pi^{-1} p_i x_i (APx)_i - \pi^{-2} p_i x_i \Sigma p_j x_j (APx)_j$$

(since the other two terms cancel)

$$= p_i x_i ((A\pi^{-1}Px)_i - xP\pi^{-1}A\pi^{-1}Px)$$

$$= \pi(v(px))_i$$

Therefore Dp maps w onto v multiplied by the scalar π, and so p maps φ_{AP}-orbits to φ_A-orbits as required.

Call a matrix central if it has an isolated fixed point at the barycentre of Δ. Suppose we are now given A with an isolated fixed point $e \in \mathring{\Delta}$ (which is then the unique fixed point in $\mathring{\Delta}$ by Lemma 1). Let E denote the diagonal matrix with e along the diagonal. Define the centralisation of A to be the matrix $\bar{A} = (n+1)AE$.

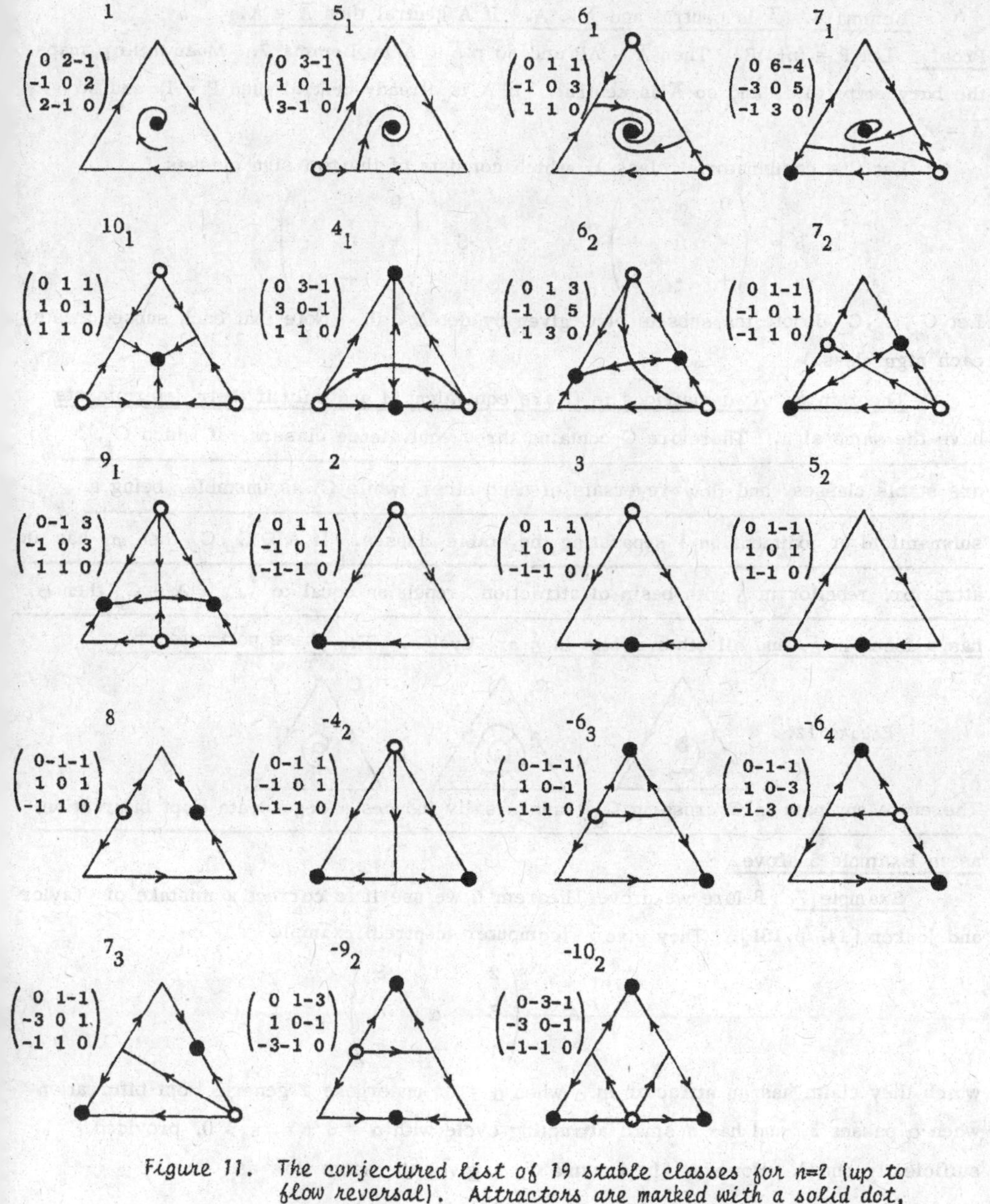

Figure 11. *The conjectured list of 19 stable classes for $n=2$ (up to flow reversal). Attractors are marked with a solid dot, repellors by an open dot, and saddles by their insets and outsets. All other orbits flow from a repellor to an attractor, except in class 1, where the α-limit is the boundary. The numbers refer to the combinatorial class in Figure 10, and a minus sign indicates flow reversal. A representative matrix is given for each class.*

Lemma 8. $\overline{A}$ is central and $\overline{A} \sim A$. If A central then $\overline{A} = A$.

Proof. Let $P = (n+1)E$. Then $\overline{A} = AP$ and so $p:\overline{A} \sim A$ by Lemma 7. Meanwhile p maps the barycentre to e, and so $\overline{A}$ is central. If A is already central then $P = I$, and so $\overline{A} = A$.

Let C = combinatorial class 1, which consists of the two sign classes

$$S = \begin{pmatrix} 0 & + & - \\ - & 0 & + \\ + & - & 0 \end{pmatrix}, \qquad -S = \begin{pmatrix} 0 & - & + \\ + & 0 & - \\ - & + & 0 \end{pmatrix}$$

Let C_+, C_0, C_- denote the subsets of C given by det $A \gtreqless 0$. (Note that each subset meets each sign class.)

Theorem 6. Two matrices in C are equivalent if and only if their determinants have the same sign. Therefore C contains three equivalence classes, of which C_+, C_- are stable classes and flow reversals of each other, while C_0 is unstable, being a submanifold of codimension 1 separating the stable classes. If $A \in C_+, C_-$ then φ_A has an attractor, repellor in $\mathring{\Delta}$ with basin of attraction, repulsion equal to $\mathring{\Delta}$. If $A \in C_0$ then φ_A has a focus in $\mathring{\Delta}$, and all other orbits in $\mathring{\Delta}$ are cycles. The phase portraits are :

Figure 12.

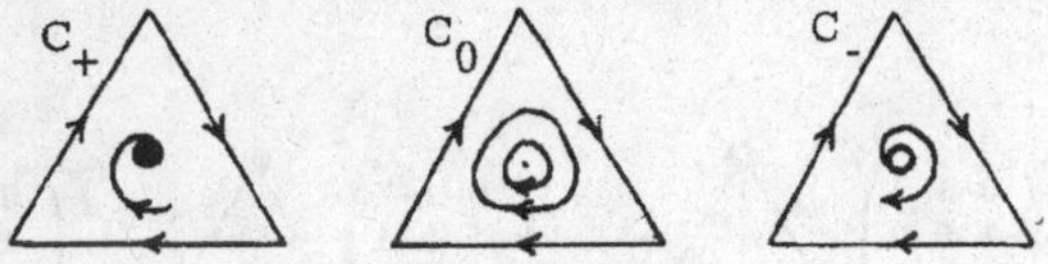

Therefore any path in C crossing C_0 transversally induces a degenerate Hopf bifurcation, as in Example 5 above.

Example 7. Before we prove Theorem 6 we use if to correct a mistake of Taylor and Jonker [14, p.151]. They give a (computer inspired) example

$$A = \begin{pmatrix} 2 & 1 & 5 \\ 5 & \alpha & 0 \\ 1 & 4 & 3 \end{pmatrix}$$

which they claim has an attractor in $\mathring{\Delta}$ when $\alpha = 3$, undergoes a generic Hopf bifurcation when α passes 3, and has a small attracting cycle with $\alpha = 3 + \varepsilon$, $\varepsilon > 0$, provided ε sufficiently small. However, by Lemma 3, A gives the same flow as

$$B = \begin{pmatrix} 0 & 1-\alpha & 2 \\ 3 & 0 & -3 \\ -1 & 4-\alpha & 0 \end{pmatrix}$$

and $\det B = 9(3-\alpha) = -9\varepsilon$. Therefore by Theorem 6 when $\alpha = 3$ the fixed point is a focus rather than an attractor, as α passes 3 the Hopf bifurcation is degenerate rather than generic, and when $0 < \varepsilon < 1$ the fixed point is a repellor with basin of repulsion $\mathring{\Delta}$, so there are no cycles.

Proof of Theorem 6. Up to equivalence it suffices to confine attention to the sign class S, because if $B \in -S$, and σ is any odd permutation, then $\sigma B \in S$, $\sigma B \sim B$ and det σB = det B. Therefore suppose $A \in S$. Then all the coefficients of adjA are positive, and hence so are those of (adjA)u. Therefore the vector subspace generated by (adjA)u meets $\mathring{\Delta}$ is a point e, which is therefore the unique fixed point of φ_A in $\mathring{\Delta}$. By Lemma 8, A is equivalent to its centralisation. Therefore up to equivalence it suffices to assume A is central, in other words e is the barycentre. Therefore, since $(Ae)_i$ is independent of i, the sum of the columns of A is a multiple of u, $= 2\theta u$ say. Therefore we can write

$$A = \begin{pmatrix} 0 & \theta+a_0 & \theta-a_0 \\ \theta-a_1 & 0 & \theta+a_1 \\ \theta+a_2 & \theta-a_2 & 0 \end{pmatrix}, \quad 0 \le |\theta| < a_i .$$

Then $\det A = 2\theta(\theta^2+\rho)$, where $\rho = a_0a_1 + a_1a_2 + a_2a_0 > 0$. Therefore $\det A \gtreqless$ as $\theta \gtreqless 0$.

We now construct a Lyapunov function V for φ_A in $\mathring{\Delta}$ as follows. For i = 0, 1, 2 let $b_i = \frac{b}{a_i}$, where $b = (\Sigma\frac{1}{a_i})^{-1}$. Then $b_i > 0$ and $\Sigma b_i = 1$. Given $x \in \mathring{\Delta}$, let

$$V = PQ, \text{ where } P = \prod x_i^{-b_i}, \; Q = \Sigma b_i x_i .$$

Then $V_i \equiv \frac{\partial V}{\partial x_i} = PQ_i + P_iQ = Pb_i - \frac{Pb_iQ}{x_i}$. By Lagrange's method V has a stationary point at x provided $V - \lambda(\Sigma x_i - 1)$ is stationary. $\therefore V_i - \lambda = 0$

$$\therefore Pb_ix_i - Pb_iQ - \lambda x_i = 0.$$

Summing over i, $\lambda = 0$. $\therefore x_i = Q$. $\therefore x = e$. Therefore e is the only stationary point of V in $\mathring{\Delta}$, and is a minimum because $V \to \infty$ as $x \to \partial\Delta$.

$$\begin{aligned}
\dot{V} = \Sigma V_i\dot{x}_i &= \Sigma Pb_i(1-\frac{Q}{x_i})x_i[(Ax)_i - xAx] \\
&= P\Sigma b_i(x_i-Q)[(Ax)_i - xAx] \\
&= P[\Sigma b_ix_i(Ax)_i - QbAx], \text{ since the other two terms cancel.}
\end{aligned}$$

$$\begin{aligned}
\text{Now } b_0x_0(Ax)_0 &= b_0x_0[(\theta+a_0)x_1 + (\theta-a_0)x_2] \\
&= \theta b_0(x_0x_1+x_0x_2) + b(x_0x_1-x_0x_2) \\
\therefore \Sigma b_ix_i(Ax)_i &= \theta\sum_{i<j}(b_i+b_j)x_ix_j . \\
b_0(Ax)_0 &= b_0[(\theta+a_0)x_1 + (\theta-a_0)x_2] \\
&= \theta b_0(x_1+x_2) + b(x_1-x_2) \\
\therefore bAx &= \theta\sum_{i\neq j} b_ix_j = \theta\Sigma_j(1-b_j)x_j , \quad \text{since } \Sigma b_i = 1 \\
\therefore QbAx &= \theta\Sigma b_ix_i\Sigma(1-b_j)x_j = \theta[\sum_i b_i(1-b_i)x_i^2 + \sum_{i\neq j} b_i(1-b_j)x_ix_j] \\
&= \theta\sum_{i<j}[b_ib_j(x_i^2+x_j^2) + (b_i+b_j-2b_ib_j)x_ix_j] \\
&= \theta\sum_{i<j}[b_ib_j(x_i-x_j)^2 + (b_i+b_j)x_ix_j] . \\
\therefore \dot{V} &= -\theta P\sum_{i<j} b_ib_j(x_i-x_j)^2
\end{aligned}$$

If $x \in \mathring{\Delta} - e$ then $x_i \neq x_j$ for some $i \neq j$, and so

$$\dot{V} \lesseqgtr 0 \text{ as } \theta \gtreqless 0, \text{ hence as } \det A \gtreqless 0.$$

If $\det A > 0$ all orbits in $\mathring{\Delta} - e$ flow towards the minimum e of V, which is therefore an attractor with basin $\mathring{\Delta}$. Similarly if $\det A < 0$ then e is a repellor with basin $\mathring{\Delta}$. If $\det A = 0$ then $\dot{V} = 0$ and the orbits of φ_A in $\mathring{\Delta} - e$ are the level curves of V, which are all simple closed curves surrounding e. We have established the phase portraits. Given two central matrices A, B whose determinants have the same sign we need to show they are equivalent, and in order to construct the required homeomorphism the following lemma is convenient.

Lemma 9. The orbits in $\mathring{\Delta} - e$ cross the rays through e transversally, going clockwise around e.

Proof. $\dot{x} \wedge (x-e)$ = multiple of u, $= \frac{1}{3}\delta u$ say. We need to show that $\delta > 0$ on $\mathring{\Delta} - e$. Throughout this proof let Σ denote the sum of the three terms obtained by permuting the suffices 012 cyclically. Adding the components of $\dot{x} \wedge (x-e)$ gives

$$\delta = \Sigma \dot{x}_0(x_1 - x_2) = \Sigma x_0(Ax)_0(x_1 - x_2).$$

When $\theta = 0$, $(Ax)_0 = a_0(x_1 - x_2)$, and so

$$\delta = \Sigma a_0 x_0 (x_1 - x_2)^2 > 0, \text{ since } x_i \neq 0, \text{ some } x_i \neq x_j.$$

When $\theta > 0$, $(Ax)_0 = (a_0 - \theta)(x_1 - x_2) + 2\theta x_1$, and so

$$\delta = \Sigma(a_0 - \theta)x_0(x_1 - x_2)^2 + 2\theta\alpha, \text{ where } \alpha = \Sigma x_0 x_1 (x_1 - x_2).$$

Since $|\theta| < a_i$ the first term > 0, and so it suffices to show $\alpha \geq 0$ in Δ. Let $\beta = \Sigma x_0(x_0 - x$
Now $\alpha, \beta \geq 0$ on $\partial\Delta$, because if $x = (s, 1-s, 0)$, $0 \leq s \leq 1$, or cyclically for the other sides, then

$$\alpha x = s(1-s)^2 \geq 0, \quad \beta x = 3s^2 - 3s + 1 > 0.$$

Therefore if $x_t = tx + (1-t)e$, $x \in \partial\Delta$, $0 \leq t \leq 1$,

$$\alpha(x_t) = t^3 \alpha x + \tfrac{1}{3} t^2 (1-t) \beta x \geq 0.$$

Finally the case $\theta < 0$ is obtained by reversing the flow and permuting 01.

Returning to the proof of Theorem 6, we can construct a radial homeomorphism of Δ keeping e and $\partial\Delta$ fixed, and throwing φ_A-orbits to φ_B-orbits by the standard technique of structural stability [10] of using the two Poincaré return maps on one particular ray, and extending orbitwise to the other rays. Hence $A \sim B$.

Finally C_0 is a submanifold of codimension 1 separating $C_{\pm}$ necause we can parametrise any matrix in C by the parameters (e, θ, a_i) where e denotes its fixed point and θ, a_i denote the parameters of its centralisation. This completes the proof of Theorem 6.

Remark. If we allow the parameters a_i to be negative, then we can use the same Lyapunov function V to determine the phase portraits of the four other stable classes in Figure 11 with an attractor (or repellor) in $\mathring{\Delta}$ provided $\rho > 0$. However, if $\rho < 0$ then V is no good because its stationary point becomes a saddle rather than a maximum or minimum, and so what is needed to complete the proof of Conjecture 3 is to find another Lyapunov function to cover the case $\rho < 0$. On the other hand we shall show that Hopf bifurcations can only occur when $\rho > 0$, and so we can at least classify those.

Suppose A has an isolated fixed point $e \in \mathring{\Delta}$, and centralisation

$$\overline{A} = 3AE = \begin{pmatrix} 0 & \theta+a_0 & \theta-a_0 \\ \theta-a_1 & 0 & \theta+a_1 \\ \theta+a_2 & \theta-a_2 & 0 \end{pmatrix}, \quad \theta, a_i \in \mathbb{R}$$

We call θ, a_i the central parameters of A. As before let $\rho = a_0a_1+a_1a_2+a_2a_0$.

Lemma 10. The eigenvalues of φ_A at e are $\frac{1}{3}(-\theta \pm \sqrt{-\rho})$. Therefore they depend only on the central parameters and not on e.

Proof. Let $x = e+y$. Then $yu = 0$ and so $yAe = 0$.

$$\therefore \dot{y}_i = (e_i+y_i)[(Ae)_i + (Ay)_i - eAe - eAy - yAe - yAy]$$
$$= e_i[(Ay)_i - eAy] + \text{higher order terms in } y.$$

Therefore the linearisation et e is

$$\dot{y} = EAy - e(eAy) = (E-ee)Ay.$$

The eigenvalues are the same as those of the matrix

$$M = A(E-ee) = AE - (Ae)e = AE - (\tfrac{2}{3}\theta u)e, \text{ since } Ae = AEu = \tfrac{2}{3}\theta u,$$

$$= \frac{1}{3}\begin{pmatrix} -2\theta e_0 & \theta+a_0-2\theta e_1 & \theta-a_0-2\theta e_2 \\ \theta-a_1-2\theta e_0 & -2\theta e_1 & \theta+a_1-2\theta e_2 \\ \theta+a_2-2\theta e_0 & \theta-a_2-2\theta e_1 & -2\theta e_2 \end{pmatrix}$$

Now $\det M = 0$ because $(E-ee)u = 0$. Therefore the eigenvalues are given by $\lambda^3-2\alpha\lambda^2+\beta\lambda = 0$, where

$$2\alpha = \text{trace } M = -\frac{2\theta}{3} \quad \therefore \alpha = -\frac{\theta}{3}.$$
$$\beta = \text{trace}(\text{adj}M) = \frac{1}{9}\Sigma[-(\theta+a_0)(\theta-a_1) + 2\theta e_0(\theta+a_0) + 2\theta e_1(\theta-a_1)]$$
$$= \frac{1}{9}(\theta^2+\rho).$$

The eigenspace corresponding to $\lambda = 0$ is transverse to Δ, and so the eigenvalues for φ_A are $\lambda = \alpha \pm \sqrt{\alpha^2-\beta} = \frac{1}{3}(-\theta \pm \sqrt{-\rho})$.

Theorem 7. When $n = 2$ all Hopf bifurcations are degenerate.

Proof. For a Hopf bifurcation to occur at a matrix A_0, it is necessary for it to have an isolated fixed point $e \in \mathring{\Delta}$ with pure imaginary eigenvalues. Therefore if θ, a_i denote the central parameters of A_0 then $\theta = 0$ and $\rho > 0$ by Lemma 10. Up to equivalence it suffices to assume A_0 central, by Lemma 8, since this does not affect the eigenvalues, by Lemma 10. There are three cases :-

(1) All a_i non-zero and the same sign.

(2) All a_i non-zero, but not all the same sign.

(3) Some $a_i = 0$.

Case (1) is covered by Theorem 6. In case (2), by permuting and reversing sign if necessary, we can assume

$$a_0, a_1 > 0, \quad a_2 < 0.$$

As before let $b = (\Sigma\frac{1}{a_i})^{-1} = \frac{a_0a_1a_2}{\rho}$, $b_i = \frac{b}{a_i}$. Then $\Sigma b_i = 1$ and $b_2 > 0$. but $b, b_0, b_1 < 0$. We use the same Lyapunov function V as in Theorem 6, and again the only stationary point of V in $\mathring{\Delta}$ is the barycentre e. However, this time V = 0 on the edges X_0X_2, X_1X_2 and on the line Q = 0, which meets those edges in points

$$Y = \left(\frac{a_0}{a_0-a_2}, 0, \frac{-a_2}{a_0-a_2}\right), \quad Z = \left(0, \frac{a_1}{a_1-a_2}, \frac{-a_2}{a_1-a_2}\right).$$

Also $V \to -\infty$ as $x \to$ interior (edge X_0X_1). The level curves of V are illustrated in the middle picture of Figure 13 below. The condition $\rho > 0$ implies that e lies inside the triangle X_2YZ, in which $V > 0$, and so e is a maximum, and the level curves inside the triangle are simple closed curves surrounding e. Meanwhile inside the complementary trapezium $V < 0$ and the level curves are arcs joining X_0, X_1. As before $\dot{V} = 0$ in $\mathring{\Delta}$ because $\theta = 0$. Therefore the orbits of φ_{A_0} are the level curves of V, with a focus at e, an attractor at X_0, a repellor at X_1, saddles at X_2, Y, Z, and a saddle-connection ZY, as shown in Figure 13.

Now consider a perturbation A of A_0. We can assume A has an isolated fixed point in $\mathring{\Delta}$, since this is a robust property by Theorem 4, and so up to equivalence we can centralise A. Therefore we can write A in the same form as in the proof of Theorem 6, with fixed point at the barycentre e, and central parameters θ, a_i satisfying

$$a_0, a_1 > 0, \quad a_2 < 0, \quad |\theta| < |a_2|, \quad \rho > 0.$$

As before

$$\dot{V} = -\theta P \sum_{i<j} b_i b_j (x_i - x_j)^2 = K\psi \text{ say, where}$$

$$K = -\frac{6Pb^2}{a_0a_1a_2} \gtrless 0 \text{ as } \theta \gtrless 0, \text{ since } a_2 < 0, \text{ and}$$

$$\psi = \text{cyclic sum } \Sigma a_i(x_j - x_k)^2 .$$

<u>Lemma 11. $\psi > 0$ on $\Delta - e$.</u>

<u>Proof.</u> Notice that the sum of any pair of a_i's is positive, for :

$$\begin{aligned} a_0+a_1 &> 0 &&, \text{ since } a_0, a_1 > 0 \\ a_0(a_1+a_2) &> -a_1a_2 &&, \text{ since } \rho > 0 \\ &> 0 &&, \text{ since } a_1 > 0 > a_2 \\ \therefore a_1+a_2 &> 0 &&, \text{ since } a_0 > 0 . \\ a_0+a_2 &> 0 &&, \text{ similarly.} \end{aligned}$$

It suffices to prove the lemma for $x \in \partial\Delta$, because if $x_t = tx+(1-t)e$ then $\psi(x_t) = t^2\psi x > 0$, $\forall t$, $0 < t \leq 1$. If x = (s, 1-s, 0) then

$$\psi x = a_0(1-s)^2 + a_1 s^2 + a_2(2s-1)^2$$
$$= (a_0+a_1+4a_2)s^2 - 2(a_0+2a_2)s + (a_0+a_2)$$
$$> 0, \ \forall s, \text{ since}$$

$a_0 + a_2 > 0$, and $(a_0+a_1+4a_2)(a_0+a_2) - (a_0+2a_2)^2 = \rho > 0$. Therefore $\psi > 0$ on X_0X_1, and similarly on the other edges. This completes the proof of Lemma 11.

Continuing with the proof of Theorem 7 we have shown $\dot{V} \gtreqless 0$ as $\theta \gtreqless 0$ on $\mathring{\Delta} - e$. Therefore if $\theta > 0$ then φ_A has an attractor at e. It also has another attractor at X_0, a repellor at X_1, and saddles at X_2, Y_θ, Z_θ, where

$$Y_\theta = \left(\frac{a_0-\theta}{a_0-a_2-2\theta}, 0, \frac{-a_2-\theta}{a_0-a_2-2\theta}\right) \in YX_0$$

$$Z_\theta = \left(0, \frac{a_1+\theta}{a_1-a_2+2\theta}, \frac{-a_2+\theta}{a_1-a_2+2\theta}\right) \in ZX_2$$

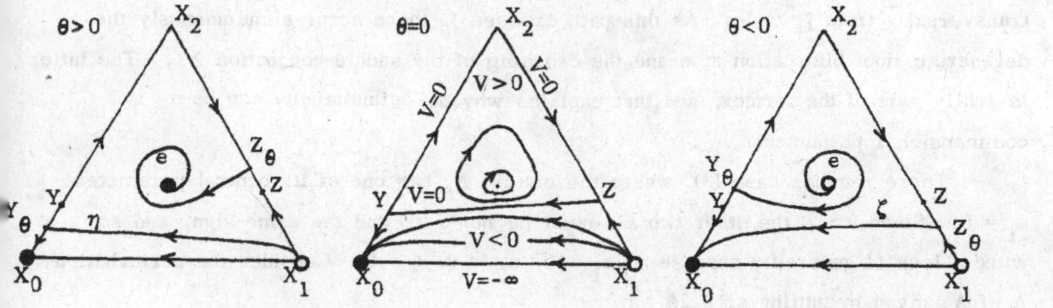

Figure 13.

Therefore, since $\dot{V} > 0$, the inset η of Y_θ must come from the repellor X_1, and the outset of Z_θ must go to the attractor e. Since $\dot{V} > 0$ there are no closed cycles in $\mathring{\Delta} - e$, and so all orbits in $\mathring{\Delta} - e$ must come from X_1, and, except for η, must go to X_0 or e. Therefore η separates the basins of attraction of X_0 and e, and the phase portrait is as in Figure 13. The numerical Example 1 in Section 1 above was obtained by putting $a_0 = 5$, $a_1 = 4$, $a_2 = -2$, $\theta = 1$.

If $\theta < 0$ then the reverse situation occurs, with X_0 the only attractor, e a repellor, and the outset ζ of Z_θ separating the basins of repulsion of e and X_1.

Let J denote the open subset of the sign class $\begin{pmatrix} 0 & + & - \\ - & 0 & + \\ - & + & 0 \end{pmatrix}$ consisting of matrices having an isolated fixed point in $\mathring{\Delta}$, and central parameters θ, a_i such that $a_1, a_2 > 0$, $a_2 < 0$, $|\theta| < |a_2|$, $\rho > 0$. We have shown that the subsets J_+, J_0, J_- of J given by $\theta \gtreqless 0$ have phase portraits as in Figure 13. Given two matrices in the same subset we show equivalence by constructing a homeomorphism of Δ throwing orbits to orbits, as follows. When $\theta \neq 0$

the flows are gradient-like, so the construction uses the standard techniques of structural stability [9], mapping fixed points to fixed points, and extending inductively to tubular neighbourhoods of their insets, starting with repellors and finishing with attractors. When $\theta = 0$ again map fixed points to fixed points and extend piecewise linearly to $\partial\Delta \cup YZ$; then map the inside of the triangle radially from e so as to preserve orbits, and use the structural stability technique inside the trapezium.

Therefore $J_\pm, J_0$ are the intersections of J with 3 equivalence classes. Since $J_\pm$ are open, they are contained in stable classes; J_+ is contained in class 7_1 of Figure 11, and J_- in the reversal. However they are not connected components of the stable classes, because the latter also contain matrices for which $\rho \leq 0$. On the other hand J_0 is a connected component of its equivalence class because by Lemma 9 a focus implies $\rho > 0$; the other 5 components are obtained by the action on the triangle of the symmetry group of order 6. Now J is a neighbourhood of J_0, and J_0 is a submanifold of codimension 1 separating $J_\pm$. Therefore to obtain a Hopf bifurcation we must take a path in J crossing J_0 transversally from J_+ to J_-. As this path crosses J_0 there occur simultaneously the degenerate Hopf bifurcation at e and the crossing of the saddle-connection ZY. The latter is really part of the former, and that explains why the simultaneity can be a codimension 1 phenomenon.

There remains case (3), where the matrix A_0 has one of its central parameters $a_i = 0$. Since $\rho > 0$ the other two a's must be non-zero and the same sign, and so without loss of generality suppose $a_0, a_1 > 0$, $a_2 = 0$, $\theta = 0$. Consider the perturbation A of A_0 given by putting $a_2 = 2\theta \neq 0$.

$$A = \begin{pmatrix} 0 & \theta+a_0 & \theta-a_0 \\ \theta-a_1 & 0 & \theta+a_1 \\ 3\theta & -\theta & 0 \end{pmatrix}, \quad |\theta| < a_0, a_1 .$$

When $\theta > 0$ the phase portrait of φ_A is as in case (1), the left-hand picture in Figure 12, and when $\theta < 0$ it is as in case (2), the right-hand picture in Figure 13. Therefore there are no small cycles when $\theta \neq 0$, and so by the Hopf bifurcation theorem [5] there is a 1-parameter family of cycles surrounding e in the phase portrait of φ_{A_0}. Therefore any path through A_0 transverse to $\theta = 0$ induces a degenerate Hopf bifurcation. This completes the proof of Theorem 7.

References.

1. M. Eigen & P. Schuster, The Hypercycle, a principal of natural self organisation. (A) Emergence of the hypercycle *Naturwissenschaften* 64 (1977) 541-565; (B) The abstract hypercycle 65 (1978), 7-41; (C) The realistic hypercycle 65 (1978) 341-369.

2. W.G.S. Hines, Strategy stability in complex populations, Preprint, Guelph, Ontario, 1979.

3. M.W. Hirsch & S. Smale, *Differential equations, dynamical systems and linear algebra*, Academic, New York, London, 1974.

4. J. Hofbauer, P. Schuster & K. Sigmund, A note on evolutionary stable strategies and game dynamics, *J. theor. Biol*, (to appear).

5. J. Marsden & M. McCracken, *The Hopf bifurcation and its applications*, Appl. Math. Sc. 19 Springer, New York, 1976.

6. J. Maynard Smith & G.R. Price, The logic of animal conflicts, *Nature* 246 (1973) 5427, 15-18.

7. J. Maynard Smith, The theory of games and the evolution of animal conflicts, *J. theor. Biol.* 47 (1974) 209-221.

8. J. Maynard Smith, Evolution and the theory of games, *Am. Scientist*, 64 (1976) 41-45.

9. J. Palis & S. Smale, Structural stability theorems, *Global Analysis, Proc. Symp. Pure Math.* 14, Am. Math. Soc. (1970) 223-231.

10. M.M. Peixoto, Structural stability on two-dimensional manifolds, *Topology* 1 (1962) 101-120.

11. P. Schuster, K. Sigmund & R. Wolff, Dynamical systems under constant organisation. I : Topological analysis of a family of non-linear differential equations, *Bull. Math. Biophys.* 40 (1978), 743-769. II (with J. Hofbauer) : Homogeneous growth of functions of degree p = 2, *SIAM J. Appl. Math.* (to appear). III : Cooperative and competitive behaviour of hypercycles, *J. Diff. Eq.* 32 (1979), 357-368. IV : Second order growth terms for mass action kinetics, Preprint, Vienna, 1979.

12. P. Schuster, K. Sigmund & R. Wolff, A mathematical model of the hypercycle, Preprint, Vienna, 1979.

13. P. Schuster & K. Sigmund, Coyness, philandering and stable strategies, Preprint, Vienna, 1979.

14. P.D. Taylor & L.B. Jonker, Evolutionarily stable strategies and game dynamics, *Math. Biosc.* 40 (1978) 145-156.

15. R. Thom, *Structural stability and morphogenesis*, (Trans. D.H. Fowler), Benjamin, Reading, Massachusetts, 1975.

16. E.C. Zeeman, Dynamics of the evolution of animal conflicts, Preprint, Warwick, 1979.

University of Warwick
Coventry, England

INDEX

Numbers refer to papers as listed in the contents

ol. 670: Fonctions de Plusieurs Variables Complexes III, Proeedings, 1977. Edité par F. Norguet. XII, 394 pages. 1978.

ol. 671: R. T. Smythe and J. C. Wierman, First-Passage Perculation n the Square Lattice. VIII, 196 pages. 1978.

ol. 672: R. L. Taylor, Stochastic Convergence of Weighted Sums f Random Elements in Linear Spaces. VII, 216 pages. 1978.

ol. 673: Algebraic Topology, Proceedings 1977. Edited by P. Hoffan, R. Piccinini and D. Sjerve. VI, 278 pages. 1978.

ol. 674: Z. Fiedorowicz and S. Priddy, Homology of Classical roups Over Finite Fields and Their Associated Infinite Loop Spaces. , 434 pages. 1978.

ol. 675: J. Galambos and S. Kotz, Characterizations of Probability istributions. VIII, 169 pages. 1978.

ol. 676: Differential Geometrical Methods in Mathematical Physics II, roceedings, 1977. Edited by K. Bleuler, H. R. Petry and A. Reetz. I, 626 pages. 1978.

ol. 677: Séminaire Bourbaki, vol. 1976/77, Exposés 489–506. IV, 64 pages. 1978.

ol. 678: D. Dacunha-Castelle, H. Heyer et B. Roynette. Ecole d'Eté e Probabilités de Saint-Flour. VII-1977. Edité par P. L. Hennequin. X, 379 pages. 1978.

ol. 679: Numerical Treatment of Differential Equations in Applicaons, Proceedings, 1977. Edited by R. Ansorge and W. Törnig. IX, 63 pages. 1978.

ol. 680: Mathematical Control Theory, Proceedings, 1977. Edited y W. A. Coppel. IX, 257 pages. 1978.

ol. 681: Séminaire de Théorie du Potentiel Paris, No. 3, Directeurs: I. Brelot, G. Choquet et J. Deny. Rédacteurs: F. Hirsch et G. Mokoodzki. VII, 294 pages. 1978.

ol. 682: G. D. James, The Representation Theory of the Symmetric Groups. V, 156 pages. 1978.

ol. 683: Variétés Analytiques Compactes, Proceedings, 1977. Edité par Y. Hervier et A. Hirschowitz. V, 248 pages. 1978.

ol. 684: E. E. Rosinger, Distributions and Nonlinear Partial Diferential Equations. XI, 146 pages. 1978.

ol. 685: Knot Theory, Proceedings, 1977. Edited by J. C. Hausmann. II, 311 pages. 1978.

ol. 686: Combinatorial Mathematics, Proceedings, 1977. Edited y D. A. Holton and J. Seberry. IX, 353 pages. 1978.

ol. 687: Algebraic Geometry, Proceedings, 1977. Edited by L. D. Olson. V, 244 pages. 1978.

ol. 688: J. Dydak and J. Segal, Shape Theory. VI, 150 pages. 1978.

ol. 689: Cabal Seminar 76–77, Proceedings, 1976–77. Edited by .S. Kechris and Y. N. Moschovakis. V, 282 pages. 1978.

ol. 690: W. J. J. Rey, Robust Statistical Methods. VI, 128 pages. 1978.

ol. 691: G. Viennot, Algèbres de Lie Libres et Monoïdes Libres. I, 124 pages. 1978.

ol. 692: T. Husain and S. M. Khaleelulla, Barrelledness in Topoogical and Ordered Vector Spaces. IX, 258 pages. 1978.

ol. 693: Hilbert Space Operators, Proceedings, 1977. Edited by . M. Bachar Jr. and D. W. Hadwin. VIII, 184 pages. 1978.

ol. 694: Séminaire Pierre Lelong – Henri Skoda (Analyse) Année 976/77. VII, 334 pages. 1978.

ol. 695: Measure Theory Applications to Stochastic Analysis, Proceedings, 1977. Edited by G. Kallianpur and D. Kölzow. XII, 261 pages. 1978.

ol. 696: P. J. Feinsilver, Special Functions, Probability Semigroups, and Hamiltonian Flows. VI, 112 pages. 1978.

ol. 697: Topics in Algebra, Proceedings, 1978. Edited by M. F. Newman. XI, 229 pages. 1978.

ol. 698: E. Grosswald, Bessel Polynomials. XIV, 182 pages. 1978.

ol. 699: R. E. Greene and H.-H. Wu, Function Theory on Manifolds Which Possess a Pole. III, 215 pages. 1979.

Vol. 700: Module Theory, Proceedings, 1977. Edited by C. Faith and S. Wiegand. X, 239 pages. 1979.

Vol. 701: Functional Analysis Methods in Numerical Analysis, Proceedings, 1977. Edited by M. Zuhair Nashed. VII, 333 pages. 1979.

Vol. 702: Yuri N. Bibikov, Local Theory of Nonlinear Analytic Ordinary Differential Equations. IX, 147 pages. 1979.

Vol. 703: Equadiff IV, Proceedings, 1977. Edited by J. Fábera. XIX, 441 pages. 1979.

Vol. 704: Computing Methods in Applied Sciences and Engineering, 1977, I. Proceedings, 1977. Edited by R. Glowinski and J. L. Lions. VI, 391 pages. 1979.

Vol. 705: O. Forster und K. Knorr, Konstruktion verseller Familien kompakter komplexer Räume. VII, 141 Seiten. 1979.

Vol. 706: Probability Measures on Groups, Proceedings, 1978. Edited by H. Heyer. XIII, 348 pages. 1979.

Vol. 707: R. Zielke, Discontinuous Čebyšev Systems. VI, 111 pages. 1979.

Vol. 708: J. P. Jouanolou, Equations de Pfaff algébriques. V, 255 pages. 1979.

Vol. 709: Probability in Banach Spaces II. Proceedings, 1978. Edited by A. Beck. V, 205 pages. 1979.

Vol. 710: Séminaire Bourbaki vol. 1977/78, Exposés 507–524. IV, 328 pages. 1979.

Vol. 711: Asymptotic Analysis. Edited by F. Verhulst. V, 240 pages. 1979.

Vol. 712: Equations Différentielles et Systèmes de Pfaff dans le Champ Complexe. Edité par R. Gérard et J.-P. Ramis. V, 364 pages. 1979.

Vol. 713: Séminaire de Théorie du Potentiel, Paris No. 4. Edité par F. Hirsch et G. Mokobodzki. VII, 281 pages. 1979.

Vol. 714: J. Jacod, Calcul Stochastique et Problèmes de Martingales. X, 539 pages. 1979.

Vol. 715: Inder Bir S. Passi, Group Rings and Their Augmentation Ideals. VI, 137 pages. 1979.

Vol. 716: M. A. Scheunert, The Theory of Lie Superalgebras. X, 271 pages. 1979.

Vol. 717: Grosser, Bidualräume und Vervollständigungen von Banachmoduln. III, 209 pages. 1979.

Vol. 718: J. Ferrante and C. W. Rackoff, The Computational Complexity of Logical Theories. X, 243 pages. 1979.

Vol. 719: Categorial Topology, Proceedings, 1978. Edited by H. Herrlich and G. Preuß. XII, 420 pages. 1979.

Vol. 720: E. Dubinsky, The Structure of Nuclear Fréchet Spaces. V, 187 pages. 1979.

Vol. 721: Séminaire de Probabilités XIII. Proceedings, Strasbourg, 1977/78. Edité par C. Dellacherie, P. A. Meyer et M. Weil. VII, 647 pages. 1979.

Vol. 722: Topology of Low-Dimensional Manifolds. Proceedings, 1977. Edited by R. Fenn. VI, 154 pages. 1979.

Vol. 723: W. Brandal, Commutative Rings whose Finitely Generated Modules Decompose. II, 116 pages. 1979.

Vol. 724: D. Griffeath, Additive and Cancellative Interacting Particle Systems. V, 108 pages. 1979.

Vol. 725: Algèbres d'Opérateurs. Proceedings, 1978. Edité par P. de la Harpe. VII, 309 pages. 1979.

Vol. 726: Y.-C. Wong, Schwartz Spaces, Nuclear Spaces and Tensor Products. VI, 418 pages. 1979.

Vol. 727: Y. Saito, Spectral Representations for Schrödinger Operators With Long-Range Potentials. V, 149 pages. 1979.

Vol. 728: Non-Commutative Harmonic Analysis. Proceedings, 1978. Edited by J. Carmona and M. Vergne. V, 244 pages. 1979.

Vol. 729: Ergodic Theory. Proceedings, 1978. Edited by M. Denker and K. Jacobs. XII, 209 pages. 1979.

Vol. 730: Functional Differential Equations and Approximation of Fixed Points. Proceedings, 1978. Edited by H.-O. Peitgen and H.-O. Walther. XV, 503 pages. 1979.

Vol. 731: Y. Nakagami and M. Takesaki, Duality for Crossed Products of von Neumann Algebras. IX, 139 pages. 1979.

Vol. 732: Algebraic Geometry. Proceedings, 1978. Edited by K. Lønsted. IV, 658 pages. 1979.

Vol. 733: F. Bloom, Modern Differential Geometric Techniques in the Theory of Continuous Distributions of Dislocations. XII, 206 pages. 1979.

Vol. 734: Ring Theory, Waterloo, 1978. Proceedings, 1978. Edited by D. Handelman and J. Lawrence. XI, 352 pages. 1979.

Vol. 735: B. Aupetit, Propriétés Spectrales des Algèbres de Banach. XII, 192 pages. 1979.

Vol. 736: E. Behrends, M-Structure and the Banach-Stone Theorem. X, 217 pages. 1979.

Vol. 737: Volterra Equations. Proceedings 1978. Edited by S.-O. Londen and O. J. Staffans. VIII, 314 pages. 1979.

Vol. 738: P. E. Conner, Differentiable Periodic Maps. 2nd edition, IV, 181 pages. 1979.

Vol. 739: Analyse Harmonique sur les Groupes de Lie II. Proceedings, 1976-78. Edited by P. Eymard et al. VI, 646 pages. 1979.

Vol. 740: Séminaire d'Algèbre Paul Dubreil. Proceedings, 1977-78. Edited by M.-P. Malliavin. V, 456 pages. 1979.

Vol. 741: Algebraic Topology, Waterloo 1978. Proceedings. Edited by P. Hoffman and V. Snaith. XI, 655 pages. 1979.

Vol. 742: K. Clancey, Seminormal Operators. VII, 125 pages. 1979.

Vol. 743: Romanian-Finnish Seminar on Complex Analysis. Proceedings, 1976. Edited by C. Andreian Cazacu et al. XVI, 713 pages. 1979.

Vol. 744: I. Reiner and K. W. Roggenkamp, Integral Representations. VIII, 275 pages. 1979.

Vol. 745: D. K. Haley, Equational Compactness in Rings. III, 167 pages. 1979.

Vol. 746: P. Hoffman, τ-Rings and Wreath Product Representations. V, 148 pages. 1979.

Vol. 747: Complex Analysis, Joensuu 1978. Proceedings, 1978. Edited by I. Laine, O. Lehto and T. Sorvali. XV, 450 pages. 1979.

Vol. 748: Combinatorial Mathematics VI. Proceedings, 1978. Edited by A. F. Horadam and W. D. Wallis. IX, 206 pages. 1979.

Vol. 749: V. Girault and P.-A. Raviart, Finite Element Approximation of the Navier-Stokes Equations. VII, 200 pages. 1979.

Vol. 750: J. C. Jantzen, Moduln mit einem höchsten Gewicht. III, 195 Seiten. 1979.

Vol. 751: Number Theory, Carbondale 1979. Proceedings. Edited by M. B. Nathanson. V, 342 pages. 1979.

Vol. 752: M. Barr, *-Autonomous Categories. VI, 140 pages. 1979.

Vol. 753: Applications of Sheaves. Proceedings, 1977. Edited by M. Fourman, C. Mulvey and D. Scott. XIV, 779 pages. 1979.

Vol. 754: O. A. Laudal, Formal Moduli of Algebraic Structures. III, 161 pages. 1979.

Vol. 755: Global Analysis. Proceedings, 1978. Edited by M. Grmela and J. E. Marsden. VII, 377 pages. 1979.

Vol. 756: H. O. Cordes, Elliptic Pseudo-Differential Operators – An Abstract Theory. IX, 331 pages. 1979.

Vol. 757: Smoothing Techniques for Curve Estimation. Proceedings, 1979. Edited by Th. Gasser and M. Rosenblatt. V, 245 pages. 1979.

Vol. 758: C. Năstăsescu and F. Van Oystaeyen; Graded and Filtered Rings and Modules. X, 148 pages. 1979.

Vol. 759: R. L. Epstein, Degrees of Unsolvability: Structure an Theory. XIV, 216 pages. 1979.

Vol. 760: H.-O. Georgii, Canonical Gibbs Measures. VIII, 190 pages 1979.

Vol. 761: K. Johannson, Homotopy Equivalences of 3-Manifolds wit Boundaries. 2, 303 pages. 1979.

Vol. 762: D. H. Sattinger, Group Theoretic Methods in Bifurcatio Theory. V, 241 pages. 1979.

Vol. 763: Algebraic Topology, Aarhus 1978. Proceedings, 1978 Edited by J. L. Dupont and H. Madsen. VI, 695 pages. 1979.

Vol. 764: B. Srinivasan, Representations of Finite Chevalley Groups XI, 177 pages. 1979.

Vol. 765: Padé Approximation and its Applications. Proceedings 1979. Edited by L. Wuytack. VI, 392 pages. 1979.

Vol. 766: T. tom Dieck, Transformation Groups and Representatio Theory. VIII, 309 pages. 1979.

Vol. 767: M. Namba, Families of Meromorphic Functions on Compac Riemann Surfaces. XII, 284 pages. 1979.

Vol. 768: R. S. Doran and J. Wichmann, Approximate Identities an Factorization in Banach Modules. X, 305 pages. 1979.

Vol. 769: J. Flum, M. Ziegler, Topological Model Theory. X, 151 pages 1980.

Vol. 770: Séminaire Bourbaki vol. 1978/79 Exposés 525-542. IV 341 pages. 1980.

Vol. 771: Approximation Methods for Navier-Stokes Problems. Pro ceedings, 1979. Edited by R. Rautmann. XVI, 581 pages. 1980.

Vol. 772: J. P. Levine, Algebraic Structure of Knot Modules. XI, 10 pages. 1980.

Vol. 773: Numerical Analysis. Proceedings, 1979. Edited by G. A Watson. X, 184 pages. 1980.

Vol. 774: R. Azencott, Y. Guivarc'h, R. F. Gundy, Ecole d'Eté d Probabilités de Saint-Flour VIII-1978. Edited by P. L. Hennequi XIII, 334 pages. 1980.

Vol. 775: Geometric Methods in Mathematical Physics. Proceedings 1979. Edited by G. Kaiser and J. E. Marsden. VII, 257 pages. 198

Vol. 776: B. Gross, Arithmetic on Elliptic Curves with Comple Multiplication. V, 95 pages. 1980.

Vol. 777: Séminaire sur les Singularités des Surfaces. Proceedings 1976-1977. Edited by M. Demazure, H. Pinkham and B. Teissie IX, 339 pages. 1980.

Vol. 778: SK1 von Schiefkörpern. Proceedings, 1976. Edited b P. Draxl and M. Kneser. II, 124 pages. 1980.

Vol. 779: Euclidean Harmonic Analysis. Proceedings, 1979. Edite by J. J. Benedetto. III, 177 pages. 1980.

Vol. 780: L. Schwartz, Semi-Martingales sur des Variétés, et Martin gales Conformes sur des Variétés Analytiques Complexes. XV, 13 pages. 1980.

Vol. 781: Harmonic Analysis Iraklion 1978. Proceedings 1978. Edite by N. Petridis, S. K. Pichorides and N. Varopoulos. V, 213 pages. 1980

Vol. 782: Bifurcation and Nonlinear Eigenvalue Problems. Proceed ings, 1978. Edited by C. Bardos, J. M. Lasry and M. Schatzmar VIII, 296 pages. 1980.

Vol. 783: A. Dinghas, Wertverteilung meromorpher Funktionen i ein- und mehrfach zusammenhängenden Gebieten. Edited by R. Ne vanlinna and C. Andreian Cazacu. XIII, 145 pages. 1980.

Vol. 784: Séminaire de Probabilités XIV. Proceedings, 1978/79 Edited by J. Azéma and M. Yor. VIII, 546 pages. 1980.

Vol. 785: W. M. Schmidt, Diophantine Approximation. X, 299 pages 1980.

Vol. 786: I. J. Maddox, Infinite Matrices of Operators. V, 122 pages 1980.